Carbon and Related Composites for Sensors and Energy Storage: Synthesis, Properties, and Application

Carbon and Related Composites for Sensors and Energy Storage: Synthesis, Properties, and Application

Guest Editors

Olena Okhay
Gil Goncalves

Basel • Beijing • Wuhan • Barcelona • Belgrade • Novi Sad • Cluj • Manchester

Guest Editors

Olena Okhay
TEMA–Centre for Mechanical
Technology and Automation
Department of Mechanical
Engineering
University of Aveiro
Aveiro
Portugal

Gil Goncalves
TEMA–Centre for Mechanical
Technology and Automation
Department of Mechanical
Engineering
University of Aveiro
Aveiro
Portugal

Editorial Office
MDPI AG
Grosspeteranlage 5
4052 Basel, Switzerland

This is a reprint of the Special Issue, published open access by the journal C (ISSN 2311-5629), freely accessible at: https://www.mdpi.com/journal/carbon/special_issues/carbon_sensor_energy.

For citation purposes, cite each article independently as indicated on the article page online and as indicated below:

Lastname, A.A.; Lastname, B.B. Article Title. *Journal Name* **Year**, *Volume Number*, Page Range.

ISBN 978-3-7258-2850-0 (Hbk)
ISBN 978-3-7258-2849-4 (PDF)
https://doi.org/10.3390/books978-3-7258-2849-4

© 2024 by the authors. Articles in this book are Open Access and distributed under the Creative Commons Attribution (CC BY) license. The book as a whole is distributed by MDPI under the terms and conditions of the Creative Commons Attribution-NonCommercial-NoDerivs (CC BY-NC-ND) license (https://creativecommons.org/licenses/by-nc-nd/4.0/).

Contents

About the Editors ix

Olena Okhay and Gil Goncalves
Carbon and Related Composites for Sensors and Energy Storage: Synthesis, Properties, and Application
Reprinted from: C **2024**, 10, 101, https://doi.org/10.3390/c10040101 1

Nastaran Ghaffari, Nazeem Jahed, Zareenah Abader, Priscilla G. L. Baker and Keagan Pokpas
Preferential Stripping Analysis of Post-Transition Metals (In and Ga) at Bi/Hg Films Electroplated on Graphene-Functionalized Graphite Rods
Reprinted from: C **2024**, 10, 95, https://doi.org/10.3390/c10040095 6

Ankit Yadav, Rajeev Kumar, Deepu Joseph, Nygil Thomas, Fei Yan and Balaram Sahoo
Impact of Dispersive Solvent and Temperature on Supercapacitor Performance of N-Doped Reduced Graphene Oxide
Reprinted from: C **2024**, 10, 89, https://doi.org/10.3390/c10040089 26

Karoline S. Nantes, Ana L. H. K. Ferreira, Marcio C. Pereira, Francisco G. E. Nogueira and André S. Afonso
A Novel Non-Enzymatic Efficient H_2O_2 Sensor Utilizing δ-FeOOH and Prussian Blue Anchoring on Carbon Felt Electrode
Reprinted from: C **2024**, 10, 82, https://doi.org/10.3390/c10030082 50

Xiaoyan Shu, Yuanjiang Yang, Zhongtang Yang, Honghui Wang and Nengfei Yu
A Nitrogen/Oxygen Dual-Doped Porous Carbon with High Catalytic Conversion Ability toward Polysulfides for Advanced Lithium–Sulfur Batteries
Reprinted from: C **2024**, 10, 67, https://doi.org/10.3390/c10030067 64

Sirine Zallouz, Bénédicte Réty, Jean-Marc Le Meins, Mame Youssou Ndiaye, Philippe Fioux and Camélia Matei Ghimbeu
FeS_2 Nanoparticles in S-Doped Carbon: Ageing Effects on Performance as a Supercapacitor Electrode
Reprinted from: C **2023**, 9, 112, https://doi.org/10.3390/c9040112 75

Xiaoyan Wang, Eng Gee Lim, Kai Hoettges and Pengfei Song
A Review of Carbon Nanotubes, Graphene and Nanodiamond Based Strain Sensor in Harsh Environments
Reprinted from: C **2023**, 9, 108, https://doi.org/10.3390/c9040108 98

Han-Wei Chang, Zong-Ying Tsai, Jia-Jun Ye, Kuo-Chuang Chiu, Tzu-Yu Liu and Yu-Chen Tsai
Synthesis and Characterization of Ni–Co–O Nanosheets on Silicon Carbide Microspheres/Graphite Composite for Supercapacitor Applications
Reprinted from: C **2023**, 9, 101, https://doi.org/10.3390/c9040101 135

Vitalii Vashchynskyi, Olena Okhay and Tetiana Boychuk
Chemical Activation of Apricot Pit-Derived Carbon Sorbents for the Effective Removal of Dyes in Environmental Remediation
Reprinted from: C **2023**, 9, 93, https://doi.org/10.3390/c9040093 147

Hsin-Ya Chiu and Chun-Pei Cho
Impacts of Mn Content and Mass Loading on the Performance of Flexible Asymmetric Solid-State Supercapacitors Using Mixed-Phase MnO_2/N-Containing Graphene Composites as Cathode Materials
Reprinted from: C **2023**, *9*, 88, https://doi.org/10.3390/c9030088 . . . . **164**

Mariana A. Vieira, Tainara L. G. Costa, Gustavo R. Gonçalves, Daniel F. Cipriano, Miguel A. Schettino, Jr., Elen L. da Silva, et al.
Phosphorus/Sulfur-Enriched Reduced Graphene Oxide Papers Obtained from Recycled Graphite: Solid-State NMR Characterization and Electrochemical Performance for Energy Storage
Reprinted from: C **2023**, *9*, 60, https://doi.org/10.3390/c9020060 . . . . **184**

Hossein Bisheh and Yasmine Abdin
Carbon Fibers: From PAN to Asphaltene Precursors; A State-of-Art Review
Reprinted from: C **2023**, *9*, 19, https://doi.org/10.3390/c9010019 . . . . **201**

J. L. Sánchez Toural, V. Marzoa, R. Bernardo-Gavito, J. L. Pau and D. Granados
Hands-On Quantum Sensing with NV^- Centers in Diamonds
Reprinted from: C **2023**, *9*, 16, https://doi.org/10.3390/c9010016 . . . . **231**

Audrey F. Adcock, Ping Wang, Elton Y. Cao, Lin Ge, Yongan Tang, Isaiah S. Ferguson, et al.
Carbon Dots versus Nano-Carbon/Organic Hybrids—Divergence between Optical Properties and Photoinduced Antimicrobial Activities
Reprinted from: C **2022**, *8*, 54, https://doi.org/10.3390/c8040054 . . . . **251**

Yuvarat Ngernyen, Thitipong Siriketh, Kritsada Manyuen, Panta Thawngen, Wipha Rodtoem, Kritiyaporn Wannuea, et al.
Easy and Low-Cost Method for Synthesis of Carbon–Silica Composite from Vinasse and Study of Ibuprofen Removal
Reprinted from: C **2022**, *8*, 51, https://doi.org/10.3390/c8040051 . . . . **264**

Lyane M. Darabian, Tainara L. G. Costa, Daniel F. Cipriano, Carlos W. Cremasco, Miguel A. Schettino, Jr. and Jair C. C. Freitas
Synthesis of Graphene Quantum Dots by a Simple Hydrothermal Route Using Graphite Recycled from Spent Li-Ion Batteries
Reprinted from: C **2022**, *8*, 48, https://doi.org/10.3390/c8040048 . . . . **278**

Sidra Ayaz, Afzal Shah and Shamsa Munir
Investigation of Electron Transfer Mechanistic Pathways of Ferrocene Derivatives in Droplet at Carbon Electrode
Reprinted from: C **2022**, *8*, 45, https://doi.org/10.3390/c8030045 . . . . **287**

Takayuki Ohta, Hiroaki Iwata, Mineo Hiramatsu, Hiroki Kondo and Masaru Hori
Power Generation Characteristics of Polymer Electrolyte Fuel Cells Using Carbon Nanowalls as Catalyst Support Material
Reprinted from: C **2022**, *8*, 44, https://doi.org/10.3390/c8030044 . . . . **304**

Giovanni G. Daniele, Daniel C. de Souza, Paulo Roberto de Oliveira, Luiz O. Orzari, Rodrigo V. Blasques, Rafael L. Germscheidt, et al.
Development of Disposable and Flexible Supercapacitor Based on Carbonaceous and Ecofriendly Materials
Reprinted from: C **2022**, *8*, 32, https://doi.org/10.3390/c8020032 . . . . **318**

Reza Moheimani, Paniz Hosseini, Saeed Mohammadi and Hamid Dalir
Recent Advances on Capacitive Proximity Sensors: From Design and Materials to Creative Applications
Reprinted from: *C* **2022**, *8*, 26, https://doi.org/10.3390/c8020026 . . . . **332**

About the Editors

Olena Okhay

Since January 2019, Dr. O. Okhay has worked at TEMA, the University of Aveiro, as an independent researcher. Her current work concentrates on materials for energy harvesting and storage. In particular, she works with 1D-2D carbon family materials and different composites, including metal oxides/hydroxide, polymers, etc. She obtained the first prize at the XXXIX Meeting of the Portuguese Society for Microscopy and Cell Biology (2004) and Young Scientist Award from the European Materials Research Society (E-MRS) at the 2008 Spring Meeting. She has participated in COST actions, FAME international networks, Graphene Flagship events, as well as many summer/winter schools and workshops on different scientific thematics in order to improve her personal research potential. She has many international collaborations, which has resulted in publications. In total, Dr. Okhay (h=14 based on SCOPUS) has published 50 papers and four book chapters. Six of her articles are published in the TOP-5 journals, and nine of them are published in the TOP-10 journals of their areas. She is a reviewer for 30 international journals. She has participated in international conferences with posters (about 30) and oral presentations (12 invited/keynotes talks), as well as being a member of the organizing committee.

Gil Goncalves

Gil Gonçalves graduated in 2003 in Industrial Chemistry at the University of Aveiro (Portugal) and obtained his Master's degree in Materials Science (EMMS, Joint European Master's Programme) in 2008. In 2012, he received his Ph.D. in Mechanical Engineering from the University of Aveiro with a thesis dedicated to nanocomposite materials for biomedical application. After obtaining a Marie Curie research fellowship in 2016, he started working at the Institute of Material Science of Barcelona, High Council of Spanish Research (ICMAB-CSIC) on the development of ultra-sensitive nanotherapeutic anticancer agents for neutron capture therapy (H2020-MSCA-IF-2015). He was a short-term visitor on various occasions at the Laboratory of Applied Nuclear Energy (LENA, University of Pavia) and the Department of Chemical and Pharmaceutical Sciences, University of Trieste. Currently, he is working at the Centre for Mechanical Technology and Automation (TEMA, University of Aveiro) as a researcher on the development of new graphene-based nanocomposite materials with applications in many different fields, such as graphene-based porous structures for heterogeneous catalysis (catalysis) and water purification (environment), three-dimensional graphene scaffolds for biomedical applications (biomaterials), nanostructured graphene substrates for selective biomolecules detection (sensors), and carbon-based nanoplatforms for the detection and therapy of cancer cells (therapeutic agent). Dr. Gil has (co-)authored numerous scientific papers (h-index 28 and >3300 citations (Scopus)) and communications at national and international conferences, and he is also a regular reviewer for high-impact scientific journals in the field of multifunctional carbon nanomaterials. He is a member of the Editorial Board of *Scientific Reports* (Nature Publishing Group) in the field of Chemical Physics.

Editorial

Carbon and Related Composites for Sensors and Energy Storage: Synthesis, Properties, and Application

Olena Okhay [1,2,*] and Gil Goncalves [1,2]

1. Centre for Mechanical Technology and Automation (TEMA), University of Aveiro, 3810-193 Aveiro, Portugal; ggoncalves@ua.pt
2. LASI-Intelligent Systems Associate Laboratory, 4800-058 Guimaraes, Portugal
* Correspondence: olena@ua.pt

In recent years, mankind's energy needs have been increasing; therefore, current research is focused on the collection and storage of energy [1]. Thus, there is a need to develop new materials with improved characteristics, such as improved collection and conversion efficiencies as well as improved energy storage properties [2]. This Special Issue, 'Carbon and Related Composites for Sensors and Energy Storage: Synthesis, Properties, and Application' of *C—Journal of Carbon Research* presents state-of-the-art contributions to the preparation and characterization techniques of carbon-related materials in the field of energy storage and sensor applications. Specifically, three review papers and sixteen research papers were included in this Special Issue.

It is well known that carbon and carbon nanomaterials, including 0D quantum dots, fullerenes, 1D carbon nanotubes, 2D graphene, reduced graphene oxide, and other carbon-related nanostructures, have shown unique morphological, electrical, thermal, mechanical, electromechanical, and electromagnetic properties for a wide range of applications [3]. One of the carbon-related materials—carbon fibers—is detailly described in the review paper written by Bisheh and Abdin entitled 'Carbon Fibers: From PAN to Asphaltene Precursors; A State-of-Art Review' [contribution 1]. This paper reviews the material properties and use of carbon fibers in various applications and industries and compares them with other existing fillers and reinforcing fibers. In addition, the processing of the carbon fibers and the main challenges in their fabrication were examined. This review demonstrates that low-cost asphaltene-based carbon fibers can be a substitute for expensive polyacrylonitrile/pitch-based carbon fibers for functional applications. The value proposition, performance/cost advantages, potential markets, market size, processing challenges, and methods for overcoming these issues are discussed.

Carbon-related materials are widely used in sensors [4], and this Special Issue contains a review article by Moheimani et al. which discusses current progress in capacitive sensor design approaches [contribution 2]. Such capacitive sensors can have different applications such as electrical capacitance tomography, capacitive voltage sensors, capacitive humidity sensors, capacitive gas sensors, displacement detection, and muscle action for interaction. Contribution 2 presents a the detailed comparison of the properties of flexible capacitance-type proximity sensors fabricated from different nanomaterials, including graphene oxide (GO), carbon microcoils, and carbon nanotubes (CNT).

Deeper state-of-the-art stain sensors for harsh environments were developed by Wang et al. [Contribution 3]. The change in capacitance with pressure or humidity and the problem of dispersion of nanofillers such as graphene and CNT in the polymer matrix are discussed, and some improvements have been proposed to overcome this problem.

Several studies have been published on the use of carbon-based materials in special sensing applications. Nantes et al. developed H_2O_2 sensor based on electrochemical Prussian blue (PB) synthesized from the acid suspension of δ-FeOOH and $K_3[Fe(CN)_6]$ using cyclic voltammetry (CV) and anchored on carbon felt (CF) [contribution 4]. The

Citation: Okhay, O.; Goncalves, G. Carbon and Related Composites for Sensors and Energy Storage: Synthesis, Properties, and Application. *C* 2024, *10*, 101. https://doi.org/10.3390/c10040101

Received: 25 November 2024
Accepted: 29 November 2024
Published: 3 December 2024

Copyright: © 2024 by the authors. Licensee MDPI, Basel, Switzerland. This article is an open access article distributed under the terms and conditions of the Creative Commons Attribution (CC BY) license (https://creativecommons.org/licenses/by/4.0/).

CF/PB-FeOOH electrode exhibits excellent selectivity for H_2O_2 in the presence of dopamine (DA), uric acid (UA), and ascorbic acid (AA). Thus, the proposed electrodes could be used in electrochemical sensing technologies for various biological and environmental applications.

Ghaffari et al. demonstrated a sensor that could effectively quantify gallium (Ga^{3+}) and indium (In^{3+}) in tap water using square-wave anodic stripping voltammetry was demonstrated by [contribution 5]. This novel electrochemical sensor combines reduced graphene oxide sheets with a bismuth–mercury (Bi/Hg) film electroplated onto pencil graphite electrodes for the high-sensitivity detection of trace amounts of gallium and indium in water samples.

The introduction of the properties of nitrogen vacancy centers in diamonds was proposed by Toural et al. [contribution 6]. Quantitative measurement of external magnetic fields with high sensitivity is possible under optical and radiofrequency excitations. In this paper, a step-by-step process for developing and testing a simple magnetic quantum sensor based on color centers with significant potential for the development of highly compact multisensor systems is described.

Four ferrocene derivatives, 4-ferrocenyl-3-methyl aniline (FMA), 3-Chloro-4-ferrocenyl aniline (CFA), 4-ferrocenyl aniline (FA), and ferrocenyl benzoic acid (FBA), were studied by Ayaz et al. to understand their biochemical actions [contribution 7]. It has been reported that the redox behavior of these compounds is sensitive to pH, concentration, scan number, and scan rate and can be used as electrochemical sensors based on ferrocene as a mediator.

The optical spectroscopic properties and photoinduced antimicrobial activities of carbon dots (CDots) and nanocarbon/organic hybrids were compared in a study by Adcock et al. [Contribution 8]. The samples were obtained by thermal processing citric acid (CA) with an oligomeric polyethylenimine (PEI) precursor mixture under different conditions, including hydrothermal and heating by microwave irradiation, and were compared with PEI–CDots (classically defined CDots of PEI-functionalized pre-existing small-carbon nanoparticles). The optical spectroscopic properties of the samples showed significant similarities; however, significant differences in their visible light-activatedd antibacterial activities were detected.

The use of carbon-based materials as sorbents to remove dyes or ibuprofen was reported by Vashchynskyi et al. [contribution 9] and Ngernyen et al. [contribution 10], respectively. Vashchynskyi et al. proposed the use of apricot-pit-derived carbon sorbents prepared using two-step carbonization and acid treatment [contribution 9]. The obtained sorbents exhibited adsorption activities of 190–235 mg g^{-1} and 210–260 mg g^{-1} for methylene blue and methyl orange, respectively. Ngernyen et al. used vinasse to synthesize carbon–silica composite with a low-cost silica source available in Thailand (sodium silicate, Na_2SiO_3) [contribution 10]. Tetraethyl orthosilicate (TEOS) was used as a silica source for comparison. A one-step sol–gel process was used to prepare these samples by varying weight ratio of vinasse (carbon source) to silica source (Na_2SiO_3 or TEOS). The maximum adsorption capacities of the Na_2SiO_3-based composites were 406 mg g^{-1} and 418 mg g^{-1}, respectively.

In addition to the aforementioned applications of carbon family materials for sensors, these materials have been widely studied for energy storage in various electrical energy storage devices, such as supercapacitors [5,6], batteries [7,8], and fuel cells [9].

In the current Special Issue, Darabian et al. [contribution 11] and Vieira et al. [contribution 12] reported on the preparation of reduced graphene oxide paper and graphene quantum dots, respectively, using graphite from spent Li-ion batteries. Darabian et al. proposed a simple and low-cost synthesis of graphene quantum dots starting from an alcoholic aqueous suspension of graphene oxide using a hydrothermal route [contribution 11]. In contrast, Vieira et al. used mixtures of sulfuric and phosphoric acids (with different H_2SO_4/H_3PO_4 ratios), leading to the production of materials with significant S and P contents, to obtain GO via the oxidation of graphite recycled from spent Li-ion batteries [contribution 12]. This material reached 155 and 110 F g^{-1} at 1 and 8 A g^{-1}, respectively, and showed good capacitance retention (~95%) after 1000 cycles.

Daniele et al. proposed a flexible supercapacitor developed from a polyethylene terephthalate substrate, reused from beverage bottles, and conductive ink based on carbon black (CB) and cellulose acetate (CA) [contribution 13]. This supercapacitor was fabricated using carbonaceous and eco-friendly materials composed of 2.45 Wh kg^{-1} and a specific energy of approximately 1000 W kg^{-1} for specific power.

The combination of nitrogen-containing graphene and MnO_2 with varying Mn contents was used by Chiu and Cho as electrode materials for flexible asymmetric solid-state supercapacitors [contribution 14]. It has been reported that excessive MnO_2 reduces the conductivity, hindering ion diffusion and charge transfer. However, overloading the electrode with active materials also negatively affects conductivity.

Zallouz et al. studied the effects of aging, thermal treatments, and activation on the structural and other properties of carbon/FeS_2 formation for a supercapacitor electrode [contribution 15]. The aging of the FeS_2 nanoparticles was investigated in air, and a progressive transformation of nano-FeS_2 into hydrated iron hydroxy sulfate with significant morphological modification was observed, resulting in a drastic decrease in capacitance (70%) and retention. In contrast, aging nano-FeS_2 during cycling led to the formation of a supplementary iron oxyhydroxide phase which contributed to the enhanced capacitance (57%) and long-term cycling (132% up to 10,000 cycles) of the device.

The critical roles of the dispersion medium and temperature during the solvothermal synthesis of nitrogen-doped reduced graphene oxide as an active material in supercapacitor electrodes were discussed by Yadav et al. [contribution 16]. Among the different solvents (THF, ethanol, acetonitrile, water, N, N-dimethylformamide, ethylene glycol, or N-Methyl-2-pyrrolidone) as the dispersive medium and different temperatures (60, 75, 95, 120, 150, 180, and 195 °C) used for the preparation of N-doped reduced graphene oxide, the highest 514 F g^{-1} at 0.5 Ag^{-1} was reported for the material synthesized with N,N-Dimethylformamide at 150 °C.

Chang et al. used silicon carbide microspheres/graphite composites to grow Ni–Co–O nanosheets for energy-storage applications [contribution 17]. Ni–Co–O prepared by electrodeposition on the silicon carbide (SiC) microspheres/graphite composite presented a larger cyclic voltammetry (CV) curve area than that of Ni–Co–O grown on graphite only, which can be attributed to the synergistic effects between the Ni–Co–O nanosheets and silicon carbide microspheres.

The utilization of porous carbon in Li-S batteries was reported by Shu et al. [contribution 18]. In this study, nitrogen/oxygen dual-doped porous carbon (N/O-PC) was synthesized by annealing a precursor of zeolitic imidazolate framework-8 grown in situ on MWCNTs (ZIF-8/MWCNTs). The interconnected porous carbon-based structure facilitates electron and ion transfer, and the S/N/O-PC cathode exhibits high cycling stability in the $LiCF_3SO_3$ electrolyte (a stable capacity of 685.9 mA h g^{-1} at 0.2 C after 100 cycles).

In addition to the use of carbon materials in supercapacitors and batteries for energy storage, Ohta et al. proposed the use of carbon nanowalls as catalyst support materials in polymer electrolyte fuel cells (PEFC) [Contribution 19]. It was concluded that voltage losses could be mitigated by increasing the height of the carbon nanowalls, which could improve the power generation of the PEFC.

In summary, this Special Issue, 'Carbon and Related Composites for Sensors and Energy Storage: Synthesis, Properties, and Application' of C—*Journal of Carbon Research*, compiles a series of 19 original research articles and review papers that provide new insights into the preparation of carbon-related materials and their possible sensing applications as well as energy storage applications. We hope that the articles published in this Special Issue will improve the readers' general understanding of the research in this rapidly developing scientific field.

Funding: O.O. thanks FCT (Fundacao para a Ciencia e a Tecnologia) in the scope of the framework contract foreseen in numbers 4, 5, and 6 of article 23 of the Decree Law 57/2016 of 29 August, UIDB/00481/2020 and UIDP/00481/2020—FCT, and the CENTRO-01-0145- FEDER-022083—Centro

Portugal Regional Operational Programme (Centro2020), under the PORTUGAL 2020 Partnership Agreement through the European Regional Development Fund.

Conflicts of Interest: The authors declare no conflicts of interest.

List of Contributions

1. Bisheh, H.; Abdin, Y. Carbon Fibers: From PAN to Asphaltene Precursors; A State-of-Art Review. *C* **2023**, *9*, 19. https://doi.org/10.3390/c9010019
2. Moheimani, R.; Hosseini, P.; Mohammadi, S.; Dalir, H. Recent Advances in Capacitive Proximity Sensors: From Design and Materials to Creative Applications. *C* **2022**, *8*, 26. https://doi.org/10.3390/c8020026
3. Wang, X.; Lim, E.G.; Hoettges, K.; Song, P. A Review of Carbon Nanotubes, Graphene and Nanodiamond Based Strain Sensor in Harsh Environments. *C* **2023**, *9*, 108. https://doi.org/10.3390/c9040108
4. Nantes, K.S.; Ferreira, A.L.H.K.; Pereira, M.C.; Nogueira, F.G.E.; Afonso, A.S. A Novel Non-Enzymatic Efficient H_2O_2 Sensor Utilizing δ-FeOOH and Prussian Blue Anchoring on Carbon Felt Electrode. *C* **2024**, *10*, 82. https://doi.org/10.3390/c10030082
5. Ghaffari, N.; Jahed, N.; Abader, Z.; Baker, P.G.L.; Pokpas, K. Preferential Stripping Analysis of Post-Transition Metals (In and Ga) at Bi/Hg Films Electroplated on Graphene-Functionalized Graphite Rods. *C* **2024**, *10*, 95. https://doi.org/10.3390/c10040095
6. Toural, L.S.; Marzoa, V.; Bernardo-Gavito, R.; Pau, J.L.; Granados, D. Hands-On Quantum Sensing with NV Centers in Diamonds. *C* **2023**, *9*, 16. https://doi.org/10.3390/c9010016
7. Ayaz, S.; Shah, A.; Munir, S. Investigation of Electron Transfer Mechanistic Pathways of Ferrocene Derivatives in Droplet at Carbon Electrode. *C* **2022**, *8*, 45. https://doi.org/10.3390/c8030045
8. Adcock, A.F.; Wang, P.; Cao, E.Y.; Ge, L.; Tang, Y.; Ferguson, I.S.; Sweilem, F.S.A.; Petta, L.; Cannon, W.; Yang, L.; et al. Carbon Dots versus Nano-Carbon/Organic Hybrids: Divergence between Optical Properties and Photoinduced Antimicrobial Activities. *C* **2022**, *8*, 54. https://doi.org/10.3390/c8040054
9. Vashchynskyi, V.; Okhay, O.; Boychuk, T. Chemical Activation of Apricot Pit-Derived Carbon Sorbents for the Effective Removal of Dyes in Environmental Remediation. *C* **2023**, *9*, 93. https://doi.org/10.3390/c9040093
10. Ngernyen, Y.; Siriketh, T.; Manyuen, K.; Thawngen, P.; Rodtoem, W.; Wannuea, K.; Knijnenburg, J.T.N.; Budsaereechai, S. Easy and Low-Cost Method for Synthesis of Carbon–Silica Composite from Vinasse and Study of Ibuprofen Removal. *C* **2022**, *8*, 51. https://doi.org/10.3390/c8040051
11. Darabian, L.M.; Costa, T.L.G.; Cipriano, D.F.; Cremasco, C.W.; Schettino, M.A., Jr.; Freitas, J.C.C. Synthesis of Graphene Quantum Dots by a Simple Hydrothermal Route Using Graphite Recycled from Spent Li-Ion Batteries. *C* **2022**, *8*, 48. https://doi.org/10.3390/c8040048
12. Vieira, M.A.; Costa, T.L.G.; Gonçalves, G.R.; Cipriano, D.F.; Schettino, M.A.; da Silva, E.L.; Cuña, A.; Freitas, J.C.C. Phosphorus/Sulfur-Enriched Reduced Graphene Oxide Papers Obtained from Recycled Graphite: Solid-State NMR Characterization and Electrochemical Performance for Energy Storage. *C* **2023**, *9*, 60. https://doi.org/10.3390/c9020060
13. Daniele, G.G.; de Souza, D.C.; de Oliveira, P.R.; Orzari, L.O.; Blasques, R.V.; Germscheidt, R.L.; da Silva, E.C.; Pocrifka, L.A.; Bonacin, J.A.; Janegitz, B.C. Development of Disposable and Flexible Supercapacitor Based on Carbonaceous and Ecofriendly Materials. *C* **2022**, *8*, 32. https://doi.org/10.3390/c8020032
14. Chiu, H.-Y.; Cho, C.-P. Effects of Mn Content and Mass Loading on the Performance of Flexible Asymmetric Solid-State Supercapacitors Using Mixed-Phase MnO_2/N-Containing Graphene Composites as Cathode Materials. *C* **2023**, *9*, 88. https://doi.org/10.3390/c9030088
15. Zallouz, S.; Réty, B.; Meins, J.-M.L.; Ndiaye, M.Y.; Fioux, P.; Ghimbeu, C.M. FeS_2 Nanoparticles in S-Doped Carbon: Ageing Effects on Performance as a Supercapacitor Electrode. *C* **2023**, *9*, 112. https://doi.org/10.3390/c9040112
16. Yadav, A.; Kumar, R.; Joseph, D.; Thomas, N.; Yan, F.; Sahoo, B. Impact of Dispersive Solvent and Temperature on Supercapacitor Performance of N-Doped Reduced Graphene Oxide. *C* **2024**, *10*, 89. https://doi.org/10.3390/c10040089
17. Chang, H.-W.; Tsai, Z.-Y.; Ye, J.-J.; Chiu, K.-C.; Liu, T.-Y.; Tsai, Y.-C. Synthesis and Characterization of Ni–Co–O Nanosheets on Silicon Carbide Microspheres/Graphite Composite for Supercapacitor Applications. *C* **2023**, *9*, 101. https://doi.org/10.3390/c9040101

18. Shu, X.; Yang, Y.; Yang, Z.; Wang, H.; Yu, N. A Nitrogen/Oxygen Dual-Doped Porous Carbon with High Catalytic Conversion Ability toward Polysulfides for Advanced Lithium-Sulfur Batteries. *C* **2024**, *10*, 67. https://doi.org/10.3390/c10030067
19. Ohta, T.; Iwata, H.; Hiramatsu, M.; Kondo, H.; Hori, M. Power Generation Characteristics of Polymer Electrolyte Fuel Cells Using Carbon Nanowalls as Catalyst Support Material. *C* **2022**, *8*, 44. https://doi.org/10.3390/c8030044

References

1. Koohi-Fayegh, S.; Rose, M.A. A review of energy storage types, applications and recent developments. *J. Energy Storage* **2020**, *27*, 101047. [CrossRef]
2. Elalfy, D.A.; Gouda, E.; Kotb, M.F.; Bureš, V.; Sedhom, B.E. Comprehensive review of energy storage systems technologies, objectives, challenges, and future trends. *Energy Strategy Rev.* **2024**, *54*, 101482. [CrossRef]
3. Waris; Chaudhary, M. S.; Anwer, A.H.; Sultana, S.; Ingole, P.P.; Nami, S.A.A.; Khan, M.Z. A Review on Development of Carbon-Based Nanomaterials for Energy Storage Devices: Opportunities and Challenges. *Energy Fuels* **2023**, *37*, 24–19433. [CrossRef]
4. Bezzon, V.D.N.; Montanheiro, T.L.A.; de Menezes, B.R.C.; Ribas, R.G.; Righetti, V.A.N.; Rodrigues, K.F.; Thim, G.P. Carbon Nanostructure-based Sensors: A Brief Review on Recent Advances. *Adv. Mater. Sci. Eng.* **2019**, *2019*, 4293073. [CrossRef]
5. Olabi, A.G.; Abbas, Q.; Abdelkareem, M.A.; Alami, A.H.; Mirzaeian, M.; Sayed, E.T. Carbon-Based Materials for Supercapacitors: Recent Progress, Challenges and Barriers. *Batteries* **2023**, *9*, 19. [CrossRef]
6. Okhay, O.; Tkach, A. Graphene/Reduced Graphene Oxide-Carbon Nanotubes Composite Electrodes: From Capacitive to Battery-Type Behaviour. *Nanomaterials* **2021**, *11*, 1240. [CrossRef] [PubMed]
7. Qiu, Z.; Cao, F.; Pan, G.; Li, C.; Chen, M.; Zhang, Y.; He, X.; Xia, Y.; Xia, X.; Zhang, W. Carbon materials for metal-ion batteries. *ChemPhysMater* **2023**, *2*, 267–281. [CrossRef]
8. Okhay, O.; Tkach, A. A comprehensive review of the use of porous graphene frameworks for various types of rechargeable lithium batteries. *J. Energy Storage* **2024**, *80*, 110336. [CrossRef]
9. Rey-Raap, N.; dos Santos-Gómez, L.; Arenillas, A. Carbons for fuel cell energy generation. *Carbon* **2024**, *228*, 119291. [CrossRef]

Disclaimer/Publisher's Note: The statements, opinions and data contained in all publications are solely those of the individual author(s) and contributor(s) and not of MDPI and/or the editor(s). MDPI and/or the editor(s) disclaim responsibility for any injury to people or property resulting from any ideas, methods, instructions or products referred to in the content.

Article

Preferential Stripping Analysis of Post-Transition Metals (In and Ga) at Bi/Hg Films Electroplated on Graphene-Functionalized Graphite Rods

Nastaran Ghaffari, Nazeem Jahed, Zareenah Abader, Priscilla G. L. Baker and Keagan Pokpas *

SensorLab, Department of Chemistry, University of the Western Cape, Robert Sobukwe Road, Bellville 7535, South Africa; 3415240@myuwc.ac.za (N.G.); njahed@uwc.ac.za (N.J.); 3438692@myuwc.ac.za (Z.A.); pbaker@uwc.ac.za (P.G.L.B.)
* Correspondence: kpokpas@uwc.ac.za; Tel.: +27-21959-4038

Abstract: In this study, we introduce a novel electrochemical sensor combining reduced graphene oxide (rGO) sheets with a bismuth–mercury (Bi/Hg) film, electroplated onto pencil graphite electrodes (PGEs) for the high-sensitivity detection of trace amounts of gallium (Ga^{3+}) and indium (In^{3+}) in water samples using square wave anodic stripping voltammetry (SWASV). The electrochemical modification of PGEs with rGO and bimetallic Bi/Hg films (ERGO-Bi/HgF-PGE) exhibited synergistic effects, enhancing the oxidation signals of Ga and In. Graphene oxide (GO) was accumulated onto PGEs and reduced through cyclic reduction. Key parameters influencing the electroanalytical performance, such as deposition potential, deposition time, and pH, were systematically optimized. The improved adsorption of Ga^{3+} and In^{3+} ions at the Bi/Hg films on the graphene-functionalized electrodes during the preconcentration step significantly enhanced sensitivity, achieving detection limits of 2.53 nmol L^{-1} for Ga^{3+} and 7.27 nmol L^{-1} for In^{3+}. The preferential accumulation of each post-transition metal, used in transparent displays, to form fused alloys at Bi and Hg films, respectively, is highlighted. The sensor demonstrated effective quantification of Ga^{3+} and In^{3+} in tap water, with detection capabilities well below the USEPA guidelines. This study pioneers the use of bimetallic films to selectively and simultaneously detect the post-transition metals In^{3+} and Ga^{3+}, highlighting the role of graphene functionalization in augmenting metal film accumulation on cost-effective graphite rods. Additionally, the combined synergistic effects of Bi/Hg and graphene functionalization have been explored for the first time, offering promising implications for environmental analysis and water quality monitoring.

Keywords: electrochemically reduced graphene oxide; pencil graphite electrode; bimetallic bismuth/mercury film; gallium; indium; anodic stripping voltammetry

Citation: Ghaffari, N.; Jahed, N.; Abader, Z.; Baker, P.G.L.; Pokpas, K. Preferential Stripping Analysis of Post-Transition Metals (In and Ga) at Bi/Hg Films Electroplated on Graphene-Functionalized Graphite Rods. *C* **2024**, *10*, 95. https://doi.org/10.3390/c10040095

Academic Editors: Olena Okhay, Gil Goncalves and Craig E. Banks

Received: 25 July 2024
Revised: 30 October 2024
Accepted: 7 November 2024
Published: 12 November 2024

Copyright: © 2024 by the authors. Licensee MDPI, Basel, Switzerland. This article is an open access article distributed under the terms and conditions of the Creative Commons Attribution (CC BY) license (https://creativecommons.org/licenses/by/4.0/).

1. Introduction

In contemporary times, there is a surging interest within environmental communities in developing methodologies for detecting, remediating, and quantifying trace metals, including Gallium and Indium. This heightened interest is primarily driven by the imperative to ensure effective wastewater treatment [1–5]. Recently, semiconductor materials like III–V semiconductors (e.g., InAs, InP, GaAs, and GaP) have emerged as crucial strategic materials extensively employed in high-tech industries for devices like light-emitting diodes, semiconductor laser diodes, solar cells, optical computers, and liquid crystal displays. Indium and Gallium, both belonging to group IIIA metallic elements, exhibit semiconductor or optoelectronic characteristics and find widespread application in thin-film transistor liquid crystal display (TFT-LCD) production for television screens, portable computer screens, and cell phone displays [6,7].

The etching wastewater discharged from semiconductor or optoelectronic plants contains trace amounts of Indium and its compounds, suspected to be carcinogenic to humans and capable of causing damage to the heart, kidney, and liver [4,8]. Consequently, there is a growing interest in developing new methods to determine trace Indium in the environment, motivated by health and environmental concerns. Similarly, trace amounts of Gallium exist in the human body through natural water, vegetables, and fruits, although its physiological function remains unknown. Gallium (III) can easily be absorbed into the blood as the transferrin–gallium complex due to its similarity to Fe (III). Some Gallium compounds have demonstrated anti-inflammatory and immunosuppressive activities, making them potential antimicrobial agents against certain pathogens [9–14]. The increasing use of Gallium and its compounds has led to a rise in Gallium levels in the environment around industrial areas, necessitating studies on the determination of trace Gallium in environmental or biological samples.

Traditionally, the determination of trace metals has been conducted using techniques such as atomic absorption spectroscopy (AAS) or inductively coupled plasma atomic emission spectroscopy (ICP-AAS) and ICP-mass spectrometry (ICP-MS) [5,15]. However, these methods are expensive, require qualified personnel, and are time-consuming. An alternative approach involves electrochemical sensors, with electrochemical stripping voltammetry emerging as a valuable technique. Among these, anodic stripping voltammetry (ASV) stands out due to its ability to accumulate metal ions at the electrode surface, resulting in high sensitivity and low detection limits [16–19].

Electrode materials are crucial in stripping voltammetry. Carbon-based electrodes have been favoured in various applications due to their favourable potential ranges and minimal contribution to background current [20–23]. Of these, pencil rods made of graphite, referred to as pencil graphite electrodes (PGE), have gained popularity in recent years owing to their favourable characteristics, including good electro-conductivity, low cost, disposability, and the absence of time-consuming pretreatments [24–26]. Additionally, the modifications to PGE surfaces and treatment steps cause changes in reactivity and active area [27–31].

The utilization of graphene and reduced graphene oxide in carbon electrode functionalization improves surface area, strength, conductivity, and ease of functionalization [32–37]. Graphene-based electrodes, particularly those modified with graphene oxide, have demonstrated superior detection capabilities, resulting in environmental and biological sensors [38–42]. Adjusting the surface area while maintaining the ratio, ease of functionalization, and improved electron-transport kinetics is leveraged for improved detection sensitivity in a range of electrochemical sensors.

Mercury electrodes have historically yielded the best analytical results due to their excellent electrochemical performance. Mercury, in its pure liquid form at ambient temperatures, can easily form a fresh surface crucial for sensitive analysis, and it boasts a wide usable potential range [43–48]. Mercury films have shown wide application in the stripping analysis of metals [49–51]. Thin bismuth-film electrodes have shown similar or comparable results to mercury-film electrodes, due to their alloy formation capabilities offering improved sensitivity, resolution, and simplicity [52–56]. The formation of Bi and Hg films has been leveraged for the past 2–3 decades and are well documented in the literature. It is believed that the high concentration of metallic film material used over the analyte (usually >1000 times) leads to film formation instead of nanoparticles. To the best of our knowledge, all work reporting the formation of bismuth nanoparticles has used chemical and other methods for the synthesis of the nanostructures with electrochemical methods used only for electrodeposition. The synergistic effects of Bi/Hg bimetallic films have been leveraged in a handful of studies for the stripping analysis of heavy metals [48,57,58]. To date, only one study has investigated its use for the determination of metalloids or transition metals, establishing a unique gap in current research for further exploration [59]. Moreover, the high affinity of the cation-π interactions facilitates the electroplating of metallic films at carbon nanostructure surfaces.

In this specific study, the pencil graphite electrode (PGE) was electrochemically modified with reduced graphene oxide. The resulting modified pencil graphite electrode, in conjunction with an in situ plated bismuth–mercury film (ERGO-Bi/HgF-PGE), was for the first time employed for the simultaneous determination of Gallium and Indium in water samples. To date, conventional and expensive solid electrode materials or hanging mercury drop electrodes have been reported in all studies for the quantitative determination of In^{3+} and Ga^{3+} in various media. This study, therefore, is one of the first reported works using low-cost and disposable alternatives for post-transition metal detection. The method's detection limit was found to be comparable to those reported in the literature, though lower detection limits have been achieved. Despite this, low detection limits remain a significant challenge for the simultaneous analysis of Gallium and Indium. The utilization of ERGO-Bi/HgF–PGE involves relatively simple modification protocols, and the method yields low detection limits.

2. Materials and Methods

2.1. Chemicals and Reagents

In this investigation, only chemicals of analytical reagent grade were employed, and they were utilized without undergoing additional purification steps. Standard stock solutions, specifically atomic absorption standard solutions with a concentration of 1000 mg L^{-1}, were procured from Sigma-Aldrich (Darmstadt, Germany) and subsequently diluted as needed for the experiments. To create the supporting electrolyte, an acetate buffer with a pH of 4.38 and a concentration of 0.1 M was utilized. The acetate buffer comprised of appropriate ratios of acetic acid and sodium acetate. All solutions were prepared with impurity-free ultra-pure distilled water obtained from MilliporeSigma (Darmstadt, Germany).

2.2. Apparatus

The Metrohm (Herisau, Switzerland), 797 VA Computrace instruments were used for all voltammetry experiments. The experimental setup involved a three-electrode system, where the electrochemically reduced graphene oxide bismuth–mercury film pencil graphite electrode (ERGO-Bi/HgF-PGE) functioned as the working electrode. Additionally, a platinum wire served as the counter electrode, and an Ag/AgCl (saturated KCl) electrode served as the reference electrode. All electrode potentials recorded herein are plotted vs. the Ag/AgCl reference electrode. The standard hydrogen electrode (SHE) potential for the Ag/AgCl reference electrode (saturated KCl) is 0.25 V. All experimental procedures took place in a single-compartment 20 mL voltammetric cell maintained at room temperature.

2.3. Electrochemical Reduction of Grapheme Oxide onto Pencil Graphite Electrodes (ERGO-PGE)

Graphene oxide (GO) dispersions were prepared in aqueous media following a conventional modified Hummer's method [17,60]. GO nanoplatelets were synthesized from graphite powder in the presence of sodium nitrate, sulfuric acid, and potassium permanganate as oxidizing agents. Alternate acid and ultra-pure water washes were used to purify the resultant GO flakes prior to drying in a desiccator. Subsequently, mechanical exfoliation (1 h) of the prepared graphite oxide was performed in an ultrasonic bath and dispersed in an acetate buffer solution (ABS).

The 6-cm pencil graphite rods (0.5 mm diameter), specifically the Pentel brand, were used as working electrodes for all experiments. Graphite rods inserted into a plastic syringe casing were affixed with copper wire through the top to ensure electrical connection, and 1-cm portions exposed for detection. Before undergoing modification, the surface of the pencil graphite electrode (PGE) was polished with 3M nitric acid and cleaned with deionized water. The electrode was then conditioned through five cycles in an acetate buffer solution (ABS) using cyclic voltammetry. Subsequently, ERGO films were prepared from 1 mg mL^{-1} graphite oxide (GO) dispersions. Voltammetric scanning between −1.4 V and +0.3 V (vs. Ag/AgCl sat) for five successive cycles resulted in the reduction and deposition of ERGO on the PGE surface. The instrumental parameters for the electrodeposition procedure included a deposition time of 120 s, frequency of 50 Hz, amplitude of 0.04 V, and voltage step of 0.004 V. Following the electrodeposition procedure, the electrochemically reduced graphene oxide-modified pencil graphite electrode (ERGO-PGE) was dried for one hour and then conditioned in an acetate buffer solution (ABS) through square wave voltammetry before each measurement.

2.4. Voltammetric Analysis of Post-Transition Metals (Ga^{3+} and In^{3+})

Square wave anodic stripping voltammetry (SWASV) measurements were carried out in a 10 mL solution of 0.1 mol L^{-1} acetate buffer solution (ABS) with a pH of 4.38. The solution included 200 ppb of bismuth (Bi), 100 ppb of mercury (Hg), and Ga^{3+} and In^{3+}. Bismuth–mercury films and the target metal ions were deposited onto the ERGO-PGE by in situ coating following fixed potential reduction at −1.2 V (vs. Ag/AgCl sat) for 120 s. An anodic scan of the ERGO-PGE from −1.2 V to −0.5 V (vs. Ag/AgCl sat) through a square-wave waveform was used for detection. At the end of the scan, the ERGO-PGE underwent an electrochemical cleaning process to remove residual metals. This was achieved by applying a potential of 0.5 V (vs. Ag/AgCl sat) for 120 s, ensuring the electrode was thoroughly cleaned from any remaining metal ions. The standard oxidation potentials of Ga^{3+} (−0.88 V), In^{3+} (−0.67 V), Bi (−0.10 V), and Hg (0.21 V) recorded at the ERGO-PGE in 0.1 M of acetate buffer solution (pH 4.38) vs. the Ag/AgCl saturated reference electrode is shown in Figure S1. An overview of the ERGO-PGE preparation and stripping procedure (deposition and stripping steps) is given in Scheme 1.

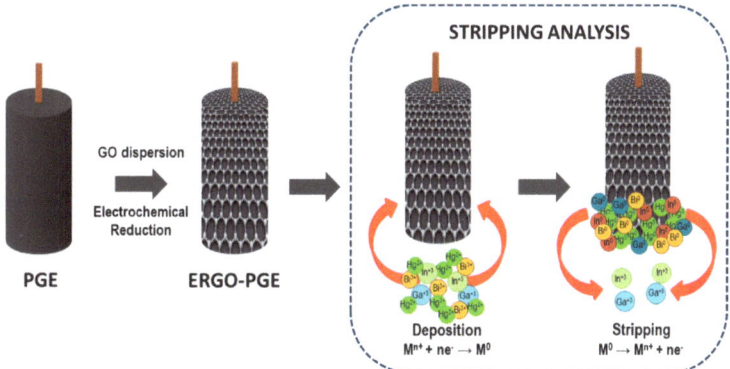

Scheme 1. Schematic illustration of the preferential stripping analysis of post-transition metals (In and Ga) at Bi/Hg films electroplated on graphene-functionalized graphite rods. Typically, electrochemically reduced graphene oxide (ERGO) nanoplatelets were electrochemically deposited on graphitic rods through successive fixed potential (−1.4 V, vs. Ag/AgCl sat) and cyclic reduction (five cycles) before the electroplating of bimetallic Bi/Hg-films. The ERGO-Bi/Hg-film-functionalized PGEs were then applied to the stripping analysis of In^{3+} and Ga^{3+} in wastewater samples.

3. Results and Discussion

3.1. Electrochemical Characterization and Optimization of ERGO-Bi/Hg-Film PGEs for the Simultaneous Stripping Analysis of Post-Transition Metals

The electrochemical reduction of graphene oxide (GO) dispersions prepared in 0.1 acetate buffer solution was performed by cyclic voltammetric reduction between −1.4 V and +0.3 V (vs. Ag/AgCl sat) for five successive cycles. The successive reduction and consequent deposition cycles are shown in Figure S2 for 20 cycles. Changes in the oxidation of the PGE surface during successive scans confirm the deposition of electrochemically reduced graphene oxide (ERGO) at the PGE surface. High-resolution scanning electron microscopy (HRSEM) images shown in Figure S3A,B confirm the deposition of ERGO sheets as irregular deposits on the spherical graphitic rod surface.

Figure 1a provides a comprehensive visualization of the electrochemical response of the PGE and ERGO-PGE in a ferricyanide solution. This evaluation involves the examination of the oxidation and reduction responses of a 5 mM $[Fe(CN)_6]^{3-/4-}$ redox couple, situated within a 0.1 M KCl solution, at both the unmodified PGE and the PGE modified with electrochemically reduced graphene oxide (ERGO-PGE). The measurements were conducted using cyclic voltammetry, employing a scan rate of 50 mV s^{-1}. Upon close inspection of the results, it is evident that the bare pencil graphite electrode exhibits broad peaks associated with the $[Fe(CN)_6]^{3-/4-}$ redox couple. Notably, at the bare PGE, the separation between the anodic and cathodic peaks measures 0.715 V (vs. Ag/AgCl sat). In stark contrast, the ERGO-PGE displays a notably different electrochemical response. The separation between the anodic and cathodic peaks at the ERGO-PGE is significantly narrower, measuring 0.115 V (vs. Ag/AgCl sat). This reduced peak separation can be attributed to the modifications involving the deposition of reduced graphene oxide (GO). The effect of ERGO modification is reflected in an enhancement of the response current, signifying a substantial improvement in the sensitivity of the ERGO-PGE electrode. This is attributed to the high electrochemical active surface area (ECSA) of the deposited ERGO. This phenomenon signifies a substantial improvement in the sensitivity of the electrode surface. The enhanced sensitivity is primarily a consequence of the higher surface-to-volume ratio associated with the modified ERGO-PGE. This alteration facilitates a more efficient and rapid transfer of electrons, resulting in superior electrochemical performance. The visual representation in Figure 1a effectively captures these changes, demonstrating the positive impact of ERGO modification on the electrochemical response of the pencil graphite electrode.

Figure 1b illustrates the relationship between peak currents and the scan rate across a range of 10 mV s^{-1} to 100 mV s^{-1}. This assessment is critical in understanding the electrochemical behavior of the system. The peak currents, corresponding to the oxidation and reduction processes, display a distinct and noteworthy pattern; they exhibit a linear increase as the square root of the scan rate increases (see Inset). The observed linearity in the relationship between peak currents and the square root of the scan rate is a significant finding. It strongly suggests that the electrochemical processes at play are predominantly diffusion-controlled. In other words, the observed changes in peak currents are primarily dictated by the rate at which analyte species can diffuse to and from the electrode surface. To further quantify this relationship, a correlation coefficient of 0.996 was obtained for both the anodic and cathodic peak currents concerning the square root of the scan rate. This high correlation coefficient serves as compelling evidence that the electrochemical behavior being studied is indeed governed by a diffusion-controlled process. Such knowledge is vital for designing and optimizing electroanalytical methods, ensuring that the system operates predictably and consistently across different conditions.

The sensitivity and overall performance of any electrode are intricately linked to the efficiency of electron transfer across the deposited film [61,62]. The figure displayed in Figure 1c underscores the influence of the number of graphene oxide (GO) electrodeposition cycles on the oxidation peak of Gallium (Ga^{3+}) and Indium (In^{3+}). Upon careful examination, it is evident that a deposition consisting of five cycles yields the most prominent peak for Ga^{3+} and In^{3+}. This observation prompts the selection of the five-cycle deposition approach for subsequent experiments aimed at detecting Ga^{3+} and In^{3+} at electrodes modified with electrochemically reduced graphene oxide (ERGO-PGE). However, it is noteworthy that the peak current for Ga^{3+} and In^{3+} displays a decline beyond the fifth cycle (Figure 1d). This decrease in peak current is primarily attributed to the thickening of the graphene film. As the film thickness increases, it progressively hinders the efficient flow of electrons to the electrode surface. In essence, while a thicker film can be advantageous in certain scenarios, it reaches a point where it starts to impede electron transfer, leading to a reduction in peak current. It is imperative to highlight that the electrodeposition of GO onto the pencil electrode, employing cycling voltammetry, offers a distinct advantage. This advantage lies in the precise control it affords over the thickness of the film deposited on the electrode surface. The ability to modulate film thickness is a key factor in tailoring electrode performance to suit specific analytical requirements.

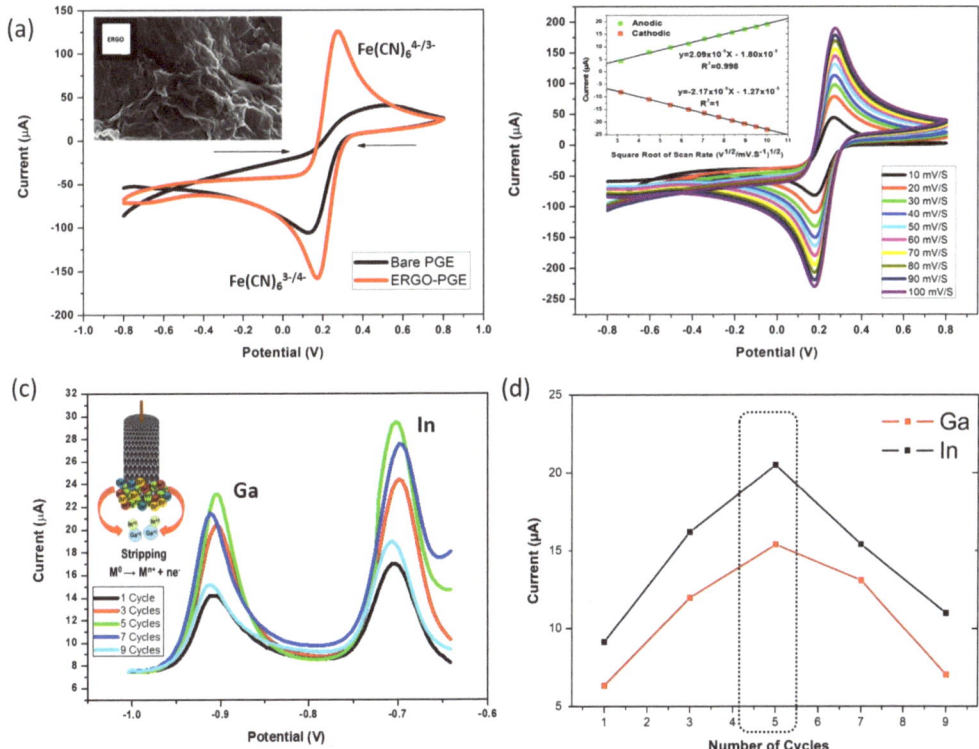

Figure 1. (**a**) Comparative CV voltammograms of bare PGE (black) and ERGO-PGE (red) recorded in 5 mM [Fe(CN)$_6$]$^{3-/4-}$ with 0.1 M KCl as supporting electrolyte. Inset: HRTEM images of ERGO-nanoplatelets deposited on PGE surfaces. (**b**) Scan rate dependence (10 to 100 mV s^{-1}) of ERGO-PGE recorded in the presence of redox probe and inset of the recorded currents vs. square root of scan rate. (**c**) SWASV voltammograms of 20 ppb Ga^{3+} and 20 ppb In^{3+} in 0.1 M acetate buffer solution (pH: 4.38) at ERGO-Bi/HgF-PGE with different numbers of GO deposition cycles from 1 to 9 cycles. (**d**) Corresponding plot of the effect of ERGO reduction cycles on peak currents of Ga^{3+} and In^{3+} at the ERGO-Bi/HgF-PGE in 0.1 M acetate buffer solution (pH 4.38).

3.2. The Synergetic Effect of Graphene Sheets and Bismuth-Mercury Film on the Electrochemical Response of Ga^{3+} and In^{3+}

Figure 2a shows changes observed in the oxidation peak heights of 30 ppb Ga^{3+} and 20 ppb In^{3+} through the use of Square Wave Anodic Stripping Voltammetry (SWASV). A comparison is made between two types of electrodes: the bare pencil graphite electrode (PGE) and the modified pencil graphite electrode. The competitive stripping analysis of Ga^{3+} and In^{3+} at the bimetallic metal-film-functionalized carbon nanostructured electrode surfaces follows the schematic shown in Figure 2b. Reduced graphene oxides have readily been studied for their ability to absorb metal cations through weak intermolecular forces. Cation-π interactions arise as a result of strong attractive forces between positively charged entities and the π-electron cloud of aromatic groups. Typically, Bi and Hg cations are accumulated at the ERGO surfaces through cation-π stacking interactions arising from the unoccupied π-orbitals in the graphene lattice structure. The weak Van der Waal's interactions and increased active surface area associated with graphene inclusion facilitate metal accumulation at the electrode surface during the deposition step. Accumulated Bi^{3+} and Hg^{2+} cations are reduced at the ERGO-functionalized PGE surfaces to create bimetallic films. While this phenomenon has not been independently studied in stripping analysis, the accumulation may follow a similar route to the work performed in membrane analysis and energy applications [63,64]. Simultaneously, target metal ions (Ga^{3+} and In^{3+}) are reduced to forms with the deposited metallic film:

$$xM^{n+} + (nx+3y+2z)e^- + yBi^{3+} + zHg^{2+} \rightarrow M_xBi_yHg_z) \qquad (1)$$

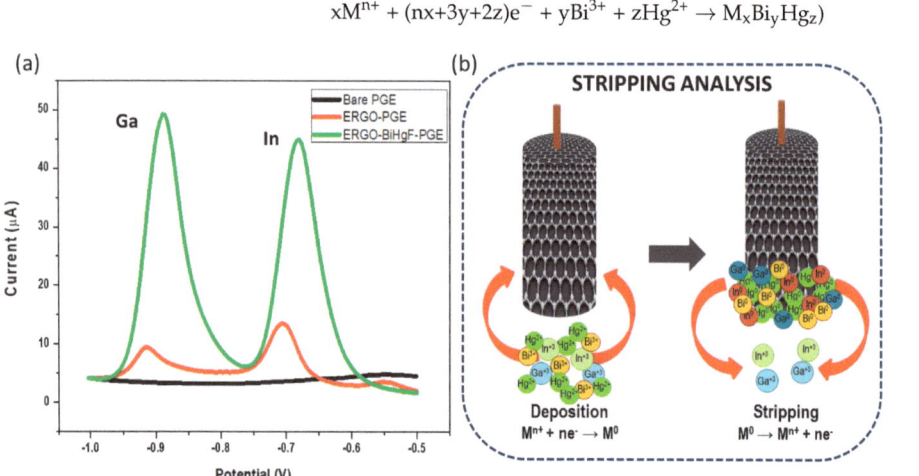

Figure 2. (a) The comparison of SWASV measurements of 0.1 mol L^{-1} ABS at pH 4.38 containing 30 ppb Ga^{3+} and 20 ppb In^{3+} at bare PGE, ERGO-PGE and ERGO-Bi/HgF-PGE. (b) Schematic illustration of anodic stripping voltammetry detection of In^{3+} and Ga^{3+}.

Bismuth and Mercury were selected for their ability to form alloys with a wide range of metal cations, facilitating Ga^{3+} and In^{3+} accumulation for reduction, thereby pre-concentrating the ERGO-PGE surface. Competition for available film sites results in preferential accumulation of In^{3+} and Ga^{3+} for the Hg^{2+} and Bi^{3+} cations, respectively. Thereafter, the subsequent re-oxidation of the metal cation at prescribed redox potentials causes its stripping, giving rise to two distinct peaks in the stripping step. Upon careful examination, it becomes evident that the oxidation peaks corresponding to Ga^{3+} and In^{3+} are distinctly visible only when the electrode is subject to modification with electrochemically reduced Graphene Oxide (ERGO). This is a crucial observation, indicating the pivotal role played by the ERGO modification in enabling the electrochemical determination of these analytes. Notably, a substantial and noteworthy enhancement in the peak height is observed when transitioning from the bare PGE to the ERGO-Bi/HgF-PGE, as illustrated in Figure 2. This enhancement in peak currents can be attributed to several key factors. First and foremost, the introduction of ERGO-Bi/HgF on the electrode's surface significantly increases the surface area-to-volume ratio. The resulting high surface area within the nanometer range (1–100 nm) is instrumental in promoting enhanced conductivity and electron transfer. The concept of quantum confinement plays a crucial role in this context. This phenomenon, occurring at the nanoscale, effectively confines electrons and other charge carriers within specific energy levels, leading to more efficient electron transfer processes. The combined synergistic effects of the ERGO cation-π stacking and alloy formation (between Ga^{3+} and In^{3+} and bimetallic Bi/Hg-films) gives rise to high sensitivity. As a result, the ERGO-Bi/HgF-PGE exhibits superior electrochemical performance, enabling the precise and sensitive determination of Ga^{3+} and In^{3+} in the tested samples. This heightened sensitivity is essential for trace analysis in various fields, making the ERGO-Bi/HgF-PGE a valuable tool in analytical chemistry and environmental monitoring. The bimetallic ERGO-Bi/HgF-PGE was compared to the ERGO-BiF-PGE and ERGO-HgF-PGE and the results are shown in Figure S4. Enhanced peak currents are shown for both Ga^{3+} and In^{3+} over the monometallic counterparts under optimized Bi and Hg concentrations.

3.3. Experimental Parameter Optimization of ERGO-Bi/Hg-Film PGE for Ga^{3+} and In^{3+} Detection

The impact of electrochemical cleaning on the restoration of the initial response of Ga^{3+} and In^{3+} at the ERGO-PGE was subject to thorough investigation using Square Wave Anodic Stripping Voltammetry (SWASV). As depicted in Figure 3a, the relationship between the oxidation peaks of Ga^{3+} and In^{3+} at the ERGO-PGE and the electrochemical cleaning process is elucidated. Figure 3a clearly illustrates that a remarkable recovery of the original response of Ga^{3+} and In^{3+} can be achieved through the systematic application of electrochemical cleaning to the electrode. The restoration of the peak height and signal intensity is particularly significant for accurate and reliable analysis, ensuring that subsequent measurements yield consistent and precise results. After careful examination, it was determined that a cleaning duration of 120 s using electrochemical cleaning delivered the most effective recovery of the electrode's initial response. This duration was established as the optimal cleaning time for further measurements, as it strikes a balance between thorough cleaning and efficient recovery. The chosen 120 s cleaning time ensures that the ERGO-PGE remains in an optimal condition for the sensitive and accurate determination of Ga^{3+} and In^{3+} in subsequent analyses. This process enhances the reliability and reproducibility of the analytical results, making it a critical step in the electroanalytical methodology employed in this study. For each experimental set, the same modified electrodes are employed and subjected to electrochemical cleaning between successive measurements. Given the low cost of pencil graphite electrodes, a freshly modified electrode is utilized for each distinct set of experiments.

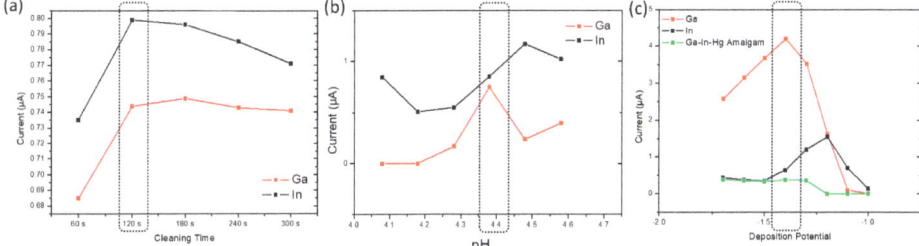

Figure 3. Effect of (**a**) electrochemically cleaning time (60–300 s), (**b**) pH (4.1–4.6), and (**c**) deposition potential (−1 to −1.7 V (vs. Ag/AgCl sat)) on the oxidation peak currents of Ga^{3+} and In^{3+} at the ERGO-Bi/HgF-PGE in a 0.1 M ABS (pH 4.38) containing 10 ppb Ga^{3+} and 2 ppb In^{3+}.

The influence of pH on the stripping peak current (I_p) was meticulously examined through a series of square wave voltammetry measurements conducted within a pH range spanning from 4.08 to 4.58. The analyte solution comprised 10.0 ppb Ga^{3+} and 2.0 ppb In^{3+}, as depicted in Figure 3b. The relationship between pH and the stripping peak current (I_p) for Ga^{3+} and In^{3+} was clearly observed. It is evident from the data that the peak current for Ga^{3+} reaches its maximum magnitude at a pH of 4.38. At this pH, two distinct and well-defined oxidation peaks, corresponding to Ga^{3+} and In^{3+}, are clearly discernible. The selection of pH 4.38 as the optimal pH setting is of paramount importance for subsequent Ga^{3+} and In^{3+} measurements. At this pH, the analytical system exhibits the highest sensitivity and provides a clear demarcation between the oxidation peaks of Ga^{3+} and In^{3+}. The use of pH 4.38 ensures that the electroanalytical methodology is operating under conditions that yield the most precise and reliable results, enhancing the overall accuracy of the measurements.

Reduction potentials influence the deposition of Ga^{3+} and In^{3+} at the ERGO-Bi/HgF-PGE. Their influence on peak currents was methodically investigated within a potential range spanning from −1.0 V to −1.7 V (vs. Ag/AgCl sat). The data presented in Figure 3c provide insights into the influence of deposition potential on the electrochemical behavior of these metal ions. At deposition potentials more negative than −1.2 V (vs. Ag/AgCl sat), a signal emerges at approximately −1.3 V (vs. Ag/AgCl sat). This signal is attributed to the formation of a Ga-In-Hg amalgam, a phenomenon that suppresses the reduction reaction responsible for the deposition of metal ions from the solution onto the electrode surface. In general, the peak currents corresponding to In and Ga exhibit an upward trend as the deposition potential becomes more negative. This increase in peak current is especially pronounced, up to -1.2 V (vs. Ag/AgCl sat) for In^{3+} and up to -1.4 V (vs. Ag/AgCl sat) for Ga^{3+}. This behavior can be attributed to the preferential reduction and deposition of In^{3+} and Ga^{3+} at the electrode surface under these conditions. However, it is noteworthy that beyond the deposition potential of -1.2 V (vs. Ag/AgCl sat), electrode saturation occurs, leading to a decline in the stripping response for In^{3+}. As a result, a deposition potential of -1.2 V (vs. Ag/AgCl sat) was judiciously selected for further analysis. This choice ensures that the electrode operates in a regime that maximizes the analytical sensitivity for In^{3+} and Ga^{3+}, contributing to more accurate and reliable measurements.

3.4. Intermetallic Interferences

Mercury and Bismuth films notably show markedly different performances for accumulating In^{3+} or Ga^{3+} ions at the electrode surface, respectively. Figure 4a,b show the effect of (a) Ga^{3+} concentration and (b) In^{3+} concentration on Hg and Bi film formation. The data clearly indicate that during the deposition of Ga^{3+} onto the electrode surface, competition ensues between Hg and Ga^{3+} ions for available surface sites on the electrode. Consequently, with increasing Ga^{3+} concentration, less Hg is plated at the surface of the ERGO-PGE. Conversely, an increase in In^{3+} concentration increases Bi plating due to fused alloy formation. The findings highlight preferential interactions between Ga^{3+} and Hg, and In^{3+} and Bi, respectively.

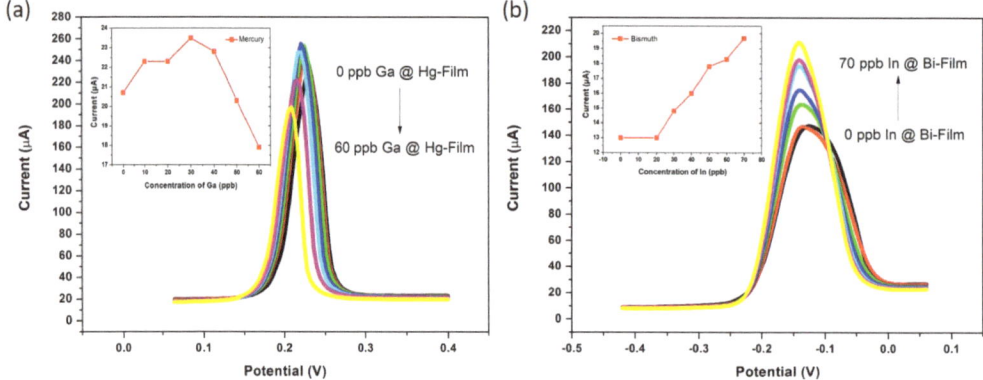

Figure 4. The effect of (**a**) Ga^{3+} concentration, and (**b**) In^{3+} concentration on Hg and Bi film formation.

Figure 5a,b provide valuable insights into the interaction between Ga^{3+} ions and the peak height of In^{3+} at a constant concentration of In^{3+}. The literature indicates that gallium demonstrates selectivity toward bismuth, while indium exhibits selectivity toward mercury [65]. As illustrated in Figure 5b, the amount of deposited gallium on the electrode surface has a direct and significant impact on the In^{3+} peak height during anodic stripping. This observation aligns with the fact that Ga^{3+} ions experience a faster rate of reduction compared to Hg ions. Consequently, the number of unoccupied sites on the electrode surface, which can be utilized for the electrodeposition of mercury from the solution, decreases as the amount of deposited gallium increases. The net result of this competition is the inhibition of deposited mercury, limiting the formation of an In^{3+} amalgam. This phenomenon becomes more pronounced with a decrease in the amount of electrodeposited mercury on the electrode surface. These insights are critical for understanding and optimizing the electrochemical behavior of In^{3+} and Ga^{3+} ions, thereby enhancing the accuracy and efficiency of subsequent analytical measurements.

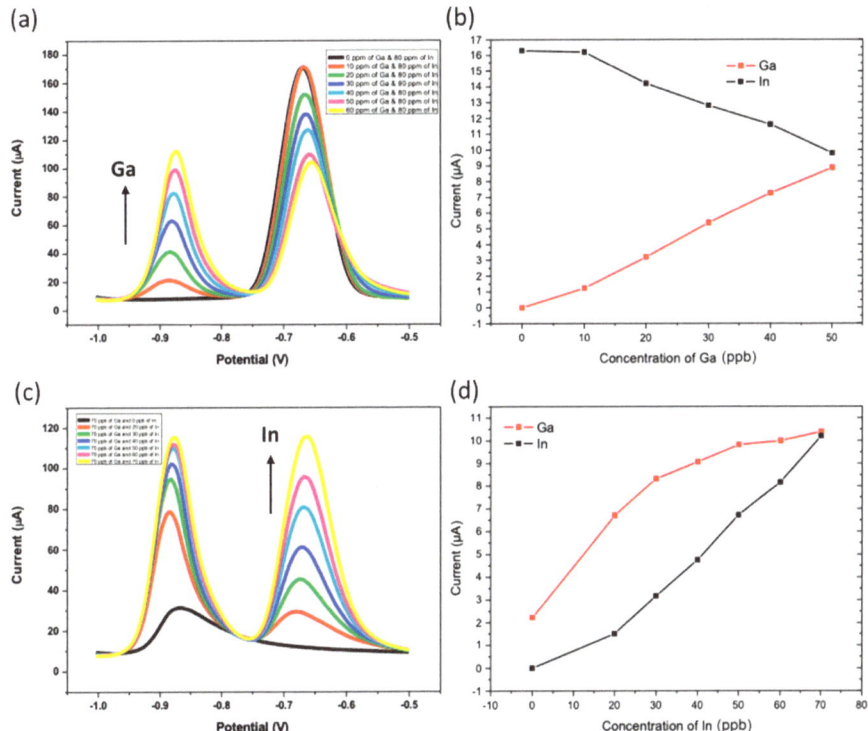

Figure 5. Voltammograms and corresponding scatter plots of Ga^{3+} and In^{3+} at the ERGO-Bi/HgF-PGE in 0.1 M acetate buffer (pH 4.38) with (**a**,**b**) 80 ppb In^{3+} concentration and Ga^{3+} concentration varied between 0 ppb to 60 ppb, and (**c**,**d**) 70 ppb Ga^{3+} concentration and In^{3+} concentration varied between 0 ppb and 70 ppb.

Figure 5c provides a clear depiction of the influence of In^{3+} ions on the oxidation response of gallium ions. The data reveal an interesting trend, as depicted in Figure 5d, where an increase in the concentration of indium results in a continuous and noticeable enhancement in the oxidation peak of a constant amount of gallium. In simpler terms, as the concentration of indium ions in the solution increases, the electrochemical response of gallium ions during oxidation becomes increasingly pronounced. This observed phenomenon indicates an accelerated electrodeposition of bismuth ions on the surface sites of the electrode, effectively outperforming indium ions. This intensified electrodeposition of bismuth, known for its stronger affinity with gallium, leading to the formation of a gallium-fused alloy, ultimately manifests in a heightened Ga^{3+} oxidation peak.

Figure S5 illustrates the investigation into interferences from other metal ions in the water samples, using standard addition and spiking techniques. Metal ions that interfere with Ga^{3+} and In^{3+} were added to the test solution. These ions can form alloys with bismuth or amalgams with mercury. They may also produce reduction peaks that overlap with or suppress the Ga^{3+} and In^{3+} peaks, thus affecting measurement accuracy. The study evaluated the impact of Zn^{2+}, Cd^{2+}, Pb^{2+}, Ni^{2+}, Co^{2+}, and Cu^{2+}. Notably, a two-fold excess of Pd^{2+} and a three-fold excess of Co^{2+} caused a 30% and 20% decrease in Ga^{3+} peak currents, respectively, and a 10% and 5% decrease in In^{3+} peak currents. The remaining metals did not interfere with Ga^{3+} and In^{3+} measurements, improving the method's selectivity and reliability.

3.5. Analytical Performance of the ERGO-Bi/Hg-Film PGEs for the Individual Determination of Ga^{3+} and In^{3+}

The analytical performance of the developed ERGO-Bi/Hg-film PGEs for the individual analysis of Ga^{3+} and In^{3+} in test samples was evaluated in test solutions using an acetate buffer as the electrolyte. The detection limit was determined by constructing a calibration curve for Ga^{3+} using the optimized SWASV conditions at the ERGO-Bi/HgF-PGE. The relationship between Ga^{3+} concentration (ranging from 30 to 70 ppb) and peak current is illustrated in Figure 6a. Increases in the stripping peak current associated with the three-electron oxidation of Ga^0 to Ga^{3+} were observed during the anodic scan, with no observable shift in peak potential. The data shown in the inset display a linear correlation between Ga^{3+} concentration and peak current, with a noteworthy correlation coefficient of 0.988 (n = 3). The linear regression equation for the calibration curves (n = 3) was employed to determine the DL. The equation is as follows:

$$I(\mu A) = 0.367 \, (\mu A/ppb) \, [Ga^{3+}] \, (ppb) - 3.08 \, (\mu A) \quad (2a)$$

$$I(\mu A) = 25.59 \, (\mu A/\mu M) \, [Ga^{3+}] \, (\mu M) - 3.08 \, (\mu A) \quad (2b)$$

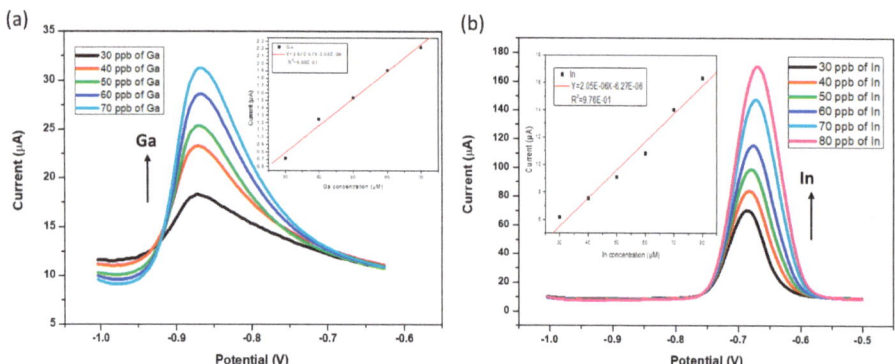

Figure 6. SWAS voltammograms of the individual analysis of (**a**) Ga^{3+} between 30 and 70 ppb and (**b**) In^{3+} between 30 and 80 ppb recorded at ERGO-Bi/HgF-PGE, under optimized parameters. The corresponding calibration curves are shown as insets.

Using the 3σ/slope ratio, where σ represents the standard deviation of the mean value from 10 voltammograms of the blank, the detection limit was determined to be 1.02 μM. This value signifies the lowest concentration of gallium that can be reliably detected under the given experimental conditions.

Similarly, the ERGO-Bi/Hg-film PGE was investigated for In^{3+} analysis. A linear correlation between In^{3+} concentration, ranging from 30 to 80 ppb, was found, and peak current was observed with a commendable correlation coefficient of 0.976, as depicted in Figure 6b. The standard deviation was determined based on the linear regression equation for the average of calibration curves (n = 6), expressed as follows:

$$I(\mu A) = 2.05 \, (\mu A/ppb) \, [In^{3+}] \, (ppb) - 3.08 \, (\mu A) \quad (3a)$$

$$I(\mu A) = 235.38 \, (\mu A/\mu M) \, [In^{3+}] \, (\mu M) - 3.08 \, (\mu A) \quad (3b)$$

The detection limit for In^{3+} using the ERGO-PG-Bi/HgFE was calculated using the following formula:

$$D.L. = 3\sigma/slope \quad (3c)$$

Here, D.L. stands for the limit of detection, 3σ denotes three times the standard deviation of the blanks, and Slope represents the gradient (slope) of the calibration curve. Blanks were recorded at the ERGO-Bi/HgF-PGE in 0.1 M acetate buffer (pH 4.38) in the absence of Ga^{3+} and In^{3+}. Currents were recorded at the Ga^{3+} and In^{3+} oxidation potentials for 10 replications. The resultant bar graphs of the ten replications are shown in Figure S6, with relative standard deviation (RSD) percentages of 0.92% and 1.4% found at Ga^{3+} and In^{3+} oxidation potentials. A summary of the analytical data used for the limit of detection calculations for both simultaneous and individual analysis is given in Table S1. The determined detection limit was 0.276 µM, based on the results of ten replications of the electrode's response in the blank solutions. This value indicates the lowest concentration of indium that can be reliably detected under the given experimental conditions. Improved current responses may be observed for In^{3+} over Ga^{3+} owing to the higher affinity of In^{3+} for Hg films, which offer improved current responses over Bi film electrodes. Moreover, the findings show the high sensitivity of both post-transition metal cations for the ERGO-Bi/Hg-film PGEs facilitated through enhanced electroplating of the bimetallic film at the ERGO-functionalized surfaces.

3.6. Simultaneous Analysis of In^{3+} and Ga^{3+} at ERGO-Bi/Hg-Film PGEs in Simulated Samples

The simultaneous analysis of Ga^{3+} and In^{3+} ions was carried out across a concentration range of 20 to 70 ppb at the ERGO-PG-Bi/HgFE, aimed at assessing the analytical performance of the electrode. The resulting peak currents from the voltammogram (as illustrated in Figure 7a) were used to create the calibration plot presented in Figure 7b. Based on the calibration curve, the detection limits for these metal ions were computed using the following formula: three times the standard deviation (3σblank) of the blank solution divided by the slope of the calibration curve. The standard deviation of the blank was determined through ten replications in the blank solutions. The calculated detection limits for Ga^{3+} and In^{3+} were 0.00253 µmol L^{-1} and 0.00727 µmol L^{-1}, respectively, with correlation coefficients of 0.977 and 0.996, respectively. These detection limits signify the lowest concentrations of Ga^{3+} and In^{3+} that can be accurately and reliably detected under the given experimental conditions, while the correlation coefficients reflect the quality of the calibration curves for these respective ions.

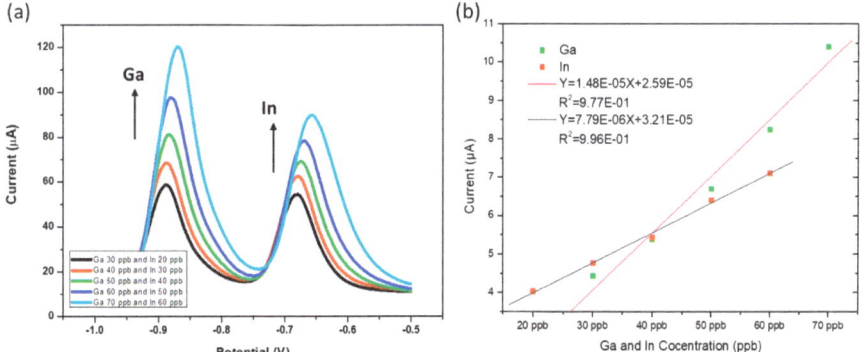

Figure 7. (**a**) SWAS voltammograms, and (**b**) corresponding calibration curves for ERGO-Bi/Hg-Film-PGE, with the optimized parameters (**a**). The Ga^{3+} concentrations range from 30 µmol L^{-1} to 80 ppb and the In^{3+} concentrations range from 20 to 70 ppb.

An overview of recent work regarding In and Ga^{3+} detection is summarized in Tables 1 and 2, respectively. Hanging mercury electrodes remain the most common support material for Ga^{3+} and In^{3+} determination to date, owing to the high-affinity amalgam formation achieved with mercury drops. Metallic films, like Bi, Pb, and Sb, are commonly employed to reduce the amount of toxic mercury. Mercury silver amalgams are also being used quite commonly. Of these, comparable detection limits are achieved for all platforms studied, indicating that metallic film surfaces are able to detect both Ga^{3+} and In^{3+} sensitively. The Bi/Hg bimetallic film, which here is studied for the first time for post-transition metals, shows good performance compared to the more conventional films. A comparison of this study with other research on the electrochemical detection of gallium and indium reveals several key distinctions. All reported studies to date have relied on costly solid electrodes or toxic hanging mercury and mercury-film electrodes, which have contributed to the high sensitivity of these works. Unlike previous studies, the proposed work is the only to investigate cost-effective, disposable electrodes utilizing pencil graphite electrodes (PGEs) to improve accessibility in resource-limited settings. Additionally, this is the first study to use graphene to enhance the low electrode sensitivity associated with disposable electrode materials for the detection of target metals, a strategy that enabled the successful application of the inexpensive PGE. To the best of our knowledge, all previous studies have performed individual detection of In^{3+} or Ga^{3+} ions. The lower sensitivity achieved herein may be a consequence of the fact that this is the first reported work regarding the simultaneous detection of gallium and indium, with only one instance in 2004 attempting this under significantly higher metal concentrations. In that study, the detection limits achieved were approximately 10^7 times higher than those obtained here. The use of a low-cost, disposable PGE, enhanced with graphene for improved sensitivity, and the application of a bimetallic mercury–bismuth film to achieve selectivity for the simultaneous detection of gallium and indium, demonstrate the novelty of this study, particularly in achieving low detection limits.

Table 1. Comparison of the proposed method with some of the previous electrochemical stripping techniques used for the determination of gallium.

Target Analyte	Electrode	Linear Range (mol. L^{-1})	Detection Limit (mol. L^{-1})	Reference
Ga	Hg(Ag)FE	2×10^{-9}–1×10^{-7}	1.0×10^{-10}	[5]
Ga	BiFe	3×10^{-10}–3×10^{-7}	1×10^{-10}	[7]
Ga	HDME	1.4×10^{-8}–2.7×10^{-7}	5×10^{-8}	[8]
Ga	BiSME	2×10^{-8}–2×10^{-6}	7×10^{-9}	[10]
Ga	PbFE	1×10^{-8}–2×10^{-7}	3.8×10^{-9}	[11]
Ga	Hg(Ag)FE	1.25×10^{-9}–9×10^{-8}	1.6×10^{-9}	[13]
Ga	Ag-HgFE	5×10^{-9}–8×10^{-8}	1.4×10^{-9}	[46]
Ga	HMDE	1×10^{-9}–1×10^{-7}	4×10^{-10}	[66]
Ga	HDME	1×10^{-8}–1.7×10^{-6}	5.7×10^{-11}	[67]
Ga	CPE	2.9×10^{-10}–8.6×10^{-8}	1.4×10^{-10}	[68]
Ga	HMDE	1×10^{-4}–1×10^{-5}	2×10^{-8}	[69]
Ga	ERGO-Bi/HgF–PGE	0.43×10^{-6}–1.14×10^{-6}	2.53×10^{-9}	This study

Table 2. Comparison of the proposed method with some of the previous electrochemical stripping techniques used to determine indium.

Target Analyte	Electrode	Linear Range (mol. L^{-1})	Detection Limit (mol. L^{-1})	Reference
In	Nafion-GCE	1.0×10^{-10}–1.0×10^{-9}	7.5×10^{-10}	[4]
In	Hg(Ag)FE	1.25×10^{-9}–9×10^{-8}	1.4×10^{-9}	[13]
In	Static Hg drop	34–340×10^{-9}	4.3×10^{-11}	[70]
In	HMDE	3.4–87×10^{-10}	1.1×10^{-12}	[71]

Table 2. Cont.

Target Analyte	Electrode	Linear Range (mol. L^{-1})	Detection Limit (mol. L^{-1})	Reference
In	HDME	3.9×10^{-7}–5×10^{-4}	1×10^{-7}	[72]
In	SbFe	8.4–84×10^{-10}	1.2×10^{-11}	[73]
In	TMFE	-	1×10^{-14}	[74]
In	HMDE	0–3.2×10^{-7}	1.7×10^{-9}	[75]
In	Metal thin-film	4.3–21.7×10^{-6}	0.012	[65]
In	SbFe	1.7–17×10^{-7}	6.9×10^{-11}	[76]
In	ERGO-Bi/HgF–PGE	0.17×10^{-6}–0.60×10^{-6}	7.27×10^{-9}	This study

3.7. Application to Tap Water Samples

The potential and applicability of the ERGO-Bi/HgF–PGE sensors were assessed for the individual and simultaneous determination of Ga^{3+} and In^{3+} in real water samples. For this assessment, the ERGO-Bi/HgF–PGE was employed to quantify Ga^{3+} and In^{3+} concentrations in tap water samples, prepared as described in the experimental procedures. The analysis of Ga^{3+} and In^{3+} in these samples was carried out in triplicate, both individually and simultaneously, using the standard addition method, as depicted in Figures 8 and 9. Recovery studies were performed in test solutions and measured the ability of the sensor to accurately detect known concentrations of Ga^{3+} and In^{3+}. Test solutions containing 2 ppb of Ga^{3+} and In^{3+} were prepared separately. The concentrations were determined via standard addition, wherein samples were spiked with fixed quantities of the target metal cation, and the concentration was determined from the obtained calibration curve (inset). Recovery percentages, calculated as the ratio of detected to known concentrations (Equation (4)), ranged from 90.1% to 103.8% for Ga^{3+} and In^{3+}, respectively, demonstrating the sensor's accuracy and reliability in test samples. A recovery percentage above or below 100%, indicates minor deviations in sensor accuracy. Recovery percentages greater than 100% can result from several factors, including matrix effects, where other substances present in the water sample may interfere with the ion detection process, or slight variations in the sensor's sensitivity/reproducibility. In such cases, the sensor may amplify the response, leading to an overestimation of ion concentration. Recovery values close to 100% are generally considered favorable, as they indicate reliable performance with minimal bias in real-world sample conditions. Moreover, the relative standard deviations (RSD) were determined to be in the range of 3–5% for the oxidation peak of Ga^{3+} when measured individually and 1.8–3.4% for In^{3+} under the same conditions. These findings are further summarized in Table 3.

$$\text{Recovery \%} = \frac{(\text{Determined Concentration})}{(\text{Spiked Concentration})} \times 100\% \quad (4)$$

The simultaneous determination of Ga^{3+} and In^{3+} in tap water samples using the ERGO-Bi/HgF-PGE also yielded favorable results, with recovery percentages ranging from 95% to 106% for Ga^{3+} and In^{3+}, respectively. The RSD for the quantification of Ga^{3+} and In^{3+} simultaneously in tap water samples remained less than 5%, indicating the precision and robustness of the electrode for concurrent analysis in a real-world sample matrix. The findings from Tables 3 and 4 show that both individual and simultaneous detection of Ga^{3+} and In^{3+} is possible with good accuracy. Simultaneous detection of In^{3+} proved to be more difficult, and higher concentrations were required.

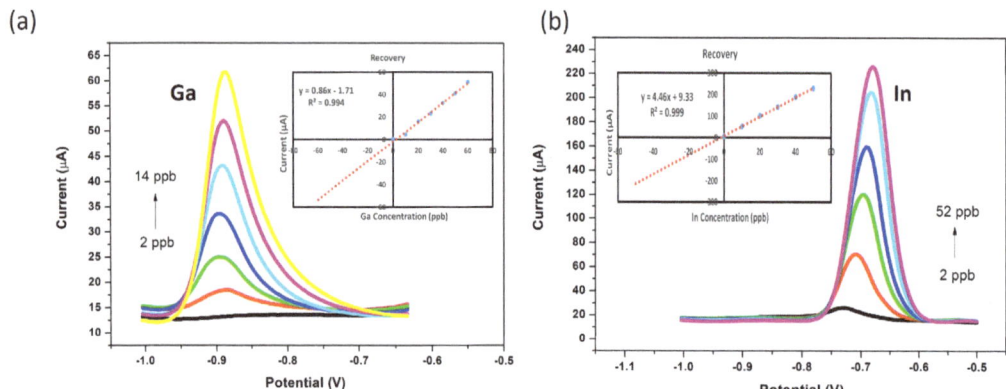

Figure 8. Analysis of 2 ppb of (**a**) Ga^{3+} and (**b**) In^{3+} in tap water (pH 4.38). The recorded voltammograms and standard addition plots are provided.

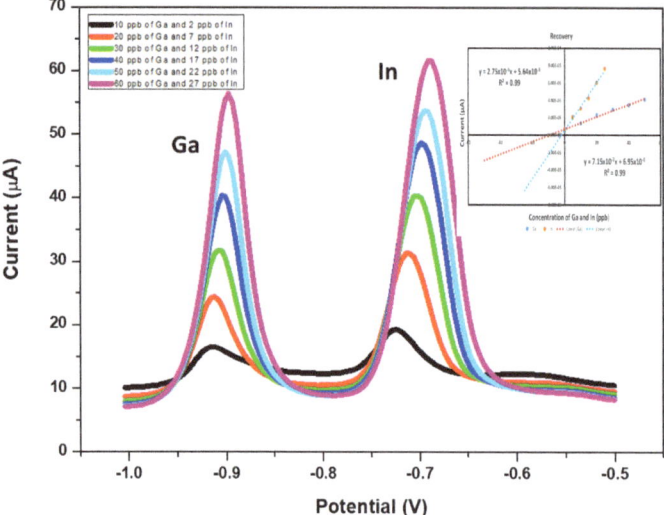

Figure 9. Voltammograms and standard addition plots observed for the simultaneous detection of 2 ppb of In^{3+} and 10 ppb Ga^{3+} in tap water.

Table 3. Recovery for the individual determination of gallium and indium in tap water samples using ERGO-Bi/HgF-PGE.

Target Analyte	Repetitive Cycles	Original (ppb)	Added (ppb)	Found (ppb)	RSD (%)	Recovery (%)
Gallium	First	ND *	2	2.07	2.43	103.5
	Second	ND *	2	1.95	1.79	97.5
	Third	ND *	2	2.1	3.45	105
Indium	First	ND *	2	1.9	3.62	95
	Second	ND *	2	2.05	1.74	102.5
	Third	ND *	2	1.95	1.79	97.5

* ND = Not detected.

Table 4. Recovery for the simultaneous determination of gallium and indium in tap water samples using ERGO-Bi/HgF-PGE.

Target Analyte	Repetitive Cycles	Original (ppb)	Added (ppb)	Found (ppb)	RSD (%)	Recovery (%)
Indium	First	ND *	2	2.08	2.77	104
	Second	ND *	2	2.11	3.78	105.5
	Third	ND *	2	1.9	3.62	95
Gallium	First	ND *	10	10.5	3.45	105
	Second	ND *	10	9.6	2.88	96
	Third	ND *	10	2.1	1.40	102

* ND = Not Detected.

4. Conclusions

The electrochemically reduced graphene oxide pencil graphite electrode with an in situ plated bismuth–mercury film (ERGO-PG-Bi/HgFe) was for the first time introduced as an effective alternative for the simultaneous quantification of trace amounts of Ga (III) and In (III) in a single measurement using square wave anodic stripping voltammetry. The improved adsorption of Ga^{3+} and In^{3+} ions at the Bi/Hg films on the graphene-functionalized electrodes, due to cation-π interactions and fused alloy formation during the preconcentration step, significantly enhanced sensitivity, achieving detection limits of 2.53 nmol L^{-1} for Ga^{3+} and 7.27 nmol L^{-1} for In^{3+}. The preferential accumulation of each post-transition metal, used in transparent displays, to form fused alloys at Bi and Hg films, respectively, is highlighted. This innovative electrode offers a more environmentally friendly and cost-effective approach for the simultaneous determination of gallium and indium, while delivering comparable performance to established techniques described in the literature. The development of these highly sensitive electrochemical sensors has enabled the achievement of well-resolved stripping voltammetric peaks, allowing for the simultaneous analysis of Ga^{3+} and In^{3+} in water samples through square wave anodic stripping voltammetry. Moreover, the synergistic effects of the bimetallic film were, for the first time, demonstrated for transition metal targets and demonstrate the improved electron transfer rates and active surface area of the disposable carbon nanostructured surface. The study is the first report to use a bimetallic Bi/Hg film for the detection of In^{3+} and Ga^{3+}. Moreover, graphite rods were used for the first time as disposable electrodes for the detection of post-transition metals, compared to the costly solid and toxic hanging mercury drop electrodes used in past studies. This is the first report on the use of electrochemically reduced graphene oxide plates for In^{3+} and Ga^{3+} detection.

Supplementary Materials: The following supporting information can be downloaded at: https://www.mdpi.com/article/10.3390/c10040095/s1, Figure S1: Stripping peaks of Ga (-0.88 V), In (-0.67 V), Bi (-0.10 V), and Hg (0.21 V) recorded at the ERGO-PGE in 0.1 M Acetate Buffer Solution (pH 4.7) vs. Ag/AgCl saturated reference electrode; Figure S2: Cyclic electrochemical reduction and deposition of Electrochemically Reduced Graphene Oxide (ERGO) onto pencil graphite electrodes (PGE) recorded between -1.5 and 0.3 V (vs. Ag/AgCl sat); Figure S3: HRSEM images of bare PGE (A), and Electrochemically Reduced Graphene Oxide PGE (B) at 100.00 times magnification; Figure S4: (a) SW voltammograms of 0.1 mol. L^{-1} ABS at pH 4.38 containing 30 ppb Ga^{3+} and 20 ppb In^{3+} at ERGO-BiF-PGE and ERGO-HgF-PGE. (b) Comparative bar graphs of the ERGO-BiF-PGE, ERGO-HgF-PGE and ERGO-Bi/HgF-PGE; Figure S5: The change of percentages of the oxidation peak current of 20 ppb of Ga and 5 ppb of In in presence of Zn^{2+}, Cd^{2+}, Pb^{2+}, Ni^{2+}, Co^{2+}, and Cu^{2+}; Figure S6: Ten replications of the response of the ERGO-PGE in the blank solution at gallium and indium oxidation potentials; Table S1: Analytical data for detection limit calculations.

Author Contributions: N.G. was responsible for the conceptualization, data acquisition, formal analysis, investigation, validation, and drafting of the manuscript. Z.A. assisted in data acquisition, formal analysis, and validation. N.J. and K.P. conceived the study and assisted in the formal analysis, review, and editing of the article, as well as the necessary funding acquisition. N.J. and P.G.L.B. assisted in the final review, editing, and clarification of the main ideas in the manuscript, and funding acquisition. All authors have read and agreed to the published version of the manuscript.

Funding: This study was supported by the National Research Foundation (NRF) of South Africa (Ref: TTK2204062362) and the DSI/NRF-SARChI chair in Analytical Systems and Processes for Priority and Emerging Contaminants (ASPPEC).

Institutional Review Board Statement: The work was conducted in accordance with the University of the Western Cape's ethical clearance guidelines.

Data Availability Statement: All research data can be found within this publication.

Conflicts of Interest: The authors declare that they have NO affiliations with or involvement in any organization or entity with any financial interest in the subject matter or materials discussed in this manuscript.

References

1. Hassanien, M.M.; Kenawy, I.M.; Mostafa, M.R.; El-Dellay, H. Extraction of gallium, indium and thallium from aquatic media using amino silica gel modified by gallic acid. *Microchim. Acta* **2011**, *172*, 137–145. [CrossRef]
2. Mortada, W.I.; Kenawy, I.M.; Hassanien, M.M. A cloud point extraction procedure for gallium, indium and thallium determination in liquid crystal display and sediment samples. *Anal. Methods* **2015**, *7*, 2114–2120. [CrossRef]
3. White, S.J.O.; Hemond, H.F. The anthrobiogeochemical cycle of indium: A review of the natural and anthropogenic cycling of indium in the environment. *Crit. Rev. Environ. Sci. Technol.* **2012**, *42*, 155–186. [CrossRef]
4. Xiang, C.; Zou, Y.; Xie, J.; Fei, X.; Li, J. Nafion-modified glassy carbon electrode for trace determination of indium. *Anal. Lett.* **2005**, *38*, 2045–2055. [CrossRef]
5. Piech, R. Novel Sensitive Voltammetric Detection of Trace Gallium(III) with Presence of Catechol Using Mercury Film Silver Based Electrode. *Electroanalysis* **2009**, *21*, 1842–1847. [CrossRef]
6. Moskalyk, R.R. Gallium: The backbone of the electronics industry. *Miner. Eng.* **2003**, *16*, 921–929. [CrossRef]
7. Kamat, J.V.; Guin, S.K.; Pillai, J.S.; Aggarwal, S.K. Scope of detection and determination of gallium(III) in industrial ground water by square wave anodic stripping voltammetry on bismuth film electrode. *Talanta* **2011**, *86*, 256–265. [CrossRef]
8. González, M.J.G.; Renedo, O.D.; Lomillo, M.A.A.; Martínez, M.J.A. Determination of gallium by adsorptive stripping voltammetry. *Talanta* **2004**, *62*, 457–462. [CrossRef]
9. Chitambar, C.R. Medical applications and toxicities of gallium compounds. *Int. J. Environ. Res. Public Health* **2010**, *7*, 2337–2361. [CrossRef]
10. Grabarczyk, M.; Wlazlowska, E. An Activated Bismuth Layer Formed In Situ on a Solid Bismuth Microelectrode for Electrochemical Sensitive Determination of Ga(III). *Membranes* **2022**, *12*, 1267. [CrossRef]
11. Grabarczyk, M.; Wasąg, J. Determination of trace amounts of Ga(III) by adsorptive stripping voltammetry with in situ plated bismuth film electrode. *Talanta* **2015**, *144*, 1091–1095. [CrossRef] [PubMed]
12. Grabarczyk, M.; Wasąg, J. Adsorptive Cathodic Stripping Voltammetric Method for Determination of Gallium Using an In Situ Plated Lead Film Electrode. *Electroanalysis* **2015**, *27*, 2596–2600. [CrossRef]
13. Grabarczyk, M.; Adamczyk, M.; Wardak, C. Simultaneous AdSV determination of Ga and In on Hg(Ag)FE electrode by AdSV in presence of cupferron. *Ionics* **2020**, *26*, 1019–1027. [CrossRef]
14. Chung, Y.; Lee, C.-W. Electrochemistry of Gallium. *J. Electrochem. Sci. Technol.* **2013**, *4*, 1–18. [CrossRef]
15. Paolicchi, I.; Renedo, O.D.; Alonso Lomillo, M.A.; Arcos Martínez, M.J. Application of an optimization procedure in adsorptive stripping voltammetry for the determination of trace contaminant metals in aqueous medium. *Anal. Chim. Acta* **2004**, *511*, 223–229. [CrossRef]
16. Gomdje, V.H.; Rosie, T.; Ngono, L.; Najih, R.; Chtaini, A. Electroanalytical Determination of Lead with Carbon Paste Modified Steel Electrode. *Acta Tech. Corviniensis-Bull. Eng.* **2013**, *6*, 139.
17. Pokpas, K.; Zbeda, S.; Jahed, N.; Mohamed, N.; Baker, P.G.; Iwuoha, E.I. Electrochemically reduced graphene oxide pencil-graphite in situ plated bismuth-film electrode for the determination of trace metals by anodic stripping voltammetry. *Int. J. Electrochem. Sci.* **2014**, *9*, 736–759. [CrossRef]
18. Wang, J.; Lu, J.; Kirgöz, Ü.A.; Hocevar, S.B.; Ogorevc, B. Insights into the anodic stripping voltammetric behavior of bismuth film electrodes. *Anal. Chim. Acta* **2001**, *434*, 29–34. [CrossRef]
19. Franke, J.P.; de Zeeuw, R.A. Differential pulse anodic stripping voltammetry as a rapid screening technique for heavy metal intoxications. *Arch. Toxicol.* **1976**, *37*, 47–55. [CrossRef]

20. Pokpas, K.; Jahed, N.; Tovide, O.; Baker, P.G.; Iwuoha, E.I. Nafion-graphene nanocomposite in situ plated bismuth-film electrodes on pencil graphite substrates for the determination of trace heavy metals by anodic stripping voltammetry. *Int. J. Electrochem. Sci.* **2014**, *9*, 5092–5115. [CrossRef]
21. Ghaffari, N.; Pokpas, K.; Iwuoha, E.; Jahed, N. Sensitive Electrochemical Determination of Bisphenol a Using a Disposable, Electrodeposited Antimony-Graphene Nanocomposite Pencil Graphite Electrode (PGE) and Differential Pulse Voltammetry (DPV). *Anal. Lett.* **2023**, *57*, 1008–1025. [CrossRef]
22. Li, F.; Li, J.; Feng, Y.; Yang, L.; Du, Z. Electrochemical behavior of graphene doped carbon paste electrode and its application for sensitive determination of ascorbic acid. *Sens. Actuators B Chem.* **2011**, *157*, 110–114. [CrossRef]
23. Dervin, S.; Ganguly, P.; Dahiya, R.S. Disposable Electrochemical Sensor Using Graphene Oxide-Chitosan Modified Carbon-Based Electrodes for the Detection of Tyrosine. *IEEE Sens. J.* **2021**, *21*, 26226–26233. [CrossRef]
24. Annu; Sharma, S.; Jain, R.; Raja, A.N. Review—Pencil Graphite Electrode: An Emerging Sensing Material. *J. Electrochem. Soc.* **2020**, *167*, 037501. [CrossRef]
25. Özcan, A. Synergistic Effect of Lithium Perchlorate and Sodium Hydroxide in the Preparation of Electrochemically Treated Pencil Graphite Electrodes for Selective and Sensitive Bisphenol A Detection in Water Samples. *Electroanalysis* **2014**, *26*, 1631–1639. [CrossRef]
26. Demetriades, D.; Economou, A.; Voulgaropoulos, A. A study of pencil-lead bismuth-film electrodes for the determination of trace metals by anodic stripping voltammetry. *Anal. Chim. Acta* **2004**, *519*, 167–172. [CrossRef]
27. David, I.G.; Buleandra, M.; Popa, D.E.; Cheregi, M.C.; David, V.; Iorgulescu, E.E.; Tartareanu, G.O. Recent Developments in Voltammetric Analysis of Pharmaceuticals Using Disposable Pencil Graphite Electrodes. *Processes* **2022**, *10*, 472. [CrossRef]
28. David, I.G.; Litescu, S.C.; Moraru, R.; Albu, C.; Buleandra, M.; Popa, D.E.; Riga, S.; Ciobanu, A.M.; Noor, H. Electroanalysis of Naringin at Electroactivated Pencil Graphite Electrode for the Assessment of Polyphenolics with Intermediate Antioxidant Power. *Antioxidants* **2022**, *11*, 2306. [CrossRef]
29. Ishtiaq, S.; Sohail, M.; Rasul, S.; Zia, A.W.; Siller, L.; Chotana, G.A.; Sharif, M.; Nafady, A. Selenium Nanoneedles Deposited on a Pencil Graphite Electrode for Hydrazine Sensing. *ACS Appl. Nano Mater.* **2022**, *5*, 14336–14346. [CrossRef]
30. Buleandră, M.; Popa, D.E.; Popa, A.; Codreanu, N.A.M.; David, I.G. Multi-Analyte Sensor Based on Pencil Graphite Electrode for Riboflavin and Pyridoxine Determination. *J. Electrochem. Soc.* **2022**, *169*, 017517. [CrossRef]
31. Kumar Naik, T.S.S.; Kesavan, A.V.; Swamy, B.E.K.; Singh, S.; Anil, A.G.; Madhavi, V.; Ramamurthy, P.C. Low cost, trouble-free disposable pencil graphite electrode sensor for the simultaneous detection of hydroquinone and catechol. *Mater. Chem. Phys.* **2022**, *278*, 125663. [CrossRef]
32. Zbeda, S.; Pokpas, K.; Titinchi, S.; Jahed, N.; Baker, P.G.; Iwuoha, E.I. Few-layer binder free graphene modified mercury film electrode for trace metal analysis by square wave anodic stripping voltammetry. *Int. J. Electrochem. Sci.* **2013**, *8*, 11125–11141. [CrossRef]
33. Bunch, J.S.; Verbridge, S.S.; Alden, J.S.; Van Der Zande, A.M.; Parpia, J.M.; Craighead, H.G.; McEuen, P.L. Impermeable atomic membranes from graphene sheets. *Nano Lett.* **2008**, *8*, 2458–2462. [CrossRef] [PubMed]
34. Noveslov, K.S.; Geim, A.K.K.; Morozov, S.V.V.; Jiang, D.; Zhang, Y.; Dubonos, S.V.V.; Grigorieva, I.V.V.; Firsov, A.A.A.; Novoselov, K.S.; Geim, A.K.K.; et al. Electric Field Effect in Atomically Thin Carbon Films. *Science* **2004**, *306*, 666–669. [CrossRef] [PubMed]
35. Geim, A.K.; Novoselov, K.S. The rise of graphene. *Nat. Mater.* **2007**, *6*, 183–191. [CrossRef]
36. Neto, A.C.; Guinea, F.; Peres, N.M. Drawing conclusions from graphene. *Phys. World* **2006**, *19*, 33–37. [CrossRef]
37. Park, H.J.; Meyer, J.; Roth, S.; Skákalová, V. Growth and properties of few-layer graphene prepared by chemical vapor deposition. *Carbon* **2010**, *48*, 1088–1094. [CrossRef]
38. Reina, A.; Thiele, S.; Jia, X.; Bhaviripudi, S.; Dresselhaus, M.S.; Schaefer, J.A.; Kong, J. Growth of large-area single- and Bi-layer graphene by controlled carbon precipitation on polycrystalline Ni surfaces. *Nano Res.* **2009**, *2*, 509–516. [CrossRef]
39. Voutilainen, M.; Seppala, E.T.; Pasanen, P.; Oksanen, M. Graphene and carbon nanotube applications in mobile devices. *IEEE Trans. Electron Devices* **2012**, *59*, 2876–2887. [CrossRef]
40. Patel, M.; Bisht, N.; Prabhakar, P.; Sen, R.K.; Kumar, P.; Dwivedi, N.; Ashiq, M.; Mondal, D.P.; Srivastava, A.K.; Dhand, C. Ternary nanocomposite-based smart sensor: Reduced graphene oxide/polydopamine/alanine nanocomposite for simultaneous electrochemical detection of Cd^{2+}, Pb^{2+}, Fe^{2+}, and Cu^{2+} ions. *Environ. Res.* **2023**, *221*, 115317. [CrossRef]
41. Rahman, H.A.; Rafi, M.; Putra, B.R.; Wahyuni, W.T. Electrochemical Sensors Based on a Composite of Electrochemically Reduced Graphene Oxide and PEDOT:PSS for Hydrazine Detection. *ACS Omega* **2023**, *8*, 3258–3269. [CrossRef] [PubMed]
42. Pichún, B.; Núñez, C.; Arancibia, V.; Martí, A.A.; Aguirre, M.J.; Pizarro, J.; Segura, R.; Flores, E. Enhanced voltammetric sensing platform based on gold nanorods and electrochemically reduced graphene oxide for As(III) determination in seafood samples. *J. Appl. Electrochem.* **2024**, *54*, 1595–1606. [CrossRef]
43. Economou, A.; Fielden, P.R. Mercury film electrodes: Developments, trends and potentialities for electroanalysis. *Analyst* **2003**, *128*, 205–213. [CrossRef] [PubMed]
44. Sánchez-Calvo, A.; Blanco-López, M.C.; Costa-García, A. Paper-based working electrodes coated with mercury or bismuth films for heavy metals determination. *Biosensors* **2020**, *10*, 52. [CrossRef] [PubMed]
45. Tefera, M.; Tessema, M.; Admassie, S.; Guadie, A. Electrochemical determination of endosulfan in vegetable samples using mercury film modified glassy carbon electrode. *Sens. Bio-Sens. Res.* **2021**, *33*, 100431. [CrossRef]

46. Piech, R.; Bas, B. Sensitive voltammetric determination of gallium in aluminium materials using renewable mercury film silver based electrode. *Int. J. Environ. Anal. Chem.* **2011**, *91*, 410–420. [CrossRef]
47. Kudr, J.; Nguyen, H.V.; Gumulec, J.; Nejdl, L.; Blazkova, I.; Ruttkay-Nedecky, B.; Hynek, D.; Kynicky, J.; Adam, V.; Kizek, R. Simultaneous automatic electrochemical detection of zinc, cadmium, copper and lead ions in environmental samples using a thin-film mercury electrode and an artificial neural network. *Sensors* **2015**, *15*, 592–610. [CrossRef]
48. Yıldız, C.; Eskiköy Bayraktepe, D.; Yazan, Z. Highly sensitive direct simultaneous determination of zinc(II), cadmium(II), lead(II), and copper(II) based on in-situ-bismuth and mercury thin-film plated screen-printed carbon electrode. *Monatshefte Fuer Chem./Chem. Mon.* **2021**, *152*, 1527–1537. [CrossRef]
49. Sanga, N.A.; Jahed, N.; Leve, Z.; Iwuoha, E.I.; Pokpas, K. Simultaneous Adsorptive Stripping Voltammetric Analysis of Heavy Metals at Graphenated Cupferron Pencil Rods. *J. Electrochem. Soc.* **2022**, *169*, 017502. [CrossRef]
50. Pokpas, K.; Jahed, N.; Iwuoha, E. Tuneable, Pre-stored Paper-Based Electrochemical Cells (μPECs): An Adsorptive Stripping Voltammetric Approach to Metal Analysis. *Electrocatalysis* **2019**, *10*, 352–364. [CrossRef]
51. Pokpas, K.; Jahed, N.; Bezuidenhout, P.; Smith, S.; Land, K.; Iwuoha, E. Nickel contamination analysis at cost-effective silver printed paper-based electrodes based on carbon black dimethyl-glyoxime ink as electrode modifier. *J. Electrochem. Sci. Eng.* **2022**, *12*, 153–164. [CrossRef]
52. Finšgar, M.; Kovačec, L. Copper-bismuth-film in situ electrodes for heavy metal detection. *Microchem. J.* **2020**, *154*, 104635. [CrossRef]
53. Albalawi, I.; Hogan, A.; Alatawi, H.; Moore, E. A sensitive electrochemical analysis for cadmium and lead based on Nafion-Bismuth film in a water sample. *Sens. Bio-Sens. Res.* **2021**, *34*, 100454. [CrossRef]
54. Li, H.; Zhao, J.; Zhao, S.; Cui, G. Simultaneous determination of trace Pb(II), Cd(II), and Zn(II) using an integrated three-electrode modified with bismuth film. *Microchem. J.* **2021**, *168*, 106390. [CrossRef]
55. Ai, Y.; Yan, L.; Zhang, S.; Ye, X.; Xuan, Y.; He, S.; Wang, X.; Sun, W. Ultra-sensitive simultaneous electrochemical detection of Zn(II), Cd(II) and Pb(II) based on the bismuth and graphdiyne film modified electrode. *Microchem. J.* **2023**, *184*, 108186. [CrossRef]
56. Rojas-Romo, C.; Aliaga, M.E.; Arancibia, V.; Gomez, M. Determination of Pb(II) and Cd(II) via anodic stripping voltammetry using an in-situ bismuth film electrode. Increasing the sensitivity of the method by the presence of Alizarin Red S. *Microchem. J.* **2020**, *159*, 105373. [CrossRef]
57. Yaman, B.; Zaman, B.T.; Bakırdere, S.; Dilgin, Y. Sensitive, Accurate and Selective Determination of Cd(II) Using Anodic Stripping Voltammetry with in-situ Hg-Bi Film Modified Pencil Graphite Electrode After Magnetic Dispersive Solid Phase Microextraction. *Electroanalysis* **2021**, *33*, 2161–2168. [CrossRef]
58. Ouyang, R.; Zhu, Z.; Tatum, C.E.; Chambers, J.Q.; Xue, Z.L. Simultaneous stripping detection of Zn(II), Cd(II) and Pb(II) using a bimetallic Hg–Bi/single-walled carbon nanotubes composite electrode. *J. Electroanal. Chem.* **2011**, *656*, 78–84. [CrossRef]
59. Sharifian, P.; Aliakbar, A. Determination of Se(IV) by adsorptive cathodic stripping voltammetry at a Bi/Hg film electrode. *Anal. Methods* **2015**, *7*, 2121–2128. [CrossRef]
60. William, S.; Hummers, J.; Offeman, R.E. Preparation of Graphitic Oxide. *J. Am. Chem. Soc.* **1958**, *80*, 1339. [CrossRef]
61. Silva, R.M.; Sperandio, G.H.; da Silva, A.D.; Okumura, L.L.; da Silva, R.C.; Moreira, R.P.L.; Silva, T.A. Electrochemically reduced graphene oxide films from Zn-C battery waste for the electrochemical determination of paracetamol and hydroquinone. *Microchim. Acta* **2023**, *190*, 273. [CrossRef] [PubMed]
62. Leve, Z.D.; Jahed, N.; Sanga, N.A.; Iwuoha, E.I.; Pokpas, K. Determination of paracetamol at the electrochemically reduced graphene oxide-antimony nanocomposite modified pencil graphite electrode using adsorptive stripping differential pulse voltammetry. *Sensors* **2022**, *22*, 5784. [CrossRef] [PubMed]
63. Chen, H.; Guo, B.; Gong, D.; Fan, J.; Li, Z.; Wang, C.; Wang, L.; Yu, C.; Wang, C.; Zeng, G. A cation–π interaction confined graphene oxide membrane for separation of light paraffins and olefins. *Chem. Commun.* **2023**, *59*, 5257–5260. [CrossRef] [PubMed]
64. Zhao, G.; Zhu, H. Cation–π Interactions in Graphene-Containing Systems for Water Treatment and Beyond. *Adv. Mater.* **2020**, *32*, 1905756. [CrossRef]
65. Medvecký, L.; Briančin, J. Possibilities of simultaneous determination of indium and gallium in binary InGa alloys by anodic stripping voltammetry in acetate buffer. *Chem. Pap.* **2004**, *58*, 93–100.
66. Qu, L.; Jin, W. Adsorption voltammetry of the gallium morin system. *Anal. Chim. Acta* **1993**, *274*, 65–70. [CrossRef]
67. Udisti, R.; Piccardi, G. Determination of gallium traces by differential pulse anodic stripping voltammetry. *Anal. Bioanal. Chem.* **1988**, *331*, 35–38. [CrossRef]
68. Li, Y.H.; Zhao, Q.L.; Huang, M.H. Cathodic Adsorptive Voltammetry of the Gallium-Alizarin Red S Complex at a Carbon Paste Electrode. *Electroanalysis* **2005**, *17*, 343–347. [CrossRef]
69. East, G.; Cofre, P. Determination of gallium by square-wave voltammetry anodic stripping, based on the electrocatalytic action of 2,2′-bipyridine in dimethylsulphoxide: Comparison with an aqueous NaSCN/NaClO4 electrolyte. *Talanta* **1993**, *40*, 1273–1281. [CrossRef]
70. Farias, P.A.M.; Martin, C.M.L.; Ohara, A.K.; Gold, J.S. Cathodic adsorptive stripping voltammetry of indium complexed with morin at a static mercury drop electrode. *Anal. Chim. Acta* **1994**, *293*, 29–34. [CrossRef]
71. Benvidi, A.; Ardakani, M.M. Subnanomolar Determination of Indium by Adsorptive Stripping Differential Pulse Voltammetry Using Factorial Design for Optimization. *Anal. Lett.* **2009**, *42*, 2430–2443. [CrossRef]

72. Nosal-Wiercińska, A.; Dalmata, G. Application of the catalytic properties of N-methylthiourea to the determination of In(III) at low levels by square wave voltammetry. *Monatshefte Fuer Chem./Chem. Mon.* **2009**, *140*, 1421–1424. [CrossRef]
73. Bobrowski, A.; Putek, M.; Zarebski, J. Antimony Film Electrode Prepared In Situ in Hydrogen Potassium Tartrate in Anodic Stripping Voltammetric Trace Detection of Cd(II), Pb(II), Zn(II), Tl(I), In(III) and Cu(II). *Electroanalysis* **2012**, *24*, 1071–1078. [CrossRef]
74. Florence, T.M.; Batley, G.E.; Farrar, Y.J. The determination of indium by anodic stripping voltammetry: Application to natural waters. *J. Electroanal. Chem. Interfacial Electrochem.* **1974**, *56*, 301–309. [CrossRef]
75. Taher, M.A. Differential pulse polarography determination of indium after column preconcentration with [1-(2-pyridylazo)-2-naphthol]-naphthalene adsorbent or its complex on microcrystalline naphthalene. *Talanta* **2000**, *52*, 301–309. [CrossRef]
76. Zhang, J.; Shan, Y.; Ma, J.; Xie, L.; Du, X. Simultaneous determination of indium and thallium ions by anodic stripping voltammetry using antimony film electrode. *Sens. Lett.* **2009**, *7*, 605–608. [CrossRef]

Disclaimer/Publisher's Note: The statements, opinions and data contained in all publications are solely those of the individual author(s) and contributor(s) and not of MDPI and/or the editor(s). MDPI and/or the editor(s) disclaim responsibility for any injury to people or property resulting from any ideas, methods, instructions or products referred to in the content.

Article

Impact of Dispersive Solvent and Temperature on Supercapacitor Performance of N-Doped Reduced Graphene Oxide

Ankit Yadav [1,*], Rajeev Kumar [1,2,*], Deepu Joseph [3], Nygil Thomas [4], Fei Yan [2] and Balaram Sahoo [1,*]

1. Materials Research Centre, Indian Institute of Science, Bengaluru 560012, Karnataka, India
2. Department of Chemistry and Biochemistry, North Carolina Central University, Durham, NC 27707, USA; fyan@nccu.edu
3. Department of Physics, Nirmalagiri College, Nirmalagiri, Kuthuparamba, Kannur 670701, Kerala, India; deepuj@nirmalagiricollege.ac.in
4. Department of Chemistry, Nirmalagiri College, Nirmalagiri, Kuthuparamba, Kannur 670701, Kerala, India; nygill@gmail.com
* Correspondence: ankit.skb02@gmail.com (A.Y.); rkumar@nccu.edu (R.K.); bsahoo@iisc.ac.in (B.S.)

Citation: Yadav, A.; Kumar, R.; Joseph, D.; Thomas, N.; Yan, F.; Sahoo, B. Impact of Dispersive Solvent and Temperature on Supercapacitor Performance of N-Doped Reduced Graphene Oxide. *C* **2024**, *10*, 89. https://doi.org/10.3390/c10040089

Academic Editor: Olena Okhay

Received: 4 September 2024
Revised: 30 September 2024
Accepted: 8 October 2024
Published: 10 October 2024

Copyright: © 2024 by the authors. Licensee MDPI, Basel, Switzerland. This article is an open access article distributed under the terms and conditions of the Creative Commons Attribution (CC BY) license (https://creativecommons.org/licenses/by/4.0/).

Abstract: This study evaluates the critical roles of the dispersion medium and temperature during the solvothermal synthesis of nitrogen-doped reduced graphene oxide (NG) for enhancing its performance as an active material in supercapacitor electrodes. Using a fixed volume of a solvent (THF, ethanol, acetonitrile, water, N,N-Dimethylformamide, ethylene glycol, or N-Methyl-2-pyrrolidone) as the dispersive medium, a series of samples at different temperatures (60, 75, 95, 120, 150, 180, and 195 °C) are synthesized and investigated. A proper removal of the oxygen moieties from their surface and an optimum number of N-based defects are essential for a better reduction of graphene oxide and better stacking of the NG sheets. The origin of the supercapacitance of NG sheets can be correlated to the inherent properties such as the boiling point, viscosity, dipole moment, and dielectric constant of all the studied solvents, along with the synthesis temperature. Due to the achievement of a suitable synthesis environment, NG synthesized using N,N-Dimethylformamide at 150 °C displays an excellent supercapacitance value of 514 F/g at 0.5 A/g, which is the highest among all our samples and also competitive among several state-of-the-art lightweight carbon materials. Our work not only helps in understanding the origin of the supercapacitance exhibited by graphene-based materials but also tuning them through a suitable choice of synthesis conditions.

Keywords: supercapacitor; reduced graphene oxide; nitrogen doping; synthesis parameters; dispersive solvents

1. Introduction

Graphene, a 2D carbon framework, has attracted marvelous consideration from experimental and theoretical scientific groups in recent times due to its enormous potential for future applications in numerous technological domains, e.g., in the Internet of Things (IoT), batteries, supercapacitors [1], and water splitting [2–4]. The surface and electrical properties of graphene can be tweaked through chemical alteration via heteroatom doping [5–7], which makes N-doped graphene more appropriate for certain applications than pure graphene [8,9]. For instance, N-doped reduced graphene oxide (NG) is used in many applications such as for energy storage devices [10,11], carbon-based biosensing devices with improved biocompatibility [12], etc. Previously, NG sheets were synthesized by the nitrogen plasma treatment of GO at ~300 °C, where precise control on the quality of the rGO is difficult to achieve. Hence, for the synthesis of NG sheets recently, many of the reports employ the hydrothermal/solvothermal method. During hydrothermal synthesis, various nitrogen sources such as dicyanadiamide [13], ammonia [14], glucosamine [15], hexamethylenetetramine [16], pyrrole [17], ammonium oxalate [18], urea [19,20], amino acids [21], and hydrazine [22] are utilized. Similarly, NG sheets are also synthesized by the solvothermal method using amitrole, urea, or ammonia as a nitrogen source in ethylene glycol or N,N-Dimethylformamide as the solvent [23–25] for their application in energy devices.

For several important electronic and energy device applications, well-dispersed NG sheets are required. However, the preparation of the homogenously dispersed form of NG sheets is not a straightforward process since its colloidal stability in various solvents at different temperatures is a critical aspect. In this perspective, the dispersibility of graphene oxide (GO)/reduced graphene oxide (rGO) in multiple solvents is examined by several groups [26–28]. The dispersion/quality of the final synthesized NG sheets would depend on the dispersibility of GO, the type of reducing agent, the nature of the particular dispersive solvent used, and the reaction temperature used during the hydrothermal/solvothermal reaction process. However, detailed studies in this regard are lacking in the literature. Furthermore, as usually reported by several researchers, only a minuscule amount of GO (i.e., <~1 mg/mL) is usually processed/reduced by the hydrothermal/solvothermal route [24,29,30]. However, for meeting the industrial-scale requirements of high-quality NG, a lacuna of proficient synthesis approaches is explored. These difficulties are mainly due to the contradicting issues of quantity and quality control by the hydrothermal process and other methods, respectively. In achieving such a feat of synthesizing high-quality graphene materials in bulk, the role of various solvents has to be explored, and hydrothermal reaction conditions should be optimized. In this direction, only a few scattered studies are available, but the results cannot be compared due to variability in the parameters. For example, Gopalkrishnan et al. and Jiang et al. studied the hydrothermal/solvothermal synthesis of NG at a few different temperatures but by using only one dispersive solvent [29,30]. Mayyas et al. have studied the in-situ synthesis of NG in different solvent media by the electrochemical method but only at room temperature (RT) [31]. Herein, we intend to explore the critical role of the solvent's intrinsic properties such as the viscosity, boiling point, dielectric constant, and dipole moment, along with the synthesis temperature's effect on the quality of the synthesized NG sheets. Our work not only offers the optimized conditions for the synthesis of large-scale and high-quality NG, but also the mechanistic details governing it. In this study, we aim to investigate how the intrinsic properties of solvents—specifically the viscosity, boiling point, dielectric constant, and dipole moment—impact the quality of synthesized nanoporous graphene sheets. These solvent characteristics can influence key aspects of the synthesis process, including the dispersion of precursors, reaction kinetics, and the overall morphology of the resulting graphenic sheets. Additionally, we will consider how varying the synthesis temperature interacts with these solvent properties to further affect the characteristics of the NG sheets. A more detailed exploration of these relationships will enhance our understanding of optimizing synthesis conditions for improved material quality.

Liu et al. [32] discussed that the high specific capacitance in carbon materials often revolves around their structural characteristics, particularly their porosity and defects. Furthermore, they have some similarities, such as the importance of surface area, the role of surface chemistry, and disorder. Apart from these, the differences are the emphasis on structural disorder, mechanism of capacitance, and the material types and application. In this work, we optimized the synthesis parameters for the hydrothermal/solvothermal reduction of GO by using urea (NH_2-CO-NH_2) as the reducing agent, which also acts as the nitrogen source for N doping in rGO. For the reduction of GO, i.e., the chemical eradication of oxygen functional groups during the reaction progress, we have used many different (aqueous and organic) solvents to dissolve the reducing agent (urea) and disperse GO and studied the effect of each solvent at different temperatures via the hydrothermal/solvothermal reaction. Urea is chosen as the nitrogen source and reducing agent because it is economical and extensively used by other researchers [20,33]. Based on the effectiveness of the reducing agent in different solvents and the dispersibility of the GO sheets, we explored the plausible reasons behind the variation in supercapacitance behavior exhibited by different samples. In this way, we figure out the optimum conditions favorable for the large-scale processing of NG, which can meet the demand of the electronic and energy device industries.

2. Materials and Methods

2.1. Synthesis of GO and N-Doped rGO (NG)

The chemicals used to synthesize graphene oxide (GO) sheets are graphite flake powder purchased from Sigma Aldrich (Bengaluru, India), conc. HCl (35.4%, SD Fine Chemicals Ltd., Mumbai, India), acetone ($\geq$99.5%, EMPLURA, Bengaluru, India), $KMnO_4$ (99%, Qualigens, Mumbai, India), H_2SO_4 (98% GR, MERCK, Bengaluru, India), H_2O_2 (30%, MERCK, Bengaluru, India), and H_3PO_4 (Sigma Aldrich, Bengaluru, India). GO is synthesized using the well-established improved Hummers' method [34,35]. Briefly, 3 g of graphite flakes were taken in a 500 mL beaker and mixed with 360 mL of H_2SO_4 and 40 mL of H_3PO_4 with continuous stirring in an ice bath, and 18 g of $KMnO_4$ was added to this mixture in small parts. Then, the reaction mixture was allowed to cool to room temperature and washed several times with deionized water, conc. HCl, and ethanol to remove excess acid and oxide impurities. The sample was then dried in an oven at 80 °C for 24 h.

The preparatory details and the synthesis procedure for the preparation of NG are given below. For the hydrothermal/solvothermal synthesis of NG sheets, urea was taken as a nitrogen source and dissolved in various solvents. To carry out the hydrothermal/solvothermal synthesis of NG, we used a Teflon-lined autoclave of 100 mL capacity. For the synthesis, 3 g of urea was dissolved in 80 mL of Tetrahydrofuran (THF) solvent, and 1 g of GO was added to this solution. The mixture was then ultrasonicated for 15 min. This ratio of 1:3 for GO/urea and the reaction time of 24 h are chosen according to our earlier work [33], where this combination shows the best supercapacitive performance. The obtained homogeneous mixture was then poured into a 100 mL Teflon-lined stainless-steel autoclave and kept in a muffle furnace at a temperature of 60 °C for 24 h for the hydrothermal/solvothermal reduction reaction to occur. After 24 h, when the furnace was cooled down to room temperature, the sample was recovered and washed several times with water and ethanol to remove the organic impurities. The sample was then dried overnight in a hot-air oven at 60 °C and labeled as NG-3-THF-60. In a similar way, the N-doped samples were prepared at 120, 150, 180, and 195 °C, keeping all other synthesis conditions identical. The samples are correspondingly named NG-3-THF-120, NG-3-THF-150, and NG-3-THF-180, respectively. Similarly, by keeping the same amount of urea and GO (i.e., 3 g of urea and 1 g of GO), the other samples were synthesized by using different dispersive solvents such as ethanol (EtOH), acetonitrile (ACN), water (H_2O), N,N-Dimethylformamide (DMF), ethylene glycol (EG), and N-Methyl-2-pyrrolidone (NMP). The hydrothermal/solvothermal reaction temperatures were chosen between 60 °C and 195 °C; in particular, the reaction temperatures of 60, 75, 95, 120, 150, 180, and 195 °C were used, depending on the boiling points of the solvents. These selected temperatures range from below the boiling point of all the solvents to that above it. As we will notice later, any temperature below 60 °C is not suitable for the removal of oxygen moieties, i.e., the hydrothermal/solvothermal reduction reaction does not occur effectively. Furthermore, there is the danger of damage to the Teflon lining of the autoclave above 200 °C. Hence, the temperature range was limited between 60 °C and 195 °C. Note that our chosen solvents not only have different boiling points but have different viscosities, dipole moments, and dielectric constants too. The details of the solvent properties and corresponding synthesis parameters used, along with the respective sample codes, are listed in Table 1. The details of the characterization methods, including the electrochemical study and the electrode preparation for supercapacitance measurement, are described in the Supplementary Materials (ESI).

Table 1. The physico-chemical parameters associated with different dispersive media used for the synthesis of nitrogen-doped reduced graphene oxide (NG) samples and the corresponding sample codes.

Solvent (80 mL)	Boiling Point (°C)	Viscosity (cP) (25 °C)	Dipole Moment (Debye)	Dielectric Constant	Temp. (°C)	Sample Name
THF	66	0.48	1.75	7.52	60	NG-3-THF-60
					120	NG-3-THF-120
					150	NG-3-THF-150
					180	NG-3-THF-180
Ethanol	78.6	0.98	1.69	24	75	NG-3-EtOH-75
					120	NG-3-EtOH-120
					150	NG-3-EtOH-150
					180	NG-3-EtOH-180
Acetonitrile (ACN)	82	0.33	3.92	36.6	75	NG-3-ACN-75
					120	NG-3-ACN-120
					150	NG-3-ACN-150
					180	NG-3-ACN-180
H_2O	100	0.89	1.85	78.4	95	NG-3-H_2O-95
					120	NG-3-H_2O-120
					150	NG-3-H_2O-150
					180	NG-3-H_2O-180
DMF	153	0.80	3.82	38.25	120	NG-3-DMF-120
					150	NG-3-DMF-150
					180	NG-3-DMF-180
Ethylene Glycol (EG)	197.6	16.1	2.31	37.7	120	NG-3-EG-120
					150	NG-3-EG-150
					180	NG-3-EG-180
					195	NG-3-EG-195
NMP	202	1.89	3.75	32	120	NG-3-NMP-120
					150	NG-3-NMP-150
					180	NG-3-NMP-180
					195	NG-3-NMP-195

2.2. Characterization Methods

The X-ray diffraction (XRD) patterns of all our synthesized samples were recorded using a PANalytical X-ray diffractometer X-ray diffractometer (operated at 40 kV and 150 mA) in a 2θ range from 10° to 55° (Cu-K$_\alpha$ radiation, λ = 1.540 Å). The scanning electron microscopy (SEM) images were obtained through a 'Supra55 Zeiss' field emission scanning electron microscope. Raman spectra were taken on 'LabRam HR' equipment (with a 532 nm laser source). The X-ray photoelectron spectroscopy (XPS) measurements were performed on an 'AXIS ULTRA system'. The excitation energy is 1486.7 eV (Al K$_\alpha$ X-ray source), while energy resolution is around 0.5 eV with a monochromated source. The background subtraction was carried out using the Shirley function in Origin software (version 8.1). Spectra fitting was performed manually in Origin software based on the available literature of similar samples.

2.3. Electrochemical Measurements

A potentiostat/galvanostat (Biologic, Model: SP-200, Software: EC-Lab v.11) was employed for all electrochemical experiments. The electrochemical measurements were performed in the three-electrode configuration by using Ag/AgCl (3.5 M KCl) as the reference electrode, a Pt wire as the counter electrode, and 3 mm diameter glassy carbon electrodes (GCEs) as the working electrodes. For exploring the performance of the supercapacitors, cyclic voltammetry (CV), galvanometric charging–discharging (GCD), and electrochemical impedance spectroscopy (EIS) experiments were performed in a 0.5 M H_2SO_4 solution at room temperature (RT) without stirring.

The working electrodes for electrochemical measurements were fabricated using the following procedure: at the start, the glassy carbon electrodes (GCEs), having a surface area of ~0.07 cm^2, were polished using different grades of alumina slurry (starting from coarse to fine), washed several times with DI water, and dried afterward. Later, by dispersing 2 mg of an as-synthesized powder (e.g., NG-3-THF-60) sample in 500 µL of ethanol through ultrasonication for 15 min, a steady ink-like fluid was prepared. This fluid was then mixed with 2.5 µL of Nafion (5% in isopropanol) as a binder and further sonicated for better dispersion. Afterward, 5 µL of this resulting fluid was drop-cast on a GCE and dried in air (at RT) [8,36]. The dried sample on the GCE works as the working electrode for all the electrochemical experiments. Similarly, using each of the other samples (synthesized by using different solvents at different temperatures), other working electrodes were prepared. Prior to all electrochemical measurements, nitrogen gas was purged onto the electrodes (at room temperature) for removing any accumulated dust on the surface of the working electrode.

From the measured CV curves, the areal capacitance values are obtained through Equation (1) as follows [16]:

$$C_{Areal} = \frac{\int IdV}{A \times v \times \Delta V} \quad (1)$$

where 'I' denotes the current obtained within a small voltage window dV in the CV curve, ΔV represents the total voltage window covering the CV loop, 'v' is the potential scanning rate, and 'A' is the surface area of the electrode (~0.07 cm^2). The specific capacitance (C_{sp}) value (F/g) is calculated from the measured GCD curve by using Equation (2) as follows [37]:

$$C_{sp} = \frac{I \times \Delta t}{m \times \Delta V} \quad (2)$$

here, 'I' is the implemented constant current (mA), 'Δt' is discharge time (seconds), 'm' is the mass (mg) of the sample used on the active electrode surface area, and 'ΔV' is the difference between the initial and the final potential. Hence, C_{sp} is the value of the net specific capacitance (supercapacitance).

3. Results and Discussion

3.1. Morphological and Structural Characterization

3.1.1. SEM Study

The SEM images for two representative samples (NG-3-DMF-150 and NG-3-H_2O-150) are shown in Figure 1. The SEM images of all other samples are given in Figures S1 and S2 (ESI file). It can be seen from the SEM images that the samples appear as sponge-like structures due to the agglomeration/stacking of the interlinked three-dimensional NG sheets, as usually reported in the literature [20]. The strongly packed lamellar and aggregated flaky texture shows the multilayered microstructure of the NG sheets.

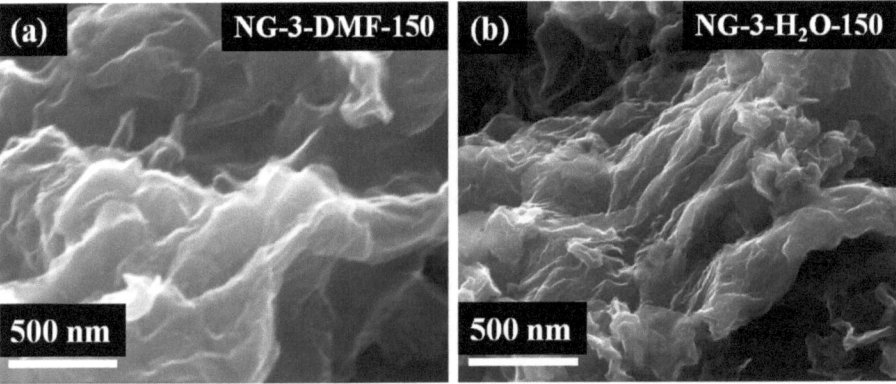

Figure 1. SEM images of (**a**) NG-3-DMF-150 and (**b**) NG-3-H$_2$O-150 samples.

3.1.2. XRD Study

The XRD patterns of all the samples synthesized at different temperatures and by using different solvents are shown in Figure 2. Figure 2a shows the XRD patterns of the synthesized GO and rGO samples. The peak at $2\theta = 10.6°$ for GO represents the (001) Bragg peak associated with an interlayer spacing of about 0.83 nm, whereas the Bragg peak observed at $2\theta = 26°$ for rGO represents the (002) peak associated with an interlayer spacing of 0.34 nm. In carbonaceous materials, both these Bragg peaks are often observed and are indexed as (001) and (002), respectively; hence, we have also indexed them accordingly. These interlayer spacing values are similar to those reported in the literature [24]; however, there is a small variation from report to report due to the concentration and size of different oxygen moieties, viz. -COOH (carboxyl), -C=O (carbonyl), -OH (hydroxyl), -C-O-C- (epoxide), etc. [34], present on the surface of the graphene sheets. Figure 2b shows the XRD patterns of the NG samples synthesized using the THF solvent (NG-3-THF samples) at different temperatures, as indicated. The sample synthesized at the lowest temperature (60 °C) shows the prominent peak at $2\theta = 10.6°$ corresponding to the (001) diffraction peak of GO, along with a broad hump at $2\theta \approx 26°$. This proves that GO is not completely reduced, but some oxygen-containing functional groups are still present on the basal planes. At higher temperatures, i.e., $\geq$~120 °C, the (001) peak at $2\theta = 10.6°$ for GO disappears from the XRD patterns, and the sharp peak corresponding to the (002) Bragg peak of rGO appears at $2\theta \approx 26°$. This illustrates that the oxygen moieties are removed from the surface of the GO sheets, which leads to lower interplanar spacing and the formation of the reduced graphene sheets that stack on each other, forming multilayers. As seen from Figure 2b, as the hydrothermal reaction temperature increases from 120 °C to 150 °C, the (002) Bragg peak (corresponding to the graphitic-type arrangement of the sheets) shifts toward higher 2θ values, indicating that the stacking of the graphenic sheets becomes better, i.e., the crystallinity improved. However, as the hydrothermal reaction temperature increases to 180 °C, a broad peak/shoulder emerges at $2\theta \approx 21°$, and the graphitic peak also becomes broad. The appearance of these broad peaks in the XRD patterns indicates lower ordering due to the breakage of the graphenic network. Hence, the high solvothermal reaction temperature of 180 °C is unfavorable/unsuitable for the stability of the graphenic sheets during their synthesis when using low-boiling solvents such as THF.

The other sets of samples synthesized using different solvents also show similar features in the XRD patterns (Figure 2), as explained above. However, the temperatures at which these features are observed vary depending on the other synthesis conditions for the different sets of samples. Overall, as the reaction temperature increases above ~60 °C, there is a gradual sharpening of the (002) XRD peak of the samples synthesized using the reaction temperatures of up to a certain value, and beyond that reaction temperature, this peak gradually becomes broad. This can be rationalized by inspecting the XRD patterns of all the samples synthesized at different temperatures by using different dispersive media.

From the above discussion, it seems that the choice of the hydro/solvothermal reaction temperature in relation to the boiling point of the solvents may be important. One can rationalize that if the synthesis temperature is below the boiling point of the solvents, the GO is not reduced properly. From Figure 2a–e, it appears that if the reaction temperature is below the boiling point of the solvent, the GO is only partially reduced. Moreover, for the hydro/solvothermal reaction temperature of ~95 °C (which is below the boiling point of water), although no GO peak is observed (Figure 2e), the partial reduction of GO is evident due to lower 2θ positions of the observed broad peaks (i.e., higher interlayer spacing caused due to the presence of some of the oxygen moieties). This discards the relationship between the boiling point and the hydro/solvothermal reaction temperature for the reduction of GO. Furthermore, it is known in the literature that the thermal reduction of GO (to form rGO) occurs at ~110 °C or lower at ambient (open furnace) conditions [38]. In case of both EG and NMP, whose boiling points are above 195 °C (Table 1), all the chosen synthesis temperatures (for these solvents) are below the boiling points of EG and NMP. This further supports our observation that the boiling point has very little role in controlling the reduction of GO. However, our observation of the partial reduction of GO even at reaction temperatures $\leq$ 95 °C (Figure 2a–e) suggests that the solvents in the hydro/solvothermal method help in the better dispersion of GO and also in lowering the reaction temperature to $\leq$~95 °C due to this better dispersion.

Closely observing the gradual sharpening of the (002) Bragg peaks for samples synthesized at different temperatures and different solvents, we can understand that the stacking of the NG sheets becomes better at different temperatures for the various solvents. This provides us with better insights into choosing a suitable solvent and optimizing the synthesis temperature for improving the functional behavior of the NG. Since there is an incomplete reduction of GO for the samples synthesized below 95 °C (i.e., NG-3-THF-60, NG-3-EtOH-75, and NG-3-ACN-75 samples), we have excluded these samples from further analysis. Furthermore, from Figure 2, we have observed high disorderliness in the NG samples (emergence of a broad shoulder at 2θ ≈ 21° along with broadening of the (002) peak) for all the different sets of samples synthesized using different solvents above ~180 °C. Hence, combining all the above discussion, we can understand that if the synthesis temperature is below ~95 °C, an incomplete reduction of GO occurs, while for temperatures above ~180 °C, the heavily disordered NG flakes are obtained. Hence, our investigations provide us a synthesis temperature window between 120 and 180 °C, which can be further optimized for obtaining perfect NG sheets with optimized supercapacitor performance.

The analysis of the supercapacitance behavior of our NG samples will be demonstrated later, but understanding the origin and optimization of supercapacitance performance shown by our NG samples is the goal of the present study. To obtain detailed insights into the origin and mechanism of the supercapacitive behavior of our synthesized samples, the structural details and the nature of stacking of the NG sheets are very important. Hence, we have least squares fitted the (002) Bragg peak of each of the NG samples and calculated their crystallite sizes (L_c) through the Debye–Scherrer equation (Equation (S3), ESI). These crystallite sizes essentially define the thickness of the NG flakes over which the stacking is proper. The broad peak present at 2θ ≈ 26° for all the samples is fitted with two sub-peaks (as given in Figures S3 and S4 (ESI)), where the subpeaks seen at lower angles (~21.3°, assumed as peak 1) and at higher angles (~25.7°, assumed as peak 2) correspond to the amorphous/disordered and graphitic-type/crystalline arrangement of the NG sheets, respectively. The detailed results of the least squares fitting of the (002) Bragg peaks for all our samples are listed in Table S1 (ESI).

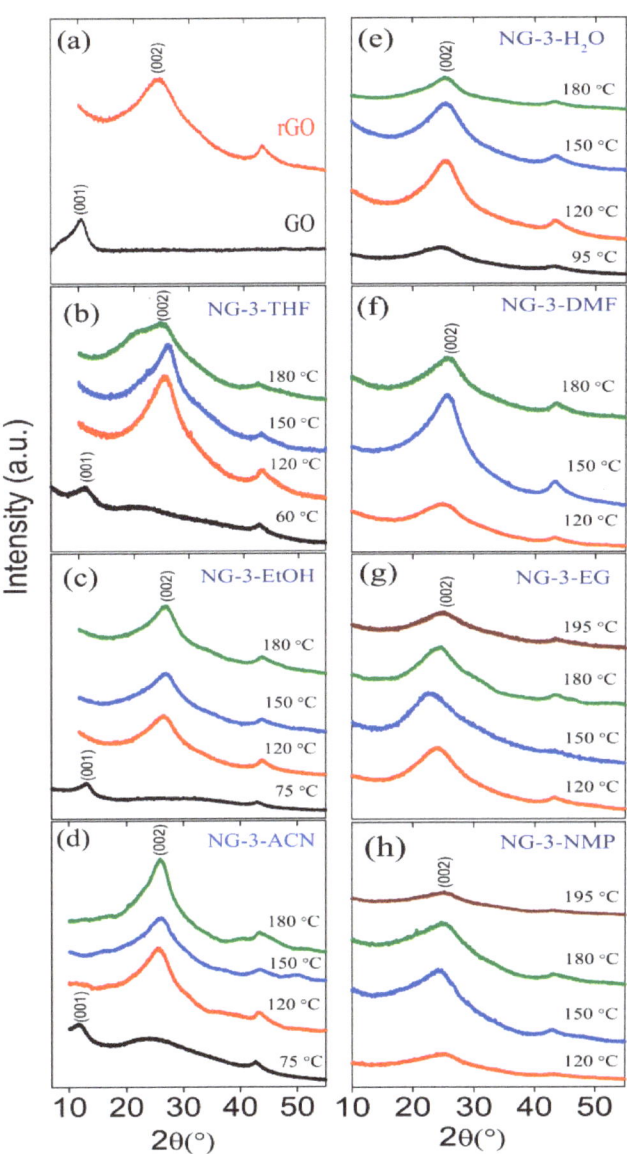

Figure 2. The XRD pattern of the (**a**) GO and rGO, and (**b**–**h**) NG-3 samples synthesized in various dispersive media and at different temperatures, as indicated.

From Table S1, it is clear that for the NG samples synthesized using THF, EtOH, ACN, water, DMF, EG, and NMP as the solvent, the best crystallinity is obtained either at 150 or 180 °C, respectively. As discussed above, the boiling point is not a very important parameter for the hydro/solvothermal reduction of GO. However, as a variation in the degree of crystallinity is observed at different reaction temperatures for the various dispersive solvents, it indicates that other inherent physical properties of the solvents such as the viscosity, dielectric constant, dipole moment, etc., can be important in this respect [39–41]. As these physical properties of the solvents depend on the temperature, the boiling point plays its role in controlling the nature of stacking/structural aspects of NG during their synthesis. For example, a low dielectric constant of the solvent may not facilitate improved activity of the NH_3 and CO/CO_2 molecules (originating from the reducing agent, urea) and as a result, the proper reduction of GO or proper dissociation of the oxygen moieties from the surface of GO does not occur. In addition, another important inherent property that may control the reduction of GO is the dipole moment of the solvent. If the solvent molecules have a high dipole moment, i.e., if the solvent is highly polar, they will facilitate better interaction with the oxygen moieties and other polar groups present on the basal plane of GO. This better GO sheet–solvent interaction can immensely help in the reduction of GO. Furthermore, a low viscosity of the solvent provides better dispersibility to GO and improves the mobility of the NH_3 and CO/CO_2 molecules present in the fluid, which helps in a better reduction of GO (along with better N doping in the basal carbon plane of the graphenic sheets). Simultaneously, a better reduction and suitable dispersibility help in a better stacking of the reduced GO sheets, which in turn decides the crystallinity of the NG flakes. The role of all the solvent parameters in controlling the structural and electrochemical aspects of the synthesized NG flakes is further explored.

3.1.3. Raman Spectroscopy Study

For examining the amount of surface defects and the nature of the disorder in the NG samples, we have performed Raman spectroscopy. As expected, each of the Raman spectra consists of two broad peaks positioned at the Raman shifts of ~1342 and ~1583 cm^{-1}, which are known as the disorder (D) and the graphitic (G) peaks, respectively. The relative amount of surface defects of the samples can be determined by the integrated intensity ratio of the G and the D peaks [35]. However, for an accurate determination of the amount and the type of defects present in each sample, we have performed a rigorous analysis of the fitting results. For this, each Raman spectrum is fitted with five sub-peaks (Figure S5) by using Gaussian–Lorentzian functions. Detailed explanations are given in Section 3.2 (ESI).

In our previous study, we have discussed the detailed analyses of Raman spectra of GO and rGO and evaluated the number of layers, which are 9.1 and 12.5, respectively [33]. From Table S3 (ESI), it is clear that the intensity of the D3 peak (which corresponds to the amorphous-type defects) for the samples prepared using a particular solvent (at different temperatures) varies in a similar trend as that observed for the amorphous carbon peak (peak 1) of the XRD patterns (Table S1). This relationship is also valid for the variation in the area under the G-peak of the Raman spectra and the crystalline graphitic peak (peak 2) of the XRD patterns of the corresponding samples prepared using a particular solvent. Notably, the D3 peak intensity is the lowest, and the G-peak intensity is the highest for the samples prepared at 150 °C (THF), 180 °C (EtOH), 180 °C (ACN), 150 °C (water), 150 °C (DMF), 180 °C (EG), and 180 °C (NMP) temperatures for the corresponding solvents given within brackets. This is also observed in the XRD patterns of the amorphous carbon peak (having a low intensity) and the graphitic peaks (having a high intensity) for the samples prepared at the same temperatures and by using the same solvents, as given above. Note that the crystallite size obtained using Scherrer's formula through the XRD linewidth (FWHM) corresponds to the thickness perpendicular to the plane, or the '*out-of-plane thickness*' of the graphene sheets of the graphitic flakes, whereas the crystallite sizes obtained using the Tuinstra–Koenig formula (Equation (S4)) through the I_G/I_D intensity ratios of Raman spectra correspond to the '*in-plane thickness*' of the graphene sheets present in the flakes. This is because the Raman spectrum sees only the vibrational modes associated with the bonds in the plane of the graphene sheet, and the graphene sheets are not bonded (no covalent bond) with the neighboring graphene sheets in the flakes. Hence, by combining the results from the XRD and Raman spectra, i.e., both the in-plane and out-of-plane crystallite size, one can rationalize the overall size of the nano-graphite crystallites. From our results, it is clear that the nano-graphite crystallites are bigger in size and better in quality for the sample prepared at an optimum temperature in a particular solvent. Depending on the nature of the solvent and synthesis temperature, the thickness of the nano-graphite crystallites varies from ~1.5 nm to ~2.2 nm (Table S1, ESI), and the in-plane sizes vary from ~2 nm to 4 nm (Table S4, ESI). Hence, our Raman study emphasizes the role of the hydro/solvothermal reaction temperature and the dispersive solvents in controlling the size and crystalline quality of the synthesized NG flakes.

3.2. Electrochemical Properties

The electrochemical techniques, such as cyclic voltammetry (CV), galvanostatic charge–discharge (GCD), and electrochemical impedance spectroscopy (EIS), were used to estimate the electrochemical performance of our NG samples. As given earlier, the NG samples were used as the working electrode in a three-electrode configuration of the measurement system. We have used 0.5 M of a H_2SO_4 aqueous electrolyte and recorded the electrochemical results in a voltage range between 0 and 0.8 V, which is the selected non-Faradic region. Figure 3 shows the CV voltammograms recorded at a scan rate of 5 mV/s for the NG samples prepared using different solvents and at various temperatures, as indicated. It can be seen that all the voltammograms have almost rectangular shapes with symmetrical current–potential characteristics. This suggests that the supercapacitance is originating due to the electrochemical double layer (EDL) formation, but not through the occurrence of any electrochemical redox reaction.

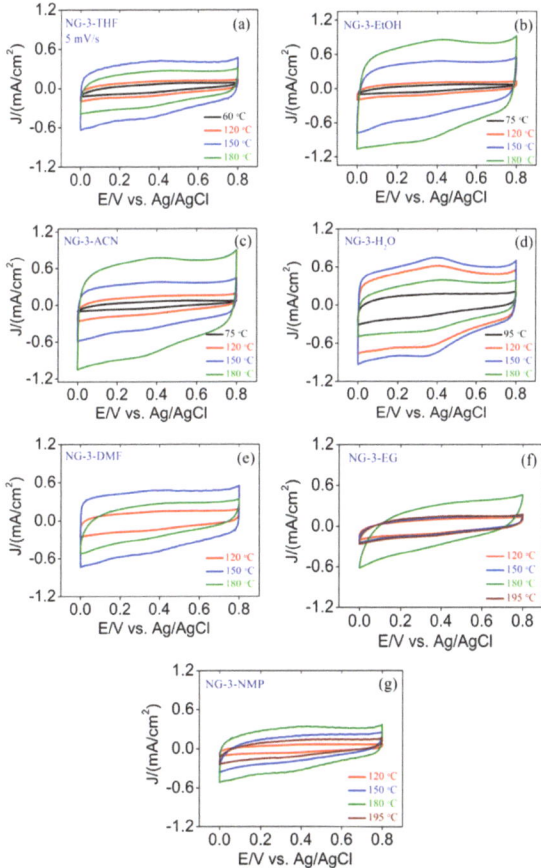

Figure 3. (**a**) Cyclic voltammograms of all the NG samples at different temperatures with various dispersive solvents, measured at a fixed scan rate of 5 mV/s. The solvents used for synthesis are (**a**) THF, (**b**) ethanol, (**c**) acetonitrile, (**d**) H_2O, (**e**) DMF, (**f**) ethylene glycol, and (**g**) NMP.

The areal capacitance values of our samples are calculated from the measured CV by using Equation (1). The obtained capacitance values are given in Table 2. It is clear from Figure 3 and Table 2 that the measured capacitance values obtained for the samples prepared using a particular solvent at different synthesis temperatures vary. Furthermore, when comparing the CV curves for the samples prepared using different solvents but at a particular temperature (with all the other experimental conditions being the same), we find variations in their features.

Table 2. The areal capacitance (C_{Areal}) calculated using the CV curves and the specific capacitance (C_{sp}) calculated using discharging of the GCD curves for our NG samples synthesized using different solvents.

Solvent	Temp (°C)	C_{areal} (mF/cm^2)	C_{sp} (F/g)
THF	60	6.38	22.28
	120	16.33	58.31
	150	72.70	259.35
	180	45.77	164.48
Ethanol	75	5.44	19.55
	120	16.88	60.82
	150	69.42	249.82
	180	75.53	268.88
Acetonitrile	75	4.79	19.05
	120	28.57	101.12
	150	51.23	185
	180	75.26	270.58
H$_2$O	95	35.72	102.64
	120	91.03	327.58
	150	103.44	371.87
	180	67.58	241.03
DMF	120	24.49	85.78
	150	141.60	513.86
	180	64.10	234.7
Ethylene Glycol	120	16.49	59.36
	150	21.13	76.08
	180	58.98	208.85
	195	23.25	83.85
NMP	120	8.60	30.32
	150	33.56	119.92
	180	59.76	214.03
	195	19.44	68.27

For example, consider all the NG samples prepared using THF as the solvent (Figure 3a). As the synthesis temperature increases from 60 °C to up to 150 °C, the current density gradually increases, but with a further increase in the synthesis temperature to 180 °C, the current density decreases. This behavior suggests that the conductivity and mobility of the charge carriers, especially in the carbonaceous layer, changes accordingly, as discussed below. This change can only be associated with the size and quality of crystallinity of the nano-graphitic regions in the sample. According to the XRD and Raman spectroscopy results (Tables S1 and S4, ESI), the nano-graphite crystallites become bigger in size and better in crystallinity as the synthesis temperature increases up to 150 °C (for the THF solvent) and it decreases beyond 150 °C. Hence, the nano-graphitic regions in the flakes have better crystallinity and are bigger in size for the sample prepared using THF as the solvent at 150 °C as compared to the other samples prepared at lower or higher synthesis temperatures. As the graphitic regions are better conductors than the amorphous carbon regions, the higher current density in the NG-3-THF-150 sample is justified.

Now, comparing the variation in current density for the samples prepared with different solvents (Figure 3), we observe that the sample synthesized using H$_2$O as the solvent has the highest current density (and the CV loops have the highest area), and the sample synthesized using EG has the lowest current density (and the loops have the lowest area). This low current density (or conductivity) of the samples prepared by EG depicts its more insulating character. This higher insulating character of the samples can originate due to the incomplete removal of the oxygen moieties from the surface of the samples (low reduction of GO), where the surface oxygen moieties block the charge transfer behavior. This aspect is further explored through the analysis of the XPS results, as discussed later.

Carefully inspecting all the CV curves (Figure 3), we realize that the CV curves for some of the samples have small redox peaks, which is an indication that these samples have a tendency for redox reactions. The redox peaks are prominent for the samples showing high current densities, i.e., for the samples synthesized using EtOH, ACN, and H_2O as solvents. This high current density in combination with the redox peaks for these samples suggests that the surface of these samples facilitates better charge accumulation and also the tendency to react with the charged species. In general, an ideally flat graphene surface, having no defects, is not prone to react with the accumulated charge species (hence, they show perfect EDLC behavior) [42]. Thus, the tendency for a redox reaction to occur for these samples indicates that the surface of these NG sheets has more defects than the samples where the redox peaks are not prominent. This aspect of the samples can be verified by comparing the XRD patterns of the samples synthesized using different solvents (at the optimum temperatures), especially by referring to the (002) XRD peaks (Figure 1). The samples synthesized using EtOH (at 180 °C), ACN (at 180 °C), and H_2O (at 150 °C) as solvents have broad (002) peaks (due to the presence of the pronounced amorphous shoulder to the left) as compared to that of the samples prepared using DMF (at 150 °C), THF (at 150 °C), NMP (at 180 °C), and EG (at 180 °C) as solvents. This clearly proves that the samples prepared using EtOH (at 180 °C), ACN (at 180 °C), and H_2O (at 150 °C) as dispersive solvents have a higher amount of surface defects. These defects create energy levels within the electronic bandgap, which leads to higher conductivity and a high current density for these samples, as observed.

From the above discussions, it can be inferred that the various solvents facilitate the reducing species (NH_3 and CO/CO_2 present in the solvent) differently to reach the site of the oxygen moieties of GO, thereby reducing it. This leads to the modification in the structural aspects of the NG sheets and in turn, the supercapacitance of the samples. Note that although the FWHM of the XRD (002) peak of the samples prepared at the lower (than the optimum) temperatures are higher, they do not show a prominent redox peak. This is because the higher peak width for the low-temperature synthesized samples is not due to the presence of the higher number of defects, but rather due to the improper reduction of the GO sheets, as discussed earlier (effect of temperature). These oxygen moieties act as a shield for the redox reaction. Hence, these samples show a lower current density and lower capacitance (Figure 3 and Table 2).

From Table 2, it is clear that the sample prepared using DMF shows the highest 'areal' and 'specific' capacitance, followed by those of the samples synthesized using H_2O and ACN. These samples demonstrate better reduced (bare, without having much oxygen moieties) surfaces of the NG sheets with higher number of defects. As there are more defects, it seems they allow the diffusion of charged species (from the solution) at the surface of the sample, originating as the oxidation peak in the CV curves.

Figure 4 shows the galvanostatic charging–discharging (GCD) curves for all our synthesized samples. In the CV curves (Figure 3), we observed a constant current density over the potential range of 0 to 0.8 V, and the same potential range is covered here in the GCD curves, i.e., all the samples are charged to a voltage of 0.8 V by application of a constant current density of 0.5 A/g. Note that the mass of the sample (electrode material) is estimated during the preparation of the electrodes. All the plots exhibit nearly linear discharge–time profiles, revealing an exemplary capacitive behavior. Generally, in all the GCD curves, four different features can be noticed, namely (1) the initial fast charging, (2) forced (slow) charging by the applied current up to 0.8 V, (3) fast discharging (also known as IR drop), and (4) slow discharging to 0 V. The fast charging and fast discharging occur through the highly mobile charged species, mostly by the electrons transiting to the conduction band (because the motion of the ions in the electrolyte is slow to make this IR drop). Here, 'I' corresponds to the electronic current and 'R' represents the resistance offered by the material to these electrons' motion. As soon as a small current is applied, these highly mobile charges (electrons) accumulate and increase the voltage (charging). Similarly, during the discharging experiment, these electrons move out rapidly and initially discharge the capacitor very fast, known as IR drop. This occurs as soon as the applied current is removed and the capacitor is allowed to discharge. As these electrons do not originate from the electrolyte or from the ions in the electrolyte, they have their origin in the electrode material (i.e., our NG sample). It is known that the structural defects in the carbonaceous materials create electronic defect levels within the energy bandgap of the materials. Depending on the concentration of these defect levels, the conductivity of the carbonaceous materials change. These defect energy levels (within the bandgap) facilitate the electrons to transit to the conduction band through hopping, requiring only a small voltage to be created within the material and pass current. Ideally, for materials having no defects, there will be no conduction until a voltage equivalent to the band gap energy is provided. However, the presence of too many of the defects in the materials (as in the case of amorphous-type materials), the electrical resistivity is enhanced through the scattering of the electrons by the defects in their path during transport.

Hence, often a moderate number of defects in the carbonaceous materials helps in obtaining better conductivity. As we have observed in the XRD patterns, when the FWHM of the (002) peak is high or has a shoulder, it signifies that a different degree of disorder exists in the samples synthesized through different solvents. These results were further supported by the results from the Raman spectroscopy study. Hence, we can understand the GCD curves through the structural and associated electronic properties of the carbonaceous (NG) materials used.

Followed by the initial fast charging and the IR drop region of the GCD curves, the slow charging and discharging region is observed. The origin of this slow charging/discharging behavior is basically governed by the diffusion/hopping-mediated motion of the charged species. Unlike the fast charging or IR drop region (where electrons are the only charge carriers), the charge carriers during this slow charging/discharging process are both electrons and ions. This includes the diffusion of electrons through the electronic defect levels and the ions diffusing into the electrode material and within the electrolyte. As we have used aq. H_2SO_4 (0.5 M) as the electrolyte during the GCD measurements, the ions present in the electrolyte are H^+ and SO_4^{2-}. Considering the faster dynamics of the ions in the solution than that of diffusion into the electrode material, we may further divide the slow charging/discharging regions of the GCD curves into two different regions. The regions that immediately follow the fast-charging region can be assigned to the diffusion of electrons inside the electrode materials and the arrangement of the ions in the electrolyte to form the double layer near the electrode surface. The opposite process happens in the region following the IR drop regions. Furthermore, the last (very slow) part of the charging and discharging curves can be assigned to the diffusion of only the ions into or out of the electrode material, respectively.

Let us understand the GCD curves obtained for different sets of samples prepared using different solvents and at different temperatures (Figure 4). Consider first the charging time observed for a set of samples prepared at different temperatures using a particular solvent; for instance, let us analyze Figure 4d. Here, we observe that as the hydrothermal synthesis temperature increases up to 150 °C, the time taken by the samples for fully charging the capacitor (up to 0.8 V) gradually increases from 256 s to 855 s (these values are obtained for the sample synthesized at 95 °C and 150 °C, respectively), whereas it decreases to 559 s for the sample synthesized at a higher temperature, 180 °C. This observation confirms that for the samples having bigger sized and high-quality nano-graphitic crystalline regions (observed from XRD and Raman spectroscopy), it takes a longer charging time to reach 0.8 V, whereas this charging time decreases for the samples synthesized at a higher temperature than the optimum temperature. This behavior is universally observed in all different sets of samples (prepared using different solvents), as seen in Figure 4.

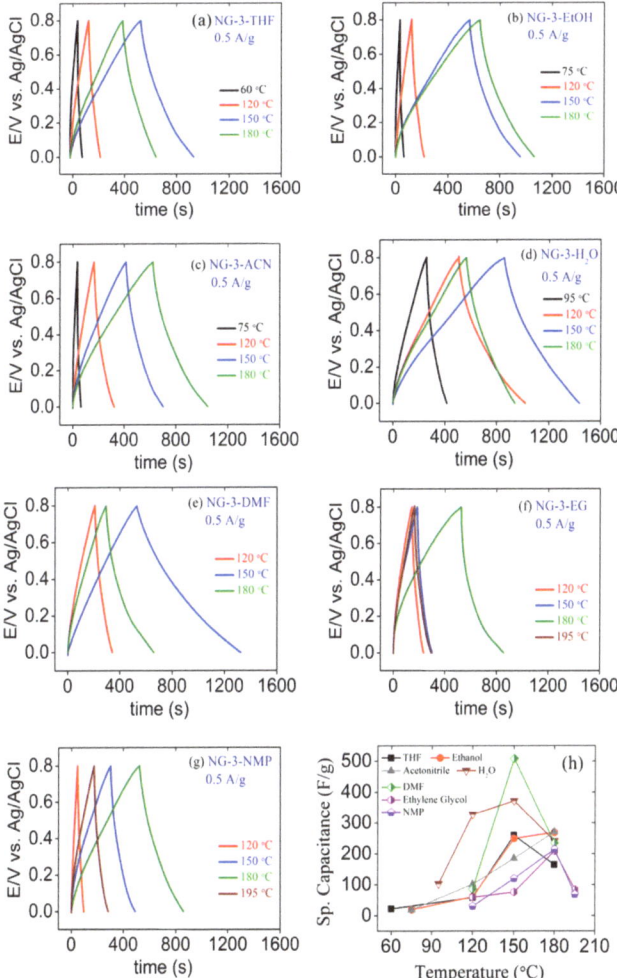

Figure 4. (**a**) The galvanostatic charging–discharging (GCD) plots for all the NG samples synthesized at different temperatures with various dispersive solvents as indicated. (**a**) THF, (**b**) ethanol, (**c**) acetonitrile, (**d**) H_2O, (**e**) DMF, (**f**) ethylene glycol, and (**g**) NMP. (**h**) The specific capacitance of all the NG samples is calculated from the discharging region of the GCD curves by using Equation (2).

As discussed earlier, for any temperature below the optimum synthesis temperature, an improper reduction of GO occurs, and the oxygen moieties (that still remain on the sheets) do not facilitate better stacking of the NG sheets. Furthermore, the presence of more oxygen moieties on the surface of the graphene sheets makes the graphene sheet more insulating, and they do not permeate the ions of the electrolytes to diffuse into the electrode material, as they act as a shield against the charge intercalation/accumulation in the electrode materials. However, for any synthesis temperature above the optimum temperature, the creation of too many defects is observed by XRD and Raman spectroscopy. This reveals the voids created in the graphene sheets due to the high-temperature synthesis of the sample, and the presence of voids in the graphene sheets makes the graphitic flakes amorphous-like. This leads to a slow permeation of the ions (present in the electrolyte) into the electrode material through these voids. The opposite process occurs during the discharging experiments. Hence, the oxygen moieties are key in providing the insulating and impermeable (to foreign ions) nature of the samples, thereby controlling the overall nature of the GCD curves. This aspect will be further explored in the XPS results.

The fast charging and discharging observed in the GCD curves for the sample prepared at low temperatures (lower than the optimum temperature) could be assigned to the fast formation/dispersion of the double layer by the ions in the electrolyte. As these samples do not conduct electronically and the ions in the electrolyte are shielded by the oxygen moieties to become intercalated in the NG sheets, they facilitate the formation of double layers. The GCD curves of the samples prepared using EG as the solvent clearly reveal this aspect. Furthermore, this aspect of the samples can also be rationalized from the CV curves (Figure 3), where these samples show better EDLC behavior than the samples prepared at the optimum temperature. Hence, as discussed in the previous paragraph, the origin of the gradually increasing charging time with an increase in the synthesis temperature of the samples (up to the optimum synthesis temperature) is due to the facilitation/permeation for the ion diffusion or electron diffusion into the graphitic flakes. With any further increase in the synthesis temperature above the optimum temperature, the amorphous-type graphitic flakes do not offer the slow motion of the ions of the electrolytes; hence, the charging and discharging time decreases again. This understanding can be universally applicable to all our samples.

When we compare the GCD curves among the samples prepared using different dispersive solvents, another interesting point to note is the difference in the charging and discharging times. A low charging time saves time by quickly charging the device, while slow discharging saves energy by decreasing leakage; hence, a better supercapacitor should have a low charging time and long discharging time. As stated earlier, the (002) XRD peak for the sample prepared by using DMF as the solvent (at 150 °C) is the sharpest among all samples, and the quality of crystallinity observed by Raman spectroscopy for this sample is the highest. This suggests that an optimum number of defects, suitable for slow discharging, is created for this sample at the given synthesis condition (i.e., at 150 °C by using DMF). This is the most plausible explanation for the long discharging time or slow diffusion during the discharging of the sample. This suggests that the choice of the dispersive solvent and the synthesis temperature are the most important parameters in obtaining high-quality NG samples for energy storage applications, such as supercapacitors.

As the discharging time is more important for the application of the device and to store energy for a long time, we have calculated the specific capacitance (C_{sp}) for all our samples through Equation (2) by taking the discharging curve into account. The obtained specific capacitance values are provided in Table 2. It is seen that among all the sample synthesized at different temperatures and in different dispersive solvents, the sample synthesized at 150 °C by using DMF as the dispersive solvent shows the highest C_{sp} value of 514 F/g (at a 0.5 A/g current density). To further explore the behavior of the samples, we have measured the CV curves at different scan rates, as shown in Figure S7(a,c,e) (ESI file). From these CV curves, it can be rationalized that as the scan rate increases from 5 mV/s to 100 mV/s, the rectangular nature of the CV curves decreases. This reveals that the diffusion time scales for the charged species are different due to the nature (activity) and the type of defects present on the surface of the NG sheets being different, and they interact differently with the charge's species. To further explore the nature of the samples, we have measured the GCD curves at different current densities, as shown in Figure S7b,d,f. As the current density increases, the charging and discharging time of our NG-based capacitors decreases. It can be understood that the ions quickly charge and discharge the capacitors, similar to that observed in the case of the CV curves.

To understand the variation in the supercapacitance values obtained for our samples and to support the XRD, Raman spectroscopy, CV, and GCD results given earlier, electrochemical impedance spectroscopy (EIS) studies were performed for all our samples in the range of frequency between 100 mHz and 200 kHz with a sine wave amplitude of 5 mV. Figure 5 demonstrates the Nyquist plots (the imaginary (Z'') versus real (Z') part of the impedance, Z^*) for all our NG samples. These plots can be broadly categorized into two regions, (1) the high-frequency (semicircular) region (near the origin) and (2) the low-frequency (linear/Warburg) region. In all the impedance plots, we observe a (not well-pronounced) semicircular region in the high-frequency part and a nearly straight line in the low-frequency part of the Nyquist plots.

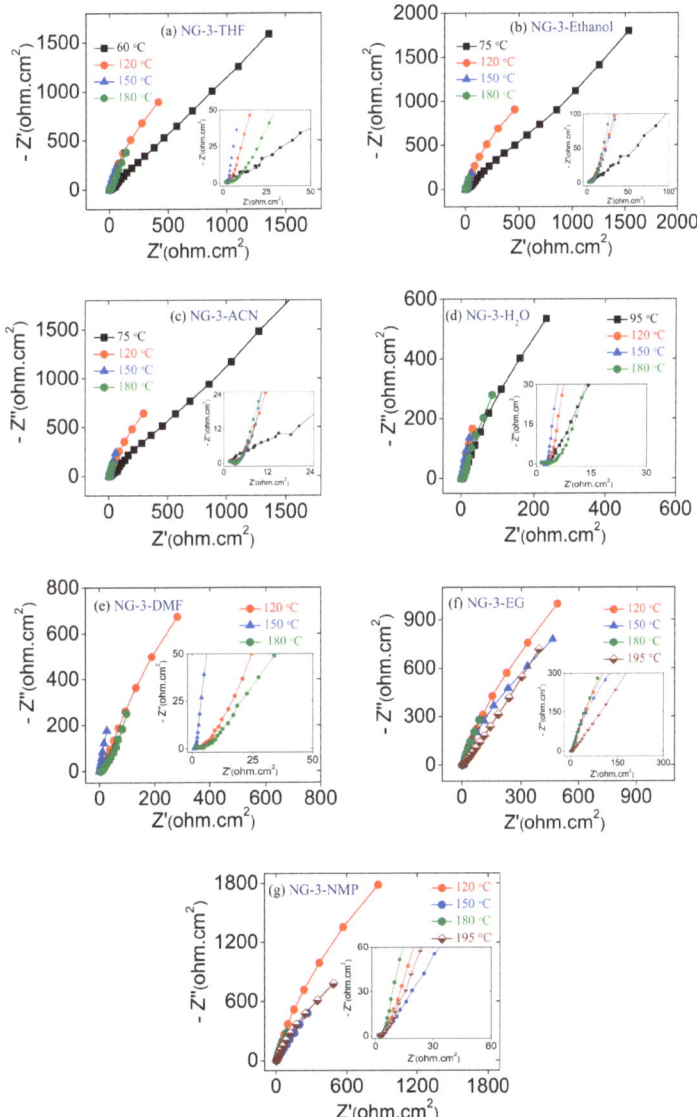

Figure 5. EIS spectra of all our synthesized NG samples at different temperatures in various dispersive solvents.

In general, the presence of semicircles in the high-frequency region is related to the electronic response within the electrode materials and also the electronic charge transfer into or out of the electrode material through diffusion. The value at which the semicircular region (if extended) would cut the Z' axis is the charge transfer resistance (R_{ct}). This charge transfer can be the electronic transport within the NG materials or the diffusion of the ions into the NG materials. The higher the value of R_{ct}, the more difficult it is for the charge transfer. In our NG materials, this could happen in the following two situations: (1) if the NG material on the electrode is more resistive to electronic motion, i.e., if the crystallinity of the graphitic flakes becomes too poor and/or the presence of a high amount of oxygen moieties, which offers an insulating character to the NG sheets, and (2) if the ions in the electrolyte are not allowed to diffuse into the electrode material, i.e., if the NG sheets are not properly reduced (the amount of the oxygen moieties is still greater on the surface, which repels the ions). The low-frequency linear (straight-up) region shows the characteristics of the ion accumulation at the surface of the materials and forms the EDL, i.e., the ions in the electrolyte forms the double layer at the electrode–electrolyte interface [43,44]. It is clear that if the low-frequency linear region is parallel to the y-axis (-Z''), it signifies the formation of a perfect EDL capacitor at the interface, whereas a decrease in the slope of this linear region occurs due to the diffusion of ions from the first ionic layer into the electrode material. Furthermore, the highest (top) value of this part (at a certain low frequency) is the maximum capacitance contributed by the ions, including EDL and diffusion behavior.

In all the sets of samples prepared in different solvents, it can be rationalized that the samples prepared at the lowest temperatures show the highest R_{ct} (Figure 5 (insets)). This suggests that these NG samples have a more insulating character due to the improper reduction of GO, leading to the presence of the oxygen moieties on the NG sheets, as discussed earlier. It can also be rationalized that for the samples prepared at the optimum temperatures, the low-frequency linear (tail) region has the highest slope (Figure 5) among the set of samples prepared using a particular solvent, i.e., at 150 °C (THF), 180 °C (EtOH), 180 °C (ACN), 150 °C (water), 150 °C (DMF), 180 °C (EG), and 180 °C (NMP) temperatures for the solvents mentioned inside the brackets. This suggests that these samples have a better flat surface and perfect EDL formation, which contributes more to the net capacitance of the samples. This can also be assessed through the CV curves for these samples. Furthermore, for these samples, the low R_{ct} values are observed, signifying that the electronic charge transfer becomes easier in these samples due to the proper reduction of GO and the formation of optimally crystalline NG sheets. Now, comparing the EDL behavior among different sets of samples, it is clear that the sample prepared by using DMF as the solvent at 150 °C has the highest slope of the low-frequency tail (Figure 5e), and it shows the best rectangular CV curve (Figure 3e), very low R_{ct}, and the highest specific supercapacitance (Table 2).

3.3. XPS Study

The XPS spectra of the samples prepared using a particular solvent and at the optimum temperature, i.e., at 150 °C (THF), 180 °C (EtOH), 180 °C (ACN), 150 °C (water), 150 °C (DMF), 180 °C (EG), and 180 °C (NMP), for the solvents mentioned inside the brackets are shown in Figures 6 and 7.

Considering the N 1s spectra of all the samples shown in the left panel of Figure 6, we assign the peaks occurring at ~398.3 ± 0.3 eV, 399.2 ± 0.3 eV, 400.1 ± 0.3 eV, 401.4 ± 0.3 eV, and 402.7 ± 0.3 eV to the pyridinic-N, amino-N, pyrrolic-N, graphitic-N, and oxidized-N species, respectively [2,45]. The fitting results (given in Table S4, ESI file) suggest that all types of N defects are present in all the samples but in different amounts. The middle panel of Figure 6 shows the C 1s spectrum of all the studied samples. The peaks occurring at 284.5 ± 0.3 eV, 285.3± 0.3 eV, 286.5 ± 0.3 eV, 287.6 ± 0.2 eV, and 289 ± 0.4 eV are assigned to the C=C aromatic/C-H bonds (sp^2), C-C aliphatic (sp^3), C-O-C/C-N bonds, and -C=O and O=C-OH/O=C-NH- bonds, respectively [46,47]. The fitting results are tabulated in Table S5 (ESI file). Similarly, the O 1s spectra of all our samples are shown in the right panel of Figure 6. The O 1s peaks are fitted to the following sub-peaks: quinone (~530.1 ± 0.3 eV), N-O (531.1 ± 0.3 eV), O=C-OH, (531.6 ± 0.3 eV),

-C=O (carbonyl oxygen, 532.3 ± 0.3 eV), C-O-C (533.1 ± 0.3 eV), C-OH (hydroxyl/phenolic oxygen, 534.01 ± 0.3 eV), and chemisorbed oxygen/adsorbed H_2O (535.32 ± 0.3 eV), respectively [2,48]. The O 1s peak of a particular oxygen species occurs almost at the same binding energy for all the samples, indicating that all the oxygenated groups are present within all samples with varying quantities. The fitting results are listed in Table S6 (ESI file). Considering the area under the C 1s, O 1s, and N 1s peaks of a particular sample, we have quantified the amount of C, O, and N present in our samples. The results are given in Table S7 (ESI file).

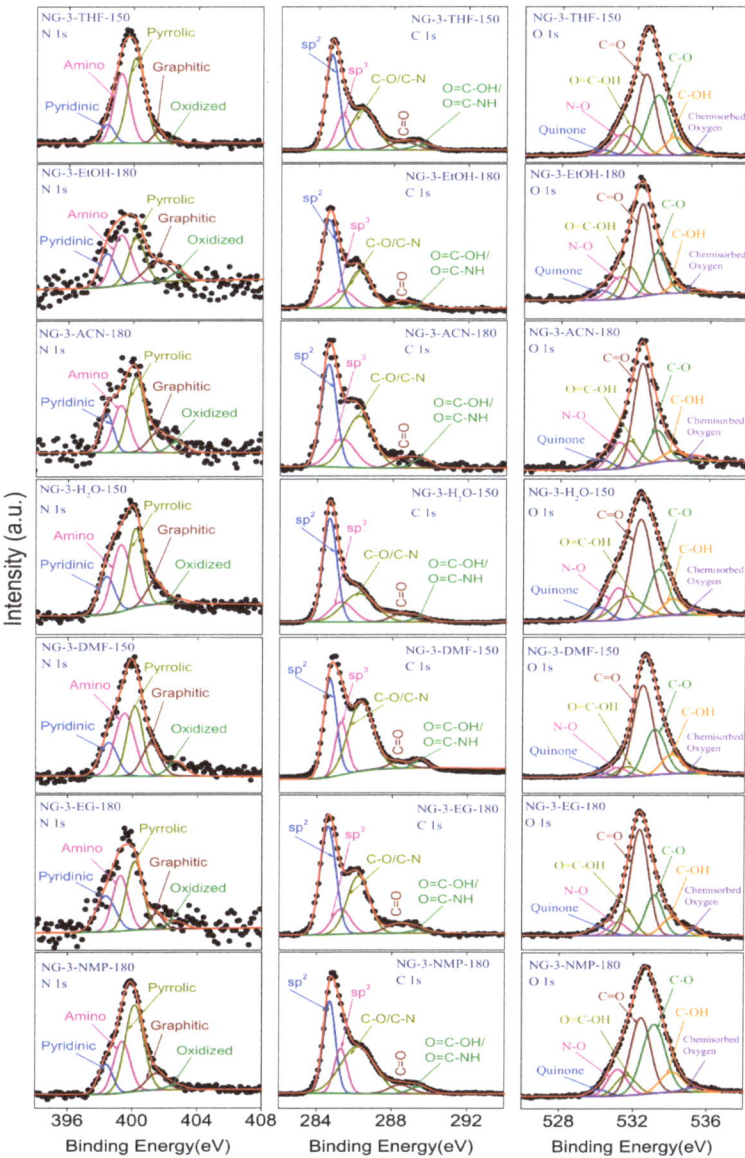

Figure 6. XPS spectra of synthesized NG-3 sample (**left panel**) N 1s, (**middle panel**) C 1s, and (**right panel**) O 1s.

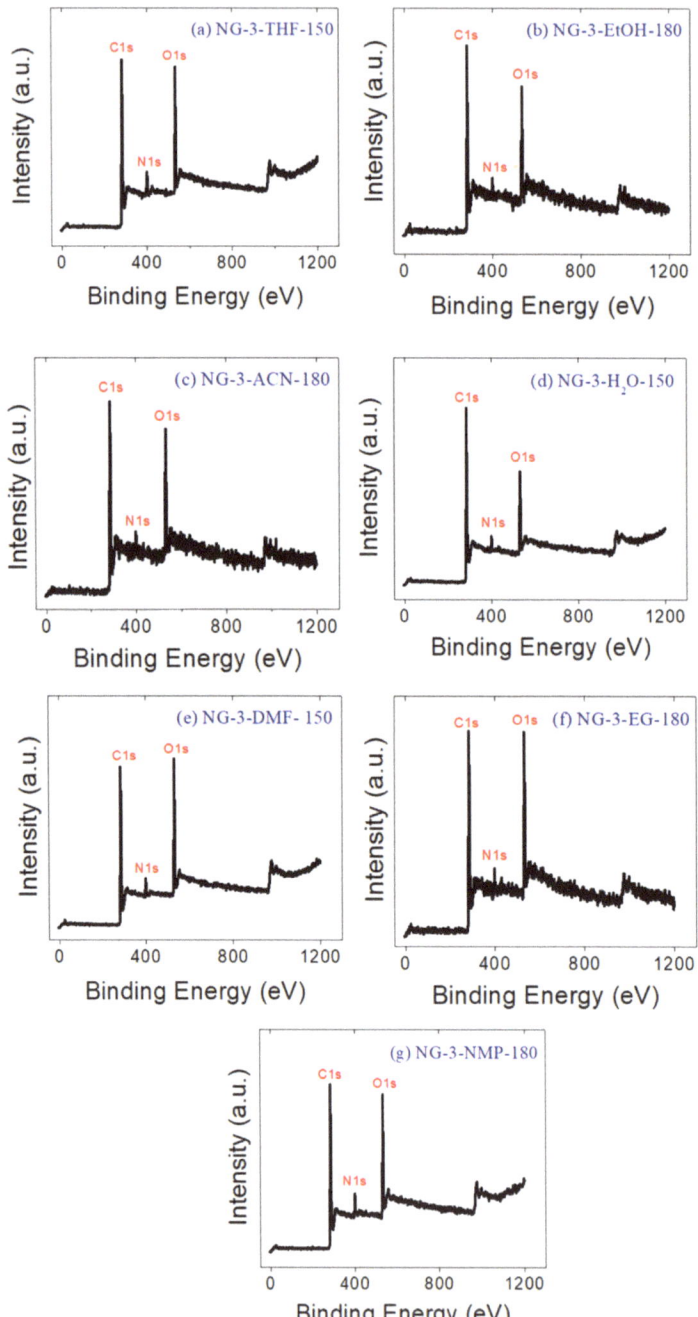

Figure 7. The wide XPS spectra of the synthesized samples in the full measured energy range.

To estimate the relative amount of N doping with respect to temperature, we have explored the XPS spectra obtained for a representative set of samples synthesized using DMF as the solvent and at three different synthesis temperatures, viz. 120 °C, 150 °C, and 180 °C (Figure S9, ESI file). Furthermore, the reason for the disparity in the specific capacitance values of the samples synthesized using different solvents can be explained through the content and type of different N environments in different samples. The number of different types of N environments (in %) obtained from the fitting of these spectra are given in Table S8 (ESI file). The important information that can be revealed from this study is that the amount of graphitic N is more in the DMF-150 sample, which shows the highest supercapacitance (Table S4, ESI). As we have mentioned previously, this graphitic N could be the main reason for the high conductivity of the sample prepared using DMF at 150 °C. In contrast, for the NG-3-THF-150 sample, the percentage of graphitic nitrogen is significantly less (Table S4, ESI). Due to this, the sample may be less conductive and shows a substantially less specific capacitance of ~259 F/g at 0.5 A/g. The results illustrate that the presence of graphitic nitrogen in the NG-3-DMF-150 sample boosts its capacitive behavior, along with other structural and electronic parameters, as discussed previously. Furthermore, the high electronegativity of nitrogen assists in generating significant interactions with the solvated charged species coming to the surface of the NG electrode.

Furthermore, from Table S4 (ESI), it can be rationalized that an optimum amount of nitrogen doping is essential in maximizing the specific capacitance of the NG samples. If the N doping is too low, it may not modify the carbon framework suitably, and it may not show a high specific capacitance value. However, according to our results (Table S4, ESI), a moderate doping of nitrogen (~5.74%) is quite fruitful in achieving high C_{sp}, and this also provides the optimum crystallinity and structural stability to the graphenic framework of the sample. Maybe this amount of nitrogen doping is just enough to allow for a good permeation path and the effortless movement of electrons across the carbon framework during the charging and discharging of the supercapacitor. As the amount of nitrogen in the graphene sheets of the NG samples increases, too many defects are formed [49], resulting in breakage of the carbon frameworks, and the electron cloud is not well localized [50]. Hence, the control of N doping in NG samples is quite essential. To compare our results with the values reported in the literature, in Table S9 (ESI), we have listed the specific capacitances of different amounts of N doping in graphene-based materials synthesized using different methods. Clearly, all our optimized samples are either equivalent or better than the reported values (compare with Table 2), and among all our samples, the NG-3-DMF-150 sample shows an excellent C_{sp} value (~514 F/g). A two-electrode symmetric device had been demonstrated to show excellent cycling stability (~82% even after 8000 cycles) with this optimized material, as shown in our earlier work [33].

It is clear that the role of the dispersive solvents and the hydrothermal/solvothermal reaction temperature are important factors that decide the structural and electronic aspects of the NG sheets. These aspects control the supercapacitative behavior of the NG samples. Furthermore, as we have used 1 g of GO, the yield is higher than that reported in most of the literature; hence, the synthesis process is industrially scalable. Here, we try to explore the nature of the dispersive solvents responsible for the supercapacitative behavior of the NG sheets. As it is known, the dispersive solvents have their own intrinsic properties such as the boiling point, dipole moment, viscosity, dielectric constant, and dispersibility of GO in these solvents [51]. Hence, all these properties of the solvents play their vital roles in creating the differences in the structural and electronic properties of the NG samples, which in turn leads to the observed discrepancy in the trends of C_{sp} values. Furthermore, it is conceivable that a proper reduction of GO depends on these intrinsic properties. Among them, the viscosity of the solvent significantly influences the conductivity of the ions and the dispersibility of GO sheets in these liquids/gels. The dielectric constant (ε), along with the viscosity, are the two principal physical properties of the solvents that directly affect the ionic conductivity of the solvents. This is because the dissociation of the salts (such as urea in the present case, which dissociates into NH_3 and CO/CO_2) is associated with the dielectric constant, and the ionic/molecular mobility is elucidated by the viscosity of the solvents [39]. A low-viscosity solvent offers a superior conductance for the ions as compared to a highly viscous solvent. However, it does not frequently happen, since viscosity is also related to the solvent's molecular interactions and dipole moment. These two features ascertain the dielectric constant (ε) of the solvent; a superior dielectric constant lessens ion coupling and ameliorates the conductivity of a known salt. Therefore, even though lower viscosity ameliorates the real mobility of the free ions, the small dielectric constant, which is generally coupled with low-viscosity solvents (due to feeble intermolecular interactions), leans to abridge the conductivity [39].

The dispersibility of GO in various solvents has been studied by several researchers, where they demonstrated that as-synthesized GO dispersions in organic solvents (THF, ethanol, acetonitrile, ethylene glycol, and NMP) were noticed to display short-term colloidal constancy compared to the dispersion of the similar materials in water and DMF [51]. Several studies examined that ethylene glycol and THF dispersions acquired a fairly superior quantity of precipitate compared to water, DMF, and NMP dispersions, signifying that the other solvents acquire a moderately lower dispersion capability. Water exhibits the most superior dispersion capability, as it delivers the utmost absorption and consequently the prevalent amount of suspended GO, pursued narrowly by DMF and NMP [26]. Ethylene glycol and THF demonstrate extremely analogous dispersion skills toward as-synthesized GO, even though they are evidently lighter compared to other solvents [27]. Several researchers observed that as-synthesized GO nanoplatelets are showing more dispersibility in *N,N*-dimethylformamide (DMF) devoid of chemical modification [52,53].

Presently, the mechanisms that permit the steady dispersion of GO in the discussed organic solvents are unambiguous. An essential but insufficient circumstance appears to be that the solvent molecules should be noticeably polar. This is sensible, as the GO sheets are noticeably and profoundly ornamented by polar oxygen-containing functional groups (hydroxyl, carbonyl, and carboxyl), which is the main cause to endorse a fine GO sheet–solvent interaction. It is known that water and four other organic solvents demonstrate noteworthy electrical dipole moment values as follows: 1.85 D (water), 3.25 D (DMF), 3.75 D (NMP), 1.75 D (THF), and 2.31 D (ethylene glycol). On the contrary, the solvents having a minute dipole moment (*n*-hexane, 0.085 D; *o*-xylene, 0.45 D) are evidently unable to disperse the as-synthesized GO sheets. Conversely, there are a variety of solvents with elevated dipole moments (mainly DMSO, 4.09 D) that are unsuccessful in affording GO dispersions with long-lasting dispersion constancy, which advocates that additional features excluding solvent polarity are imperative for examining superior dispersibility [40,41].

Hence, from the above observations, we concluded that the role of the solvents and temperature is very important for the synthesis of nitrogen-doped reduced graphene oxide materials for achieving high supercapacitance. Furthermore, it is clear that all the different physical properties (such as the boiling point, dipole moment, viscosity, dielectric constant, and dispersibility of GO) of the solvents, along with the synthesis temperature, are significant in describing the discrepancy in specific supercapacitance values observed for the synthesized samples at different temperatures in various solvents. Overall, DMF is an ideal solvent for the proper reduction of GO and for the perfect stacking of NG sheets.

4. Conclusions

We have used different solvents such as THF, ethanol, acetonitrile, H_2O, DMF, ethylene glycol, and NMP for the solvothermal synthesis of N-doped reduced graphene oxide (NG) at different synthesis temperatures between 60 °C and 195 °C. Our results show that in these solvents, GO is differently reduced and differently exfoliated into individual flakes of a few layers of N-doped reduced graphene oxide-like sheets with lateral dimensions of a few nanometers. The suitable conditions for the reduction and simultaneous N doping in the reduced GO sheets, along with proper stacking of the NG sheets, are identified in several organic/aqueous solvents. Our results demonstrated that among all solvents, the NG sample prepared using DMF exhibits the best supercapacitive performance (with a specific capacitance of 514 F/g at 0.5 A/g). Moreover, the appropriate stacking of the NG sheets is achieved with an optimum number of defects required for appreciably high conductivity in comparison to the other solvents. This supercapacitance value is the highest among all our samples and also a competitive value among all the known state-of-the-art lightweight materials. Hence, our results reveal the optimum synthesis conditions and suitable solvent parameters required for the synthesis of high-quality NG sheets showing excellent supercapacitor performance for their direct use in industrial applications.

Supplementary Materials: The following supporting information can be downloaded at: https://www.mdpi.com/article/10.3390/c10040089/s1. References [54–60] are cited in the supplementary materials.

Author Contributions: A.Y.: conceptualization, physical and electrochemical measurements, data analysis, writing—original draft, writing—review and editing. R.K.: conceptualization, writing—review and editing. D.J.—data analysis, writing—review and editing. N.T.—data analysis, writing—review and editing. F.Y.—writing—review and editing, project funding acquisition. B.S.: project administration, project funding acquisition, data analysis, writing—review and editing. All authors have read and agreed to the published version of the manuscript.

Funding: The funding for this work was received from the ISRO-IISc STC (project code- ISTC/CMR/BS/431) and the U.S. National Science Foundation (NSF) under grant #DMR-2122044 and the U.S. Army Research Office (ARO) under grant #W911NF2210109.

Data Availability Statement: The raw/processed data required to reproduce these findings cannot be shared at this time as the data also form part of an ongoing study.

Acknowledgments: The authors acknowledge SSCU, IPC, AFMM and CeNSE-MNCF at IISc Bengaluru for using the characterization facilities.

Conflicts of Interest: There are no conflicts of interest.

References

1. Wang, Z.F.; Tang, C.; Sun, Q.; Han, Y.L.; Wang, Z.J.; Xie, L.; Zhang, S.C.; Su, F.Y.; Chen, C.M. Effect of N-Doping-Derived Solvent Adsorption on Electrochemical Double Layer Structure and Performance of Porous Carbon. *J. Energy Chem.* **2023**, *80*, 120–127. [CrossRef]
2. Ajravat, K.; Pandey, O.P.; Brar, L.K. Significance of N Bonding Configurations in N-Doped Graphene for Enhanced Supercapacitive Performance: A Comparative Study in Aqueous Electrolytes. *FlatChem* **2024**, *43*, 100588. [CrossRef]
3. Li, X.; Cao, J.; Chen, J.; Zhu, Y.; Xia, H.; Xu, Z.; Gu, C.; Xie, J.; Jones, M.; Lyu, C.; et al. UV-Induced Synthesis of Graphene Supported Iridium Catalyst with Multiple Active Sites. *Adv. Func. Mater.* **2024**, *34*, 2313530. [CrossRef]
4. Fan, Y.F.; Yi, Z.L.; Song, G.; Wang, Z.F.; Chen, C.J.; Xie, L.J.; Sun, G.H.; Su, F.Y.; Chen, C.M. Self-Standing Graphitized Hybrid Nanocarbon Electrodes towards High-Frequency Supercapacitors. *Carbon* **2021**, *185*, 630–640. [CrossRef]
5. Wang, H.; Maiyalagan, T.; Wang, X. Review on Recent Progress in Nitrogen-Doped Graphene: Synthesis, Characterization, and Its Potential Applications. *ACS Catal.* **2012**, *2*, 781–794. [CrossRef]
6. Lu, Y.; Huang, Y.; Zhang, M.; Chen, Y. Nitrogen-Doped Graphene Materials for Supercapacitor Applications. *J. Nanosci. Nanotechnol.* **2014**, *14*, 1134–1144. [CrossRef]
7. Jing, M.; Wu, T.; Zhou, Y.; Li, X.; Liu, Y. Nitrogen-Doped Graphene via In-Situ Alternating Voltage Electrochemical Exfoliation for Supercapacitor Application. *Front. Chem.* **2020**, *8*, 428. [CrossRef]
8. Nolan, H.; Mendoza-Sanchez, B.; Ashok Kumar, N.; McEvoy, N.; O'Brien, S.; Nicolosi, V.; Duesberg, G.S. Nitrogen-Doped Reduced Graphene Oxide Electrodes for Electrochemical Supercapacitors. *Phys. Chem. Chem. Phys.* **2014**, *16*, 2280–2284. [CrossRef]
9. Deng, D.; Pan, X.; Yu, L.; Cui, Y.; Jiang, Y.; Qi, J.; Li, W.X.; Fu, Q.; Ma, X.; Xue, Q.; et al. Toward N-Doped Graphene via Solvothermal Synthesis. *Chem. Mater.* **2011**, *23*, 1188–1193. [CrossRef]
10. Jeong, H.M.; Lee, J.W.; Shin, W.H.; Choi, Y.J.; Shin, H.J.; Kang, J.K.; Choi, J.W. Nitrogen-Doped Graphene for High-Performance Ultracapacitors and the Importance of Nitrogen-Doped Sites at Basal Planes. *Nano Lett.* **2011**, *11*, 2472–2477. [CrossRef]
11. Qu, L.; Liu, Y.; Baek, J.-B.; Dai, L. Nitrogen-Doped Reduced-Graphene Oxide as an Efficient Metal-Free Electrocatalyst for Oxygen Reduction in Fuel Cells. *ACS Nano* **2010**, *4*, 1321–1326. [CrossRef] [PubMed]
12. Wang, Y.; Shao, Y.; Matson, D.W.; Li, J.; Lin, Y. Nitrogen-Doped Graphene and Its Application in Electrochemical Biosensing. *ACS Nano* **2010**, *4*, 1790–1798. [CrossRef] [PubMed]
13. Yadav, A.; Kumar, R.; Yadav, K.; Thomas, N.; Mishra, M.; Sahoo, B. Synthesis, Characterization and Insights into the Supercapacitive and Electrocatalytic (OER) Bi-Functional Properties of Nitrogen-Doped Reduced Graphene Oxide Using Dicyandiamide Precursor. *Solid State Sci.* **2024**, *147*, 107377. [CrossRef]
14. Hasan, S.A.; Tsekoura, E.K.; Sternhagen, V.; Strømme, M. Evolution of the Composition and Suspension Performance of Nitrogen-Doped Graphene. *J. Phys. Chem. C* **2012**, *116*, 6530–6536. [CrossRef]
15. Fan, X.; Yu, C.; Yang, J.; Ling, Z.; Qiu, J. Hydrothermal Synthesis and Activation of Composite for High-Performance Supercapacitors. *Carbon* **2014**, *70*, 130–141. [CrossRef]

16. Lee, J.W.; Ko, J.M.; Kim, J.D. Hydrothermal Preparation of Nitrogen-Doped Graphene Sheets via Hexamethylenetetramine for Application as Supercapacitor Electrodes. *Electrochim. Acta* **2012**, *85*, 459–466. [CrossRef]
17. Zhao, Y.; Hu, C.; Hu, Y.; Cheng, H.; Shi, G.; Qu, L. A Versatile, Ultralight, Nitrogen-Doped Graphene Framework. *Angew. Chem. Int. Ed.* **2012**, *51*, 11371–11375. [CrossRef]
18. Wang, D.; Min, Y.; Yu, Y.; Peng, B. A General Approach for Fabrication of Nitrogen-Doped Graphene Sheets and Its Application in Supercapacitors. *J. Colloid Interface Sci.* **2014**, *417*, 270–277. [CrossRef]
19. Lei, Z.; Lu, L.; Zhao, X.S. The Electrocapacitive Properties of Graphene Oxide Reduced by Urea. *Energy Environ. Sci.* **2012**, *5*, 6391–6399. [CrossRef]
20. Lee, Y.H.; Chang, K.H.; Hu, C.C. Differentiate the Pseudocapacitance and Double-Layer Capacitance Contributions for Nitrogen-Doped Reduced Graphene Oxide in Acidic and Alkaline Electrolytes. *J. Power Sources* **2013**, *227*, 300–308. [CrossRef]
21. Wang, T.; Wang, L.; Wu, D.; Xia, W.; Zhao, H.; Jia, D. Hydrothermal Synthesis of Nitrogen-Doped Graphene Hydrogels Using Amino Acids with Different Acidities as Doping Agents. *J. Mater. Chem. A* **2014**, *2*, 8352–8361. [CrossRef]
22. Long, D.; Li, W.; Ling, L.; Miyawaki, J.; Mochida, I.; Yoon, S.H. Preparation of Nitrogen-Doped Graphene Sheets by a Combined Chemical and Hydrothermal Reduction of Graphene Oxide. *Langmuir* **2010**, *26*, 16096–16102. [CrossRef] [PubMed]
23. Sun, L.; Wang, L.; Tian, C.; Tan, T.; Xie, Y.; Shi, K.; Li, M.; Fu, H. Nitrogen-Doped Graphene with High Nitrogen Level via a One-Step Hydrothermal Reaction of Graphene Oxide with Urea for Superior Capacitive Energy Storage. *RSC Adv.* **2012**, *2*, 4498–4506. [CrossRef]
24. Śliwak, A.; Grzyb, B.; Díez, N.; Gryglewicz, G. Nitrogen-Doped Reduced Graphene Oxide as Electrode Material for High Rate Supercapacitors. *Appl. Surf. Sci.* **2017**, *399*, 265–271. [CrossRef]
25. Lai, L.; Chen, L.; Zhan, D.; Sun, L.; Liu, J.; Lim, S.H.; Poh, C.K.; Shen, Z.; Lin, J. One-Step Synthesis of NH2-Graphene from in Situ Graphene-Oxide Reduction and Its Improved Electrochemical Properties. *Carbon* **2011**, *49*, 3250–3257. [CrossRef]
26. Paredes, J.I.; Villar-Rodil, S.; Martínez-Alonso, A.; Tascón, J.M.D. Graphene Oxide Dispersions in Organic Solvents. *Langmuir* **2008**, *24*, 10560–10564. [CrossRef]
27. Park, S.; An, J.; Jung, I.; Piner, R.D.; An, S.J.; Li, X.; Velamakanni, A.; Ruoff, R.S. Colloidal Suspensions of Highly Reduced Graphene Oxide in a Wide Variety of Organic Solvents. *Nano Lett.* **2009**, *9*, 1593–1597. [CrossRef]
28. Hernandez, Y.; Lotya, M.; Rickard, D.; Bergin, S.D.; Coleman, J.N. Measurement of Multicomponent Solubility Parameters for Graphene Facilitates Solvent Discovery. *Langmuir* **2010**, *26*, 3208–3213. [CrossRef]
29. Gopalakrishnan, K.; Moses, K.; Govindaraj, A.; Rao, C.N.R. Supercapacitors Based on Nitrogen-Doped Reduced Graphene Oxide and Borocarbonitrides. *Solid State Commun.* **2013**, *175–176*, 43–50. [CrossRef]
30. Jiang, B.; Tian, C.; Wang, L.; Sun, L.; Chen, C.; Nong, X.; Qiao, Y.; Fu, H. Highly Concentrated, Stable Nitrogen-Doped Graphene for Supercapacitors: Simultaneous Doping and Reduction. *Appl. Surf. Sci.* **2012**, *258*, 3438–3443. [CrossRef]
31. Mayyas, M.; Li, H.; Kumar, P.; Ghasemian, M.B.; Yang, J.; Wang, Y.; Lawes, D.J.; Han, J.; Saborio, M.G.; Tang, J.; et al. Liquid-Metal-Templated Synthesis of 2D Graphitic Materials at Room Temperature. *Adv. Mater.* **2020**, *32*, 2001997. [CrossRef] [PubMed]
32. Liu, X.; Lyu, D.; Merlet, C.; Leesmith, M.J.A.; Hua, X.; Xu, Z.; Grey, C.P.; Forse, A.C. Structural Disorder Determines Capacitance in Nanoporous Carbons. *Science* **2024**, *384*, 321–325. [CrossRef] [PubMed]
33. Yadav, A.; Kumar, R.; Sahoo, B. Exploring Supercapacitance of Solvothermally Synthesized N-RGO Sheet: Role of N-Doping and the Insight Mechanism. *Phys. Chem. Chem. Phys.* **2022**, *24*, 1059–1071. [CrossRef] [PubMed]
34. Marcano, D.C.; Kosynkin, D.V.; Berlin, J.M.; Sinitskii, A.; Sun, Z.; Slesarev, A.; Alemany, L.B.; Lu, W.; Tour, J.M. Improved Synthesis of Graphene Oxide. *ACS Nano* **2010**, *4*, 4806–4814. [CrossRef]
35. Yadav, A.; Kumar, R.; Kumar, S.; Sahoo, B. Mechanistic Insights into the Roles of Precursor Content, Synthesis Time, and Dispersive Solvent in Maximizing Supercapacitance of N-RGO Sheets. *J. Alloys Compd.* **2024**, *971*, 172648. [CrossRef]
36. Saraf, M.; Natarajan, K.; Mobin, S.M.; Natarajan, K.; Mobin, S.M. Robust Nanocomposite of Nitrogen-Doped Reduced Graphene Oxide and MnO2 Nanorods for High-Performance Supercapacitors and Nonenzymatic Peroxide Sensors. *ACS Sustain. Chem. Eng.* **2018**, *6*, 10489–10504. [CrossRef]
37. Sun, Q.; Yi, Z.; Fan, Y.; Xie, L.; Wang, Z.; Sun, G.; Wang, Z.; Huang, X.; Liu, Z.; Su, F.; et al. Whole Landscape of the Origin and Evolution of Gassing in Supercapacitors at a High Voltage. *ACS Appl. Mater. Interfaces* **2023**, *15*, 54386. [CrossRef]
38. Chen, W.; Yan, L. Preparation of Graphene by a Low-Temperature Thermal Reduction at Atmosphere Pressure. *Nanoscale* **2010**, *2*, 559–563. [CrossRef]
39. Pal, B.; Yang, S.; Ramesh, S.; Thangadurai, V.; Jose, R. Electrolyte Selection for Supercapacitive Devices: A Critical Review. *Nanoscale Adv.* **2019**, *1*, 3807–3835. [CrossRef]
40. Furtado, C.A.; Kim, U.J.; Gutierrez, H.R.; Pan, L.; Dickey, E.C.; Eklund, P.C. Debundling and Dissolution of Single-Walled Carbon Nanotubes in Amide Solvents. *J. Am. Chem. Soc.* **2004**, *126*, 6095–6105. [CrossRef]
41. Ausman, K.D.; Piner, R.; Lourie, O.; Ruoff, R.S.; Korobov, M. Organic Solvent Dispersions of Single-Walled Carbon Nanotubes: Toward Solutions of Pristine Nanotubes. *J. Phys. Chem. B* **2000**, *104*, 8911–8915. [CrossRef]
42. Li, X.; Zang, X.; Li, Z.; Li, X.; Li, P.; Sun, P.; Lee, X. Large-Area Flexible Core–Shell Graphene / Porous Carbon Woven Fabric Films for Fiber Supercapacitor Electrodes. *Adv. Funct. Mater.* **2013**, *23*, 4862–4869. [CrossRef]
43. Qu, D. Studies of the Activated Mesocarbon Microbeads Used in Double-Layer Supercapacitors. *J. Power Sources* **2002**, *109*, 403–411. [CrossRef]

44. Chen, W.-C.; Wen, T.-C.; Teng, H. Polyaniline-Deposited Porous Carbon Electrode for Supercapacitor. *Electrochim. Acta* **2003**, *48*, 641–649. [CrossRef]
45. Mandal, B.; Saha, S.; Das, D.; Panda, J.; Das, S.; Sarkar, R.; Tudu, B. Supercapacitor Performance of Nitrogen Doped Graphene Synthesized via DMF Assisted Single-Step Solvothermal Method. *FlatChem* **2022**, *34*, 100400. [CrossRef]
46. Ajravat, K.; Rajput, S.; Brar, L.K. Microwave Assisted Hydrothermal Synthesis of N Doped Graphene for Supercapacitor Applications. *Diam. Relat. Mater.* **2022**, *129*, 109373. [CrossRef]
47. Sobaszek, M.; Brzhezinskaya, M.; Olejnik, A.; Mortet, V.; Alam, M.; Sawczak, M.; Ficek, M.; Gazda, M.; Weiss, Z.; Bogdanowicz, R. Highly Occupied Surface States at Deuterium-Grown Boron-Doped Diamond Interfaces for Efficient Photoelectrochemistry. *Small* **2023**, *19*, 2208265. [CrossRef]
48. Brzhezinskaya, M.; Mishakov, I.V.; Bauman, Y.I.; Shubin, Y.V.; Maksimova, T.A.; Stoyanovskii, V.O.; Gerasimov, E.Y.; Vedyagin, A.A. One-Pot Functionalization of Catalytically Derived Carbon Nanostructures with Heteroatoms for Toxic-Free Environment. *Appl. Surf. Sci.* **2022**, *590*, 153055. [CrossRef]
49. Wang, M.; Duong, L.D.; Mai, N.T.; Kim, S.; Kim, Y.; Seo, H.; Kim, Y.C.; Jang, W.; Lee, Y.; Suhr, J.; et al. All-Solid-State Reduced Graphene Oxide Supercapacitor with Large Volumetric Capacitance and Ultralong Stability Prepared by Electrophoretic Deposition Method. *ACS Appl. Mater. Interfaces* **2015**, *7*, 1348–1354. [CrossRef]
50. Kumar, M.P.; Kesavan, T.; Kalita, G.; Ragupathy, P.; Narayanan, T.N.; Pattanayak, D.K. On the Large Capacitance of Nitrogen Doped Graphene Derived by a Facile Route. *RSC Adv.* **2014**, *4*, 38689–38697. [CrossRef]
51. Konios, D.; Stylianakis, M.M.; Stratakis, E.; Kymakis, E. Dispersion Behaviour of Graphene Oxide and Reduced Graphene Oxide. *J. Colloid Interface Sci.* **2014**, *430*, 108–112. [CrossRef] [PubMed]
52. Cai, D.; Song, M.; Xu, C. Highly Conductive Carbon-Nanotube/Graphite-Oxide Hybrid Films. *Adv. Mater.* **2008**, *20*, 1706–1709. [CrossRef]
53. Cai, D.; Song, M. Preparation of Fully Exfoliated Graphite Oxide Nanoplatelets in Organic Solvents. *J. Mater. Chem.* **2007**, *17*, 3678–3680. [CrossRef]
54. Pawlyta, M.; Rouzaud, J.N.; Duber, S. Raman Microspectroscopy Characterization of Carbon Blacks: Spectral Analysis and Structural Information. *Carbon* **2015**, *84*, 479–490. [CrossRef]
55. Cançado, L.G.; Takai, K.; Enoki, T.; Endo, M.; Kim, Y.A.; Mizusaki, H.; Jorio, A.; Coelho, L.N.; Magalhães-Paniago, R.; Pimenta, M.A. General Equation for the Determination of the Crystallite Size La of Nanographite by Raman Spectroscopy. *Appl. Phys. Lett.* **2006**, *88*, 163106. [CrossRef]
56. Hassan, F.M.; Chabot, V.; Li, J.; Kim, B.K.; Ricardez-Sandoval, L.; Yu, A. Pyrrolic-Structure Enriched Nitrogen Doped Graphene for Highly Efficient Next Generation Supercapacitors. *J. Mater. Chem. A* **2013**, *1*, 2904–2912. [CrossRef]
57. Wang, C.; Zhou, Y.; Sun, L.; Zhao, Q.; Zhang, X.; Wan, P.; Qiu, J. N/P-Codoped Thermally Reduced Graphene for High-Performance Supercapacitor Applications. *J. Phys. Chem. C* **2013**, *117*, 14912–14919. [CrossRef]
58. Wang, P.; He, H.; Xu, X.; Jin, Y. Significantly Enhancing Supercapacitive Performance of Nitrogen-Doped Graphene Nanosheet Electrodes by Phosphoric Acid Activation. *ACS Appl. Mater. Interfaces* **2014**, *6*, 1563–1568. [CrossRef]
59. Lu, Y.; Zhang, F.; Zhang, T.; Leng, K.; Zhang, L.; Yang, X.; Ma, Y.; Huang, Y.; Zhang, M.; Chen, Y. Synthesis and Supercapacitor Performance Studies of N-Doped Graphene Materials Using o-Phenylenediamine as the Double-N Precursor. *Carbon* **2013**, *63*, 508–516. [CrossRef]
60. Wen, Z.; Wang, X.; Mao, S.; Bo, Z.; Kim, H.; Cui, S.; Lu, G.; Feng, X.; Chen, J. Crumpled Nitrogen-Doped Graphene Nanosheets with Ultrahigh Pore Volume for High-Performance Supercapacitor. *Adv. Mater.* **2012**, *24*, 5610–5616. [CrossRef]

Disclaimer/Publisher's Note: The statements, opinions and data contained in all publications are solely those of the individual author(s) and contributor(s) and not of MDPI and/or the editor(s). MDPI and/or the editor(s) disclaim responsibility for any injury to people or property resulting from any ideas, methods, instructions or products referred to in the content.

Article

A Novel Non-Enzymatic Efficient H$_2$O$_2$ Sensor Utilizing δ-FeOOH and Prussian Blue Anchoring on Carbon Felt Electrode

Karoline S. Nantes [1], Ana L. H. K. Ferreira [1], Marcio C. Pereira [1], Francisco G. E. Nogueira [2] and André S. Afonso [1,*]

[1] Institute of Science, Engineering, and Technology, Federal University of Jequitinhonha and Mucuri Valleys (UFVJM), Teófilo Otoni 39803-371, Brazil
[2] Department of Chemical Engineering, Federal University of São Carlos (UFSCar), São Carlos 13565-905, Brazil; nogueira@ufscar.br
* Correspondence: andre.afonso@ufvjm.edu.br

Abstract: In this study, an efficient H$_2$O$_2$ sensor was developed based on electrochemical Prussian blue (PB) synthesized from the acid suspension of δ-FeOOH and K$_3$[Fe(CN)$_6$] using cyclic voltammetry (CV) and anchored on carbon felt (CF), yielding an enhanced CF/PB-FeOOH electrode for sensing of H$_2$O$_2$ in pH-neutral solution. CF/PB-FeOOH electrode construction was proved by scanning electron microscopy (SEM), energy-dispersive X-ray spectroscopy (EDS), and X-ray diffraction (XRD), and electrochemical properties were verified by impedance electrochemical and CV. The synergy of δ-FeOOH and PB coupled to CF increases electrocatalytic activity toward H$_2$O$_2$, with the sensor showing a linear range of 1.2 to 300 μM and a limit of detection of 0.36 μM. Notably, the CF/PB-FeOOH electrode exhibited excellent selectivity for H$_2$O$_2$ detection in the presence of dopamine (DA), uric acid (UA), and ascorbic acid (AA). The calculated H$_2$O$_2$ recovery rates varied between 93% and 101% in fetal bovine serum diluted in PBS. This work underscores the potential of CF/PB-FeOOH electrodes in progressing electrochemical sensing technologies for various biological and environmental applications.

Keywords: carbon felt; hydrogen peroxide; δ-FeOOH; Prussian blue; non-enzymatic sensor

1. Introduction

Hydrogen peroxide is a potent oxidizing agent widely applied in the food, cosmetic, and textile industries [1]. Additionally, it is a vital molecule within the human body, implicated in immunological responses, inflammation, and cell signaling [2]. The scientific literature characterizes hydrogen peroxide as a reactive oxygen species, with a concentration exceeding 50 μM in biological fluids, signaling metabolic dysfunction to diseases like Alzheimer's, Parkinson's, and diabetes, among other health conditions [3,4]. In the environmental context, hydrogen peroxide concentration in aquatic ecosystems, including rivers, lakes, and oceans, indicates the health of fauna, flora, and overall water and air quality. In the food and beverages industry, hydrogen peroxide is a technological aid for diminishing the presence of bacteria [2]. Nonetheless, following processes such as pasteurization, sterilization, and packaging, the H$_2$O$_2$ content in the final product must not exceed 147 μM. Thus, it is crucial for health, safety, and the environment to ensure that the level of hydrogen peroxide remains within a safe range [2,5,6].

Considering the significance and wide-ranging applications of H$_2$O$_2$ in industrial and biological contexts, several techniques, including spectrophotometry, chemiluminescence, fluorescence, and chromatography, are employed to measure its concentration. Nonetheless, these techniques necessitate expert operators, rendering them unsuitable for point-of-care scenarios. Moreover, the extensive analysis duration and the equipment's cumbersome nature preclude their use in remote locations [7].

Conversely, electrochemical methods present a promising alternative due to their numerous benefits, such as sensitivity, portability, rapid response time, cost-effectiveness, and an environmentally friendly analytical approach; however, they often lack specificity [8]. The electrochemical sensor is an accessible tool that non-specialist technicians can utilize for point-of-care detection and continuous monitoring. Since H_2O_2 is a reactive molecule, its detection is enhanced by its interaction with the working electrode's surface. Modifying this surface with an electrocatalytic material can substantially increase the sensor's specificity [9].

Electrochemical sensors can be constructed by immobilizing biological materials on working electrodes, such as enzymes or inorganic materials, including carbon nanomaterials, metal complexes, metal oxide hydroxides, etc. [10–12]. These materials act as catalysts for H_2O_2 reactions on surface electrodes. While enzymatic sensors exhibit selectivity and sensitivity, their catalytic activity is affected by pH, temperature, and the reaction medium, which impact sensor performance quality and limit these sensors' shelf life [10]. In contrast, non-enzymatic sensors are more attractive because of their low cost and stability. Different groups of scientists have demonstrated various non-enzymatic sensors for H_2O_2 using organic molecules and other inorganic compounds. Ranni et al. demonstrated a sensor for H_2O_2 with gold nanoparticle-decorated copper cross-linked pectin on a glassy carbon electrode [13]. Yin et al. developed a biochar H_2O_2 sensor using crab shells as the carbon precursor through pyrolysis [14]. Wang et al. constructed a photoelectrochemical H_2O_2 sensor based on pillar [5] arene-functionalized Au nanoparticles and MWNTs hybrid BiOBr heterojunction [15]. An electrochemical sensor modified with cerium oxide nanocubes and carbon black as a highly active non-enzymatic H_2O_2 catalyst was proposed by Shen et al. [16]. Researchers have demonstrated that semiconductor δ-FeOOH, a material with good catalytic performance and practical application in photocatalysis [17,18], produces non-enzymatic sensors with excellent electrocatalytic activity and electronic conduction properties. For instance, Afonso and colleagues constructed an all-plastic disposable carbon electrochemical cell modified with δ-FeOOH and silver nanoparticles, which exhibited an electrocatalytic response for H_2O_2 with a limit of detection of 71 µM [19]. Further research demonstrated an improvement in the same system with the introduction of carbon black, achieving a limit of detection of 22 µM [20].

Another important inorganic compound used to modify electrodes for electrocatalytic purposes is Prussian blue (PB). PB is known as an artificial enzyme capable of catalyzing H_2O_2 reactions. Extensive research involving PB-based composite materials for electrochemical sensing of H_2O_2 has focused on synthesizing and depositing PB nanoparticles on various conductive surfaces to enhance their electrocatalytic and conductivity properties [21–24]. However, enhancing individual properties does not guarantee the material's overall performance, necessitating better structural integration of the components.

Moreover, PB, as an electrocatalyst for H_2O_2 reaction, faces several limitations. The penetration of H_2O_2 and counter ions into PB structure can lead to volume changes and mechanical stress, resulting in reduced stability [25]. Side reactions caused by lower electrical conductivity and decomposition of PB structure in a medium with pH-neutral or basic are effects that reduce the stability of PB [26–28]. Therefore, it is essential to develop methods that improve the intimate contact between components of the sensing surface and enhance the physical and electrochemical stability.

Scholtz and colleagues demonstrated that Prussian blue (PB) strongly interacts with the goethite surface (α-FeOOH) due to the prevalence of positive charge groups in a medium with a pH below 9 [29]. This interaction facilitates PB binding to the surface's negative charges and hexacyanoferrate anions, which provides remarkable stability for Prussian blue. Significantly, this property can be extended to other iron oxide hydroxides, making them effective traps for Prussian blue and its analogs. By integrating sensitive components for H_2O_2 detection with suitable conductive substrates, the resulting sensor's sensitivity, selectivity, and practicality can be significantly enhanced.

Flexible electrodes, particularly carbon felt, stand out due to their adaptability to substrate shapes, robust mechanical properties, and superior electrical conductivity [30,31]. These characteristics are instrumental in elevating the efficacy of electrochemical sensors. Carbon felt is especially noteworthy among flexible electrodes for its extensive electrochemical surface area, chemical stability, ease of regeneration, swift charge and ion transport, and affordability [32]. While various electrochemical sensors employing carbon felt have been utilized to detect H_2O_2 and other analytes using diverse nanomaterials [23,33–36], the development of a non-enzymatic sensor that synergizes the properties of PB and δ-FeOOH remains an area not yet studied. δ-FeOOH is a relatively new material in sensor development, and its unique properties and potential applications are still being explored.

In this study, we have developed a novel non-enzymatic and accurate sensor for the electrochemical sensing of H_2O_2. PB was electrochemically synthesized and deposited on CF from a δ-FeOOH acid suspension, which also was adsorbed on CF, resulting in the formation of the CF/PB-FeOOH electrode. The synergistic effect of PB and δ-FeOOH enhances the electrode's performance, as demonstrated by its excellent electrochemical reduction in H_2O_2, rapid response, high sensitivity, low limit of detection, and strong applicability in biological samples.

2. Materials and Methods

2.1. Reagents and Apparatus

The experiment used only chemicals of analytical grade. Ammonium iron (II) sulfate hexahydrate $(NH_4)_2Fe(SO_4)_2 \cdot 6H_2O$, uric acid, dopamine hydrochloride, ascorbic acid, and fetal bovine serum (FBS) were purchased from Sigma-Aldrich (St. Louis, MO, USA); potassium hexacyanoferrate (III) ($K_3[Fe(CN)_6]$), potassium hexacyanoferrate (II) trihydrate ($K_4[Fe(CN)_6]$), and iron chloride (III) ($FeCl_3$) were purchased from A.C.S. scientific (Sumaré, SP, Brazil); potassium phosphate monobasic (KH_2PO_4) and potassium phosphate dibasic (K_2HPO_4) were obtained from Dinâmica Química (Indaiatuba, SP, Brazil); 37% v/v hydrogen chloride (HCl) was supplied by Labsynth (Diadema, SP, Brazil); sodium hydroxide (NaOH) was purchased from Anidrol (Diadema, SP, Brazil); 30% v/v hydrogen peroxide (H_2O_2) was purchased from Alphatec (Santo André, SP, Brazil); and potassium chloride was obtained from C.R.Q. (Diadema, SP, Brazil). Ultrapure water from a Millipore Direct Q® system (Billerica, MA, USA) was used in this research.

The electrochemical experiments were conducted on a potentiostat/galvanostat Autolab PGSTAT128N (Utrecht, The Netherlands) connected to a computer running NOVA 2.0 software. The 50 mL one-compartment electrochemical cell was equipped with a working electrode made of carbon felt (CF) (Fuel Cell Store, TX, USA) cut into small ribbons with 2.5 cm² areas, a platinum counter electrode, and Ag|AgCl immersed in 3 M KCl reference electrode. The morphology of the PB nanoparticles was evaluated by scanning electron microscopy using a Philips XL-30 FEG microscope equipped with an energy-dispersive X-ray spectrometer (EDS, Bruker, Singapore). The structural analysis of PB-FeOOH film was carried out by XRD measurements using an X-ray diffractometer with Rigaku Miniflex 600 diffractometer, which emitted Cu Kα radiation (λ = 1.54056 Å) at a scanning speed of 2° min^{-1} at 30 kV.

2.2. Synthesis of δ-FeOOH

The synthesis of δ-FeOOH followed the procedure reported by Pereira et al. [37]. δ-FeOOH was prepared by adding 200 mL of 2 M NaOH to 200 mL of 0.71 mM $(NH_4)_2Fe(SO_4)_2 \cdot 6H_2O$ solution with stirring. After forming the green precipitate, 5 mL of H_2O_2 was added and magnetically stirred for 30 min. The reddish-brown solid formed indicated the attainment of the δ-FeOOH phase. The as-made material was washed with deionized water, centrifuged at 3600 rpm for 5 min, and dried under vacuum at room temperature.

2.3. Carbon Felt Electrode Preparation

We assessed three different approaches to preparing the CF surface for PB synthesis: (i) the CF was exposed to an air atmosphere in the oven, maintained at 100 °C for 2 h; (ii) the CF was immersed in 1 M H_2SO_4 solution for 2 h, followed by a thorough rinse with distilled water and subsequent drying in the oven at 100 °C for 2 h; (iii) we cleansed the CF with distilled water and left to dry at room temperature.

2.4. Synthesis of PB-FeOOH Film

PB-FeOOH film was formed by electrochemically synthesizing PB nanoparticles on CF using cyclic voltammetry at a scan rate of 50 mV·s^{-1} and a voltage range of –0.2 V to 0.7 V vs. Ag | AgCl. The synthesis was performed by placing CF in a precursor solution containing 5 mM of $K_3[Fe(CN)_6]$, 22 mg of δ-FeOOH, and 0.1 M KCl as a supporting electrolyte, forming the PB-FeOOH film on CF. The quantity of δ-FeOOH in precursor solution was varied in 11, 22, and 44 mg of δ-FeOOH. The conditions for forming the PB-FeOOH film, such as the pH values of the precursor solution and the number of cyclic voltammetric scans, were also assessed. The effect of the pH value of the precursor solution from 2.0 to 6.0 was studied by adjusting the solution with 0.1 M HCl. Finally, CV scan numbers were evaluated using 15, 30, 50, 100, and 150 scans. For comparison, PB nanoparticles using a CF surface previously optimized were synthesized by applying 100 cyclic voltammetric in 2 mM $FeCl_3$, 2 mM $K_3[Fe(CN)_6]$, 0.1 M KCl, and 0.1 M HCl solution based on the modified method published by De Mattos [38].

The XRD analysis was performed using conducting fluorine-doped tin oxide (FTO)-coated glass where PB–FeOOH was deposited. Before deposition, the substrate was washed with acetone and deionized water in an ultrasonic bath for 15 min.

2.5. Electrochemical Measurement

CF/PB–FeOOH electrodes were electrochemically characterized in a 0.1 M KCl solution at pH 2.0 and a 1 mM equimolar mixture of $K_3[Fe(CN)_6]$/$K_4[Fe(CN)_6]$ containing 0.1 M KCl as a supporting electrolyte by cyclic voltammetry. The solution used for amperometric measurement was 0.1 M PBS at pH 7.0, under stirring, with or without adding H_2O_2. All solutions were N_2-saturated for 15 min. The electrochemical impedance spectroscopy was conducted in the hexacyanoferrate solution, applying an open-circuit potential with a 5 mV amplitude and a frequency range from 100 kHz to 0.01 Hz.

3. Results and Discussion

According to the previous literature, CF surface treatments using acid solutions or heating in an air atmosphere have been shown to improve the electrochemical activity of the CF surface [34,35]. Therefore, we evaluated three approaches to prepare the CF surface by studying the magnitude of redox peaks of PB nanoparticles on CF. Figure 1A shows typical CV curves of PB nanoparticle synthesis, which occurs during the potential scanning in 5 mM of $K_3[Fe(CN)_6]$, 22 mg of δ-FeOOH, and 0.1 M KCl at pH 2.0, applying 30 cycles. We observed increasing redox peaks at 0.39 V and 0.12 V, along with a slight potential shift with the number of scans, suggesting the growth of stable PB nanoparticles [39] on the CF electrode. Furthermore, δ-FeOOH was also deposited on CF due to electrostatic interaction [29]. Therefore, the electrode was designated CF/PB-FeOOH. CV tests were conducted in a 0.1 M KCl solution pH 2.0 to evaluate the electrochemical properties of CF/PB-FeOOH electrodes (Figure 1B). The characterization was maintained in a solution with pH 2 to ensure the structure and properties of the PB. The CF/PB-FeOOH electrodes exhibit two pairs of sharp redox peaks related to Prussian white (PW)/PB near 0.3/0.2 V with a potential separation of 100 mV; this was observed for three different CF surfaces prepared, indicating a fast charge transfer in CF (Reaction (R1)). Among the approaches used to prepare the CF surface, a better signal was achieved by washing CF with distilled water and drying it at room temperature (curve c). Contrary to the literature, which claims that the electrochemical activity of CF electrodes improves with chemical and thermal treatments—treatments that increase the functional groups as hydroxyl and carbonyl—we did not observe this in our findings. The interaction of FeOOH with CF was more effective; therefore, we conducted the following experiments with CF prepared by method (iii).

$$\underset{\text{PB}}{Fe_4^{III}[Fe^{II}(CN)_6]_3} + 4K^+ + 4e^- \rightleftharpoons \underset{\text{PW}}{K_4Fe_4^{II}[Fe^{II}(CN)_6]_3} \tag{R1}$$

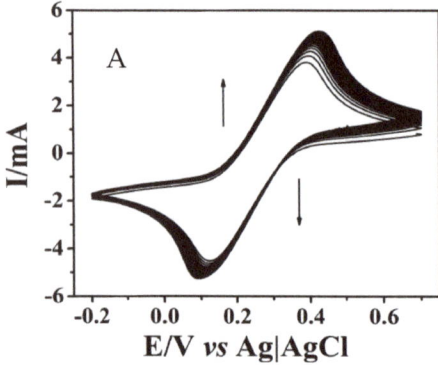 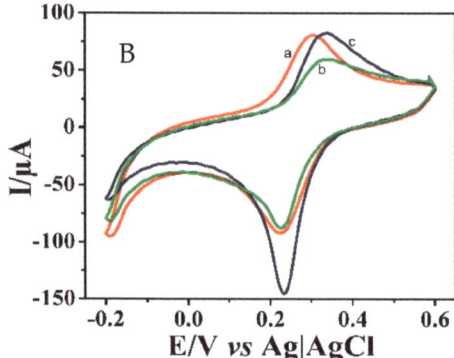

Figure 1. (**A**) Potentiodynamic growth of PB nanoparticles on CF electrode in N_2-saturated 0.1 M KCl solution at pH 2.0 with 5 mM $K_3Fe(CN)_6$ and 22 mg of δ-FeOOH. The potential window was set from −0.2 V to 0.7 V at a scan rate of 50 mV s^{-1}, applying 30 scans. The arrows indicate scans increase. (**B**) CV of FC/PB-FeOOH electrode in N_2-saturated 0.1 M KCl at pH 2.0 at a scan rate of 50 mV s^{-1}. The CF was subjected to the following treatments: (a) i, (b) ii, and (c) iii.

The electrochemical synthesis of PB on electrodes involves several controllable factors to achieve PB deposited with good electrocatalytic activity, such as deposition time, concentration of reagents, and effect of solution pH [40]. Figure 2A shows the CV of CF/PB-FeOOH electrodes in 0.1 M KCl solution pH 2.0 constructed in precursor solution with pH ranging from 2 to 6. The CF/PB-FeOOH electrode constructed under pH 2.0 exhibited notably more significant two pairs of redox peaks than those constructed under other pH conditions, indicating the successful electrochemical formation of PB from δ-FeOOH in an acid solution. Subsequently, we assessed the effect of δ-FeOOH mass in the precursor solution. Figure 2B shows the electrochemical performance of the CF/PB-FeOOH electrode in 0.1 M KCl prepared with 11, 22, and 44 mg of FeOOH. The optimal amount of FeOOH was 22 mg, which was used in the subsequent experiments. To analyze the electrochemical behavior of the CF/PB-FeOOH electrode obtained with different deposition cycles, we obtained the CV curves of the CF/PB-FeOOH electrodes in 0.1 M KCl generated with 15, 30, 50, 100, and 150 cycles. The peak current enhanced as the number of scans increased until 100 cycles but decreased with 150 cycles, indicating that the electrochemical properties of PB deteriorated with a high number of scans (Figure 2C). Based on the best electrochemical performance of the optimized CF/PB-FeOOH electrode, we proceed with the upcoming experiments under these conditions.

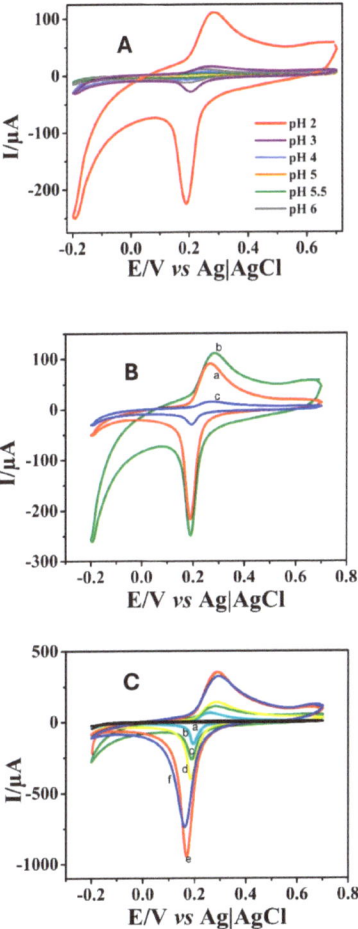

Figure 2. (**A**) Evaluation by the CV of PB synthesized in different pH levels of the precursor solution. (**B**) PB synthesized with (a) 44 mg, (b) 22 mg, or (c) 11 mg of δ-FeOOH. (**C**) Effect of the number of scans on the synthesis of PB: (a) background, (b) 15 scans, (c) 30 scans, (d) 50 scans, (e) 100 scans, and (f) 150 scans. N_2-saturated 0.1 M KCl solution at pH 2.0 was the medium in which CVs were performed. The potential range was from −0.2 V to 0.7 V, and the scan rate was 50 mV s^{-1}.

The amount of PB deposited on the CF/PB-FeOOH electrode following 100 electrodeposition cycles was determined by CV collected in 0.1 M KCl using Equation (1).

$$\Gamma_\Gamma = \frac{Q}{nFA} \qquad (1)$$

Q is the charge equivalent to the reduction peak at 0.16 V, n represents the number of electrons in the redox process (n = 1), F stands for the Faraday constant, A is the electroactive area (3.0 cm^2), and Γ_Γ is the amount of PB on the active surface. The quantity of PB deposited was calculated to be 4.4×10^{-9} mol/cm^2.

The structural property of PB-FeOOH nanoparticles was evaluated by X-ray diffraction using FTO as substrate. Figure 3 shows the XRD pattern with only a characteristic strong peak (JCPDS#1-239) at 17.65° corresponding PB phase (100). This result suggests that the PB-FeOOH film has a predominantly amorphous structure [41], indicating no crystalline peaks other than the characteristic peak of PB.

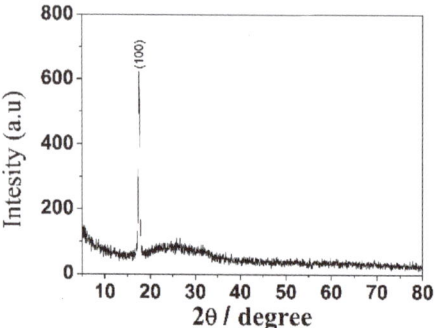

Figure 3. The XRD pattern for the PB-FeOOH film was deposited on FTO.

Figure 4 shows the SEM images of the surface morphology of CF and CF/PB-FeOOH electrodes, respectively. The bare CF electrode (Figure 4A,B) has a fibrous and smooth surface, while the CF/PB-FeOOH electrode (Figure 4C,D) has a rough surface with nanometric cubic structures [40]. In addition, some irregular microstructures can be seen on the surface. The EDS analysis in Figure 4E,F shows C, O, Fe, N, K, and Cl elements, confirming the successful electrosynthesis of the PB on CF.

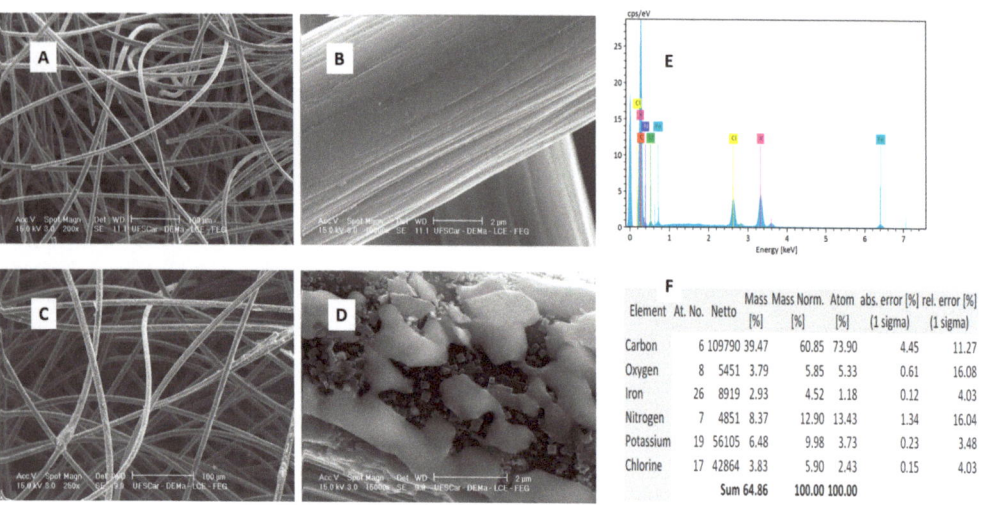

Element	At. No.	Netto	Mass [%]	Mass Norm. [%]	Atom [%]	abs. error [%] (1 sigma)	rel. error [%] (1 sigma)
Carbon	6	109790	39.47	60.85	73.90	4.45	11.27
Oxygen	8	5451	3.79	5.85	5.33	0.61	16.08
Iron	26	8919	2.93	4.52	1.18	0.12	4.03
Nitrogen	7	4851	8.37	12.90	13.43	1.34	16.04
Potassium	19	56105	6.48	9.98	3.73	0.23	3.48
Chlorine	17	42864	3.83	5.90	2.43	0.15	4.03
Sum			64.86	100.00	100.00		

Figure 4. SEM micrographs of (**A,B**) CF electrode and (**C,D**) CF/PB-FeOOH electrode. (**E**) EDS spectrum of CF/PB-FeOOH electrode and (**F**) the weight % of EDS analysis.

We use CV and EIS to assess the electrocatalytic activity of bare CF and CF/PB-FeOOH in a ferri-ferro cyanide solution. In Figure 5A, CV was carried out by cycling the potential between -0.2 and 0.7 V at a scan rate of 100 mV·s^{-1}. The peak current in the redox probe at CF/PB-FeOOH (curve b) was higher than that at CF (curve a) due to the good electrochemical properties of the modified electrode. For the CF/PB-FeOOH electrode, the I_a/I_c has an average of 0.86, and ΔE_p is +0.21 V, showing a quasi-reversible electrochemical response. However, for the bare CF electrode, ΔEp is 0.12 V. This better value is probably due to the different thermodynamic conditions of the surface.

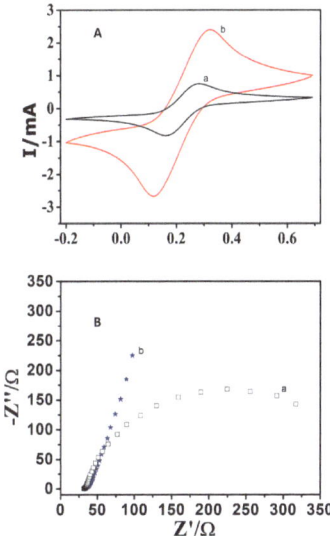

Figure 5. (**A**) CV curves at 100 mV·s^{-1} and (**B**) EIS plots at open-circuit potential using N$_2$-saturated 0.1 M KCl solution at pH 2.0 containing 1 mM K$_3$Fe(CN)$_6$ and 1 mM K$_4$Fe(CN)$_6$, (a) CF, and (b) CF/PB-FeOOH.

Furthermore, the effect of scan rate was studied. Figure S1 shows CVs obtained with different scan rates from 10 to 300 mV·s^{-1} for CF and CF/PB-FeOOH electrodes and the magnitudes of CV peak current plotted against the square root $v^{1/2}$. The results exhibited a linear relationship for both anodic and cathodic current enhancement along with an increase in scan rate for the CF and CF/PB-FeOOH, indicating a diffusion-controlled process at the surface electrode. From these data, we experimentally determined the electroactive area for CF and CF/PB-FeOOH electrodes using the Randles–Ševčík equation. The electroactive areas were 3.0 ± 0.1 cm^2 and 8.2 ± 0.4 cm^2 for CF and CF/PB-FeOOH electrodes, respectively. CF modified with PB exhibited a large surface area and improved CF's electroactive properties. These findings were aligned with EIS experiments. Figure 5B shows the Nyquist plot for CF (curve a) and CF/PB-FeOOH (curve b). The circuits used to obtain the EIS data were R$_s$(R$_{ct}$[C$_{dl}$]) for CF and R$_s$(CPE[R$_{ct}$W]) for CF/PB-FeOOH, where R$_s$ is solution resistance, R$_{ct}$ is electron transfer kinetics as charge transfer resistance, C$_{dl}$ is double layer capacitance, CPE is constant phase element, and W is mass transfer element. The R$_{ct}$ for CF and CF/PB-FeOOH were 324.2 Ω and 1.28 Ω, respectively. From the EIS data, the value of the apparent heterogeneous electron transfer constant (k$_{app}$) was calculated using Equation (2).

$$k_{app} = \frac{RT}{n^2 F^2 A R_{ct} C} \qquad (2)$$

where R stands for the ideal gas constant (8.414 Joule/(mol·K)), T is the temperature absolute (298 K), n is the number of electrons transferred during reaction redox (n = 1), F is the constant of Faraday (96,485 C/mol), A stand for the active area of the electrode (cm^2); R$_{ct}$ is electron transfer kinetics as charge transfer resistance (Ω); C is the concentration of the probe redox (mol/cm^3). In the ferri-ferro solution, the K$_{app}$ calculated were 0.25 × 10^{-4} and 1 × 10^{-7} cm·s^{-1} for CF/PB-FeOOH and CF, respectively. The K$_{app}$ value obtained for CF/PB-FeOOH is 254 folds higher than that for CF; this improvement agrees with other K$_{app}$ values for modified electrodes reported in the literature [20,42].

Cyclic voltammograms were recorded for the CF/PB-FeOOH-modified electrode in a 0.1 M PBS solution to simulate the pH of the biology environment, with and without 1.0 mM of H$_2$O$_2$, at a scan rate of 50 mV/s. For comparison, we also evaluated the performance of PB deposited on CF using iron (III) chloride and ferricyanide as precursor solution (CF/PB). Figure 6A,B display CF/PB-FeOOH and FC/PB CVs, revealing redox peaks at 0.28 and 0.21 V and 0.46 and 0.43 V in PBS, respectively. The capacitive current of CF/PB was higher than that of CF/PB-FeOOH in PBS. This behavior is likely due to the larger CF/PB film thickness than CF/PB-FeOOH. PB thick films provide easier ion exchange between PB nanoparticles and the solution interface, improving the electroactivity of the electrode [43,44]. However, this property did not enhance PB's electrochemical catalytic activity toward H_2O_2. After adding H_2O_2, CF/PB and CF/PB-FeOOH voltammetric profiles changed, especially for the CF/PB-FeOOH electrode. Evident alterations in current values were observed at 0.0 V, −0.1 V, and −0.2 V, amounting to 35%, 28%, and 29% for CF/PB, and 31%, 66%, and 85% for CF/PB-FeOOH, respectively. The difference between the current response in PBS and PBS with H_2O_2 was more pronounced for CF/PB-FeOOH, demonstrating greater sensitivity to H_2O_2. The H_2O_2 reduction by CF/PB-FeOOH can be explained by considering the synergistic effect of the Fenton-like reaction at the FeOOH (Reaction (2a–c)) surface and the PB electrocatalytic process according to Reaction (2d) [19,45]. Based on this comparison, we demonstrate the positive influence of FeOOH in the electrochemical properties of PB deposited on CF to the reduction in H_2O_2 and its potential application in electroanalytical. Therefore, we evaluated the analytical performance of the CF/PB-FeOOH electrode by amperometry at 0.0 V, −0.1 V, and −0.2 V, which were assessed for various H_2O_2 concentrations. The current response increased with H_2O_2 concentration. Figure S2 shows the calibration curve constructed taken from the three potentials. The linear regression equations were as follows: Y (µA) = 0.122 × C (µM) + 0.800 (R^2 = 0.977); Y (µA) = 0.250 × C (µM) + 2.33 (R^2 = 0.940); and Y (µA) = 0.164 × C (µM) + 0.150 (R^2 = 0.998), and the limits of detection (LOD) calculated were 0.91 µM, 0.55 µM, and 0.36 µM (S/N = 3) for 0.0 V, −0.1 V, and −0.2 V, respectively. We observed that the current value at −0.2 V exhibited outstanding linearly and showed low LOD. The current CF/PB-FeOOH electrode values for H_2O_2 detection showed a linear range of 1.2 to 300 µM, and the limit of quantification (LOQ) was 1.19 µM (Figure 7A,B). Our results show that the CF/PB-FeOOH electrode exhibits a low limit of detection comparable to those previously reported. Additionally, it is constructed with inexpensive materials (see Table S1 [19,20,38,46–50]).

(a) $FeOOH + H_2O_2 \rightarrow (FeOOH)H_2O_2$

(b) $Fe(III)_{(surf)}(H_2O_2) + e^- \rightarrow Fe(II)_{(surf)} + HOO\bullet + H^+$

(c) $Fe(II)_{(surf)} + HOO\bullet + H^+ \rightarrow Fe(III)_{(surf)} + H_2O + 1/2\, O_2$

(d) $H_2O_2 + 2e^- \rightarrow 2OH^-$

(R2)

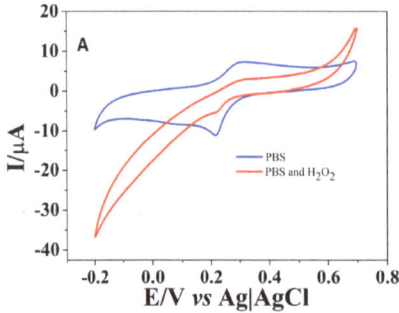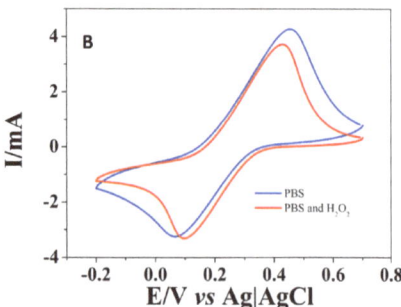

Figure 6. CV curves at a potential range from −0.2 V to 0.7 V at a scan rate of 100 mV·s^{-1} in 0.1 M of PBS without or with 1.0 mM of H_2O_2. (**A**) CF/PB-FeOOH electrode and (**B**) CF/PB electrode.

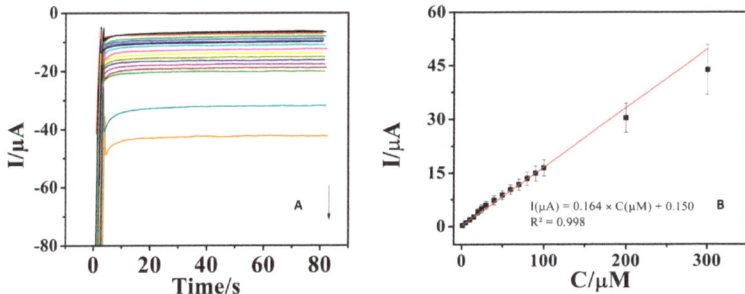

Figure 7. (**A**) Amperometric response of the CF/PB–δ-FeOOH electrode in 0.1 M PBS at −0.2 V versus Ag|AgCl with successive additions of H_2O_2 under stirring. (**B**) Corresponding calibration curve. The error bars show the standard deviation for n = 3. The arrow indicates the increase in concentration of H_2O_2.

The reproducibility of the three independent CF/PB-FeOOH electrodes was tested by measuring the reduction current of 80 μM H_2O_2 under the same experimental conditions. Figure S3 shows no change in reduction current, and the relative standard deviation (RSD) is calculated to be 3.3%. This value typically represents the acceptable reproducibility of the CF/PB-FeOOH electrode. The selectivity of the CF/PB-FeOOH electrode was evaluated in PBS in the presence of electroactive species usually found in biological samples, such as dopamine (DA), uric acid (UA), and ascorbic acid (AA). According to the literature, (DA) is found in blood plasma in a concentration range of 0.24 to 23 μM [51], (UA) from 200 to 390 μM [52], and (AA) at 30 μM [53]. Therefore, we used 100 μM of ascorbic acid, 400 μM of uric acid, 40 μM of dopamine, and 80 μM of H_2O_2 in the selectivity assay. Figure 8 displays the amperometric response in PBS to adding H_2O_2 and electroactive interferents. A significant reduction current was observed upon introducing H_2O_2 at the initial and final stages of the experiment. In contrast, a weak current was detected upon adding AA despite the concentration being three times higher than usually found in blood plasma. Moreover, no discernible reduction current is detected for DA and UA. These results demonstrate that the CF/PB-FeOOH electrode possesses high selectivity.

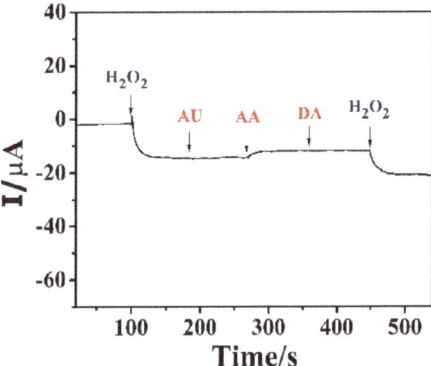

Figure 8. Amperometric response of CF/PB-FeOOH electrode in 0.1 M PBS at −0.2 V using 80 μM of H_2O_2 (initial and final additions), 400 μM of uric acid, 100 μM of ascorbic acid, and 40 μM of dopamine.

The long-term stability of the CF/PB-FeOOH electrode was estimated by storing it at 25 °C in a room. After being stored and tested eighteen days later, the sensor's current response maintained 93% of its initial response. This characteristic, combined with the steady electrochemical response of the CF/PB-FeOOH electrode, makes it useful in sensing H_2O_2 levels in biological samples.

Living cells can be induced to release H_2O_2 into a culture medium enriched with FBS, which supplies a range of nutrients essential for cell growth [54]. Different concentrations of H_2O_2 standard solution were spiked into PBS with 10% FBS to assess the practical application potential of the CF/PB-FeOOH electrode in biological samples. The calculated H_2O_2 recovery rates varied between 93% and 101%, as shown in Table 1. These results underscore the suitability of the CF/PB-FeOOH electrode for detecting H_2O_2 in biological samples.

Table 1. Spinking and recovery of H_2O_2 in PBS with 10% of FBS.

Sample	Added (μM)	Found (μM)	Recovery (%)
1	5	4.66	93
2	15	14.30	95
3	30	30.28	101

4. Conclusions

We have developed a novel non-enzymatic electrochemical sensor utilizing PB and δ-FeOOH to determine H_2O_2. PB was successfully synthesized electrochemically from δ-FeOOH acid suspension and deposited on CF, yielding a CF/PB-FeOOH electrode. XRD analysis suggests that PB-FeOOH film has a predominantly amorphous structure, and SEM images revealed nanocubes deposited on CF. The synergic effect of PB and δ-FeOOH enhanced the electroactivity of the CF/PB-FeOOH electrode, showing sensitivity, selectivity, and reproducibility to determine H_2O_2. CF/PB-FeOOH achieved greater sensitivity to H_2O_2 than the CF/PB electrode, which was used for comparison. Moreover, the CF/PB-FeOOH electrode detected H_2O_2 in 10% fetal bovine serum. This work underscores the potential of CF/PB-FeOOH electrodes in progressing electrochemical sensing technologies for various biological and environmental applications.

Supplementary Materials: The following supporting information can be downloaded at: https://www.mdpi.com/article/10.3390/c10030082/s1, Figure S1: CVs recorded at different scan rates (10–300 mV·s^{-1}) and linear plot current versus square root of scan rate at CF (A, B) and CF/PB-FeOOH (C, D) in 1 mM $K_3Fe(CN)_6$, 1 mM $K_4Fe(CN)_6$ and 1 mM KCl solution at pH 2.0; Figure S2: Calibration curve plot of CF/PB-FeOOH at three different potentials with successive addition of H_2O_2 in 0.1 M PBS; Figure S3: Amperometric response of the three independent CF/PB-FeOOH electrodes at −0.2 V in 0.1 M PBS after adding 80 μM H_2O_2; Table S1: The detection of limit, applied potential and pH for different sensors for H_2O_2 sensing. Reference [55] is cited in the supplementary materials.

Author Contributions: K.S.N.: conceptualization, methodology, experimental work, data analysis, and writing—original draft; A.L.H.K.F.: methodology and experimental work; M.C.P.: methodology; F.G.E.N.: methodology; A.S.A.: conceptualization, supervision, methodology, data analysis, resources, and writing—review and editing. All authors have read and agreed to the published version of the manuscript.

Funding: This research was funded by FAPEMIG (APQ-00607-22).

Data Availability Statement: The data presented in this study are available on request from the corresponding author.

Acknowledgments: We thank FAPEMIG, CNPq, and FINEP/MCTI for their support. We also acknowledge the Institute of Science, Engineering, and Technology of the Federal University of Jequitinhonha and Mucuri Valleys for supporting this work.

Conflicts of Interest: The authors declare no conflict of interest.

References

1. Mattos, I.L.d.; Shiraishi, K.A.; Braz, A.D.; Fernandes, J.R. Peróxido de hidrogênio: Importância e determinação. *Quim. Nova* **2003**, *26*, 373–380. [CrossRef]
2. Giaretta, J.E.; Duan, H.W.; Oveissi, F.; Farajikhah, S.; Dehghani, F.; Naficy, S. Flexible Sensors for Hydrogen Peroxide Detection: A Critical Review. *ACS Appl. Mater. Interfaces* **2022**, *14*, 20491–20505. [CrossRef]
3. Cheung, E.C.; Vousden, K.H. The role of ROS in tumour development and progression. *Nat. Rev. Cancer* **2022**, *22*, 280–297. [CrossRef] [PubMed]
4. Geraskevich, A.V.; Solomonenko, A.N.; Dorozhko, E.V.; Korotkova, E.I.; Barek, J. Electrochemical Sensors for the Detection of Reactive Oxygen Species in Biological Systems: A Critical Review. *Crit. Rev. Anal. Chem.* **2022**, *54*, 742–774. [CrossRef]
5. Xing, L.J.; Zhang, W.G.; Fu, L.J.; Lorenzo, J.M.; Hao, Y.J. Fabrication and application of electrochemical sensor for analyzing hydrogen peroxide in food system and biological samples. *Food Chem.* **2022**, *385*, 132555. [CrossRef]
6. Thatikayala, D.; Ponnamma, D.; Sadasivuni, K.K.; Cabibihan, J.-J.; Al-Ali, A.K.; Malik, R.A.; Min, B. Progress of Advanced Nanomaterials in the Non-Enzymatic Electrochemical Sensing of Glucose and H_2O_2. *Biosensors* **2020**, *10*, 151. [CrossRef]
7. Thwala, L.N.; Ndlovu, S.C.; Mpofu, K.T.; Lugongolo, M.Y.; Mthunzi-Kufa, P. Nanotechnology-Based Diagnostics for Diseases Prevalent in Developing Countries: Current Advances in Point-of-Care Tests. *Nanomaterials* **2023**, *13*, 1247. [CrossRef]
8. Baranwal, J.; Barse, B.; Gatto, G.; Broncova, G.; Kumar, A. Electrochemical Sensors and Their Applications: A Review. *Chemosensors* **2022**, *10*, 363. [CrossRef]
9. Hasanzadeh, M.; Shadjou, N.; Guardia, M.d.l. Current advancement in electrochemical analysis of neurotransmitters in biological fluids. *TrAC Trends Anal. Chem.* **2017**, *86*, 107. [CrossRef]
10. Chen, X.M.; Wu, G.H.; Cai, Z.X.; Oyama, M.; Chen, X. Advances in enzyme-free electrochemical sensors for hydrogen peroxide, glucose, and uric acid. *Microchim. Acta* **2014**, *181*, 689–705. [CrossRef]
11. Zhao, C.L.; Zhang, H.F.; Zheng, J.B. Synthesis of silver decorated sea urchin-like FeOOH nanocomposites and its application for electrochemical detection of hydrogen peroxide. *J. Mater. Sci. Mater. Electron.* **2017**, *28*, 14369–14376. [CrossRef]
12. Lim, H.C.; Cho, Y.J.; Han, D.H.; Kim, T.H. Enhancing electrocatalytic performance and Stability: A novel Prussian Blue-Graphene quantum dot nanoarchitecture for H_2O_2 reduction. *Appl. Surf. Sci.* **2024**, *646*, 158920. [CrossRef]
13. Rani, K.K.; Liu, Y.X.; Devasenathipathy, R.; Yang, C.; Wang, S.F. Simple preparation of gold nanoparticle-decorated copper cross-linked pectin for the sensitive determination of hydrogen peroxide. *Ionics* **2019**, *25*, 309–317. [CrossRef]
14. Yin, J.W.; Zhang, H.T.; Wang, Y.; Laurindo, M.B.J.; Zhao, J.F.; Hasebe, Y.; Zhang, Z.Q. Crab gill-derived nanorod-like carbons as bifunctional electrochemical sensors for detection of hydrogen peroxide and glucose. *Ionics* **2024**, *30*, 3541–3552. [CrossRef]
15. Wang, J.; Zhou, Q.X.; Fan, C.; Guo, X.; Bei, J.L.; Chen, T.T.; Yang, J.; Yao, Y. Ultrasensitive and specific photoelectrochemical sensor for hydrogen peroxide detection based on pillar 5 arene-functionalized Au nanoparticles and MWNTs hybrid BiOBr heterojunction. *Microchim. Acta* **2024**, *191*, 266. [CrossRef] [PubMed]
16. Shen, W.; Shi, J.C.; Wang, X.F.; Xu, P.C.; Li, X.X. A Non-Enzymatic Electrochemical Sensor Based on Cerium Oxide Nanocubes for the Rapid Detection of Hydrogen Peroxide Residues in Food Samples. *IEEJ Trans. Electr. Electron. Eng.* **2024**, *19*, 1573–1578. [CrossRef]
17. Chen, Y.H.; Zhao, D.; Sun, T.Q.; Cai, C.R.; Dong, Y.M. The preparation of MoS2/8-FeOOH and degradation of RhB under visible light. *J. Environ. Chem. Eng.* **2023**, *11*, 110353. [CrossRef]
18. Campos, P.T.B.; Vaiss, V.S.; Ramalho, T.C. The high potential of the pure and Nb-doped 6-FeOOH (001) surface in the adsorption and degradation of a neurotoxic agent. *Surf. Sci.* **2024**, *746*, 122491. [CrossRef]
19. de Meira, F.H.A.; Resende, S.F.; Monteiro, D.S.; Pereira, M.C.; Mattoso, L.H.C.; Faria, R.C.; Afonso, A.S. A Non-enzymatic Ag/δ-FeOOH Sensor for Hydrogen Peroxide Determination using Disposable Carbon-based Electrochemical Cells. *Electroanalysis* **2020**, *32*, 2231–2236. [CrossRef]
20. Melo, W.E.R.d.; Nantes, K.S.; Ferreira, A.L.H.K.; Pereira, M.C.; Mattoso, L.H.C.; Faria, R.C.; Afonso, A.S. A Disposable Carbon-Based Electrochemical Cell Modified with Carbon Black and Ag/δ-FeOOH for Non-Enzymatic H_2O_2 Electrochemical Sensing. *Electrochem* **2023**, *4*, 523–536. [CrossRef]
21. Jin, E.; Lu, X.F.; Cui, L.L.; Chao, D.M.; Wang, C. Fabrication of graphene/prussian blue composite nanosheets and their electrocatalytic reduction of H_2O_2. *Electrochim. Acta* **2010**, *55*, 7230–7234. [CrossRef]
22. Farah, A.M.; Shooto, N.D.; Thema, F.T.; Modise, J.S.; Dikio, E.D. Fabrication of Prussian Blue/Multi-Walled Carbon Nanotubes Modified Glassy Carbon Electrode for Electrochemical Detection of Hydrogen Peroxide. *Int. J. Electrochem. Sci.* **2012**, *7*, 4302–4313. [CrossRef]
23. Han, L.J.; Tricard, S.; Fang, J.; Zhao, J.H.; Shen, W.G. Prussian blue @ platinum nanoparticles/graphite felt nanocomposite electrodes: Application as hydrogen peroxide sensor. *Biosens. Bioelectron.* **2013**, *43*, 120–124. [CrossRef] [PubMed]
24. Husmann, S.; Nossol, E.; Zarbin, A.J.G. Carbon nanotube/Prussian blue paste electrodes: Characterization and study of key parameters for application as sensors for determination of low concentration of hydrogen peroxide. *Sens. Actuator B Chem.* **2014**, *192*, 782–790. [CrossRef]
25. Ma, L.T.; Cui, H.L.; Chen, S.M.; Li, X.L.; Dong, B.B.; Zhi, C.Y. Accommodating diverse ions in Prussian blue analogs frameworks for rechargeable batteries: The electrochemical redox reactions. *Nano Energy* **2021**, *81*, 105632. [CrossRef]
26. Pajerowski, D.M.; Watanabe, T.; Yamamoto, T.; Einaga, Y. Electronic conductivity in Berlin green and Prussian blue. *Phys. Rev. B* **2011**, *83*, 153202. [CrossRef]

27. Yi, H.C.; Qin, R.Z.; Ding, S.X.; Wang, Y.T.; Li, S.N.; Zhao, Q.H.; Pan, F. Structure and Properties of Prussian Blue Analogues in Energy Storage and Conversion Applications. *Adv. Funct. Mater.* **2021**, *31*, 2006970. [CrossRef]
28. Wang, Z.W.; Yang, H.K.; Gao, B.W.; Tong, Y.; Zhang, X.J.; Su, L. Stability improvement of Prussian blue in nonacidic solutions via an electrochemical post-treatment method and the shape evolution of Prussian blue from nanospheres to nanocubes. *Analyst* **2014**, *139*, 1127–1133. [CrossRef]
29. Scholz, F.; Schwudke, D.; Stösser, R.; Bohácek, J. The interaction of Prussian blue and dissolved hexacyanoferrate ions with goethite (α-FeOOH) studied to assess the chemical stability and physical mobility of Prussian blue in soils. *Ecotox. Environ. Saf.* **2001**, *49*, 245–254. [CrossRef]
30. Yang, Y.R.; Gao, W. Wearable and flexible electronics for continuous molecular monitoring. *Chem. Soc. Rev.* **2019**, *48*, 1465–1491. [CrossRef]
31. Kim, J.; Campbell, A.S.; de Avila, B.E.F.; Wang, J. Wearable biosensors for healthcare monitoring. *Nat. Biotechnol.* **2019**, *37*, 389–406. [CrossRef] [PubMed]
32. Moreira, F.C.; Boaventura, R.A.R.; Brillas, E.; Vilar, V.J.P. Electrochemical advanced oxidation processes: A review on their application to synthetic and real wastewaters. *Appl. Catal. B Environ.* **2017**, *202*, 217–261. [CrossRef]
33. Guo, C.A.Y.; Chen, C.F.; Lu, J.Y.; Fu, D.; Yuan, C.Z.; Wu, X.L.; Hui, K.N.; Chen, J.R. Stable and recyclable Fe3C@CN catalyst supported on carbon felt for efficient activation of peroxymonosulfate. *J. Colloid Interface Sci.* **2021**, *599*, 219–226. [CrossRef]
34. Sun, B.; Skyllas-Kazacos, M. Chemical modification of graphite electrode materials for vanadium redox flow battery application—Part II. Acid treatments. *Electrochim. Acta* **1992**, *37*, 2459–2465. [CrossRef]
35. Sun, B.; Skyllas-Kazacos, M. Modification of graphite electrode materials for vanadium redox flow battery application—I. Thermal treatment. *Electrochim. Acta* **1992**, *37*, 1253–1260. [CrossRef]
36. Zhao, Y.G.; Ying, M.; Fu, Y.B.; Chen, W. Improving Electrochemical Performance of Carbon Felt Anode by Modifying with Akaganeite in Marine Benthic Microbial Fuel Cells. *Fuel Cells* **2019**, *19*, 190–199. [CrossRef]
37. Pereira, M.C.; Garcia, E.M.; da Silva, A.C.; Lorençon, E.; Ardisson, J.D.; Murad, E.; Fabris, J.D.; Matencio, T.; Ramalho, T.D.; Rocha, M.V.J. Nanostructured δ-FeOOH: A novel photocatalyst for water splitting. *J. Mater. Chem.* **2011**, *21*, 10280–10282. [CrossRef]
38. de Mattos, I.L.; Gorton, L.; Ruzgas, T.; Karyakin, A.A. Sensor for hydrogen peroxide based on Prussian blue modified electrode: Improvement of the operational stability. *Anal. Sci.* **2000**, *16*, 795–798. [CrossRef]
39. Vieira, T.A.; Souza, J.R.; Gimenes, D.T.; Munoz, R.A.A.; Nossol, E. Tuning electrochemical and morphological properties of Prussian blue/carbon nanotubes films through scan rate in cyclic voltammetry. *Solid State Ion.* **2019**, *338*, 5–11. [CrossRef]
40. Matos-Peralta, Y.; Antuch, M. Review-Prussian Blue and Its Analogs as Appealing Materials for Electrochemical Sensing and Biosensing. *J. Electrochem. Soc.* **2019**, *167*, 037510. [CrossRef]
41. Elshorbagy, M.H.; Ramadan, R.; Abdelhady, I. Preparation and characterization of spray-deposited efficient Prussian blue electrochromic thin film. *Optik* **2017**, *129*, 130–139. [CrossRef]
42. Coros, M.; Varodi, C.; Pogacean, F.; Gal, E.; Pruneanu, S.M. Nitrogen-Doped Graphene: The Influence of Doping Level on the Charge-Transfer Resistance and Apparent Heterogeneous Electron Transfer Rate. *Sensors* **2020**, *20*, 1815. [CrossRef]
43. Fu, Z.Y.; Wei, Y.X.; Liu, W.M.; Li, J.Y.; Li, J.M.; Ma, Y.B.; Zhang, X.F.; Yan, Y. Investigation of electrochromic device based on multi-step electrodeposited PB films. *Ionics* **2021**, *27*, 4419–4427. [CrossRef]
44. Lu, S.Y.; Chen, Y.H.; Fang, X.F.; Feng, X. Hydrogen peroxide sensor based on electrodeposited Prussian blue film. *J. Appl. Electrochem.* **2017**, *47*, 1261–1271. [CrossRef]
45. Karyakin, A.A.; Karyakina, E.E.; Gorton, L. On the mechanism of H2O2 reduction at Prussian blue modified electrodes. *Electrochem. Commun.* **1999**, *1*, 78–82. [CrossRef]
46. Du, S.; Ren, Z.; Wu, J.; Xi, W.; Fu, H. Vertical α-FeOOH nanowires grown on the carbon fiber paper as a free-standing electrode for sensitive H_2O_2 detection. *Nano Res.* **2016**, *9*, 2260–2269. [CrossRef]
47. Rattanopas, S.; Schulte, A.; Teanphonkrang, S. Prussian Blue/Carbon Nanotube Sensor Spread with Gelatin/Zein Glaze: A User-Friendly Modification for Stable Interference-Free H_2O_2 Amperometry. *Anal. Chem.* **2022**, *94*, 4919–4923. [CrossRef] [PubMed]
48. Liu, X.L.; Zhang, X.J.; Zheng, J.B. One-pot fabrication of AuNPs-Prussian blue-Graphene oxide hybrid nanomaterials for non-enzymatic hydrogen peroxide electrochemical detection. *Microchem. J.* **2021**, *160*, 105595. [CrossRef]
49. Uzunçar, S.; Özdogan, N.; Ak, M. Amperometric detection of glucose and H_2O_2 using peroxide selective electrode based on carboxymethylcellulose/polypyrrole and Prussian Blue nanocomposite. *Mater. Today Commun.* **2021**, *26*, 101839. [CrossRef]
50. Li, N.; Zhou, H.Y.; Liu, Y.H.; Yu, X.J.; Cao, L.; Xu, Y.J.; Xi, L.X.; Zhao, G.; Ban, X.X. Prussian blue@Au nanoparticles/SiO2 cavity/ITO electrodes: Application as a hydrogen peroxide sensor. *New J. Chem.* **2024**, *48*, 1300–1306. [CrossRef]
51. Phung, V.D.; Jung, W.S.; Nguyen, T.A.; Kim, J.H.; Lee, S.W. Reliable and quantitative SERS detection of dopamine levels in human blood plasma using a plasmonic Au/Ag nanocluster substrate. *Nanoscale* **2018**, *10*, 22493–22503. [CrossRef] [PubMed]
52. Li, Q.; Yang, Z.; Lu, B.; Wen, J.; Ye, Z.; Chen, L.L.; He, M.; Tao, X.M.; Zhang, W.W.; Huang, Y.; et al. Serum uric acid level and its association with metabolic syndrome and carotid atherosclerosis in patients with type 2 diabetes. *Cardiovasc. Diabetol.* **2011**, *10*, 72. [CrossRef] [PubMed]
53. Da Costa, L.A.; García-Bailo, B.; Borchers, C.H.; Badawi, A.; El-Sohemy, A. Association between the plasma proteome and serum ascorbic acid concentrations in humans. *J. Nutr. Biochem.* **2013**, *24*, 842–847. [CrossRef] [PubMed]

54. van der Valk, J.J.S. Fetal bovine serum—A cell culture dilemma. *Science* **2022**, *375*, 143–144. [CrossRef]
55. Jiang, T.; Zhan, D.P.; Chen, Y. Preparation of carbon nanotube arrays nanocomposites filled with Prussian blue and electrochemical sensing of hydrogen peroxide. *Ferroelectrics* **2021**, *580*, 42–54. [CrossRef]

Disclaimer/Publisher's Note: The statements, opinions and data contained in all publications are solely those of the individual author(s) and contributor(s) and not of MDPI and/or the editor(s). MDPI and/or the editor(s) disclaim responsibility for any injury to people or property resulting from any ideas, methods, instructions or products referred to in the content.

Article

A Nitrogen/Oxygen Dual-Doped Porous Carbon with High Catalytic Conversion Ability toward Polysulfides for Advanced Lithium–Sulfur Batteries

Xiaoyan Shu [1], Yuanjiang Yang [1], Zhongtang Yang [1], Honghui Wang [2,*] and Nengfei Yu [1,*]

[1] School of Energy Sciences and Engineering, Nanjing Tech University, Nanjing 211816, China; shuxiaoyan@njtech.edu.cn (X.S.); 202261108028@njtech.edu.cn (Y.Y.); 202161208042@njtech.edu.cn (Z.Y.)
[2] Key Laboratory of Estuarine Ecological Security and Environmental Health (Fujian Province University), Xiamen University Tan Kah Kee College, Zhangzhou 363105, China
* Correspondence: hhwang@xujc.com (H.W.); yunf@njtech.edu.cn (N.Y.)

Abstract: Lithium–sulfur batteries (LSBs) have attracted widespread attention due to their high theoretical energy density and low cost. However, their development has been constrained by the shuttle effect of lithium polysulfides and their slow reaction kinetics. In this work, a nitrogen/oxygen dual-doped porous carbon (N/O-PC) was synthesized by annealing the precursor of zeolitic imidazolate framework-8 grown in situ on MWCNTs (ZIF-8/MWCNTs). Then, the N/O-PC composite served as an efficient host for LSBs through chemical adsorption and providing catalytic conversion sites of polysulfides. Moreover, the interconnected porous carbon-based structure facilitates electron and ion transfer. Thus, the S/N/O-PC cathode exhibits high cycling stability (a stable capacity of 685.9 mA h g^{-1} at 0.2 C after 100 cycles). It also demonstrates excellent rate performance with discharge capacities of 1018.2, 890.2, 775.1, 722.7, 640.4, and 579.6 mAh g^{-1} at 0.2, 0.5, 1.0, 2.0, 3.0, and 5.0 C, respectively. This work provides an effective strategy for designing and developing high energy density, long cycle life LSBs.

Keywords: lithium–sulfur batteries; metal-organic frameworks; shuttling effect

Citation: Shu, X.; Yang, Y.; Yang, Z.; Wang, H.; Yu, N. A Nitrogen/Oxygen Dual-Doped Porous Carbon with High Catalytic Conversion Ability toward Polysulfides for Advanced Lithium–Sulfur Batteries. *C* **2024**, *10*, 67. https://doi.org/10.3390/c10030067

Academic Editors: Olena Okhay and Gil Goncalves

Received: 4 June 2024
Revised: 20 July 2024
Accepted: 26 July 2024
Published: 30 July 2024

Copyright: © 2024 by the authors. Licensee MDPI, Basel, Switzerland. This article is an open access article distributed under the terms and conditions of the Creative Commons Attribution (CC BY) license (https:// creativecommons.org/licenses/by/ 4.0/).

1. Introduction

In order to achieve the goal of "carbon peak" by 2030 and "carbon neutrality" by 2060, it is critical to develop renewable green energy and sustainable energy storage technologies. To date, various rechargeable energy storage devices such as nickel–cadmium, lead–acid, or lithium–ion batteries (LIBs) have been developed [1]. Compared to other rechargeable battery technologies, LIBs have many advantages. They have a high energy density of 300 Wh kg^{-1} compared to roughly 75 Wh kg^{-1} for alternative technologies. In addition, LIBs can deliver up to 3.6 V, 2–3 times the voltage of alternatives, which makes them suitable for high-power applications like transportation [2]. LIBs have no memory effect, a detrimental process where repeated partial discharge/charge cycles can cause a battery to 'remember' a lower capacity. LIBs also have a low self-discharge rate of around 1.5–2% per month, and do not contain toxic lead or cadmium. Thus, LIBs have been widely used in portable electronic devices such as smartphones, laptops, digital cameras, smart grids, and 5G base stations. However, LIBs cannot meet the demand for batteries in electric vehicles due to their low energy density. LIBs are still around a hundred times less energy-dense than gasoline, which contains 12,700 Wh kg^{-1} by mass or 8760 Wh L^{-1} by volume. Therefore, there is an urgent need to develop high energy density rechargeable energy storage devices [3].

Lithium–sulfur batteries (LSBs), which convert energy through reversible electrochemical reactions between lithium and sulfur, have a high theoretical energy density (2600 Wh kg^{-1}) and high theoretical specific capacity (1675 mAh g^{-1}). Unlike the expensive and rare elements (nickel, manganese, and cobalt) commonly found in LIB cathodes, sulfur is plentiful and can be found almost anywhere on Earth. Moreover, LSBs require much less production energy since sulfur only requires 112 °C to melt into crystal form [4–6]. Therefore, LSBs are considered to be the most promising next-generation energy storage technology. Although LSBs possess many advantages, LSBs also have some challenges to overcome [6,7]. The main problem is that current LSBs possess poor cycling performance. It is all in the internal chemistry. The electrochemical reaction between lithium and sulfur proceeds via a complicated multi-step mechanism and a variety of intermediate polysulfides (Li_2S_n) are formed. The polysulfides, however, are often very soluble in electrolytes. The dissolution of polysulfides depletes electrodes from active materials. Furthermore, when polysulfides diffuse into the electrolyte, they can easily travel between the cathode and anode and instead of useful oxidation and reduction cycles, parasitic reduction and oxidation of intermediate polysulfides occurs on the electrodes (shuttle effect), resulting in capacity fading. And the conductivity of sulfur is low. Moreover, the volume change of sulfur during cycling is huge.

To address these problems, suitable cathode host materials are effective approaches [8–13]. Porous carbon materials hold great potential as S cathode hosts due to their high specific surface area, high conductivity, and excellent chemical stability [14,15]. Li et al. prepared a hybrid graphene album structure (g-C_3N_4@n-G) as an S host for LSBs. The g-C_3N_4@n-G provides a hierarchical porous structure, easy mass transport and excellent conductivity for the S cathode. Thus, S/C_3N_4@n-G showed a high initial capacity of 1252 mA h g^{-1} at 0.2 C and an extremely lower capacity decay of 0.028% per cycle at 0.2 C. Although porous carbon materials can immobilize polysulfides through physical adsorption, the surface affinity between carbon materials and sulfur is poor.

Doping heteroatoms (such as nitrogen, oxygen, and boron) into carbon substrates can promote covalent bonding between carbon materials and sulfur, thus improving the electrochemical performance of the sulfur cathode. MOFs are a new type of porous organic–inorganic hybrid material formed by the coordination of metal ions and organic ligands [5,16,17]. MOFs have many advantages such as rich doping of heterogeneous atoms (such as N, P, S, and B), a porous structure, and a high porosity, which makes them ideal precursors for carbon materials. MOF-derived carbon materials retain the precursors' pore structure and high specific surface area and possess high conductivity and high stability [18–20]. Therefore, MOF-derived porous carbon materials show great potential as S cathode hosts [21–24]. However, MOF-derived porous carbon materials agglomerate in the process of pyrolysis of MOFs [25,26]. The introduction of two-dimensional materials such as graphene, carbon nanotubes, and Mxene can effectively address the agglomeration issue [27]. Herein, zeolitic imidazolate framework-8 (ZIF-8) nanoparticles in situ grow on multiwalled carbon nanotubes (ZIF-8/MWCNTs) [28]. Then, the ZIF-8/MWCNT composite was annealed to obtain nitrogen/oxygen dual-doped porous carbon (N/O-PC). The N/O-PC composite serves as an efficient host for the S cathode (S/N/O-PC), which not only effectively traps soluble lithium polysulfides but also offers more catalytic conversion sites for polysulfides. Additionally, the MWCNTs in the S/N/O-PC mainly interconnect the N/O-doped porous carbon and provide richly continuous electron transport channels. Thus, the S/N/O-PC cathode exhibits high cycling stability.

2. Materials and Methods

2.1. Materials

Polyvinylpyrrolidone (PVP, K23-27, M_W = 24,000), zinc nitrate hexahydrate ($Zn(NO_3)_2 \cdot 6H_2O$, AR)), 2-methylimidazole (2-MeIm, 96%), multiwalled carbon nanotubes (MWCNTs, >98%), N-methyl-pyrrolidone (NMP, AR), and concentrated nitric acid (GR) were purchased from Aladdin. The electrolyte used in our experiments is the LS6903 electrolyte (0.5 M lithium bis(trifluoromethanesulfonyl)imide (LiTFSI) and 0.5 M lithium nitrate ($LiNO_3$) dissolved in a mixture of 1,3-dioxolane (DME) and 1,2-dimethoxyethane (DME, DME:DOL = 1:1 vol%) was purchased from Acros Organics). All chemicals were used without further purification.

2.2. Synthesis of N/O-PC Composite

The N/O-PC composite was constructed by the MWCNTs interconnecting N/O-doped porous carbon (~100 nm) via annealing the precursor of large-size zeolitic imidazolate framework-8 (ZIF-8) grown in situ on MWCNTs. The MWCNTs need to undergo activation treatment before use to increase their surface active sites and promote ZIF-8 growth on their surface. In the typical activation process, the MWCNTs were pre-activated in concentrated HNO_3 solution at 110 °C for 2 h to generate hydrophilic surface structures with oxygen-containing functional groups. Subsequently, the MWCNTs were washed three times with deionized water and then dried at 80 °C for 12 h to obtain the activated MWCNTs. After that, 300 mg activated MWCNTs and 6.0 mmol 2-MeIm were added into 400 mL methanol solution containing 200 mg 2-MeIm under vigorous stirring for 2 h to form homogeneous ink. Next, 800 mL methanol containing 3.0 mmol Zn $(NO_3)_2 \cdot 6H_2O$ was transferred into the above-mentioned MWCNTs solution and the resulting mixture was allowed to react for 24 h at room temperature to form the ZIF-8/MWCNTs precursors. The obtained ZIF-8/MWCNTs precursors were collected by centrifugation at 9000 rpm for 10 min, washed three times with methanol and deionized water and then dried in vacuum at 80 °C overnight. Finally, annealing the as-obtained ZIF-8/MWCNTs precursors under Ar atmosphere in a tube furnace at 950 °C for 3 h with a ramp rate of 2 °C min^{-1} yielded the N/O-PC composite.

2.3. Fabrication of S/N/O-PC Composite

The S/N/O-PC composite was prepared by a simple method of solution infiltration and heat treatment. First, the N/O-PC composite was activation treated at 100 °C maintained for 8 h. Then, 200 mg sulfur powder was dissolved in 10 mL CS_2 solution. Subsequently, N/O-PC composite was immersed in sulfur/CS_2 solution for 96 h, in which the mass ratio of N/O-PC composite and sulfur was 1 to 4. Then, the mixture was dried at 60 °C for 2 h at room temperature. The mixture was heated at 155 °C for 12 h in a sealed autoclave under Ar, generating the S/N/O-PC composite. The fabrication processes of S/N/O-PC are illustrated in Figure 1. Also, the S/carbon black (S/C) composite was prepared by using a method similar to that used for constructing the S/N/O-PC composite.

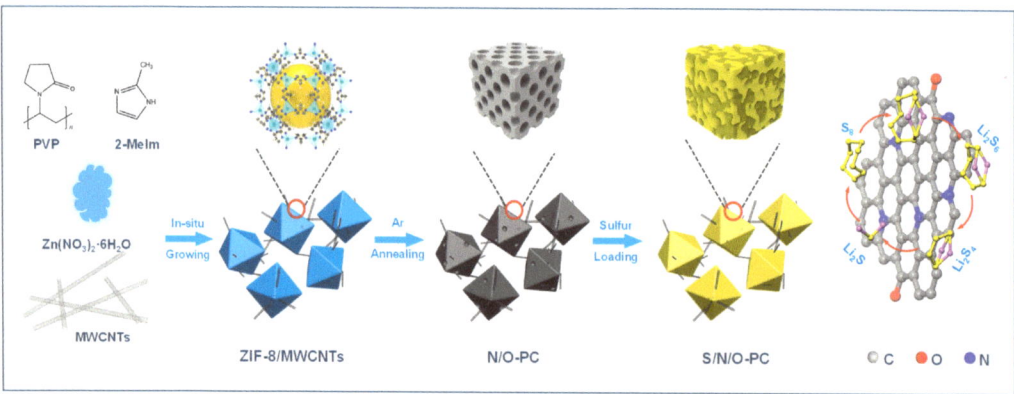

Figure 1. Schematic illustration of fabrication processes of S/N/O–PC.

2.4. Materials Characterization

X-ray diffraction (XRD) patterns were collected using a SmartLab3KW X-ray diffractometer equipped with Cu-Kα radiation (λ = 0.1540 nm). The measurements were conducted at 40 kV and 40 mA, with a scanning rate of 10° min^{-1} in a 2θ range of 10–80°. Scanning electron microscopy (SEM) and transmission electron microscopy (TEM) experiments were performed using a Phenom ProX SEM and a JEM1200EX high-resolution TEM, respectively. Thermal gravimetric analysis (TGA) and differential scanning calorimetry (DSC) were carried out using a TGA2 thermogravimetric analyzer in an argon atmosphere. The scan rate was set at 10 °C min^{-1}, covering a temperature range from room temperature to 800 °C. The Brunauer–Emmett–Teller (BET) surface area was determined by analyzing nitrogen adsorption isotherms using an Autosorb iQ3 automated gas sorption system.

2.5. Electrochemical Measurement

For the S/N/O-PC cathodes, S/N/O-PC composite, super P, and PVDF in a weight ratio of 8:1:1 were mixed in NMP to form a uniform slurry that was then coated onto Al foil. For comparison, the S/C cathodes were also prepared by using a process similar to fabricating the S/N/O-PC cathodes. Sulfur cathodes, Li foil anodes, commercial PP separators, and 40 µL electrolyte (0.5 M LiCF$_3$SO$_3$, 0.5 M LiNO$_3$ in DME:DOL = 1:1 vol%) were used to assemble CR2025 coin-type cells for electrochemical measurements. Cyclic voltammetry (CV) measurements were performed using a CHI760e electrochemical workstation (Shanghai Chenhua Instrument, Shanghai, China) with a scan rate of 0.1 mV s^{-1} and a voltage range of 1.7–2.8 V. Electrochemical impedance spectroscopy (EIS) was conducted using a PARSTAT 2273 within a frequency range of 10^{-1}–10^5 Hz. Galvanostatic charge/discharge tests were carried out using a battery test system (LAND CT2001A). The S/N/O-PC cathodes were cycled by using galvanostatic charge-discharge at a current density of 0.2 C with a voltage range of 1.7 V to 2.8 V. The cycling stability of the S/N/O-PC cathode is reflected by plotting the specific capacity versus the number of cycles. The rate performance of the S/N/O-PC cathode was investigated by using galvanostatic charge–discharge at gradual charge/discharge rates from 0.2, 0.5, 1, 2, 3 to 5 C with each current density cycled for 5 cycles. The rate performance of the S/N/O-PC cathode was demonstrated by the plotted curve of the specific capacity versus the current density.

3. Results and Discussion

MOFs possess a porous structure and high porosity, which make them ideal precursors for porous carbon materials [19]. In the process of pyrolysis of MOFs, MOF-derived porous carbon materials agglomerate. Multiwalled carbon nanotubes (MWCNTs) have a unique structure in which carbon atoms are bonded together in a hexagonal lattice, forming tubes with a diameter of nanometers. Thus, carbon nanotubes can effectively address the agglomeration issues of MOF-derived porous carbon. In this paper, we used MWCNTs to support ZIF-8 [29,30]. The morphologies of MWCNTs were characterized by SEM and TEM techniques. The MWCNTs feature an even surface (Figure 2a–c) and a diameter distribution from 10 to 30 nm with average diameter of 19.9 nm (Figure S1).

After in situ growth of ZIF-8 nanoparticles on MWCNTs and pyrolysis of the precursor, their morphologies have been changed significantly. Figure 2d,e shows the SEM images of N/O-PC. A large number of carbon nanoparticles (particle sizes ranging from 100 to 200 nm with an average particle size of 150.5 nm (Figure S2) are interconnected in series by the MWCNTs, resulting in the formation of necklace-like structures. TEM images (Figure 2f) clearly indicate that the MWCNTs penetrate through the carbon nanoparticle. The N/O-PC necklace-like structure not only provides three-dimensional interconnected electron transport paths, but also effectively alleviates the detachment of active materials caused by the volume change of sulfur during the charge–discharge process [31–35].

Figure 3a shows the XRD patterns of the ZIF-8, ZIF-8/MWCNT and N/O-PC composites. The ZIF-8 peaks are similar to that in previous work, indicating the successful synthesis of ZIF-8 crystals [36]. The peaks are sharp, indicating a high crystallinity degree of ZIF-8. For the XRD pattern of ZIF-8/MWCNTs, it is similar to that of the ZIF-8 crystal, which indicates the successful synthesis of ZIF-8 on MWCNTs. The XRD pattern of the N/O-PC composite shows a broad diffraction peak around 26°, which is characteristic of amorphous carbon. Additionally, a weak peak can be observed at around 44°, which belongs to graphite [37,38]. This weak peak suggests that the NCPs exhibit some level of graphitization. Figure 3b shows the N$_2$ adsorption–desorption isotherm of the N/O-PC composite. It can be observed that the N/O-PC composite exhibits typical Type I isotherms. At a relative low pressure ($P/P_0 < 0.1$), the adsorption shows a steep increase, indicating the presence of a large number of micropores. Between the relative pressures of 0.8 to 1.0, a hysteresis loop appears, which is characteristic of mesopores [39]. These results indicate the co-existence of micropores and mesopores in the N/O-PC composite. The pore size distribution is typically quantified using the Barrett–Joyner–Halenda (BJH) method, which is applied to the desorption branch of the nitrogen adsorption isotherm. Figure S3 illustrates the pore size distribution of the N/O-PC composite. The pore size of the N/O-PC composite is mainly at 1.2 nm, corresponding to micropores. Additionally, the specific surface area of the N/O-doped porous carbon (N/O-PC) can be measured using the Brunauer–Emmett–Teller (BET) method, which is applied to the isotherm data in the relative pressure range to calculate the surface area. The surface area and the pore volume of the N/O-PC composite were measured to be 504.7 $m^2\,g^{-1}$ and 0.18 $cm^3\,g^{-1}$, respectively [40]. As a sulfur host, the N/O-PC composite exhibits a high specific surface area and pore volume, which allows more sulfur accommodation and reduces the dissolution and diffusion of the polysulfides and provides pathways for electron and ion transport [41,42]. The XPS technique was applied in the identification of the chemical composition and electronic states of the N/O-PC composite. As shown in Figure S4, the full spectrum clearly reveals the presence of C, N, and O, and a small amount of un-gasified Zn on the surface of the derivatives. Based on the XPS data, the atomic ratio of the doped N to O was found to be 1.43. Figure 3c shows the electronic states of O in N/O-PC. Two deconvoluted peaks correspond to O-O (533.3 eV) and C-O (532.4 eV), respectively. Figure 3d shows the electronic states of N in N/O-PC. Four deconvoluted peaks correspond to pyridinic N (398.5 eV), Zn-N (399.4 eV), graphitic N (401.0 eV), and pyrrolic N (401.5 eV), respectively [43–45]. Lithium polysulfides can be adsorbed and catalyzed through nitrogen and oxygen heteroatoms of the carbon matrix.

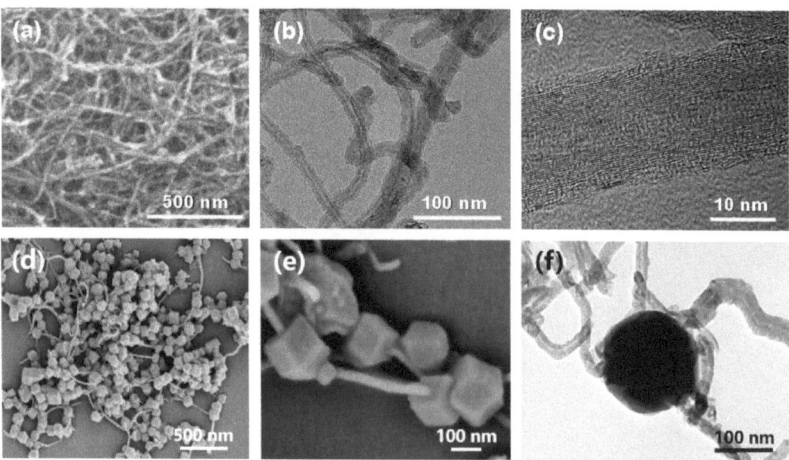

Figure 2. (a) SEM image, (b,c) TEM images of MWCNTs. (d) Low magnification, and (e) high magnification SEM image of N/O–PC, and (f) TEM image of S/N/O–PC.

To investigate the structure of S/N/O-PC, we performed XRD analysis. Figure 4a shows the XRD patterns of the pure S and S/N/O-PC composite. For the S/N/O-PC composite, no obvious crystalline S peaks are observed, indicating that sulfur is filled in the micropores of the N/O-PC composite. To determine the sulfur content in the S/N/O-PC composite, we conducted TGA [44,46]. As shown in Figure 4b, the sulfur content was 63.1 wt%. It is worth noting that there was no sulfur evaporation below 200 °C due to the sulfur nanoparticles filled in the micropores of the N/O-PC composite. We further performed nitrogen adsorption–desorption measurements for S/N/O-PC. As shown in

Figure 4c, the surface area decreased from 504.70 m² g⁻¹ of the N/O-PC to 61.4 m² g⁻¹ of the S/N/O-PC. Additionally, the micropore volume also reduced dramatically from 0.18 cm³ g⁻¹ of the N/O-PC to 0.01 cm³ g⁻¹ of the S/N/O-PC, based on changes in the pore size distribution before and after sulfur loading (Figure 4d). These findings confirm that almost all sulfur is filled into the micropores of the N/O-PC. In addition, the EDS elemental mapping images (Figure S5) unambiguously illustrate that the S, C, N and O elements are evenly distributed over the entire architecture at the nano level.

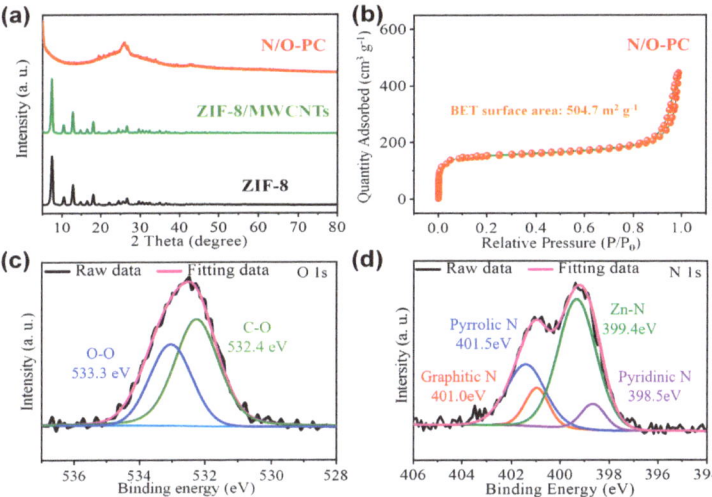

Figure 3. (**a**) XRD patterns of ZIF-8, ZIF-8/MWCNTs and N/O–PC. (**b**) Nitrogen adsorption–desorption isotherms for N/O–PC. (**c**) XPS high resolution of O 1s. (**d**) XPS high resolution of N 1s.

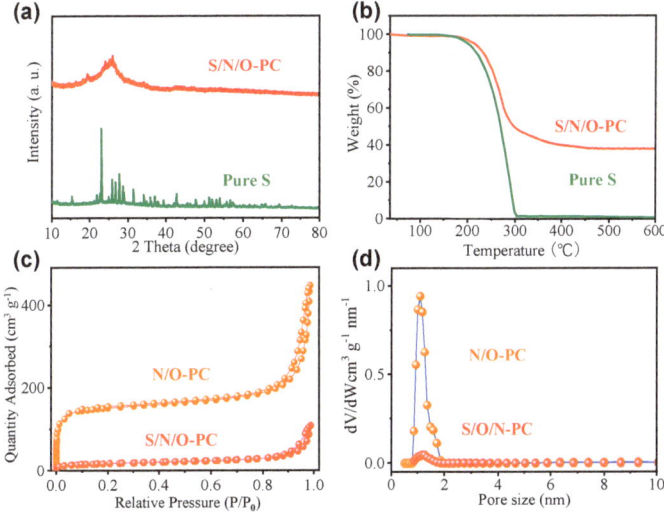

Figure 4. (**a**) XRD patterns of crystalline S, S/N/O–PC. (**b**) TGA curves of crystalline S, S/N/O–PC. (**c**) Nitrogen adsorption–desorption isotherms for the S/N/O–PC. (**d**) The pore size distribution for S/N/O-PC.

To evaluate the electrochemical performance of S/N/O-PC, cyclic voltammetry (CV) was tested first. Figure 5a shows the CV curves of the S/N/O-PC and S/C cathodes at a scan rate of 0.1 mV s^{-1}. It can be observed that the S/N/O-PC cathode exhibits two reduction peaks and one oxidation peak. The first reduction peak occurs around 2.3 V, corresponding to the reduction of S to soluble long-chain polysulfides (Li$_2$S$_{4-8}$). The second reduction peak is around 2.0 V, corresponding to the reduction from Li$_2$S$_{4-8}$ to insoluble product Li$_2$S$_2$/Li$_2$S. The oxidation peak occurs around 2.41 V, which belongs to the conversion of Li$_2$S$_2$/Li$_2$S into elemental sulfur and lithium. Although the CV curve of the S/C cathode also shows two reduction peaks and one oxidation peak, the current of the oxidation/reduction peaks are much smaller and the voltage polarization is much larger than those of the S/N/O-PC cathode, indicating the excellent electrocatalytic conversion of the polysulfides on S/N/O-PC. The galvanostatic charge/discharge profiles of the S/N/O-PC and S/C cathodes at 0.2 C (1 C = 1675 mA g^{-1}) are presented in Figure 5b, in which a charge plateau and two discharge plateaus can be apparently observed, consistent with the CV result [47]. There are also noticeable differences in the voltage gap between the charge and discharge voltage plateaus, which are associated with the reaction kinetics and the reversibility of the polysulfide redox reactions. Rather than the S/C cathode, the S/N/O-PC cathode delivers a voltage gap, implying the kinetically efficient reaction process in the S/N/O-PC cathode. Figure 5c shows the cycling performance of the S/N/O-PC and S/C cathodes at 0.2 C. The initial discharge capacity of the S/N/O-PC cathode is 1032.8 mA h g^{-1}, which is higher than that of the S/C cathode (862.4 mA h g^{-1}). After 100 cycles, the discharge capacity of S/N/O-PC maintained 685.9 mA h g^{-1}, corresponding to a 66.4% capacity retention rate. In contrast, the S/C cathode displayed a low capacity retention of 38.6%. Meanwhile, the S/N/O-PC kept a relatively high Coulombic efficiency of above 99% over the 100 cycles at 0.2 C, which is significantly superior to that of the S/C, implying efficient kinetics of the polysulfide redox reactions in the S/N/O-PC cathode. Figure 5d shows that the S/N/O-PC has small peak of negative potential at the beginning of the second discharge voltage plateau, suggesting a low overpotential for the reaction of Li$_2$S$_4$ to Li$_2$S$_2$. Also, Figure 5e shows that the S/N/O-PC has a small peak of positive potential at the start of the first charge voltage plateau, indicating a low overpotential for the reaction of Li$_2$S to Li$_2$S$_{4-8}$. These show that S/N/O-PC displays a low overpotential during both the discharge and charge processes, which is in line with the enhanced kinetics of the polysulfide redox reactions in the S/N/O-PC cathode.

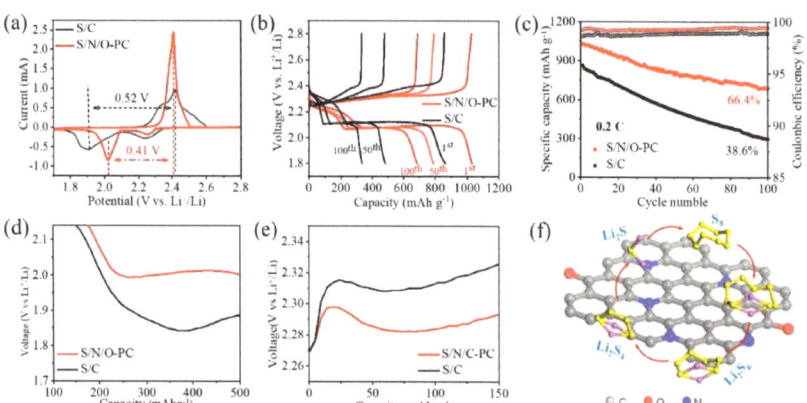

Figure 5. (**a**) CV curve of S/N/O-PC and S/C cathodes at a scan rate of 0.1 mV s^{-1}. (**b**) Galvanostatic charge/discharge profiles of S/N/O−PC and S/C cathodes at 0.2 C. (**c**) Cycle performances of S/N/O−PC and S/C cathodes at 0.2 C. (**d**) Discharge and (**e**) charge profiles of S/N/O−PC and S/C cathodes at a scan rate of 0.1 mV s^{-1}, indicating the overpotentials of solid–liquid phase conversion between soluble LiPSs and insoluble Li$_2$S$_2$/Li$_2$S. (**f**) Schematic illustration of nitrogen and oxygen heteroatoms on N/O−PC catalyzed lithium polysulfides.

The improvement in the performance of our S/N/O-PC cathode can be attributed to the advantages of the N/O-PC host. First, the N/O-PC composite has a high specific surface area and porous structure, which provide more sites for the physical adsorption of polysulfides, preventing their diffusion and dissolution in the electrolyte. Secondly, doping high electronegativity N and O into the carbon lattice can effectively disrupt the electrical neutrality of the adjacent carbon atoms and form active sites for adsorbing lithium polysulfides by forming chemical bonds, immobilizing the lithium polysulfides in the cathode region and inhibiting the shuttle of lithium polysulfides. Additionally, oxygen-doped carbon has surface functional groups such as hydroxyl, carbonyl, and carboxyl groups. These groups can interact with polysulfides through hydrogen bonds or other chemical bonds, further enhancing the chemical adsorption of polysulfides. More importantly, a stronger interaction between doped N/O and carbon atoms can promote the charge redistribution to form catalytic active sites for redox reactions of adsorbing lithium polysulfides. Then, since MWCNTs are flexible and have a relatively low volume expansion, they can effectively accommodate the volume expansion of S during cycling and improve the cycling performance of the S/N/O-PC cathode. Moreover, NCPs and MWCNTs are chemically stable in the operating conditions of LSBs, which contributes to the overall stability of the battery. Finally, lithium polysulfides can be adsorbed and catalyzed through nitrogen and oxygen heteroatoms of the carbon matrix (Figure 5f).

The electrode kinetics were studied by using electrochemical impedance spectroscopy (EIS) technology. Figure S6 shows the Nyquist plots of the cells with the S/N/O-PC collected before cycling and after 100 cycles, which are composed of a depressed semicircle and a linear section in the high- and low-frequency regions, respectively. The depressed semicircle corresponds to charge-transfer resistance when the electrons transfer from confined sulfur to the NCPs hosts and then to the MWNTs network. The linear section is associated with the Warburg impedance (W) related to the ionic diffusion. The charge-transfer resistance of the S/N/O-PC decreases after 100 cycles, which further confirms that the sulfur-loaded N/O-PC with high conductivity can significantly accelerate the kinetics of polysulfide redox reactions by reducing the charge-transfer resistance of electrochemical reactions [48]. The rate performance of the S/N/O-PC cathode was evaluated by cycling at gradual rates from 0.2 C to 5 C. As shown in Figure 6a, discharge capacities of 1018.2, 890.2, 775.1, 722.7, 640.4, and 579.6 mAh g^{-1} were obtained for the S/N/O-PC at rates of 0.2, 0.5, 1.0, 2.0, 3.0 and 5.0 C, respectively, which were always higher than that of the S/C at the same rate. When the rate was set back to 0.2 C, the discharge capacity of the S/N/O-PC was still stable, about 1017.9 mAh g^{-1}, which is significantly higher than that of the S/C cathode. More notably, when the rate exceeded 1 C, the second discharge voltage plateau corresponding to the reduction from Li_2S_{4-8} to the insoluble product Li_2S_2/Li_2S completely disappeared (Figure 6c), indicating poor conversion efficiency. In contrast, the S/N/O-PC cathode displayed an obvious second discharge voltage plateau even at a large rate of 5 C (Figure 6b), suggesting efficient catalytic performance toward conversion of Li_2S_{4-8} to the insoluble product Li_2S_2/Li_2S.

Additionally, compared with the S/C cathode, the S/N/O-PC cathode has a small voltage gap between the charge and discharge voltage plateaus as well as a low polarization. The N/O-PC composite has excellent electrical conductivity, which helps in efficient electron transport during the charge and discharge cycles, reducing the internal resistance of the battery and improving its rate capabilities. Thus, these LSBs can deliver and accept high currents, making them suitable for applications that require rapid energy release or recharge. Additionally, the performances of the S/N/O-PC cathodes can also compete with various reported sulfur composite cathodes in the literature (Table S1). Especially for high current discharge, the S/N/O-PC cathode still delivers high discharge capacity. This is due to the high catalytic activity of N/O-PC, which can accelerate the redox reaction kinetics of lithium polysulfides.

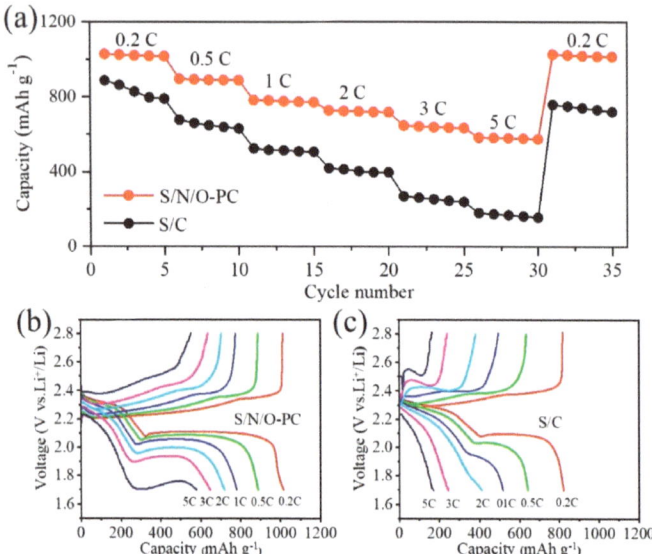

Figure 6. (**a**) The rate performance of S/N/O−PC. (**b**,**c**) Charge/discharge curves of the S/N/O−PC and S/C cathodes at various rates, respectively.

4. Conclusions

In summary, we successfully designed a novel structure that embeds O/N-doped porous carbon into a multiwalled carbon nanotubes network (N/O-PC). The structure possesses a three-dimensional conductive network, a high specific surface area, and abundant active sites. These characteristics enable effective adsorption of lithium polysulfides, thereby efficiently suppressing the "shuttle effect" of lithium polysulfides and improving the redox kinetics of the sulfur cathode. As a result, the S/N/O-PC showed a high discharge capacity of 685.9 mA h g^{-1} at 0.2 C after 100 cycles. Moreover, the S/N/O-PC showed excellent rate performance (discharge capacities of 1018.2, 890.2, 775.1, 722.7, 640.4, and 579.6 mAh g^{-1} at gradual increasing rates of 0.2, 0.5, 1.0, 2.0, 3.0, and 5.0 C). This unique structural design provides an effective approach for constructing high-performance lithium–sulfur batteries with high energy density.

Supplementary Materials: The following supporting information can be downloaded at: https://www.mdpi.com/article/10.3390/c10030067/s1.

Author Contributions: Conceptualization, X.S. and N.Y.; Methodology, X.S., H.W. and N.Y.; Software, X.S.; Validation, Y.Y. and Z.Y.; Formal analysis, X.S., H.W. and N.Y.; Investigation, X.S., Y.Y., Z.Y., H.W. and N.Y.; Resources, X.S., H.W. and N.Y.; Writing—original draft preparation, X.S., H.W. and N.Y.; Writing—review and editing, N.Y.; Visualization, X.S. and N.Y.; Project administration, X.S.; Funding acquisition, H.W. and N.Y. All authors have read and agreed to the published version of the manuscript.

Funding: This work was financially supported by the Natural Science Foundation of Xiamen City (grant number 3502Z20227321).

Data Availability Statement: The data supporting the findings of this study are available within the article.

Acknowledgments: The author acknowledges fruitful discussion with Z.Y. and H.Q. from Nanjing Tech University.

Conflicts of Interest: The authors declare no conflict of interest.

References

1. Zhang, Z.; Zhao, D.; Xu, Y.; Liu, S.; Xu, X.; Zhou, J.; Gao, F.; Tang, H.; Wang, Z.; Wu, Y.; et al. A review on electrode materials of fast-charging lithium-ion batteries. *Chem. Rec.* **2022**, *22*, e202200127. [CrossRef] [PubMed]
2. Ao, S.; Gouda, S.P.; Selvaraj, M.; Boddula, R.; Al-Qahtani, N.; Mohan, S.; Rokhum, S.L. Active sites engineered biomass-carbon as a catalyst for biodiesel production: Process optimization using RSM and life cycle assessment. *Energy Convers. Manag.* **2024**, *300*, 117956. [CrossRef]
3. Ren, J.; Wang, Z.; Xu, P.; Wang, C.; Gao, F.; Zhao, D.; Liu, S.; Yang, H.; Di, W.; Niu, C.; et al. Porous Co2VO4 nanodisk as a high-energy and fast-charging anode for lithium-ion batteries. *Nano Micro Lett.* **2021**, *14*, 5. [CrossRef] [PubMed]
4. Zhang, F.; Zhou, Y.; Zhang, Y.; Li, D.; Huang, Z. Facile synthesis of sulfur@titanium carbide Mxene as high performance cathode for lithium-sulfur batteries. *Nanophotonics* **2020**, *9*, 2025–2032. [CrossRef]
5. Baumann, A.E.; Han, X.; Butala, M.M.; Thoi, V.S. Lithium thiophosphate functionalized zirconium MOFs for Li-S batteries with enhanced rate capabilities. *J. Am. Chem. Soc.* **2019**, *141*, 17891–17899. [CrossRef] [PubMed]
6. Yin, C.; Li, Z.; Zhao, D.; Yang, J.; Zhang, Y.; Du, Y.; Wang, Y. Azo–branched covalent organic framework thin films as active separators for superior sodium–sulfur batteries. *ACS Nano* **2022**, *16*, 14178–14187. [CrossRef] [PubMed]
7. Zhao, D.; Ge-Zhang, S.; Zhang, Z.; Tang, H.; Xu, Y.; Gao, F.; Xu, X.; Liu, S.; Zhou, J.; Wang, Z.; et al. Three–dimensional honeycomb–like carbon as sulfur host for sodium–sulfur batteries without the shuttle effect. *ACS Appl. Mater. Interfaces* **2022**, *14*, 54662–54669. [CrossRef]
8. Zhao, Q.; Wang, R.; Wen, J.; Hu, X.; Li, Z.; Li, M.; Pan, F.; Xu, C. Separator engineering toward practical Li-S batteries: Targeted electrocatalytic sulfur conversion, lithium plating regulation, and thermal tolerance. *Nano Energy* **2022**, *95*, 106982. [CrossRef]
9. Zhou, C.; Li, Z.; Xu, X.; Mai, L. Metal-organic frameworks enable broad strategies for lithium-sulfur batteries. *Natl. Sci. Rev.* **2021**, *8*, nwab055. [CrossRef]
10. Cao, G.; Bi, D.; Zhao, J.; Zheng, J.; Wang, Z.; Lai, Q.; Liang, Y. Transformation of ZIF-8 nanoparticles into 3D nitrogen-doped hierarchically porous carbon for Li–S batteries. *RSC Adv.* **2020**, *10*, 17345–17352. [CrossRef]
11. Du, X.-L.; You, Y.; Yan, Y.; Zhang, D.; Cong, H.-P.; Qin, H.; Zhang, C.; Cao, F.-F.; Jiang, K.-C.; Wang, Y.; et al. Conductive carbon network inside a sulfur-impregnated carbon sponge: A bioinspired high-performance cathode for li–s battery. *ACS Appl. Mater. Interfaces* **2016**, *8*, 22261–22269. [CrossRef] [PubMed]
12. Yoshie, Y.; Hori, K.; Mae, T.; Noda, S. High-energy-density Li-S battery with positive electrode of lithium polysulfides held by carbon nanotube sponge. *Carbon* **2021**, *182*, 32–41. [CrossRef]
13. Gueon, D.; Yoon, J.; Hwang, J.T.; Moon, J.H. Microdomain sulfur-impregnated CeO2-coated CNT particles for high-performance Li-S batteries. *Chem. Eng. J.* **2020**, *390*, 124548. [CrossRef]
14. Li, H.; Shao, F.; Wen, X.; Ding, Y.; Zhou, C.; Zhang, Y.; Wei, H.; Hu, N. Graphene/Mxene fibers-enveloped sulfur cathodes for high-performance Li-S batteries. *Electrochim. Acta* **2021**, *371*, 137838. [CrossRef]
15. Hong, X.; Liang, J.; Tang, X.; Yang, H.; Li, F. Hybrid graphene album with polysulfides adsorption layer for Li-S batteries. *Chem. Eng. Sci.* **2019**, *194*, 148–155. [CrossRef]
16. Qiu, S.; Zhang, J.; Liang, X.; Li, Y.; Cui, J.; Chen, M. Tunable MOFs derivatives for stable and fast sulfur electrodes in Li-S batteries. *Chem. Eng. J.* **2022**, *450*, 138287. [CrossRef]
17. Chen, Y.; Zhang, L.; Pan, H.; Zhang, J.; Xiang, S.; Cheng, Z.; Zhang, Z. Pore-space-partitioned MOF separator promotes high-sulfur-loading Li-S batteries with intensified rate capability and cycling life. *J. Mater. Chem. A* **2021**, *47*, 26929–26938. [CrossRef]
18. Chenxi, Y.; Liping, W.; Bo, W.; Hulin, T.; Xinyue, L.; Jian, C.; Mengyao, G.; Naiqiang, L. NiCoSe4@CNFs derived from MOF compounds enabling robust polysulfide adsorption and catalysis in Li-S batteries. *J. Alloys Compd.* **2023**, *957*, 170449.
19. Xiaohua, C.; Mi, Z.; Jin, Z.; Juan, W.; Zengbao, J.; Yong, L. Boosting electrochemical performance of Li-S batteries by cerium-based MOFs coated with polypyrrole. *J. Alloys Compd.* **2022**, *901*, 163649.
20. Zhang, H.; Zhao, W.; Zou, M.; Wang, Y.; Chen, Y.; Xu, L.; Wu, H.; Cao, A. 3D, mutually embedded MOF@Carbon nanotube hybrid networks for high—performance lithium—sulfur batteries. *Adv. Energy Mater.* **2018**, *819*, 1614–6832. [CrossRef]
21. Bi, T.; Zheng, W.; Zhou, Y.; Lin, Q.; Zhang, X. Three-dimensional N, S co-doped ultra-thin-walled hierarchical porous carbon as sulfur-loading host for high-performance lithiumsulfur battery. *J. Energy Storage* **2024**, *93*, 112428. [CrossRef]
22. Haina, C.; Jingyu, S.; Menglei, W.; Yan, H.; Zixiong, S. A review in rational design of graphene toward advanced Li-S batteries. *Nano Res. Energy* **2023**, *2*, E9120054.
23. Ye, H.; Li, Y. Towards practical lean-electrolyte Li-S batteries: Highly solvating electrolytes or sparingly solvating electrolytes? *Nano Res. Energy* **2022**, *1*, e9120012. [CrossRef]
24. Liu, H.; Liu, F.; Qu, Z.; Chen, J.; Liu, H.; Tan, Y.; Guo, J.; Yan, Y.; Zhao, S.; Zhao, X.; et al. High sulfur loading and shuttle inhibition of advanced sulfur cathode enabled by graphene network skin and N, P, F-doped mesoporous carbon interfaces for ultra-stable lithium sulfur battery. *Nano Res. Energy* **2023**, *2*, E912049. [CrossRef]
25. Xue, H.; Gong, H.; Yamauchi, Y.; Sasaki, T.; Ma, R. Photo-enhanced rechargeable high-energy-density metal batteries for solar energy conversion and storage. *Nano Res. Energy* **2022**, *1*, E9120007. [CrossRef]
26. Wu, Z.; Wang, L.; Chen, S.; Zhu, X.; Deng, Q.; Wang, J.; Zeng, Z.; Deng, S. Facile and low-temperature strategy to prepare hollow ZIF-8/CNT polyhedrons as high-performance lithium-sulfur cathodes. *Chem. Eng. J.* **2020**, *404*, 126579. [CrossRef]

27. Jeon, Y.; Lee, J.; Jo, H.; Hong, H.; Lee, L.Y.S.; Piao, Y. Co/Co3O4-embedded n-doped hollow carbon composite derived from a bimetallic MOF/ZnO core-shell template as a sulfur host for li-s batteries. *Chem. Eng. J.* **2020**, *407*, 126967. [CrossRef]
28. Fan, B.; Zhao, D.; Zhou, W.; Xu, W.; Liang, X.; He, G.; Wu, Z.; Li, L. Nitrogen—doped hollow carbon polyhedrons with carbon nanotubes surface layers as effective sulfur hosts for high—rate, long—lifespan lithium–sulfur batteries. *ChemElectroChem* **2020**, *724*, 4990–4998. [CrossRef]
29. Chen, P.; Wang, T.; Tang, F.; Chen, G.; Wang, C. Elaborate interface design of CoS2/Fe7S8/NG heterojunctions modified on a polypropylene separator for efficient lithium-sulfur batteries. *Chem. Eng. J.* **2022**, *446*, 136990. [CrossRef]
30. Liu, G.; Zeng, Q.; Fan, Z.; Tian, S.; Li, X.; Lv, X.; Zhang, W.; Tao, K.; Xie, E.; Zhang, Z. Boosting sulfur catalytic kinetics by defect engineering of vanadium disulfide for high-performance lithium-sulfur batteries. *Chem. Eng. J.* **2022**, *448*, 137683. [CrossRef]
31. Feng, H.; Zhang, M.; Kang, J.; Su, Q.; Du, G.; Xu, B. Nitrogen and oxygen dual-doped porous carbon derived from natural ficus microcarpas as host for high performance lithium-sulfur batteries. *Mater. Res. Bull.* **2019**, *113*, 70–76. [CrossRef]
32. Yang, D.; Li, M.; Zheng, X.; Han, X.; Zhang, C.; Biendicho, J.J.; Llorca, J.; Wang, J.; Hao, H.; Li, J.; et al. Phase engineering of defective copper selenide toward robust lithium-sulfur batteries. *ACS Nano* **2022**, *16*, 11102–11114. [CrossRef] [PubMed]
33. Yu, H.; Zhang, B.; Sun, F.; Jiang, G.; Zheng, N.; Xu, C.; Li, Y. Core-shell polyhedrons of carbon nanotubes-grafted graphitic carbon@nitrogen doped carbon as efficient sulfur immobilizers for lithium-sulfur batteries. *Appl. Surf. Sci.* **2018**, *450*, 364–371. [CrossRef]
34. Geng, M.; Yang, H.; Shang, C. The multi-functional effects of CuS as modifier to fabricate efficient interlayer for Li-S batteries. *Adv. Sci.* **2022**, *9*, 202204561. [CrossRef] [PubMed]
35. Yang, X.-X.; Li, X.-T.; Zhao, C.-F.; Fu, Z.-H.; Zhang, Q.-S.; Hu, C. Promoted deposition of three-dimensional Li2S on catalytic Co phthalocyanine nanorods for stable high-loading lithium-sulfur batteries. *ACS Appl. Mater. Interfaces* **2020**, *1229*, 32752–32763. [CrossRef] [PubMed]
36. Yin, Y.; Xin, S.; Guo, Y.; Wan, L. Lithium-sulfur batteries: Electrochemistry, materials, and prospects. *Angew. Chem. Int. Ed.* **2013**, *52*, 13186–13200. [CrossRef] [PubMed]
37. Uppugalla, S.; Pothu, R.; Boddula, R.; Desai, M.A.; Al-Qahtani, N. Nitrogen and sulfur co-doped activated carbon nanosheets for high-performance coin cell supercapacitor device with outstanding cycle stability. *Emergent Mater.* **2023**, *6*, 1167–1176. [CrossRef]
38. Li, H.; Sun, L.; Zhao, Y.; Tan, T.; Zhang, Y. A novel CuS/graphene-coated separator for suppressing the shuttle effect of lithium/sulfur batteries. *Appl. Surf. Sci.* **2019**, *466*, 309–319. [CrossRef]
39. Ma, F.; Srinivas, K.; Zhang, X.J.; Zhang, Z.H.; Wu, Y.; Liu, D.W.; Zhang, W.L.; Wu, Q.; Chen, Y.F. Mo2N quantum dots decorated n-doped graphene nanosheets as dual-functional interlayer for dendrite-free and shuttle-free lithium-sulfur batteries. *Adv. Funct. Mater.* **2022**, *32*, 2206113. [CrossRef]
40. Chen, X.; Huang, Y.; Li, J.; Wang, X.; Zhang, Y.; Guo, Y.; Ding, J.; Wang, L. Bifunctional separator with sandwich structure for high-performance lithium-sulfur batteries. *J. Colloid Interface Sci.* **2020**, *559*, 13–20. [CrossRef]
41. Fang, R.; Zhao, S.; Sun, Z.; Wang, D.; Cheng, H.; Li, F. More reliable lithium-sulfur batteries: Status, solutions and prospects. *Adv. Mater.* **2017**, *29*, 1606823. [CrossRef]
42. Zou, K.; Zhou, T.; Chen, Y.; Xiong, X.; Jing, W.; Dai, X.; Shi, M.; Li, N.; Sun, J.; Zhang, S.; et al. Defect engineering in a multiple confined geometry for robust lithium-sulfur batteries. *Adv. Energy Mater.* **2022**, *12*, 2103981. [CrossRef]
43. Wang, W.; Xi, K.; Li, B.; Li, H.; Liu, S.; Wang, J.; Zhao, H.; Li, H.; Abdelkader, A.M.; Gao, X.; et al. A Sustainable Multipurpose Separator Directed Against the Shuttle Effect of Polysulfides for High-Performance Lithium-Sulfur Batteries. *Adv. Energy Mater.* **2022**, *12*, 2200160. [CrossRef]
44. Wang, M.; Bai, Z.; Yang, T.; Nie, C.; Xu, X.; Wang, Y.; Yang, J.; Dou, S.; Wang, N. Advances in high sulfur loading cathodes for practical lithium-sulfur batteries. *Adv. Energy Mater.* **2022**, *12*, 2201585. [CrossRef]
45. Yang, M.; Liu, P.; Qu, Z.; Sun, F.; Tian, Y.; Ye, X.; Wang, X.; Liu, X.; Li, H. Nitrogen-vacancy-regulated Mo2N quantum dots electrocatalyst enables fast polysulfides redox for high-energy-density lithium-sulfur batteries. *Nano Energy* **2022**, *104*, 107922. [CrossRef]
46. Chen, S.; Li, D.; Liu, Y.; Huang, W. Morphology-dependent defect structures and photocatalytic performance of hydrogenated anatase TiO2 nanocrystals. *J. Catal.* **2016**, *341*, 126–135. [CrossRef]
47. Fan, F.Y.; Chiang, Y.-M. Electrodeposition kinetics in Li-S batteries: Effects of low electrolyte/sulfur ratios and deposition surface composition. *J. Electrochem. Soc.* **2017**, *164*, A917–A922. [CrossRef]
48. Yang, M.; Wang, X.; Wu, J.; Tian, Y.; Huang, X.; Liu, P.; Li, X.; Li, X.; Liu, X.; Li, H. Dual electrocatalytic heterostructures for efficient immobilization and conversion of polysulfides in li-s batteries. *J. Mater. Chem. A* **2021**, *9*, 18477–18487. [CrossRef]

Disclaimer/Publisher's Note: The statements, opinions and data contained in all publications are solely those of the individual author(s) and contributor(s) and not of MDPI and/or the editor(s). MDPI and/or the editor(s) disclaim responsibility for any injury to people or property resulting from any ideas, methods, instructions or products referred to in the content.

Article

FeS$_2$ Nanoparticles in S-Doped Carbon: Ageing Effects on Performance as a Supercapacitor Electrode

Sirine Zallouz [1,2], Bénédicte Réty [1,2,3], Jean-Marc Le Meins [1,2], Mame Youssou Ndiaye [1,2], Philippe Fioux [1,2] and Camélia Matei Ghimbeu [1,2,3,*]

[1] CNRS, Institut de Science des Matériaux de Mulhouse (IS2M) UMR 7361, Université de Haute-Alsace, F-68100 Mulhouse, France; benedicte.rety@uha.fr (B.R.); jean-marc.le-meins@uha.fr (J.-M.L.M.); mameyoussoun@gmail.com (M.Y.N.); philippe.fioux@uha.fr (P.F.)

[2] CNRS, Institut de Science des Matériaux de Mulhouse (IS2M) UMR 7361, Université de Strasbourg, F-67081 Strasbourg, France

[3] Réseau sur le Stockage Electrochimique de l'Energie (RS2E), FR CNRS 3459, F-80039 Amiens, France

* Correspondence: camelia.ghimbeu@uha.fr; Tel.: +33-(0)-3-89-60-87-43

Abstract: Although transition metal sulfides have prodigious potential for use as electrode materials because of their low electronegativities, their large volume changes inhibit broad application. Moreover, there is only limited knowledge of the ageing processes of these materials at the nanoscale. Herein, nano-C/FeS$_2$ materials were prepared via one-pot syntheses from green biodegradable carbon precursors, followed by activation and sulfidation. The increased activation/sulfidation time led to an increase in the size of the nanoparticles (7 to 17 nm) and their aggregation, as well as in an increase in the specific surface area. The materials were then used as electrodes in 2-electrode symmetric supercapacitors with 2 M KOH. The activation process resulted in improved capacitance (60 F g^{-1} at 0.1 A g^{-1}) and rate capability (36%) depending on the composite porosity, conductivity, and size of the FeS$_2$ particles. The ageing of the FeS$_2$ nanoparticles was investigated under air, and a progressive transformation of the nano-FeS$_2$ into hydrated iron hydroxy sulfate with a significant morphological modification was observed, resulting in drastic decreases in the capacitance (70%) and retention. In contrast, the ageing of nano-FeS$_2$ during cycling led to the formation of a supplementary iron oxyhydroxide phase, which contributed to the enhanced capacitance (57%) and long-term cycling (132% up to 10,000 cycles) of the device.

Keywords: C/pyrite composite; CO$_2$ activation; mesoporous carbon; energy storage; ageing; sulfidation

1. Introduction

Carbon has been extensively studied as an electrode material for supercapacitors. However, it has limitations due to the nonuniformly distributed pores that hamper the accessibility of ions to all active sites [1]. Other materials, such as transition metal oxides, are constantly proposed due to their high theoretical capacities. However, this family suffers from modest electronic conductivities [2]. Recently, other groups, such as transition metal sulfides (TMSs), have received attention. TMSs are generally semiconductors with electrical conductivity. The reason for this is that the sulfur atom has a lower electronegativity than oxygen, which facilitates electron transfer and therefore increases the capacitance [3]. Among the different TMS materials, pyrite FeS$_2$ appears to be an interesting choice due to its natural abundance, low toxicity, and different active sites. It has been used in various energy storage systems, such as lithium-ion batteries [4], sodium-ion batteries [5], and supercapacitors [6]. However, it might exhibit large volume changes during cycling that can cause electrode degradation during long-term use [7].

The synthesis methodology has a substantial impact on the structure, morphology, and characteristics of the material. Therefore, the formation of particles with tuned properties and the syntheses of nanostructures are often seen as ways to obtain satisfactory electrochemical performance closer to the theoretical level [8]. For example, Venkateshalu et al. [9] synthesized octahedral-shaped FeS_2 particles to sizes of ~200 nm via a solvothermal method. The as-prepared material was tested in a three-electrode system in both 3.5 M KOH and 1 M Na_2SO_4 with low mass loadings on the working electrode of 1.9 mg and 1.7 mg, respectively. The material exhibited capacitances of 269 F g^{-1} at 1 A g^{-1} and 120 F g^{-1} in alkaline and neutral electrolytes, respectively [9].

To increase the electrochemical performance, the mixing of carbon materials with TMS is an encouraging strategy because it enhances the electrical conductivity of the TMS [10]. In the literature, the most commonly used synthesis method is preparation of the carbon material, generally graphene oxide (GO) or reduced graphene oxide (rGO), followed by incorporation of the metal sulfide precursors (iron salts for Fe and different sources for sulfur, e.g., sodium thiosulfate [7], ammonium thiocyanate [11], cysteine [12], sulfur powder [5], or sulfidation under H_2S [13]) to synthesize the carbon/FeS_2 composite used for electrodes in energy storage devices [14]. This synthesis route is acceptable but requires multiple steps, which increases the overall cost. Pei et al. synthesized flower-like FeS_2 on a graphene aerogel using a two-step self-assembly method. After preparing GO using the Hummer method, iron sulfate ($FeSO_4$) and sodium thiosulfate ($Na_2S_2O_3$) were added to the GO solution, and graphene aerogel/FeS_2 composites were prepared via hydrothermal synthesis at 180 °C for 16 h. The FeS_2 particles had sizes ranging from 250 nm to 2 µm, and most of them were severely agglomerated. The material exhibited a capacitance of 313.6 F g^{-1} at 0.5 A g^{-1} in a three-electrode cell containing 6 M KOH [7]. In another study, FeS_2/carbon microspheres were synthesized by first preparing Fe_3O_4/carbon microspheres that were later mixed with sulfur powder and annealed at 500 °C under argon for 5 h to obtain the FeS_2/carbon microspheres. The material showed a capacitance of 278 F g^{-1} at 1 A g^{-1} with 1 M KOH in a three-electrode system [13].

In fact, it has been shown before that the confinement of particles in carbon inhibits particle growth, thus maintaining small particles with limited aggregation throughout thermal treatments or electrochemical cycling [15,16]. Furthermore, the carbon matrix is also a useful scaffold to improve electronic conductivity and porosity [17]. In a recent report, graphene-wrapped FeS_2-$FeSe_2$ core-shell spheres with sizes of ~1 µm were synthesized and used as the negative electrode in an asymmetric supercapacitor. The device exhibited a capacitance of 352.3 F g^{-1} at 1 A g^{-1}. This was attributed to the protection of the particles through encapsulation in carbon, which reduced the dissolution of the metal ions during cycling and thus preserved the cycling lifetime of the device [18].

This state-of-the-art overview shows that synthesized FeS_2 has large particle sizes, and in some reports, the particle sizes are not mentioned at all. No systematic study has been conducted on nanoparticles with sizes smaller than 20 nm confined in carbon. Consequently, new synthesis approaches are required to ensure that the small particles in the carbon matrix are distributed homogeneously [17]. Moreover, despite its high potential as an electrode material, there are fewer reports on FeS_2 than there are on other TMSs and TMOs; therefore, it should be thoroughly investigated. Furthermore, it should also be mentioned that most previous reports [12,19] on TMSs determined the capacitances in half-cell or three-electrode cell configurations, which measured the electrochemical performance of a single electrode in a specific electrolyte. This configuration gives capacitance readings that are higher than those of two-electrode devices and is not representative of a real two-electrode working system. The use of a symmetric two-electrode configuration for transition metal sulfides in general, and FeS_2 in particular, has been poorly reported until now, as noted in a recent review article [10]. To our knowledge, the only example was provided in a very recent report of a symmetric solid supercapacitor composed of FeS_2/carbon nanofibers and PVA-KOH gel electrolyte, which exhibited a specific capacitance of 203.4 F g^{-1} at 1 A g^{-1} [5].

Another important aspect that needs to be mentioned is the ageing of the nanoparticles. To our knowledge, pyrite nanoparticles have not been reported, although pyrite in the nanosize is more prone to oxidation than their oxide counterparts [20]. This ageing effect was not studied with supercapacitor electrodes, and there is a lack of knowledge on the structural and morphological evolution during cycling. The ageing of carbonaceous materials during cycling led to modifications in the porosity, structure, and surface chemistry, in particular for the positive electrode [21]. This caused a decrease in the capacitance of the device. These aspects of ageing are rarely studied, and the new interest in these issues shows their importance. Investigating the ageing processes of nanoparticle electrodes could prevent later problems that may occur during the life of a supercapacitor.

It is clear from the above literature that there is a lack of studies on FeS_2 at the nanoscale. Moreover, the electrochemical tests are limited to 3-electrode cells, although 2-electrode system tests should be reported for better comparison with different material families. Nevertheless, ageing phenomena are crucial for electrochemical behaviour and device life, therefore, their understanding is of prime importance, and has not been investigated before. All of these problems mentioned above motivate this work to provide a comprehensive systematic study on FeS_2 nanoparticles as supercapacitor electrodes.

Herein, a novel synthesis route of carbon-containing nanoparticles of FeS_2 starting from green biosourced precursors, followed by pyrolysis, sulfidation, and activation is reported. Cysteine containing sulfur and nitrogen is used for a double purpose, i.e., as a sulfur source for FeS_2 formation and to dope carbon with these elements. The optimization of the sulfidation temperature and activation time were investigated to obtain materials with tuned particle size and porosity. The capacitance, rate capability, and cycling life could be improved depending on the material properties (particle size, porosity, and conductivity). The ageing of the materials was studied in air and during electrochemical cycling and evidenced progressive modifications in the morphology and structure. Ageing in air is detrimental to performance, while ageing in an electrolyte promotes redox reactions and improves performance. The results described herein provide a comprehensive understanding of the phenomena occurring at the interfaces of the electrode with both the air and the electrolyte, which have never been addressed to the best of our knowledge.

2. Materials and Methods

2.1. Synthesis of C/FeS$_2$ Composites

The C/FeS$_2$ composites were synthesized via a new one-pot synthetic method, as shown in Figure 1. The green carbon precursors of phloroglucinol (1.23 g), glyoxylic acid (1.2 g), and L-cysteine (1.76 g) were used as the sulfur precursors and Pluronic F127 (1.21 g) was used as the pore/structure-directing agent, and all were mixed together in 60 mL of a solvent mixture (with a water:ethanol volume ratio of 75:25) in a Teflon beaker. The reaction mixture was stirred until all precursors were completely dissolved (approximately 20 min). Then, 4.88 g of iron(III) nitrate nonahydrate ($Fe(NO_3)_3 \cdot 9H_2O$) was added to the reaction mixture as the iron source. The solution was kept in the fume hood and continuously stirred until the solvent was completely evaporated (generally 48 h). The dry polymer/iron salt mixture was then placed in an alumina tube and pyrolyzed under flowing argon (8 L h^{-1}) at 750 °C for 1 h with a heating rate of 2 °C min^{-1}. The products were kept in the furnace until the temperature dropped below 45 °C.

This synthesis afforded the carbon/iron sulfide (FeS) composites. Two approaches have been considered to convert the FeS into FeS$_2$. In the first route (1), the as-prepared C/FeS composites were directly subjected to sulfidation at different temperatures (250–600 °C) under a mixture of H$_2$S (2 L h^{-1}) and argon (8 L h^{-1}) for 3 h. The obtained composite was denoted as C/FeS$_2$ NA (NA means nonactivated).

In the second route (2), physical activation was performed before the sulfidation step. CO$_2$ was used in an activation step carried out at 750 °C for different times (1–4 h), which was then followed by the same sulfidation step used in route 1. The sulfidation temperature was 400 °C, and the time of exposure to H$_2$S was 3 h for most materials but was adjusted to 6 h for the material activated for 4 h. The resulting composites were noted as C/FeS$_2$ A t with t being the time of activation in hours (h). It should be highlighted that during activation under CO$_2$ and sulfidation under H$_2$S, the heating and cooling steps were carried out under an inert atmosphere. The specific gases were introduced only at the indicated temperatures.

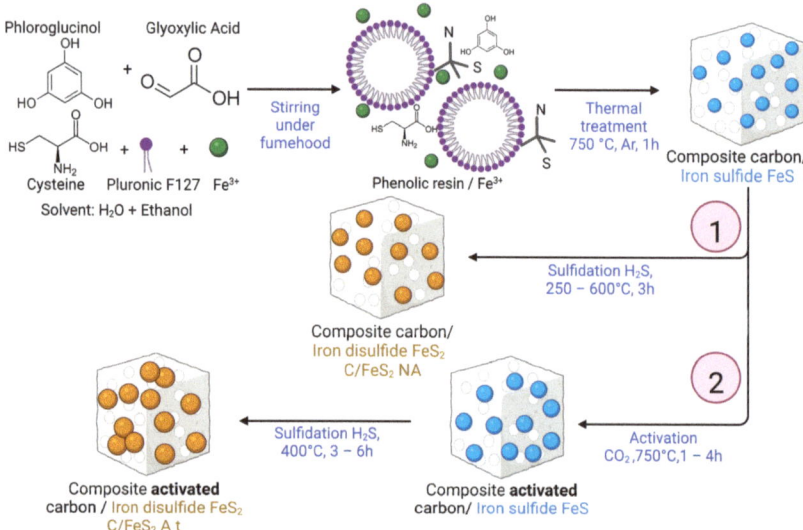

Figure 1. Synthesis scheme for the C/FeS$_2$ composites; (1) nonactivated carbon/iron disulfide composites (C/FeS$_2$ NA) and (2) activated carbon/iron disulfide composites (C/FeS$_2$ A t, where A t means activation time). The scheme was created by the authors using BioRender software.

For comparison purposes, unsupported FeS$_2$ (named nano-FeS$_2$) was prepared from commercially nanosized Fe$_2$O$_3$ (Sigma Aldrich, Saint-Louis, MO, USA) according to the procedure used for the C/FeS$_2$ composites (400 °C for 3 h under a mixture of H$_2$S/Ar).

2.2. Physicochemical Characterisation

The nanocomposites were characterised via powder X-ray diffraction (XRD) using a Bruker, D8 ADVANCE A25 instrument (Bruker, Billerica, MA, USA) in θ-θ configuration. The diffractometer is equipped with a LynxEye XE-T high-resolution energy dispersive 1-D detector and operates at <380 eV with Cu K$\alpha_{1,2}$. Data were collected over the 2θ range of 10 to 90°.

A semiquantitative study of composite ageing was performed for a period of ~4 months, and diffractograms were recorded each week. The as-collected diffraction data allowed us to evaluate the weight percentage (concentration) of a given phase with a semiquantitative analysis using peak heights and a reference intensity ratio (RIR). The semiquantitative (SQ) analysis was conducted as described in the Supporting Materials, SI. The use of a Crystallography Data Base (COD) is preferred in this study and the reasons are explained in the SI part.

Thermogravimetric analyses were carried out using a Mettler Toledo instrument under a gas flow of 100 mL min^{-1} and by heating in an alumina (Al$_2$O$_3$) crucible from 25 to 900 °C at a rate of 2 °C min^{-1}. The purpose of this study was to estimate the amounts of FeS$_2$ in the composites, as described in the SI.

The surface morphologies and microstructures of the samples were investigated using transmission electron microscopy (TEM) and scanning transmission electron microscopy (STEM) using an ARM-200F instrument (Jeol, Tokyo, Japan). ImageJ [22] software was used to treat the STEM images and to determine the particle size distributions by counting ~500 particles from several images obtained at different magnifications. The chemical composition, nature, and amounts of metal and functional groups were determined using an XPS SES-2002 spectrometer (Scienta Omicron, Uppsala, Sweden) equipped with a monochromatic X-ray source (Al Kα = 1486.6 eV) and a VG Scienta XM780 monochromator.

The textural properties were determined using a ASAP 2020 porosimetry system (Micromeritics Instrument Corporation, Norcross, GA, USA) with N_2 as the adsorbate at 77 K. Prior to the adsorption measurements, the samples (~100 mg) were placed into a glass tube and outgassed under a vacuum at 90 °C for 12 h. Furthermore, a second outgassing was performed at 90 °C under a vacuum for 2 h. The specific surface areas (S_{BET}), total pore volumes (V_T), micropore volumes (V_{micro}), and mesopore volumes (V_{meso}) were determined as described elsewhere [15]. The pore size distributions were determined from the adsorption isotherms and the 2D-NLDFT (two-dimensional nonlocal density functional theory) heterogeneous surface for carbon materials explored using SAIEUS software (Version 3.2).

2.3. Electrochemical Studies

The electrodes were prepared by mixing an 80 wt% of the active material (C/FeS$_2$-NA and C/FeS$_2$-A t and nano-FeS$_2$), with a 10 wt% of a conductive additive (ENSACO 350G, from IMERYS, Paris, France) and a 10 wt% of a binder, which was polytetrafluoroethylene PTFE (60% wt dispersion in water), dissolved in ethanol. The mixture was blended manually in an agate mortar until the solvent evaporated and a homogeneous paste was formed. The manual grinding enables a better homogeneity of the electrode. The paste was then spread between two plastic foil sheets to achieve a paste thickness of 200–250 µm. Then, self-standing electrodes with 10 mm diameters and 10–17 mg loadings were cut using a round punch and then dried under a vacuum at 120 °C for 12 h. The materials were tested in 2-electrode Swagelok cells with a symmetric configuration. All measurements were carried out using a multichannel VSP300 Potentiostat (Biologic, France). Cyclic voltammetry (CV), galvanostatic charge with potential limitation (GCPL), and electrochemical impedance spectroscopy (EIS) were conducted using a 2-electrode cell assembly containing 2 M KOH as the electrolyte, qualitative filter paper as the separator, and stainless-steel current collectors. The current collectors were covered with a layer of conductive glue (Loctite ElectroDag PF 407C) to ensure better contact and conductivity with the electrode. To characterise the materials further and observe the redox peaks, a 3-electrode cell was built. In this case, Hg/HgO was used as the reference electrode, a Pt mesh was used as the counter electrode, and the same electrolyte and current collector were used. The specific capacitance was calculated from the GCPL of the 2-electrode measurement by using the mass of the active material in one electrode, according to Equation (1) [23]:

$$C_{electrode} = \frac{2 \times I \times S_{dis}}{U_{max}^2} \qquad (1)$$

where $C_{electrode}$ is the specific capacitance of the active material in the working electrode in F g^{-1}, I is the applied current density of the active material in the working electrode in A g^{-1} reported to the mass of active material in the electrode, S_{dis} is the area under the discharge curve in V·s, and U_{max} is the electrochemical window in V. The energy density E and power density P were determined following these equations [21]:

$$E = \frac{1}{2} C_{electrode} U^2 \qquad (2)$$

$$P = \frac{E}{t_{dis}} \qquad (3)$$

where E is the energy density in W h kg^{-1}, t_{dis} is the discharge time in s, and P is the power density in W kg^{-1}.

Cycled electrodes were washed in 50 mL of distilled water and then dried at 80 °C for 8 h before postmortem characterisation with XRD and TEM.

3. Results

3.1. Carbon/FeS$_2$ Formation

The first step of the synthesis involved a mixture of the carbon precursors of phloroglucinol-glyoxylic acid and cysteine (the source of S), the Pluronic F127 surfactant, and iron nitrate in a water/ethanol mixture. A phenolic resin based on the phloroglucinol-glyoxylic acid-cysteine was formed through polymerization [24]. The abundant –OH and –COOH groups present in the phenolic resin allowed self-assembly with the ethylene oxide units of surfactant micelles through H-bonding [25]. Depending on the interactions involved, the iron nitrate can be located either in the vicinity of the micelles and/or within the phenolic resin structure [26]. The continuous mixing of the precursors until the solvent evaporated ensured the complete integration of the iron precursor into the polymer matrix. The obtained nanoassembly was pyrolyzed under Ar, leading to the decomposition of the phenolic resin to form the carbon network. Simultaneously, the surfactant micelles resulted in the creation of mesopores through a soft-template synthesis method [25]. The thermal degradation of cysteine had a double function in the synthesis: first, it ensured the doping of the carbon with S and N, and second, it provided sulfur as a source for FeS$_2$ formation. The iron(III) nitrate decomposed with increasing temperature, and it formed Fe$_2$O$_3$ at approximately 180 °C [27]. Then, the temperature was increased (up to 750 °C), and the iron oxide present in the material was sulfidated by the sulfur from the cysteine through the following reaction:

$$2\ Fe_2O_3 + 4\ S\ (s) \rightarrow 4\ FeS + 3\ O_2\ (g) \qquad (4)$$

During this step, the carbon reduced the iron oxide, thus simplifying the process of FeS formation, as reported for the WS$_2$ formation [28].

XRD data for the material was obtained after the pyrolysis step, denoted C/FeS ref, as shown in Figure 2. Large distinguishable reflections were observed, and identification of the phases by XRD using COD indicated a mixture of different FeS phases with two hexagonal crystallographic forms according to COD 1504400 (centrosymmetric NiAs-type structure, with space Group P6$_3$/mmc) and COD 9004036 for Troilite-type FeS [29], which was not centrosymmetric (space group P$\bar{6}$ 2c). Fe$_7$S$_8$ pyrrhotite 4C, COD 2104739, which contains a superstructure related to the NiAs-type, was required to describe the distribution of iron vacancies within the unit cell [30].

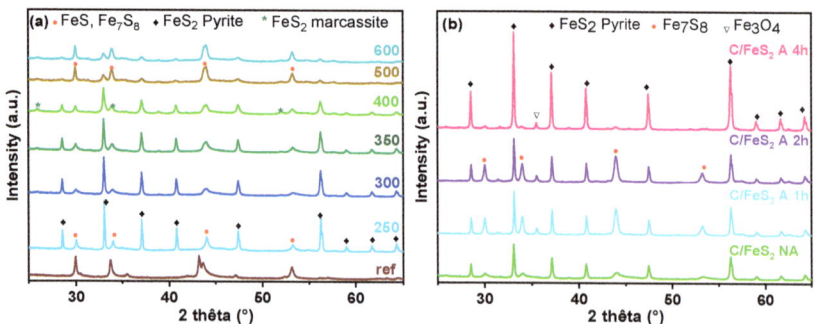

Figure 2. (a) XRD diffractograms for the C/FeS composites obtained via sulfidation at different temperatures (250–600 °C) and (b) XRD data obtained after activation and sulfidation.

In a second step, different thermal treatments were performed with a mixture of H$_2$S/Ar to determine the optimal temperature for the formation of pyrite. Figure 2a,b shows the XRD data obtained at 250 to 600 °C for the different materials. For comparison purposes, the C/FeS ref composite pyrolyzed at 750 °C (without H$_2$S treatment) is provided in the same figure.

After sulfidation at the lowest temperature (250 °C), a significant fraction of FeS had already been transformed to FeS$_2$ via Equation (5) [31]:

$$2\,Fe_2O_3 + 4\,S\,(s) \rightarrow 4\,FeS + 3\,O_2\,(g) \tag{5}$$

This is shown through the decreased intensities of the FeS reflections and the appearance of several additional intense peaks (2θ = 28.5, 33.0, 37.0, 40.7, 47.4, and 56.2°). These newly formed peaks were indexed as two coexisting polymorphs of FeS$_2$: pyrite (COD 9000594) with cubic symmetry (Pa$\bar{3}$) and marcasite (COD 9013067), which is orthorhombic (Pnnm). Marcasite is metastable with respect to pyrite from 270 to 430 °C, and above 430 °C, marcasite undergoes an irreversible exothermic transformation compared to pyrite [32]. In contrast, pyrite is stable at higher temperatures and is thought to be the predominant phase.

Therefore, increasing the reaction temperature to 400 °C increased the proportion of the pyrite FeS$_2$ phase and eliminated the FeS phase. The other peaks associated with the undesirable phase (FeS$_2$ marcasite, FeS troilite, and Fe$_7$S$_8$ pyrrhotite 4C) also weakened. At this temperature, FeS$_2$ pyrite was the predominant phase in the material, and only small peaks for the FeS$_2$ marcasite were detected. Above 500 °C, decomposition of the pyrite with the reformation of FeS was observed. Therefore, 400 °C was chosen as the optimal sulfidation temperature for the rest of the composites.

Physical activation is known to increase the porosities of carbon materials, which is important for supercapacitor performance [33]. However, most of the studies were performed with carbon materials, and composites were rarely reported, as was the case for the carbon/FeS composites. Therefore, the C/FeS material was activated under CO$_2$ for 1 h, 2 h, and 4 h prior to the sulfidation treatment. During this process, some of the carbon was burnt off due to the exposure to an oxidizing environment (CO$_2$).

Furthermore, it was shown that at temperatures between 750 and 800 °C, the oxidation reaction was slow [33]; therefore, 750 °C was chosen for this study. The activation process took place according to the reverse Boudouard equation [34]:

$$C_{(s)} + CO_{2\,(g)} \rightarrow 2\,CO_{(g)} \tag{6}$$

The resulting activated materials were sulfided at 400 °C to obtain the C/FeS$_2$ composites. Then, the composites were characterised in-depth using several techniques. The crystallographic structures of the activated C/FeS$_2$ materials were identified by XRD and compared to that of the nonactivated material (C/FeS$_2$ NA) (Figure 2b). Overall, pyrite FeS$_2$ was the predominant phase in the three activated materials, as it was in the nonactivated material. A tendency toward narrower peaks was observed with increasing activation time, indicating an increase in the size of the coherent diffraction domain during the thermal treatments (activation and sulfidation).

Small traces of Fe$_7$S$_8$ were still present with more or less distinguishable peaks observed for all materials. Small amounts of iron oxide, Fe$_3$O$_4$ (COD 1513304), were also detected in all of the activated C/FeS$_2$ samples due to the activation with CO$_2$ (Figure S1, SI). The amounts increased with increasing activation times and reached the highest level at 4 h of activation, which is why the reductions with H$_2$S/Ar required 6 h instead of 3 h for this material. To conclude, after 4 h of activation followed by 6 h of sulfidation, the diffractogram indicated the highest pyrite FeS$_2$ content among all materials.

The STEM images for the C/FeS$_2$ composites are provided in Figure 3. The C/FeS$_2$ materials presented an average size of 7.0 nm for C/FeS$_2$ NA, which increased to 12.5 nm for C/FeS$_2$ A 1 h, 14.5 nm for C/FeS$_2$ A 2 h, and 17.1 nm for C/FeS$_2$ A 4 h. Therefore, as the activation time was increased, the particles became larger and tended to coalesce, as shown in Figure S2, SI.

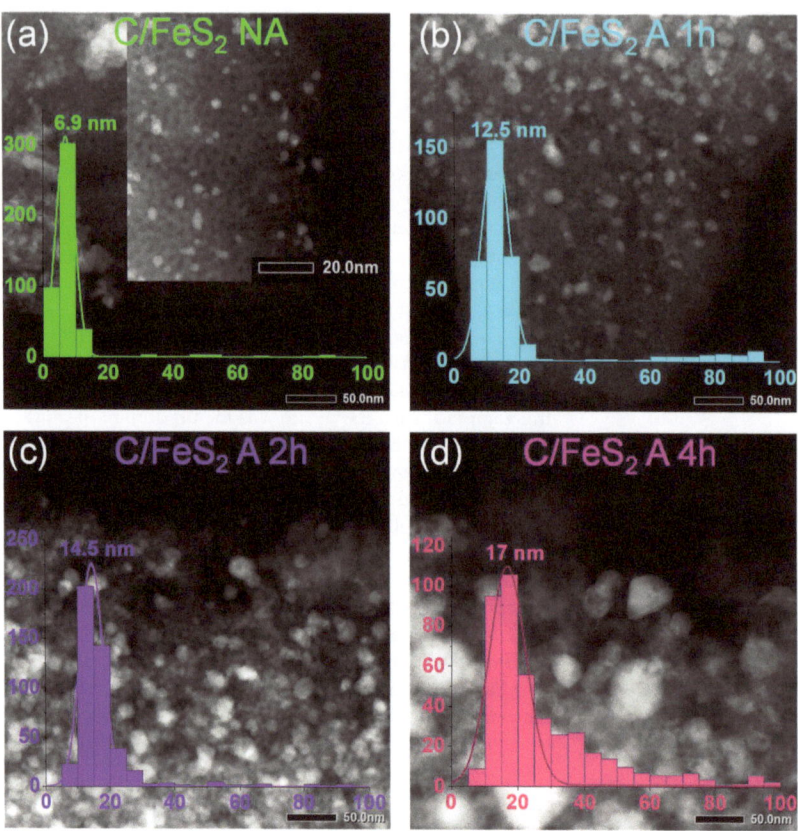

Figure 3. STEM images and (inset) particle size distributions of the C/FeS$_2$ composites; (**a**) nonactivated, (**b**) activated for 1 h, (**c**) activated for 2 h, and (**d**) activated for 4 h, then sulfidized at 400 °C.

The average sizes of the particles were very small and did not exceed 20 nm, which was smaller than those reported in the literature [35,36]. This was particularly interesting, as it offered the possibility to evaluate the electrochemical activity of FeS$_2$ on the nanoscale. Moreover, the particles were well-dispersed in the carbon matrix after short activation times. This was due to the one-pot synthesis that ensured the efficient confinement of the particles within the carbon walls [37]. However, as activation proceeded longer, the particles became more aggregated. This behaviour was previously described as the Oswald ripening mechanism [38] that occurs spontaneously and is explained thermodynamically by the fact that large particles are energetically favoured because larger particles are more stable than smaller particles. This was confirmed through the activation, which created the pores that provided more space for the particles to aggregate. In some cases, very large particles with different sizes varying up to 400 nm (Figure S3, SI) were observed. This was particularly true for C/FeS$_2$ A 4 h, as the distribution showed particles with sizes of ~40–60 nm.

The amounts of FeS$_2$ in the composites were estimated from TGA data (Figure S4, SI, and Table 1). Details of the calculation method and the equation used are provided in the SI.

Table 1. Textural properties of the synthesized C/FeS_2 materials determined through nitrogen adsorption, percentage of iron disulfide determined through TGA under air, and particle sizes (d) obtained using STEM.

Material	S_{BET} ($m^2\ g^{-1}$)	V_T ($cm^3\ g^{-1}$)	V_{micro} ($cm^3\ g^{-1}$)	V_{meso} ($cm^3\ g^{-1}$)	d STEM (nm)	% FeS_2 TGA (wt%)
C/FeS_2 NA	161	0.10	0.06	0.054	7.00	15
C/FeS_2 A 1 h	185	0.17	0.07	0.10	12.50	18
C/FeS_2 A 2 h	296	0.28	0.11	0.17	14.50	22
C/FeS_2 A 4 h	214	0.20	0.08	0.13	17.00	23

Previous works have used the same technique to determine the FeS_2 amount in carbon/FeS_2 composites [7,36]. The FeS_2 percentages in the composites were ~15, 18, 22, and 23% for C/FeS_2 NA, C/FeS_2 A 1 h, C/FeS_2 A 2 h, and C/FeS_2 A 4 h, respectively. The activation process led to slight increases in the FeS_2 percentages in the composites. The XRD data for the resulting powder confirmed the formation of the Fe_2O_3 phase, as shown in Figure S5, SI. It should be mentioned that the presence of some traces of Fe_7S_8 and Fe_3O_4 in the different composites contributes to the resulting FeS_2 value, therefore, it is only an estimation of the real value.

The textural properties impact the electrochemical performance of a composite. Therefore, nitrogen adsorption/desorption experiments were performed with the composites, and Figure 4a,b presents the corresponding isotherms and pore size distributions. The isotherms (Figure 4a) presented mixed shapes between those of type I and type IV isotherms [39], with H4 hysteresis loops for C/FeS_2 NA and C/FeS_2 A 1 h, and H3 hysteresis loops for C/FeS_2 A 2 h and C/FeS_2 A 4 h [40]. This indicated the presence of both micro- and mesopores. During the activation process, the volume of micropores increased until 2 h of activation and then decreased after 4 h of activation, which was observed at the relative pressures of $P/P_0 < 0.1$. This resulted in specific surface areas (SSAs) of 161, 185, 296, and 214 $m^2\ g^{-1}$ for C/FeS_2 NA, C/FeS_2 A 1 h, C/FeS_2 A 2 h, and C/FeS_2 A 4 h, respectively, as seen in Table 1. The decrease in SSA after 4 h of activation was caused by several factors: (i) the FeS_2 particles were larger and may have blocked access to some pores, (ii) the amount of metal particles in the material increased slightly with increasing activation times while the carbon amount decreased, and (iii) the long activation times may have caused pore enlargement, which does not account for the micropore contributions to the SSA.

Mesopores were also present, but with smaller volumes. Two types of mesopores were observed for all of the materials: (i) intrinsic mesopores with average sizes of ~4 nm (inset of Figure 4b) and (ii) intergrain mesopores measuring ~23 nm, which probably arose from the spaces between the FeS_2 nanoparticles that grew outside of the carbon for 2 h and 4 h. The presence of intergrain porosity was sustained by the shape of the isotherm, which showed an increase in the adsorption volume at high P/P_0 (>0.8).

The presence of micropores and mesopores is important for electrochemical evaluations because it enables the efficient flow of the electrolyte ions during charge and discharge. The pore size distribution determined by 2D-NLDFT indicated the presence of many micropores with sizes of ~0.7 nm in all materials. These micropore sizes are very interesting for a supercapacitor electrode material, since a match of micropore size with the electrolyte ion size leads to the enhancement of the capacitance, as has been shown elsewhere [41].

XPS is a surface analysis technique, and it was used to determine the chemical compositions, chemical bonds, and functional groups in the composites (C/FeS_2 NA, C/FeS_2 A 1 h, and C/FeS_2 A 4 h). The results are gathered in Figure 4c,d and Figure S6. Based on the XPS survey spectra, the C 1s, O 2p, S 2p, Fe 2p, and N 1s amounts were determined (Figure 4c).

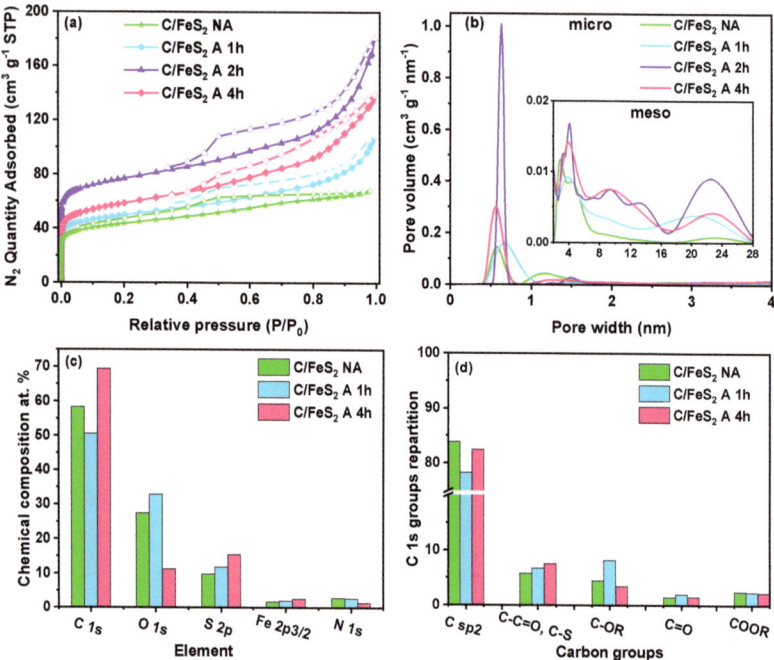

Figure 4. (**a**) Nitrogen adsorption/desorption isotherms, (**b**) pore size distributions determined with the 2D–NLDFT model for the C/FeS$_2$ composites (nonactivated and activated for different times), main graph (0–4 nm), and inset mesopore region (2–28 nm), (**c**) chemical compositions determined from XPS survey spectra showing the different elements, and (**d**) partitioning of the C 1s groups obtained through deconvolution of the C 1s peaks for the synthesized C/FeS$_2$ composites.

The carbon functional groups exhibited the highest proportions in all of the composites; (58.36 at.%) for C/FeS$_2$ NA, (50.63 at.%) for C/FeS$_2$ A 1 h, and (69.36 at.%) for C/FeS$_2$ A 4 h. Then, oxygen showed rather similar values (27.48 at.%) for C/FeS$_2$ NA and (32.93 at.%) for C/FeS$_2$ A 1 h, while for C/FeS$_2$ A 4 h (11.24 at.%), the quantity decreased significantly. This might be due to the longer activation time (6 h vs. 3 h), which caused the removal of more oxygen groups. The reduced oxygen groups are in favor of S groups, which increased continuously with the activation time from 9.74 to 15.47 at.%.

The amount of nitrogen was low (2.76 at.% maximum) as a result of cysteine decomposition. There was a slight decrease in the N 1s peak intensities (Figure 4c), which was caused by the removal of NO$_x$ groups at longer activation times [42]. In the literature, S-doped carbon [43] and N-doped carbon [44] showed superior electrochemical activities in supercapacitors. This doping improved the pseudocapacitance and electronic conductivity, and hence the rate capability and stability during long-term cycling. With increased activation times, the Fe 2p$_{3/2}$ peak for the extreme surface also increased (Figure 4c).

Deconvolution of the high-resolution C 1s spectrum showed that the main contribution came from sp^2-hybridized C, and several other smaller contributions came from oxygen-based functional groups (C-C=O, C-OR, C=O, and COOR) and C-S bonds (Figure 4d). In general, oxygen groups present on the carbon surface were reported to improve the wettability, increase electrolyte diffusion, and contribute to some pseudocapacitive reactions, hence improving the capacitance [45]. However, large amounts might diminish the electronic conductivity.

Deconvoluted high-resolution N 1s and S 2p spectra are presented in Figure S6a, SI. For nitrogen, the observed groups included pyridine (C=N) at 398.15 eV, pyrrolic groups (C-NH$_x$) at 400.29 eV, and quaternary groups (C-N$^+$) at 401.51 eV. The proportion of pyrrolic groups increased slightly with increasing activation time, and reduced the proportion of quaternary groups due to the removal of more carbon atoms, and hence the formation of C=N double bonds. Thus, during pyrolysis, nitrogen was incorporated into the carbon structure. This approach was previously used to dope carbon with other nitrogen sources [46].

In Figure S6b, SI, the high-resolution sulfur spectra showed differences for the different materials. For C/FeS$_2$ NA and C/FeS$_2$ A 1 h, the spectra contained peaks at 168–170 eV attributed to oxidized sulfur (sulfates and sulfoxides) [47]. Interestingly, the proportions of these compounds decreased significantly for C/FeS$_2$ 4 h as FeS$_2$ was formed, probably because the longer treatment with H$_2$S (6 h) caused the reduction of oxygen and sulfur. The larger particle sizes of this material (~17 nm) inhibited oxidation compared to that of the other materials containing smaller particles (7 to 12.5 nm).

For the Fe comparison, the XPS spectra of Fe 2p$_{3/2}$ are provided in Figure S7a, SI. As can be observed, the material activated for the longest duration (C/FeS$_2$ 4 h) shows the highest peak of FeS$_2$ contribution (707 eV), attesting to the presence of the richest pyrite phase in this material. Iron oxides with a high oxidation state (Fe$^{2+,3+}$ and Fe^{3+}) coming from Fe$_2$O$_3$ or Fe$_3$O$_4$ are less significant due to their reduction into Fe^{2+} (FeO). These results are in agreement with the other characterisation techniques.

3.2. Electrochemical Characterisation of Activated C/FeS$_2$ Materials

Electrochemical tests were performed with 2 M KOH in a 3-electrode cell to obtain CV data illustrating the redox activity of FeS$_2$, and then in a symmetric 2-electrode cell to evaluate the supercapacitor performance.

Figure 5 shows the CVs of the materials obtained with scan rates of 1 mV s^{-1} over a potential window ranging from −1.2 to −0.2 V vs. Hg/HgO (Figure 5a). It should be noted that 50 activation cycles were performed at 1 mV s^{-1} before recording the voltammograms (Figure 5a). The aim of the repetitive cycling was to stabilize the material until it gave a reproducible CV [48]. However, the phenomena that occurred during this step have not been reported in the literature. Therefore, to understand the structural and morphological changes occurring in the material, XRD and STEM were applied to the pristine electrode and again after 50 CV cycles (Figure 5b–d). Compared to the pristine electrode, the FeS$_2$ was preserved, while the Fe$_7$S$_8$ content diminished substantially. Interestingly, a new phase was identified after cycling, i.e., the iron(III) oxide hydroxide Fe(OOH) (COD 9010406). The observed graphite peak (COD 9011577) comes from the conductive glue used on the current collector on the cell. The formation of Fe(OOH) was beneficial because it may enhance the wettability and reversibility during cycling [49]. The High-resolution XPS spectra of the Fe 2p$_{3/2}$ pristine electrode and of the same electrode after 50 cycles (Figure S7b,c, SI) confirmed the presence of more FeOOH. According to the literature [50], the binding energy of FeOOH is in the same position as that of Fe$_3$O$_4$ (~710 eV), therefore, the contribution of the latter phase cannot be excluded in the pristine electrode. However, the amount of FeOOH after cycling is significantly higher than in pristine material (5.8% vs. 1.1%). In addition, based on the deconvoluted XPS S2p peak (not shown here), the FeS$_2$ amount decreases while Fe$_7$S$_8$ disappears after cycling, in line with XRD observations. Therefore, a part of FeS$_2$/Fe$_7$S$_8$ is transformed in FeOOH, indicating the reason for the increased pseudocapacitive activity, since with more hydrated phases, better reversibility resulted in more developed CV scans (66% increase in the area) (Figure 5a). The STEM images (Figure 5c,d) showed that the particles tended to aggregate during cycling, and distinguishable nanoparticles were no longer observed. This suggested that in contact with the electrolyte, the FeS$_2$ tended to form larger particles that completely covered the surface of the carbon. This will be discussed in detail later.

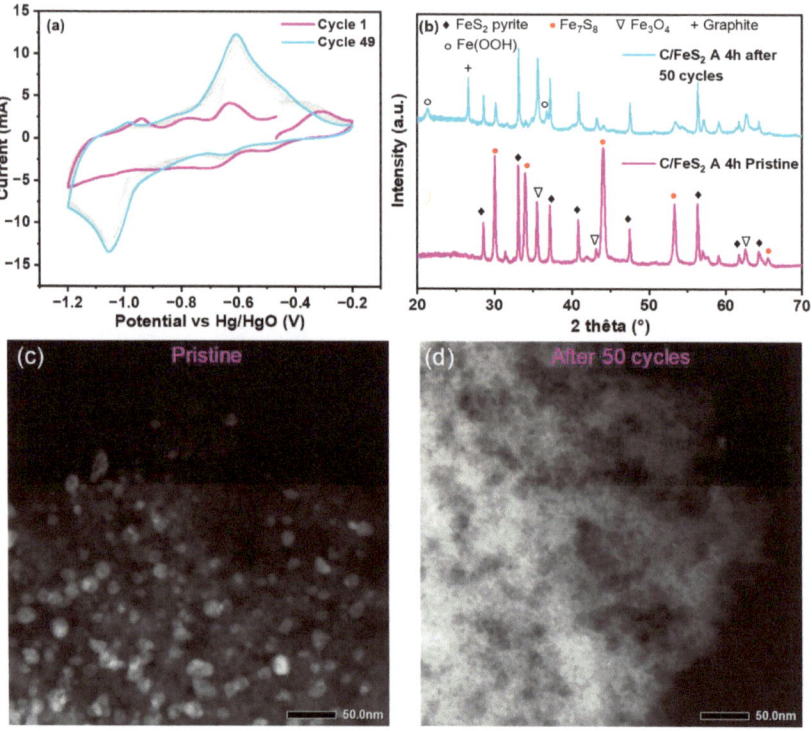

Figure 5. (**a**) CVs from cycle 1 to cycle 49 for C/FeS$_2$ A 4 h in a 3-electrode cell using 2 M KOH. (**b**) XRD data for the pristine electrode and for the same electrode after 50 cycles in a 3-electrode cell. STEM images of the (**c**) pristine electrode and (**d**) the same electrode after 50 cycles in a 3-electrode system.

As shown by the electrochemical data for the materials, (Figure S8, SI) the oxidation and reduction peaks were clearly visible for all materials and occurred at an approximate potential of −0.58 V vs. Hg/HgO for the oxidation peak and at −1.1 V vs. Hg/HgO for the reduction peak. Small shifts in the potentials for oxidation and reduction were seen for some of the materials. According to the literature [19], the redox couple involved in charge storage by FeS$_2$ was the Fe^{2+}/Fe^{3+} couple, and the reaction proposed for charge storage by pyrite FeS$_2$ is shown in Equation (5) [12]. Moreover, additional redox activity might come from FeOOH (Equation (6)).

$$FeS_2 + OH^- \leftrightarrow FeS_2OH + H_2O + e^- \tag{7}$$

$$FeOOH + H_2O + e^- \leftrightarrow Fe(OH) + OH^- \tag{8}$$

The areas under the CV curves varied for the different materials (Figure S9, SI); C/FeS$_2$ NA showed the smallest area, probably because it had the lowest SSA. The increased area under the voltammogram was consistent with the textural properties (SSA, pore volume, and width). After 2 h of activation, the SSA and the micropore volume were the highest, and the CV of the material showed the largest area, indicating good electrochemical behaviour. After 4 h, the material had a smaller surface area, but it was still larger than that of C/FeS$_2$ NA. In addition to the textural properties, other parameters influenced the electrochemical behaviour, such as the particle sizes and different types/amounts of crystalline phases and functional groups, which varied from one material to another.

Evaluations were performed with a 2-electrode supercapacitor, as shown in Figure 6. In Figure 6a, the CV obtained for C/FeS$_2$ NA with a scan rate of 5 mV s^{-1} showed obvious redox peaks that contributed to the capacitance. For the other materials, the shapes were mainly rectangular, indicating capacitive-like behaviour. This suggested that the smallest particle sizes manifested as pseudocapacitance in the cyclic voltammograms and made the peaks more visible. C/FeS$_2$ A 2 h showed the largest areas under the peaks, probably because it had the highest SSA of all the materials.

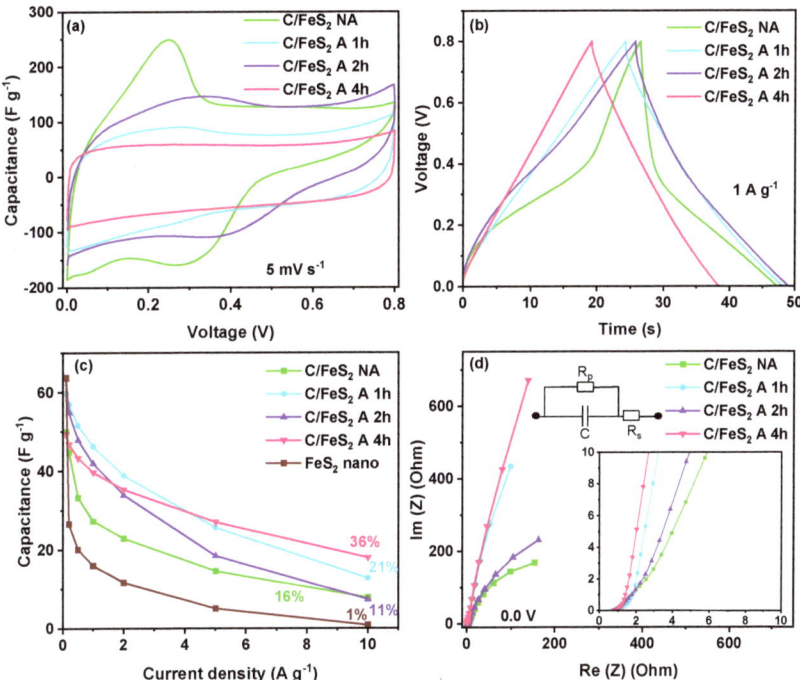

Figure 6. Electrochemical data obtained for the C/FeS$_2$ composites in symmetric 2–electrode cells with 2 M KOH used as the electrolyte; (**a**) CV at 5 mV s^{-1} (**b**) charge–discharge cycles at 1 A g^{-1} (**c**) capacitance from 0.1 A g^{-1} to 10 A g^{-1} and (**d**) EIS at 0.0 V (inset: equivalent circuit diagram). The percentage values represent the rate capabilities.

For a better evaluation at different sweep rates, the CVs of all materials are presented in Figure S9, SI, and the behaviour differed from one material to another. For the material with small particles (C/FeS$_2$ NA, Figure S9a, SI), the behaviour indicated a mixture of capacitive and faradaic contributions. However, with increasing particle sizes and specific surface areas, the redox contribution slowly vanished (Figure S9b,c, SI) until pure capacitive behaviour was observed for C/FeS$_2$ A 4 h (Figure S9d, SI).

To gain a deeper understanding of the different energy storage mechanisms in this system, the power law (I = aV^b) was used. This indicates the relationship between the peak current registered (I) and the scan rate (V) and provides insights into the capacitive and pseudocapacitive contributions [51]. If b = 1, the storage mechanism is mainly governed by surface adsorption, i.e., it is capacitive, while for b = 0.5, diffusion mechanisms originating from pseudocapacitance are involved. For this purpose, the value of b was determined by calculating the slope of a linear fit for a plot of Log (I) vs. Log(V), as shown in Figure S10, SI. The intensity of the cathodic peak in the range of 0.2 to 0.4 V was used for this calculation. It can be seen that b increased from 0.7 to 0.93 with an increase in the activation time from 1 h to 4 h.

The pristine C/FeS$_2$ NA had a low b value of 0.7, suggesting a combination of both energy storage mechanisms but with a more significant pseudocapacitive contribution. In contrast, for C/FeS$_2$ A 4 h, b = 0.93, and a rather predominant capacitive contribution was observed with only a small contribution from pseudocapacitive behaviour. This probably arose from the larger pore particles formed that did not block the pores as the small particles did; therefore, the porosity was more accessible. These results were consistent with the STEM and 3-electrode cell results (Figure 5).

The charge discharge curves determined at 1 A g^{-1} (Figure 6b) exhibited symmetric lines. Once again, the shapes were identical, and triangular, for all materials except C/FeS$_2$ NA, which showed behaviour to that of battery-type electrodes. This behaviour was already reported for graphene/FeS$_2$ rods with lengths of 2 μm and diameters of 50 nm and was attributed to battery-type behaviour [52].

The shapes of the CV curves changed with changes in the particle sizes and electrode thicknesses, as demonstrated, for example, with Nb$_2$O$_5$ [53]. Carbon/FeS$_2$ nanoparticles with bimodal size distributions, i.e., 20–50 nm and 100–200 nm, exhibited faradaic behaviour in a nonaqueous electrolyte [36]. In the present case, although C/FeS$_2$ NA had very small particles (<10 nm), it also had aggregates and exhibited similar behaviour. However, the precise effect of the particle sizes on the CV signatures must be studied further with mono-modal particles to determine their contributions.

Therefore, the capacitances for the various materials were calculated using the GCD technique; in 2 M KOH, the values were 50, 60, 63, and 49 F g^{-1} at 0.1 A g^{-1} for C/FeS$_2$ NA, C/FeS$_2$ A 1 h, C/FeS$_2$ A 2 h, and C/FeS$_2$ A 4 h, respectively. The capacitance was determined at different current densities to evaluate the material's behaviour at high current densities up to 10 A g^{-1} (Figure 6c). The resulting capacitance retention rates were 16%, 22%, 11%, and 36% for C/FeS$_2$ NA, C/FeS$_2$ A 1 h, C/FeS$_2$ A 2 h, and C/FeS$_2$ A 4 h, respectively. For comparison purposes, unsupported nanosized FeS$_2$ synthesized from nano-Fe$_2$O$_3$ was tested in a symmetric 2-electrode system. This material had a predominant FeS$_2$ marcasite phase, particle sizes of ~50–100 nm, and a surface area of 10 m^2 g^{-1} (Figure S11, SI) and was taken as a representative example for comparison with the C/FeS$_2$ composites. The rate capability (Figure 6c) for nano-FeS$_2$ showed a capacitance of 64 F g^{-1} at 0.1 A g^{-1}, which was comparable to those for the C/FeS$_2$ materials, but dropped drastically to 27 F g^{-1} at 0.2 A g^{-1}, and only 1% of the rate capability was retained at 10 F g^{-1}. These results showed that in the absence of carbon, the capacitance loss for FeS$_2$ was more drastic, probably due to the limited electron transport through the material at high current rates. This confirmed that the FeS$_2$ particles supported in carbon improved the supercapacitor performance since the porosity and electronic conductivity are provided.

EIS was used to gain insight into the resistance of the device. The Nyquist plots (Figure 6d) showed two behaviours in the low-frequency region, depending on the material. C/FeS$_2$ A 1 h and C/FeS$_2$ A 4 h exhibited almost vertical, straight lines indicating capacitive materials with high conductivities [54]. However, C/FeS$_2$ NA and C/FeS$_2$ A 2 h showed lines with angles of 45° and shapes similar to half semicircles. This probably due to internal cell resistance, but also indicated some resistance in the bulk material at low frequencies [55]. In the inset, the high-frequency regime showed similar EIS values of 0.76, 0.86, 0.75, and 0.72 Ω for C/FeS$_2$ NA, C/FeS$_2$ A 1 h, C/FeS$_2$ A 2 h, and C/FeS$_2$ A 4 h, respectively. These values were lower than those for nano-FeS$_2$ (1.02 Ω) and FeS$_2$ electrodes reported in other work (2.88 Ω) [56]. Therefore, the presence of carbon decreased the resistance and enhanced the electronic conductivity.

To develop valid metrics with which to compare supercapacitors, a Ragone plot (energy density vs. power density) is presented in Figure 7a. The energy and power were calculated from the mass of the two electrodes in the symmetric device. C/FeS$_2$ NA had the lowest profile, and the energy vs. power line dropped quickly. C/FeS$_2$ A 1 h and C/FeS$_2$ A 2 h showed higher energies of 1.32 and 1.41 W h kg^{-1}, but their profiles dropped at high power densities. Finally, C/FeS$_2$ A 4 h had the longest horizontal line with an energy density of 1.09 W h kg^{-1} at 16.32 W kg^{-1}, and it reached a maximum power of 1.3 kW kg^{-1}. A comparison with other types of metal-based electrodes is interesting, and

C/FeS$_2$ showed values comparable to those of nickel sulfide [57], copper sulfide [58], and manganese-cobalt sulfides [59].

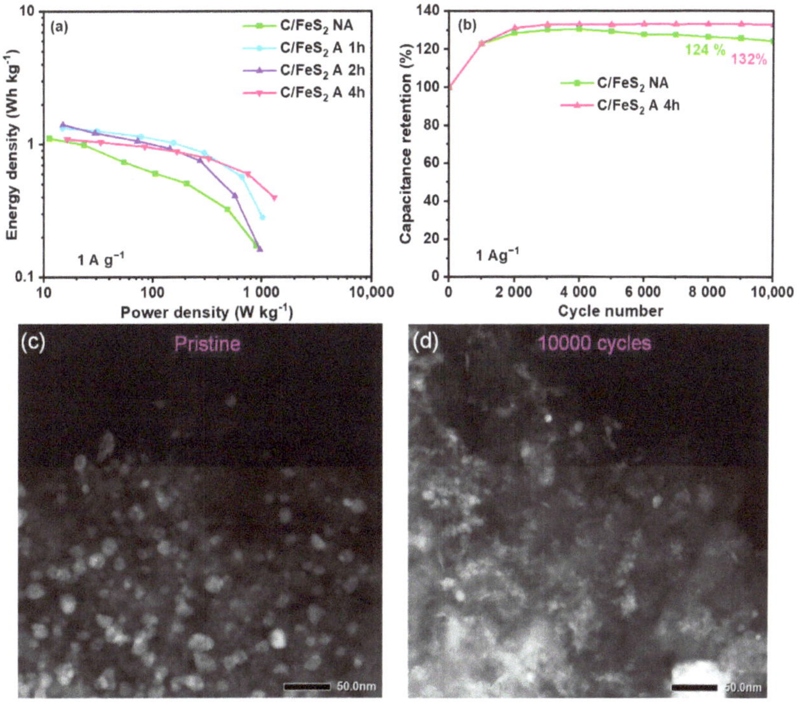

Figure 7. (**a**) Ragone plots for all C/FeS$_2$ materials and (**b**) long–term cycling of C/FeS$_2$ NA and C/FeS$_2$ A 4 h at 1 A g^{-1} with voltages of 0.8 V; STEM images of C/FeS$_2$ A 4 h; (**c**) the pristine electrode and (**d**) the same electrode after 10,000 cycles.

Since cycling stability is another important criterion for the application of supercapacitors, the nonactivated reference material C/FeS$_2$ NA and the best-performing material, i.e., that activated at 4 h, C/FeS$_2$ A 4 h, were selected for this purpose. The long-term cycling performance was evaluated at up to 0.8 V with a current density of 1 A g^{-1} in the 2-electrode cell. In the first 2000 cycles, the capacitances of C/FeS$_2$ 1 h and C/FeS$_2$ A 4 h increased. Similar behaviour was reported in previous studies [60] and was attributed to the activation of the redox species [7], as discussed above for the 3-electrode cell, but no other insights were provided.

After the capacitance increased during the first 2000 cycles, nearly identical capacitance was observed until cycle 10,000 (Figure 7b). This stable behaviour for more than 8000 cycles led us to predict that the cycle lives of these materials were still very high. To gain a better understanding of the cycling behaviour, STEM and XRD were applied to the pristine electrode and the same electrode after 10,000 cycles (Figure 7c,d and Figure S12, SI). As with the 3-electrode test, there was a progressive modification in the morphology during cycling. After 10,000 cycles, the particles looked like needles or sheets, and their sizes increased slightly but remained rather small. Moreover, the pyrite phase in the material was preserved, while the Fe$_7$S$_8$ phase was drastically diminished (Figure S12, SI). This increased the capacitance and was beneficial for long-term cycling.

The various electrochemical tests allowed us to conclude that C/FeS$_2$ A 4 h showed the best behaviour. First, it had the best rate capability of 36%, which was highly encouraging for a TMS material, and it showed values between 15 and ~40% [60]. The energy vs. power profile was steady, demonstrating efficient energy storage. Additionally, long-term cycling was steady for at least 10,000 cycles. All of these good results can be attributed to the characteristics of the material. First, it had a high proportion of the pyrite FeS_2 phase, which was not the case for the other materials. Second, the pyrite particles were the largest of the materials, but their average size did not exceed 20 nm. Finally, this material showed the highest percentage of sulfur, which may have increased the pseudocapacitance. Therefore, the synthesis procedure and thermal treatment conditions shaped and tuned the structure, functional groups, conductivity, and textural properties of the material. Table S1 presents a comparison with some composites reported in the literature. The comparisons were not absolute, as many parameters were different, including the type of cell (most of them used 3 electrode cells), the voltage window, and the current density at which cycling was performed. Therefore, it can be said that comparisons of the synthesized C/FeS_2 should only be conducted using 2-electrode cells and similar conditions.

3.3. Ageing of C/FeS_2 over Time

The stability of FeS_2 in its different allotropic forms (marcasite or pyrite) is a research topic of interest. Therefore, attempts have been made to prepare air-stable FeS_2 [61]; however, most studies were focused on large-scale pyrite for photovoltaics or environmental applications [61,62]. No reports have been found on nanosized FeS_2 (<20 nm) in the literature, and possible modifications of the structural, morphological, and electrochemical properties triggered by oxidation, hydration, etc., are not known, which motivated this work.

Therefore, the emphasis herein was on the evolution of nano-FeS_2 over time. In this case, the C/FeS_2 NA composite powder, which had the smallest particle sizes, was selected for detailed evaluation. In the first step, the sample was placed in a sample holder under ambient conditions and characterised by XRD and TEM weekly over a period of two months. Aggregation and volume expansion were observed over time; therefore, the material was realigned to perform another XRD experiment after two additional months (110d in total). After this period of 110d, the material was immediately tested in a symmetric 2-electrode cell to compare its performance with that of the initial material (1d).

Figure 8 contains the diffractograms and electrochemical data generated after the selected time intervals. The XRD data (Figure 8a) show that in the initial state (1d), the material contained pyrite FeS_2 (COD 9000594) as the main phase and Fe_7S_8 pyrrhotite 4C (COD 2104739). With increased ageing (49d and 110d), new crystalline phases emerged, i.e., hydrated iron sulfates such as jarosite $(H_3O)Fe_3(SO_4)_2(OH)_6$ (COD 9007285) and butlerite $Fe(SO_4)(OH) \cdot 2H_2O$ (COD 9000225). A possible chemical reaction pathway is shown in Equation (9) [63]:

$$FeS_2 + 8\ H_2O + 7\ O_2 \rightarrow FeSO_4\ 7\ H_2O + SO_4^{2-} + 2\ H^+ \qquad (9)$$

Density functional theory studies indicated that pyrite oxidation may be related to different environmental issues. Depending on the activation energy, either iron oxides or iron sulfates are formed [64]. The evolutions of these species were followed over time with XRD spectra (not shown for simplicity), as shown in Figure 8b.

The initial composition (75% pyrite and 25% pyrrhotite 4c) remained stable for approximately 20d, and then the jarosite phase appeared after ~18d, followed by butlerite after ~24d. At 30d, butlerite evolution increased significantly with time until it reached 80%, while the jarosite proportion remained rather stable (~20%). The proportion of the pyrrhotite 4C phase decreased slowly and continuously from 24 to 2% on 110d.

The STEM was performed on the final material to provide a better understanding of the morphological modifications (Figure S13, SI), and there were drastic modifications to the initial material, presented in Figure 3a. After ageing, there was no sign of individual particles, and the previous small particles all aggregated to form structures that looked like long fibres or filaments. This may have been due to the growth of hydrated particles along a preferred direction in the porous carbon. Wiese et al. observed the spontaneous formation of hydrated iron sulfates from pyrite samples [63]. This was accompanied by a significant change in the morphology and formation of a new material shape. The modification depended on many factors, including the degree of crystallization, since less crystalline materials are more prone to oxidation. From the TEM images (Figure S13, SI), it is difficult to distinguish between the carbon matrix and the new forms of hydrated sulfate, as these new particles were spread everywhere on the carbon surface. Therefore, the broad dispersion of the particles in the initial material was no longer present after 110d. The morphological evolution and the changes in distribution may have impacted the electrochemical performance because ion diffusion and transport will have different pathways.

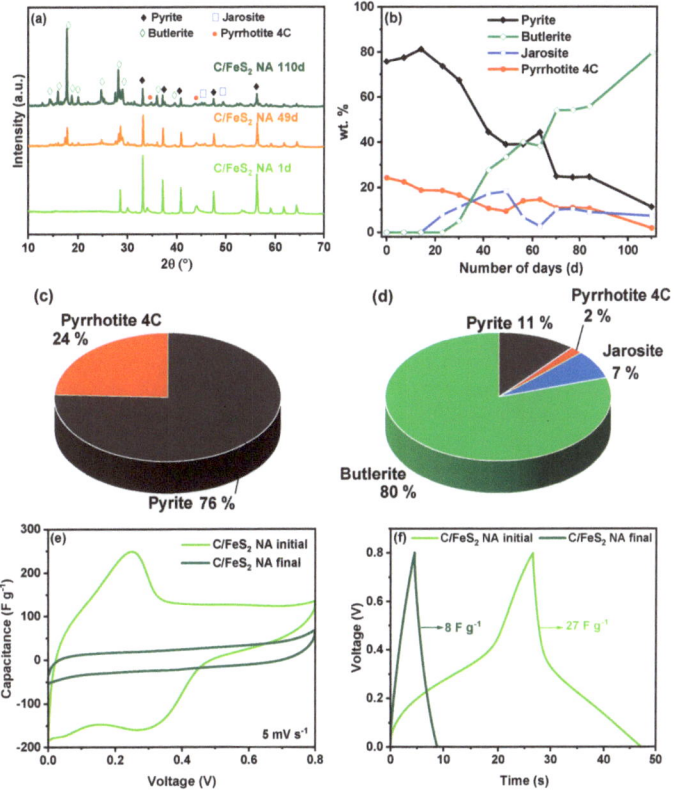

Figure 8. (a) XRD ageing study of C/FeS$_2$ NA at different time scales, (b) percentages of the crystalline phases present in C/FeS$_2$ NA materials after different time periods, (c) initial and (d) final proportions of the crystalline phases in the C/FeS$_2$ NA composite, (e) CVs run at 5 mV s^{-1} and (f) charge discharge curves at 1 A g^{-1}.

The initial materials contained 75% pyrite (Figure 8c), while only 11% was preserved after 110d (Figure 8d). A significant decrease in the amount of FeS$_2$ could have affected the performance. In Figure 8e, a CV generated for the aged material showed different behaviour. After ageing, no redox activity was observed even at low scan rates, and the shape of the CV curve was rectangular, originating only a carbon contribution. This suggested that the newly formed hydrated phases were not electrochemically active. Furthermore, this showed that a significant part of the capacitance in the initial material arose from FeS$_2$, even if the composite was not purely composed of FeS$_2$ (75%). Nevertheless, it is also possible that the growth of these structures blocked the pores; therefore, carbon contributed less to the CV as well. The charge–discharge curves (Figure 8f) also showed differences after ageing. The curves exhibited triangular shapes, attesting to the presence of carbon and its capacitive contribution, whereas there was no redox activity. Moreover, the capacitance decreased by almost two-thirds after ageing (8 vs. 27 F g^{-1}).

4. Discussions

The synthesis of the C/FeS$_2$ composites is advantageous due to the one-pot approach using cysteine that allows a direct FeS formation. Further H$_2$S treatment was required to convert FeS to FeS$_2$. The optimisation of the sulfidation temperature between 250 and 600 °C for 3 h allowed to determine the optimal temperature for FeS$_2$ formation (400 °C), which is a compromise between the FeS conversion to FeS$_2$ (at low temperatures) and the FeS$_2$ decomposition to FeS (at high temperatures). In a previous report, FeS$_2$ pyrite was synthesized at 200 °C on graphene sheets through a 24 h thermal treatment with H$_2$S [65]. However, in the present case, the FeS particles are confined in carbon, which changes the kinetics of their transformation into FeS$_2$. Consequently, the synthesis of pyrite by sulfidation is complex. Depending on the initial properties of the materials, there is a compromise between kinetics and thermodynamics, as discussed elsewhere [62].

The as-obtained C/FeS$_2$ NA composite presented a small FeS$_2$ particle size (7 nm) and a BET surface area of 161 m^2 g^{-1}. To enhance the specific surface area in order to improve the capacitance, an activation step was implemented prior to the sulfidation. CO$_2$ activation is a known pathway to increase the specific surface area [66], and the effect of activation time (1 to 4 h) on C/FeS$_2$ formation was studied. This allowed to enhance the specific surface area, in particular, the BET surface was almost doubled (296 m^2 g^{-1}) after 2 h of activation, then a decrease was observed. The increase in activation time has also led to an increase in the size of the FeS$_2$ particles, from 7 to 17 nm, and to the forming of aggregates of particles, in agreement with other works [67]. The size and location of these particles in the carbon network could affect not only the porosity of the composites but also the interactions with the electrolyte. Additionally, the duration of activation induced the FeS oxidation to Fe$_3$O$_4$, which inhibits FeS$_2$ formation. For this reason, the C/FeS$_2$ 4 h material contained the highest amount of Fe$_3$O$_4$ after activation, therefore, it was sulfidated for a longer time (6 h) compared to that of the other composites (3 h). Consequently, this material presented the highest amount of FeS$_2$ and a lower content of residual phases such as Fe$_7$S$_8$ and Fe$_3$O$_4$. Due to the presence of the latter phases, the precise quantification of the amount of FeS$_2$ in the composites was complex and estimated to be between 15 and 23%, depending on the composite.

The various electrochemical tests showed a general trend for an increase in the capacitance (50 to 63 F g^{-1} at 0.1 A g^{-1}) with the increase of BET surface area (161 to 296 m^2 g^{-1}) for the different C/FeS$_2$ composites, from NA to 2 h of activation. Then, for the 4 h activated material, there is a decrease in both BET surface area (214 m^2 g^{-1}) and the capacitance (49 F g^{-1}). This suggests that the capacitance follows the variation of the BET surface area, in line with the known knowledge on pure carbon materials [68]. However, for the latter material, there is a greater amount of S in the carbon network, larger particles, and more amount of FeS$_2$. Therefore, data should be carefully analysed since other parameters might affect the capacitance in composite materials [69]. In particular, the pseudo-capacitive contribution from FeS$_2$ nanoparticles should be considered. This depends on the amount of FeS$_2$, but also on particle size and distribution. Indeed, different electrochemical behaviours were observed depending on the particle size, i.e., the small particles (C/FeS$_2$ NA) revealed a rather faradic (battery-like) behaviour, while the larger particles (the other composites) showed pseudo-capacitive behaviour. The modification of the capacitance with the particle size has already been observed, but the charge storage mechanism behind it remains a challenge to be fully understood. Therefore, the capacitance is related to the properties of both carbon and FeS$_2$ [53].

If the other metrics are analysed, the C/FeS$_2$ A 4 h appears to present the best behaviour. First, it has the best rate capability of 36%, which is highly encouraging for a TMS material, that showed values between 15 and ~40% [70]. The energy vs. power profile was steady, demonstrating efficient energy storage. Furthermore, long-term cycling was stable for at least 10,000 cycles. All of these good results can be attributed to the characteristics of the material. First, it has a high proportion of the pyrite FeS_2 phase, which was not the case for the other materials. Second, the pyrite particles were the largest of the materials, but their average size did not exceed 20 nm. Finally, this material showed the highest percentage of sulfur, which may have increased the pseudocapacitance. Therefore, the synthesis procedure and the thermal treatment conditions shaped and tuned the structure, functional groups, conductivity, and textural properties of the material.

Table S1 presents a comparison to some C/FeS_2 composites reported in the literature. The comparisons were not absolute, as many parameters were different, including the type of cell, the voltage window, and the current density at which cycling was performed [7,35]. Most of the works reported the performance of only 3-electrode cells leading to overestimated values compared to synthesized C/FeS_2. Moreover, such tests in 3-electrodes are not representative of supercapacitor performance and should be conducted with 2-electrode cells. Furthermore, if the performance is compared to the unsupported FeS_2 nanoparticle, the C/FeS_2 composites behave much better. In particular, the carbon network improved the capacitance retention of composite materials, probably due to its higher conductivity and porosity compared to FeS_2. Nevertheless, the capacitance of the C/FeS_2 composites is still lower compared to that of pure carbon [71]. Strategies to improve the performance of C/FeS_2 composites might include the increase of the FeS_2 content in the composites, the formation of a pure FeS_2 phase, better control of particle size, and the modification of particle shape. More importantly, an understanding of FeS_2 storage mechanisms should also allow one to enhance performance. In this direction, this study showed an atypical electrochemical behavior during the cycling of the materials. A progressive increase in capacitance over cycling occurred, and post-mortem analyses evidenced the formation of FeOOH. This newly born phase from the interaction with the electrolyte with iron-based species (most probably Fe_7S_8 or/and iron oxide) positively impacted the electrochemical performance. This was associated with the redox activity of the FeOOH material [72]. Therefore, both the FeS_2 and FeOOH contributed to the capacitance through pseudo-capacitive reactions, whereas carbon did so through electrostatic interactions (porosity) and pseudocapacitance (S-doping).

Another important aspect that strongly affected the performance was the chemical state of the FeS_2 particle. On the basis of the results obtained herein, one may say that the storage of nanoscale pyrite particles (7.0 nm for C/FeS_2 NA) under ambient laboratory conditions led to the substantial modification of the structure and morphology of the material. Pyrite was converted in the hydrated iron hydroxy sulphate phase over time. The particles, initially present in the nanoscale range and highly dispersed in the carbon matrix, were no longer ensured overtime. The surface of carbon was covered by the newly formed phase, blocking, most likely, the carbon porosity. These modifications directly impacted the electrochemical behaviour and led to the formation of inactive phases and low performance. In summary, very small pyrite nanoparticles (C/FeS_2 NA) are not air-stable and should be kept in an anaerobic atmosphere, i.e., a glove box. The activated C/FeS_2 A 4 h composite showed different behaviour, i.e., it was much more stable and exhibited less phase evolution after ageing for days, as shown in Figure S14a, probably due to the larger and more stable particles. Unsupported nano-FeS_2 showed fast conversion to hydrated sulfate phases $FeSO_4 \cdot H_2O$ and $FeSO_4 \cdot 4H_2O$ beginning on 7d (from 87% to 33% FeS_2) and then gradually converted (only 12% of the FeS_2 remained after 70d), as observed in Figure S14b,c, SI. This distinct degradation pathway for nano-FeS_2 compared to C/FeS_2 NA might suggest that, in the absence of carbon, the particles were even more prone to degradation. Therefore, both the sizes of the FeS_2 particles and the surrounding environment affected the stability in air.

5. Conclusions

The current study described the syntheses of C/FeS_2 nanocomposites via one-pot routes, their evaluations as electrodes in supercapacitors, and the effects of ageing on their properties and performance. First, the mixing of all precursors followed by a carbonization step formed the C/FeS nanocomposites, whereas sulfidation under H_2S provided C/FeS_2. The sulfidation process was optimized, and the optimal temperature was 400 °C. In addition, the physical activation was conducted with CO_2 for 1 to 4 h and led to increases in the porosity up to 2 h then a slight decrease was observed while the FeS_2 nanoparticle sizes increased (from 7 to 17 nm). The evaluation of the nano-C/FeS_2 materials for use as supercapacitor electrodes was performed in a symmetric device. The capacitance generally improved with increasing activation time and active surface area, although the particle sizes also affected the performance. The C/FeS_2 A 4 h material exhibited the best compromise among the improved capacitance, rate capability, energy and power density, as well as cycling stability. This occurred because it had the richest pyrite phase with the largest nanoparticles, along with the highest amount of sulfur in the carbon. The ageing of the nanopyrite composites during electrochemical cycling and under atmospheric conditions was first studied using XRD and TEM. After cycling, hydrated forms of iron oxide were formed and contributed to the capacitance by expanding the CV shapes generated with the 3-electrode cell. The morphology changed progressively during cycling but was not detrimental to the electrochemical behaviour, as observed with long-term cycling. However, under air, the progressive conversion of the iron sulfide into hydrated iron hydroxy sulfate occurred, which was accompanied by a morphology change from spherical to fibrous. These new phases were electrochemically inactive and probably had obstructed carbon pores leading to a very weak performance. For future investigations, controlling the phase crystallinity during the activation/sulfidation can be beneficial. The size/shape of the particles can also be optimized for better stability during ageing. Finally, the content of FeS_2 in the composites can be increased, with special attention given to porosity conservation. In summary, this work shows the importance of understanding the fundamental phenomena occurring at the interface between the electrode and the air/electrolyte to enhance the performance of C/FeS_2 materials in supercapacitors.

Supplementary Materials: The following supporting information can be downloaded at: https://www.mdpi.com/article/10.3390/c9040112/s1, XRD and TGA experimental details; XRD C/FeS composites; particle size of C/FeS after activation; STEM and TGA of C/FeS_2; XRD of residue obtained from TGA; XPS of C/FeS_2; CV of C/FeS_2 in three- and two-electrode cells; correlation of peak current vs scan rate for C/FeS_2; characterization of nano FeS_2; XRD before and after long cycling of C/FeS_2; literature table comparing the performance of C/FeS_2; STEM of C/FeS_2 after aging in air; XRD semi-quantitative analyses [7,11–13,29,35].

Author Contributions: Conceptualization, S.Z. and J.-M.L.M.; methodology, S.Z., B.R., J.-M.L.M. and C.M.G.; software, S.Z. and P.F.; validation, C.M.G.; formal analysis, S.Z. and M.Y.N.; investigation, S.Z., B.R., M.Y.N. and P.F.; resources, C.M.G.; data curation, S.Z. and P.F.; writing—original draft preparation, S.Z.; writing—review and editing, J.-M.L.M. and C.M.G.; visualization, S.Z., J.-M.L.M. and C.M.G.; supervision, J.-M.L.M. and C.M.G.; project administration, S.Z. and C.M.G.; funding acquisition, C.M.G. All authors have read and agreed to the published version of the manuscript.

Funding: The authors thank the Université de Haute Alsace (UHA) and the French National Research Agency (STORE-EX Labex Project ANR-10-LABX-76-01) for the financial support of this work.

Data Availability Statement: Data are contained within the article and Supplementary Materials.

Acknowledgments: We thank L. Vidal for their technical help with the STEM analysis and Samar Hajjar-Garreau for their technical assistance with the XPS analyses.

Conflicts of Interest: The authors declare no conflict of interest.

References

1. Frąckowiak, E.; Płatek-Mielczarek, A.; Piwek, J.; Fic, K. Advanced Characterization Techniques for Electrochemical Capacitors. *Adv. Inorg. Chem.* **2022**, *79*, 151–207. [CrossRef]
2. An, C.; Zhang, Y.; Guo, H.; Wang, Y. Metal Oxide-Based Supercapacitors: Progress and Prospectives. *Nanoscale Adv.* **2019**, *1*, 4644–4658. [CrossRef] [PubMed]

3. Gao, Y.; Zhao, L. Review on Recent Advances in Nanostructured Transition-Metal-Sulfide-Based Electrode Materials for Cathode Materials of Asymmetric Supercapacitors. *Chem. Eng. J.* **2022**, *430*, 132745. [CrossRef]
4. Yu, S.; Ng, V.M.H.; Wang, F.; Xiao, Z.; Li, C.; Kong, L.B.; Que, W.; Zhou, K. Synthesis and Application of Iron-Based Nanomaterials as Anodes of Lithium-Ion Batteries and Supercapacitors. *J. Mater. Chem. A* **2018**, *6*, 9332–9367. [CrossRef]
5. Huang, Y.; Zhao, H.; Bao, S.; Yin, Y.; Zhang, Y.; Lu, J. Hollow FeS_2 Nanospheres Encapsulated in N/S Co-Doped Carbon Nanofibers as Electrode Material for Electrochemical Energy Storage. *J. Alloys Compd.* **2022**, *905*, 164184. [CrossRef]
6. Balakrishnan, B.; Balasingam, S.K.; Sivalingam Nallathambi, K.; Ramadoss, A.; Kundu, M.; Bak, J.S.; Cho, I.H.; Kandasamy, P.; Jun, Y.; Kim, H.-J. Facile Synthesis of Pristine FeS_2 Microflowers and Hybrid rGO-FeS_2 Microsphere Electrode Materials for High Performance Symmetric Capacitors. *J. Ind. Eng. Chem.* **2019**, *71*, 191–200. [CrossRef]
7. Pei, L.; Yang, Y.; Chu, H.; Shen, J.; Ye, M. Self-Assembled Flower-like FeS_2/Graphene Aerogel Composite with Enhanced Electrochemical Properties. *Ceram. Int.* **2016**, *42*, 5053–5061. [CrossRef]
8. Lai, C.-H.; Lu, M.-Y.; Chen, L.-J. Metal Sulfide Nanostructures: Synthesis, Properties and Applications in Energy Conversion and Storage. *J. Mater. Chem.* **2012**, *22*, 19–30. [CrossRef]
9. Venkateshalu, S.; Goban Kumar, P.; Kollu, P.; Jeong, S.K.; Grace, A.N. Solvothermal Synthesis and Electrochemical Properties of Phase Pure Pyrite FeS_2 for Supercapacitor Applications. *Electrochim. Acta* **2018**, *290*, 378–389. [CrossRef]
10. Iqbal, M.F.; Ashiq, M.N.; Zhang, M. Design of Metals Sulfides with Carbon Materials for Supercapacitor Applications: A Review. *Energy Technol.* **2021**, *9*, 2000987. [CrossRef]
11. Wang, Y.; Zhang, M.; Ma, T.; Pan, D.; Li, Y.; Xie, J.; Shao, S. A High-Performance Flexible Supercapacitor Electrode Material Based on Nano-Flowers-like FeS_2/NSG Hybrid Nanocomposites. *Mater. Lett.* **2018**, *218*, 10–13. [CrossRef]
12. Sridhar, V.; Park, H. Carbon Nanofiber Linked FeS_2 Mesoporous Nano-Alloys as High Capacity Anodes for Lithium-Ion Batteries and Supercapacitors. *J. Alloys Compd.* **2018**, *732*, 799–805. [CrossRef]
13. Liu, X.; Deng, W.; Liu, L.; Wang, Y.; Huang, C.; Wang, Z. Passion Fruit-like Microspheres of FeS_2 Wrapped with Carbon as an Excellent Fast Charging Material for Supercapacitors. *New J. Chem.* **2022**, *46*, 11212–11219. [CrossRef]
14. Wang, F.; Li, G.; Zhou, Q.; Zheng, J.; Yang, C.; Wang, Q. One-Step Hydrothermal Synthesis of Sandwich-Type $NiCo_2S_4$@reduced Graphene Oxide Composite as Active Electrode Material for Supercapacitors. *Appl. Surf. Sci.* **2017**, *425*, 180–187. [CrossRef]
15. Zallouz, S.; Réty, B.; Vidal, L.; Le Meins, J.-M.; Matei Ghimbeu, C. Co_3O_4 Nanoparticles Embedded in Mesoporous Carbon for Supercapacitor Applications. *ACS Appl. Nano Mater.* **2021**, *4*, 5022–5037. [CrossRef]
16. Jahel, A.; Ghimbeu, C.M.; Monconduit, L.; Vix-Guterl, C. Confined Ultrasmall SnO_2 Particles in Micro/Mesoporous Carbon as an Extremely Long Cycle-Life Anode Material for Li-Ion Batteries. *Adv. Energy Mater.* **2014**, *4*, 1400025. [CrossRef]
17. Theerthagiri, J.; Senthil, R.A.; Nithyadharseni, P.; Lee, S.J.; Durai, G.; Kuppusami, P.; Madhavan, J.; Choi, M.Y. Recent Progress and Emerging Challenges of Transition Metal Sulfides Based Composite Electrodes for Electrochemical Supercapacitive Energy Storage. *Ceram. Int.* **2020**, *46*, 14317–14345. [CrossRef]
18. Zardkhoshoui, A.M.; Davarani, S.S.H.; Ashtiani, M.M.; Sarparast, M. Designing an Asymmetric Device Based on Graphene Wrapped Yolk–Double Shell $NiGa_2S_4$ Hollow Microspheres and Graphene Wrapped FeS_2–$FeSe_2$ Core–Shell Cratered Spheres with Outstanding Energy Density. *J. Mater. Chem. A* **2019**, *7*, 10282–10292. [CrossRef]
19. Chen, Y.-C.; Shi, J.-H.; Hsu, Y.-K. Multifunctional FeS_2 in Binder-Independent Configuration as High-Performance Supercapacitor Electrode and Non-Enzymatic H_2O_2 Detector. *Appl. Surf. Sci.* **2020**, *503*, 144304. [CrossRef]
20. Dong, H.; Li, L.; Wang, Y.; Ning, Q.; Wang, B.; Zeng, G. Aging of Zero-Valent Iron-Based Nanoparticles in Aqueous Environment and the Consequent Effects on Their Reactivity and Toxicity. *Water Environ. Res.* **2020**, *92*, 646–661. [CrossRef]
21. Zallouz, S.; Meins, J.-M.L.; Ghimbeu, C.M. Alkaline Hydrogel Electrolyte from Biosourced Chitosan to Enhance the Rate Capability and Energy Density of Carbon-Based Supercapacitors. *Energy Adv.* **2022**, *1*, 1051–1064. [CrossRef]
22. Beda, A.; Villevieille, C.; Taberna, P.-L.; Simon, P.; Ghimbeu, C.M. Self-Supported Binder-Free Hard Carbon Electrodes for Sodium-Ion Batteries: Insights into Their Sodium Storage Mechanisms. *J. Mater. Chem. A* **2020**, *8*, 5558–5571. [CrossRef]
23. Bujewska, P.; Gorska, B.; Fic, K. Redox Activity of Selenocyanate Anion in Electrochemical Capacitor Application. *Synth. Met.* **2019**, *253*, 62–72. [CrossRef]
24. Wickramaratne, N.P.; Perera, V.S.; Ralph, J.M.; Huang, S.D.; Jaroniec, M. Cysteine-Assisted Tailoring of Adsorption Properties and Particle Size of Polymer and Carbon Spheres. *Langmuir* **2013**, *29*, 4032–4038. [CrossRef] [PubMed]
25. Ghimbeu, C.M.; Vidal, L.; Delmotte, L.; Meins, J.-M.L.; Vix-Guterl, C. Catalyst-Free Soft-Template Synthesis of Ordered Mesoporous Carbon Tailored Using Phloroglucinol/Glyoxylic Acid Environmentally Friendly Precursors. *Green Chem.* **2014**, *16*, 3079–3088. [CrossRef]
26. Ghimbeu, C.M.; Sopronyi, M.; Sima, F.; Delmotte, L.; Vaulot, C.; Zlotea, C.; Paul-Boncour, V.; Meins, J.-M.L. One-Pot Laser-Assisted Synthesis of Porous Carbon with Embedded Magnetic Cobalt Nanoparticles. *Nanoscale* **2015**, *7*, 10111–10122. [CrossRef]
27. Sun, S.; Matei Ghimbeu, C.; Janot, R.; Le Meins, J.-M.; Cassel, A.; Davoisne, C.; Masquelier, C.; Vix-Guterl, C. One-Pot Synthesis of $LiFePO_4$–Carbon Mesoporous Composites for Li-Ion Batteries. *Microporous Mesoporous Mater.* **2014**, *198*, 175–184. [CrossRef]
28. Kiener, J.; Girleanu, M.; Ersen, O.; Parmentier, J. Direct Insight into the Confinement Effect of WS_2 Nanostructures in an Ordered Carbon Matrix. *Cryst. Growth Des.* **2020**, *20*, 2004–2013. [CrossRef]
29. Gražulis, S.; Chateigner, D.; Downs, R.T.; Yokochi, A.F.T.; Quirós, M.; Lutterotti, L.; Manakova, E.; Butkus, J.; Moeck, P.; Le Bail, A. Crystallography Open Database—An Open-Access Collection of Crystal Structures. *J. Appl. Cryst.* **2009**, *42*, 726–729. [CrossRef]
30. Elliot, A.D. Structure of Pyrrhotite 5C (Fe_9S_{10}). *Acta Cryst. B* **2010**, *66*, 271–279. [CrossRef]

31. Li, Y.; van Santen, R.A.; Weber, T. High-Temperature FeS–FeS$_2$ Solid-State Transitions: Reactions of Solid Mackinawite with Gaseous H$_2$S. *J. Solid State Chem.* **2008**, *181*, 3151–3162. [CrossRef]
32. Grønvold, F.; Westrum, E.F. Heat Capacities of Iron Disulfides Thermodynamics of Marcasite from 5 to 700 K, Pyrite from 300 to 780 K, and the Transformation of Marcasite to Pyrite. *J. Chem. Thermodyn.* **1976**, *8*, 1039–1048. [CrossRef]
33. Zhang, T.; Walawender, W.P.; Fan, L.T.; Fan, M.; Daugaard, D.; Brown, R.C. Preparation of Activated Carbon from Forest and Agricultural Residues through CO$_2$ Activation. *Chem. Eng. J.* **2004**, *105*, 53–59. [CrossRef]
34. De Groote, A.M.; Froment, G.F.; Kobylinski, T. Synthesis Gas Production from Natural Gas in a Fixed Bed Reactor with Reversed Flow. *Can. J. Chem. Eng.* **1996**, *74*, 735–742. [CrossRef]
35. Huang, Y.; Bao, S.; Yin, Y.; Lu, J. Three-Dimensional Porous Carbon Decorated with FeS$_2$ Nanospheres as Electrode Material for Electrochemical Energy Storage. *Appl. Surf. Sci.* **2021**, *565*, 150538. [CrossRef]
36. Tung Pham, D.; Paul Baboo, J.; Song, J.; Kim, S.; Jo, J.; Mathew, V.; Hilmy Alfaruqi, M.; Sambandam, B.; Kim, J. Facile Synthesis of Pyrite (FeS$_2$/C) Nanoparticles as an Electrode Material for Non-Aqueous Hybrid Electrochemical Capacitors. *Nanoscale* **2018**, *10*, 5938–5949. [CrossRef]
37. Jahel, A.; Ghimbeu, C.M.; Darwiche, A.; Vidal, L.; Hajjar-Garreau, S.; Vix-Guterl, C.; Monconduit, L. Exceptionally Highly Performing Na-Ion Battery Anode Using Crystalline SnO$_2$ Nanoparticles Confined in Mesoporous Carbon. *J. Mater. Chem. A* **2015**, *3*, 11960–11969. [CrossRef]
38. Voorhees, P.W. The Theory of Ostwald Ripening. *J. Stat. Phys.* **1985**, *38*, 231–252. [CrossRef]
39. Sing, K.S.W. Reporting Physisorption Data for Gas/Solid Systems with Special Reference to the Determination of Surface Area and Porosity (Recommendations 1984). *Pure Appl. Chem.* **1985**, *57*, 603–619. [CrossRef]
40. Sing, K.S.W.; Williams, R.T. Physisorption Hysteresis Loops and the Characterization of Nanoporous Materials. *Adsorpt. Sci. Technol.* **2004**, *22*, 773–782. [CrossRef]
41. Chmiola, J. Anomalous Increase in Carbon Capacitance at Pore Sizes Less Than 1 Nanometer. *Science* **2006**, *313*, 1760–1763. [CrossRef] [PubMed]
42. Fu, Y.; Zhang, Y.; Li, G.; Zhang, J.; Guo, Y. NO Removal Activity and Surface Characterization of Activated Carbon with Oxidation Modification. *J. Energy Inst.* **2017**, *90*, 813–823. [CrossRef]
43. Shaheen Shah, S.; Abu Nayem, S.M.; Sultana, N.; Saleh Ahammad, A.J.; Abdul Aziz, M. Preparation of Sulfur-Doped Carbon for Supercapacitor Applications: A Review. *ChemSusChem* **2022**, *15*, e202101282. [CrossRef] [PubMed]
44. Lee, W.; Moon, J.H. Monodispersed N-Doped Carbon Nanospheres for Supercapacitor Application. *ACS Appl. Mater. Interfaces* **2014**, *6*, 13968–13976. [CrossRef]
45. Li, X.; Jiang, Y.; Wang, P.; Mo, Y.; Lai, W.; Li, Z.; Yu, R.; Du, Y.; Zhang, X.; Chen, Y. Effect of the Oxygen Functional Groups of Activated Carbon on Its Electrochemical Performance for Supercapacitors. *New Carbon Mater.* **2020**, *35*, 232–243. [CrossRef]
46. Moussa, G.; Hajjar-Garreau, S.; Taberna, P.-L.; Simon, P.; Matei Ghimbeu, C. Eco-Friendly Synthesis of Nitrogen-Doped Mesoporous Carbon for Supercapacitor Application. *C—J. Carbon Res.* **2018**, *4*, 20. [CrossRef]
47. Kiciński, W.; Dziura, A. Heteroatom-Doped Carbon Gels from Phenols and Heterocyclic Aldehydes: Sulfur-Doped Carbon Xerogels. *Carbon* **2014**, *75*, 56–67. [CrossRef]
48. Chen, J.; Zhou, X.; Mei, C.; Xu, J.; Zhou, S.; Wong, C.-P. Pyrite FeS$_2$ Nanobelts as High-Performance Anode Material for Aqueous Pseudocapacitor. *Electrochim. Acta* **2016**, *222*, 172–176. [CrossRef]
49. Zhao, S.; Song, Z.; Qing, L.; Zhou, J.; Qiao, C. Surface Wettability Effect on Energy Density and Power Density of Supercapacitors. *J. Phys. Chem. C* **2022**, *126*, 9248–9256. [CrossRef]
50. Biesinger, M.C.; Lau, L.W.M.; Gerson, A.R.; Smart, R.S.C. Resolving Surface Chemical States in XPS Analysis of First Row Transition Metals, Oxides and Hydroxides: Sc, Ti, V, Cu and Zn. *Appl. Surf. Sci.* **2010**, *257*, 887–898. [CrossRef]
51. Nita, C.; Zhang, B.; Dentzer, J.; Matei Ghimbeu, C. Hard Carbon Derived from Coconut Shells, Walnut Shells, and Corn Silk Biomass Waste Exhibiting High Capacity for Na-Ion Batteries. *J. Energy Chem.* **2021**, *58*, 207–218. [CrossRef]
52. Sun, Z.; Li, F.; Ma, Z.; Wang, Q.; Qu, F. Battery-Type Phosphorus Doped FeS$_2$ Grown on Graphene as Anode for Hybrid Supercapacitor with Enhanced Specific Capacity. *J. Alloys Compd.* **2021**, *854*, 157114. [CrossRef]
53. Come, J.; Augustyn, V.; Kim, J.W.; Rozier, P.; Taberna, P.-L.; Gogotsi, P.; Long, J.W.; Dunn, B.; Simon, P. Electrochemical Kinetics of Nanostructured Nb$_2$O$_5$ Electrodes. *J. Electrochem. Soc.* **2014**, *161*, A718. [CrossRef]
54. Diard, J.-P.; Le Gorrec, B.; Montella, C. Linear Diffusion Impedance. General Expression and Applications. *J. Electroanal. Chem.* **1999**, *471*, 126–131. [CrossRef]
55. Mei, B.-A.; Munteshari, O.; Lau, J.; Dunn, B.; Pilon, L. Physical Interpretations of Nyquist Plots for EDLC Electrodes and Devices. *J. Phys. Chem. C* **2018**, *122*, 194–206. [CrossRef]
56. Wang, Q.; Liu, Q.; Ni, Y.; Yang, Y.; Zhu, X.; Song, Y. N-Doped FeS$_2$ Achieved by Thermal Annealing of Anodized Fe in Ammonia and Sulfur Atmosphere: Applications for Supercapacitors. *J. Electrochem. Soc.* **2021**, *168*, 080522. [CrossRef]
57. Wang, Y.; Zhang, W.; Guo, X.; Liu, Y.; Zheng, Y.; Zhang, M.; Li, R.; Peng, Z.; Zhang, Y.; Zhang, T. One-Step Microwave-Hydrothermal Preparation of NiS/rGO Hybrid for High-Performance Symmetric Solid-State Supercapacitor. *Appl. Surf. Sci.* **2020**, *514*, 146080. [CrossRef]
58. Zhang, J.; Feng, H.; Yang, J.; Qin, Q.; Fan, H.; Wei, C.; Zheng, W. Solvothermal Synthesis of Three-Dimensional Hierarchical CuS Microspheres from a Cu-Based Ionic Liquid Precursor for High-Performance Asymmetric Supercapacitors. *ACS Appl. Mater. Interfaces* **2015**, *7*, 21735–21744. [CrossRef]

59. Pramanik, A.; Maiti, S.; Sreemany, M.; Mahanty, S. Carbon Doped MnCo$_2$S$_4$ Microcubes Grown on Ni Foam as High Energy Density Faradaic Electrode. *Electrochim. Acta* **2016**, *213*, 672–679. [CrossRef]
60. Liu, W.; Niu, H.; Yang, J.; Cheng, K.; Ye, K.; Zhu, K.; Wang, G.; Cao, D.; Yan, J. Ternary Transition Metal Sulfides Embedded in Graphene Nanosheets as Both the Anode and Cathode for High-Performance Asymmetric Supercapacitors. *Chem. Mater.* **2018**, *30*, 1055–1068. [CrossRef]
61. Bi, Y.; Yuan, Y.; Exstrom, C.L.; Darveau, S.A.; Huang, J. Air Stable, Photosensitive, Phase Pure Iron Pyrite Nanocrystal Thin Films for Photovoltaic Application. Available online: https://pubs.acs.org/doi/pdf/10.1021/nl202902z (accessed on 24 September 2022).
62. Fan, D.; Lan, Y.; Tratnyek, P.G.; Johnson, R.L.; Filip, J.; O'Carroll, D.M.; Nunez Garcia, A.; Agrawal, A. Sulfidation of Iron-Based Materials: A Review of Processes and Implications for Water Treatment and Remediation. *Environ. Sci. Technol.* **2017**, *51*, 13070–13085. [CrossRef] [PubMed]
63. Wiese, R.G.; Powell, M.A.; Fyfe, W.S. Spontaneous Formation of Hydrated Iron Sulfates on Laboratory Samples of Pyrite- and Marcasite-Bearing Coals. *Chem. Geol.* **1987**, *63*, 29–38. [CrossRef]
64. Dos Santos, E.C.; de Mendonça Silva, J.C.; Duarte, H.A. Pyrite Oxidation Mechanism by Oxygen in Aqueous Medium. *J. Phys. Chem. C* **2016**, *120*, 2760–2768. [CrossRef]
65. Golsheikh, A.M.; Huang, N.M.; Lim, H.N.; Chia, C.H.; Harrison, I.; Muhamad, M.R. One-Pot Hydrothermal Synthesis and Characterization of FeS$_2$ (Pyrite)/Graphene Nanocomposite. *Chem. Eng. J.* **2013**, *218*, 276–284. [CrossRef]
66. Turmuzi, M.; Daud, W.R.W.; Tasirin, S.M.; Takriff, M.S.; Iyuke, S.E. Production of Activated Carbon from Candlenut Shell by CO$_2$ Activation. *Carbon* **2004**, *42*, 453–455. [CrossRef]
67. Zallouz, S.; Pronkin, S.N.; Le Meins, J.-M.; Pham-Huu, C.; Matei Ghimbeu, C. Chapter 9—New Development in Carbon-Based Electrodes and Electrolytes for Enhancement of Supercapacitor Performance and Safety. In *Renewable Energy Production and Distribution*; Advances in Renewable Energy Technologies; Jeguirim, M., Dutournié, P., Eds.; Academic Press: Cambridge, MA, USA, 2023; Volume 2, pp. 353–408. ISBN 978-0-443-18439-0.
68. Barbieri, O.; Hahn, M.; Herzog, A.; Kötz, R. Capacitance Limits of High Surface Area Activated Carbons for Double Layer Capacitors. *Carbon* **2005**, *43*, 1303–1310. [CrossRef]
69. Nguyen, T.; Montemor, M.d.F. Metal Oxide and Hydroxide–Based Aqueous Supercapacitors: From Charge Storage Mechanisms and Functional Electrode Engineering to Need-Tailored Devices. *Adv. Sci.* **2019**, *6*, 1801797. [CrossRef]
70. Geng, P.; Zheng, S.; Tang, H.; Zhu, R.; Zhang, L.; Cao, S.; Xue, H.; Pang, H. Transition Metal Sulfides Based on Graphene for Electrochemical Energy Storage. *Adv. Energy Mater.* **2018**, *8*, 1703259. [CrossRef]
71. Platek, A.; Nita, C.; Ghimbeu, C.M.; Frąckowiak, E.; Fic, K. Electrochemical Capacitors Operating in Aqueous Electrolyte with Volumetric Characteristics Improved by Sustainable Templating of Electrode Materials. *Electrochim. Acta* **2020**, *338*, 135788. [CrossRef]
72. Owusu, K.A.; Qu, L.; Li, J.; Wang, Z.; Zhao, K.; Yang, C.; Hercule, K.M.; Lin, C.; Shi, C.; Wei, Q.; et al. Low-Crystalline Iron Oxide Hydroxide Nanoparticle Anode for High-Performance Supercapacitors. *Nat. Commun.* **2017**, *8*, 14264. [CrossRef]

Disclaimer/Publisher's Note: The statements, opinions and data contained in all publications are solely those of the individual author(s) and contributor(s) and not of MDPI and/or the editor(s). MDPI and/or the editor(s) disclaim responsibility for any injury to people or property resulting from any ideas, methods, instructions or products referred to in the content.

Review

A Review of Carbon Nanotubes, Graphene and Nanodiamond Based Strain Sensor in Harsh Environments

Xiaoyan Wang [1,2], Eng Gee Lim [1,2], Kai Hoettges [2] and Pengfei Song [1,2,*]

[1] School of Advanced Technology, Xi'an Jiaotong—Liverpool University, Suzhou 215123, China; xiaoyan.wang22@student.xjtlu.edu.cn (X.W.); enggee.lim@xjtlu.edu.cn (E.G.L.)

[2] Department of Electrical Engineering and Electronics, University of Liverpool, Liverpool L69 7ZX, UK; k.hoettges@liverpool.ac.uk

* Correspondence: pengfei.song@xjtlu.edu.cn; Tel.: +86-512-81889039

Abstract: Flexible and wearable electronics have attracted significant attention for their potential applications in wearable human health monitoring, care systems, and various industrial sectors. The exploration of wearable strain sensors in diverse application scenarios is a global issue, shaping the future of our intelligent community. However, current state-of-the-art strain sensors still encounter challenges, such as susceptibility to interference under humid conditions and vulnerability to chemical and mechanical fragility. Carbon materials offer a promising solution due to their unique advantages, including excellent electrical conductivity, intrinsic and structural flexibility, lightweight nature, high chemical and thermal stability, ease of chemical functionalization, and potential for mass production. Carbon-based materials, such as carbon nanotubes, graphene, and nanodiamond, have been introduced as strain sensors with mechanical and chemical robustness, as well as water repellency functionality. This review reviewed the ability of carbon nanotubes-, graphene-, and nanodiamond-based strain sensors to withstand extreme conditions, their sensitivity, durability, response time, and diverse applications, including strain/pressure sensors, temperature/humidity sensors, and power devices. The discussion highlights the promising features and potential advantages offered by these carbon materials in strain sensing applications. Additionally, this review outlines the existing challenges in the field and identifies future opportunities for further advancement and innovation.

Keywords: carbon nanotubes; graphene; nanodiamond; strain sensor; harsh environment

1. Introduction

With the development of modern technology and the rapid rise of artificial intelligence technology, the research on various flexible devices has received widespread attention [1–5]. The types of strain sensors include pressure, torsional, tensile, and bending strain sensors, and the strain sensors can be divided into capacitance-type, piezoelectric-type, and resistance-type based on their working principles [6]. Among them, flexible strain sensors have attracted great interest from the industry and academia due to their promising applications in wearable devices [7,8], electronic skin [9,10], intelligent robots [11], healthcare monitoring [12], and human–machine interfaces [13,14]. As an important component in determining the performance of strain sensors, conductive materials generally include metallic materials (silver particles (AgNPs) [15,16], silver nano-wires (AgNWs) [16], gold nanowires (AuNWs) [17]), carbon-based materials (carbon black (CB) [18], carbon nanotubes (CNTs) [19,20], graphene [21,22] and nano diamond [23,24]), and intrinsically conducting polymers [25], etc. The combination of conductive materials with appropriate elastic polymer materials (silicone rubber [26], natural rubber (NR) [27], polyurethane(PU) [28]) can be fabricated strain sensors with high sensitivity and high stretchability.

The exploration of strain sensors with mechanical and chemical robustness, or functionality of water repellency enables the sensing coverage with a broader application scenario to revolutionize our future. The importance of strain sensors in harsh environments cannot be overstated, as these sensors play a critical role in ensuring the safety, reliability, and performance of essential components and structures. In harsh environments, mechanical stress, extreme temperatures, and corrosive substances can lead to material fatigue and structural degradation. Strain sensors provide real-time data on the mechanical deformation and stress experienced by components, allowing for the early detection of potential failures. This timely information enables preventive measures to be taken, preventing catastrophic incidents and ensuring the safety of personnel and assets.

As technology continues to advance, strain sensors will play an increasingly vital role in various industries, such as aerospace, oil and gas, and civil engineering and health care. In space missions, deep-sea exploration, and mining operations, where human presence is limited or not feasible, strain sensors play a crucial role [29–32]. These sensors provide critical data on the structural integrity and performance of remote systems and equipment, ensuring the success and safety of these missions. By analyzing the strain data, engineers can assess the integrity of bridges, pipelines, aircraft, and other infrastructures. Moreover, strain sensors are employed in research and experiments conducted in extreme environments, such as high-temperature environments, high-pressure conditions, or corrosive atmospheres. They enable scientists to study materials and phenomena under these conditions, advancing our understanding of materials science and engineering. In industries with stringent safety regulations, such as nuclear power or aerospace, strain sensors are essential for compliance and certification. These sensors provide valuable data to ensure that components and structures meet safety standards and can withstand harsh conditions. In addition, as the population grows and the demand for healthcare increases, remote and personalized health monitoring gradually becomes the focus of attention, which can improve the accuracy of health diagnosis and also identify problems in time. Wearable strain sensors have been used extensively to monitor human physiological information, such as heartbeat, pulse, monitoring of respiratory and arterial signals, etc. [33,34].

In recent years, the advancement of materials science and nanotechnology has led to the development of cutting-edge sensing technologies that can withstand and operate in extreme environmental conditions [35–43]. Among these innovative sensors, strain sensors based on carbon materials have emerged as promising candidates for a wide range of applications in harsh environments [44–46]. Carbon-based strain sensors possess unique properties such as mechanical robustness, chemical stability, excellent electrical conductivity, and high sensitivity, making them ideal for monitoring structural integrity, deformation, and mechanical stress in extreme conditions [47–50].

The term "carbon materials" encompasses a diverse group of substances, including carbon nanotubes (CNTs), graphene, carbon black, and nanodiamonds, among others. These materials have been used in various aspects [51–53]. Each of these carbon-based materials has specific structural and electrical properties that can be harnessed to design strain sensors suitable for various harsh environments. Traditional strain sensors have been widely used [54–64]; however, their limitations in extreme conditions have driven the need for advanced sensing technologies. Carbon-based strain sensors offer several advantages over conventional sensors, including their ability to withstand harsh environments, high mechanical strength, and compatibility with diverse materials and substrates.

Previously, a few review articles for flexible strain sensors based on carbon materials have been reported [6,54,65–70], but most of them are only focused on the strain sensor used in normal environments. Here, we focus on strain sensors based on carbon nanotubes, graphene, and nanodiamond in harsh environments. Some typical strain sensors based on carbon nanotubes, graphene, and nanodiamond will be summarized in this review due to the unique property of these three kinds of carbon materials (Figure 1). This review will introduce the properties of carbon nanotubes, graphene, and nanodiamond (Table 1) and then show how the fabrication of carbon-based strain sensors withstand extreme conditions; their sensitivity, durability, and response time in Table 2. After that, the application was introduced in detail, including strain/pressure sensors, temperature/humidity sensors, and power devices that are used in motion sensing, health monitors, electronic skin, and other applications (Figure 1). The discussion highlights the promising features and potential advantages offered by these carbon materials in strain sensing applications. This review will explore the advantages of carbon materials strain sensors in harsh environments, highlighting their unique characteristics and potential applications across various industries. From aerospace and automotive to healthcare and disaster rescue, these carbon-based strain sensors pave the way for enhanced safety, predictive maintenance, and performance optimization in the most demanding settings.

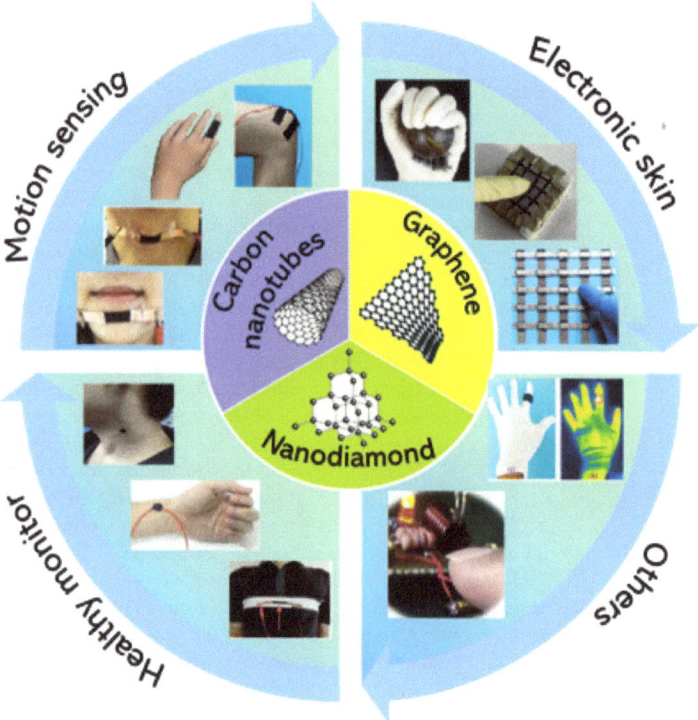

Figure 1. Application of strain sensors based on carbon nanotubes, graphene, and nanodiamond in harsh environments.

Table 1. The properties of carbon nanotubes, graphene, and nanodiamond.

Carbon Material	Electrical Conductivity	Thermal Conductivity	Young's Modulus	Crystal Lattice Parameters	Band Gap
CNT	$0.17\text{–}2.0 \times 10^7$ S m^{-1}	6600 Wm^{-1}K^{-1}	270 to 950 GPa	sp^2 1.7 nm	~0.2 eV
Graphene	224810^8 S m^{-1}	5000 Wm^{-1}K^{-1}	853.3 ± 0.9 Gpa	sp^2 0.25 nm	1.25 eV
Nanodiamond	400 S m^{-1}	2000 Wm^{-1}K^{-1}	880 ± 90 GP	sp^3 2 to 3 nm	~5.5 eV

2. The Property of Carbon Nanotubes, Graphene and Nanodiamond

In harsh and demanding environments, the need for reliable and resilient sensors to monitor mechanical strain becomes paramount. Carbon materials, including carbon nanotubes, graphene, and nanodiamonds, offer cutting-edge solutions as strain sensors in such challenging conditions. In Table 1, a comparison of the electrical conductivity, thermal conductivity, Young's modulus, crystal lattice parameters, and band gap of carbon nanotubes (CNTs), graphene, and nanodiamonds is listed [71–83]. Those carbon materials possess unique superiorities such as good electrical conductivity, high chemical and thermal stability, excellent mechanical property, low toxicity, as well as ease to be functionalized, endowing them with great potentiality for applications as strain sensors (Figure 1). These advanced carbon-based materials possess exceptional properties that enable them to withstand extreme temperatures, corrosive substances, and mechanical stress while accurately detecting and measuring strain.

Carbon nanotubes are one-dimensional nanomaterials with exceptional electrical conductivity and mechanical strength [73]. There are two classes of CNTs, single-walled carbon nanotubes (SWCNTs) and multiwalled carbon nanotubes (MWCNTs). Carbon nanotubes, tubular structures composed of rolled-up graphene sheets, are renowned for their exceptional mechanical strength. This property empowers them to endure high levels of mechanical stress in harsh environments without deformation or failure. Additionally, carbon nanotubes exhibit high sensitivity to mechanical strain, detecting even minute deformations with high accuracy. This is because carbon nanotubes have an exceptionally high aspect ratio, with lengths up to micrometers and diameters at the nanoscale, allowing for efficient load transfer and strain distribution within the material, enhancing their sensitivity to mechanical deformations even in harsh environments. Furthermore, the lightweight and small size of carbon nanotubes allow for seamless integration into various structures without compromising overall mass. By capitalizing on their electrical conductivity (0.17–2.0×10^7 S m^{-1}), carbon nanotubes serve as piezoresistive sensors, directly measuring electrical resistance changes in response to mechanical strain, thus ensuring precise and real-time strain monitoring. The nanoscale dimensions of carbon nanotubes make them suitable for sensing strain at small scales, such as in microelectromechanical systems (MEMS) or nanodevices, enabling applications in miniaturized and complex systems. Importantly, they have good chemical stability. Their chemical inertness shields them from corrosion and degradation when exposed to aggressive chemicals or harsh substances.

Graphene, a single layer of carbon atoms arranged in a two-dimensional honeycomb lattice, possesses extraordinary mechanical strength and superior electronic property [84]. Graphene, graphene oxide (GO), and reduced graphene oxide (RGO) have attracted much attention because of easy operation and high yields of bulk-quantity and they have been demonstrated for usages of strain sensors [85]. Graphene's atomic thickness ensures minimal interference with the mechanical properties of the host material, making it suitable for surface-mounted and embedded sensing. This intrinsic sensitivity to mechanical strain enables graphene to detect and measure even the most subtle deformations with precision. Graphene's exceptional mechanical strength, combined with its high electrical conductivity (10^8 S m^{-1}), results in superior sensitivity and fast response times when used as a piezoresistive strain sensor. These attributes are crucial for real-time monitoring of dynamic events and rapid mechanical changes in harsh environments. Similar to carbon nanotubes, graphene is chemically inert, resistant to corrosion, and stable in the presence of harsh substances. Its thermal stability allows it to operate effectively across a broad range of temperatures, making it an ideal candidate for applications in extreme thermal environments. Graphene can be synthesized in various forms, such as flakes, films, or as part of composites, allowing for versatile strain sensor designs catering to specific application requirements [86].

Nanodiamonds, nanometer-sized diamond particles composed of carbon atoms in a diamond crystal lattice, boast extreme hardness and durability. This exceptional mechanical strength renders them highly resistant to wear and damage in harsh environments. Like carbon nanotubes and graphene, nanodiamond (ND) is from a family of carbon materials with unique properties including extreme hardness, low friction coefficient, high corrosion resistance, high thermal conductivity, along with excellent mechanical and chemical stability. The extreme hardness of nanodiamonds (Young's moudulus 880 ± 90 GP) makes them resistant to wear and degradation, allowing for long-term and reliable strain monitoring, even in abrasive environments. ND also has a large surface area and possesses a versatile surface chemistry that can be tailored for each application via surface termination and functionalization [87,88]. Synthetic diamond, due to its chemical inertness and resistance, has the potential to be a perfect passivating material [89]. It is very cost-effective because of its production at bulk quantity using a detonation method [90,91]. Nanodiamonds demonstrate chemical inertness, preserving their stability and integrity in the face of corrosive substances. Furthermore, nanodiamonds can withstand high temperatures, making them well-suited for applications in harsh thermal environments. Their small size and surface functionalization capabilities offer the flexibility to tailor their properties for specific strain sensing applications. Nanodiamonds have excellent biocompatibility, making them suitable for strain sensing applications in biological or medical environments, including monitoring the mechanical strain of tissues or implants. Nanodiamonds can also serve as strain sensors based on stress-induced changes in their optical properties, providing an alternative sensing mechanism.

3. Fabrication Methods and Performance of Strain Sensor on Extreme Conditions

3.1. Fabrication Method of Strain Sensor on Extreme Condition

Creating a carbon-based strain sensor for harsh environments involves designing a robust and durable device capable of measuring mechanical stress or strain while withstanding extreme conditions. Various manufacturing techniques, such as coating, 3D printing, chemical vapor deposition, transfer, and spinning methods, are utilized in the production of carbon materials strain sensors. Coating technology is widely applied to manufacture thin film and fabric sensors, while 3D printing is employed for pattern generation [92]. Chemical vapor deposition is primarily used for carbon nanotubes growth. Additionally, the grown carbon materials can be transferred using a flexible substrate combined with a transfer method [93]. The integration of spinning [94] and coating processes enables the fabrication of high-performance fiber optic strain sensors. Furthermore, other methods like layer-by-layer (LBL) assembly technology, photolithography, and solution mixing [95] are also employed to manufacture flexible strain sensors. The fabrication methods of carbon nanotubes-, graphene-, and nanodiamond-based strain sensors reviewed are listed in Table 2.

3.1.1. Fabrication of Carbon Nanotubes Strain Sensors

Several synthesis methods have been reported to incorporate CNTs into various polymeric matrices, aiming to prevent CNT agglomeration and achieve uniform dispersion within the polymer matrix. However, there is no one-size-fits-all approach to attain perfect dispersion of different CNTs in various types of polymer matrices. The selection of an appropriate dispersion technique depends on several factors, including the physical state of the polymer (solid or liquid), the chemical nature (thermoplastic or thermoset), the dimensions and content of the introduced CNTs, the availability of fabrication processes and techniques, ease of synthesis from suitable monomers, desired performance characteristics of the composites, cost constraints, and the choice of solvent parameters. Li et al. [96] and Zhang et al. [97] chose the one-pot method to fabricate the CNT strain senor. Under UV light, the reaction occurred very rapidly and efficiently within a few minutes, showing fast and pollution-free characteristics [97]. In another method, the PVA was dissolved by heating in the water/glycerol(Gly) binary solvent, and the hydrogels were formed after several freeze-thaw cycles, with the hydrogen-bonded microcrystalline region acting as the crosslinking point [96]. A facile spraying method was used to prepare superhydrophobic and conductive coatings on elastic tape with a hierarchical fluorinated carbon nanotubes (FCNT)/SiO_2 nanoparticle structure in Figure 2a [98]. The flexible and super-hydrophobic elastic tape covered with FCNT/SiO2 was successfully prepared (Figure 2b), and the hierarchical FCNT/nanoparticle structure can be observed in Figure 2c.

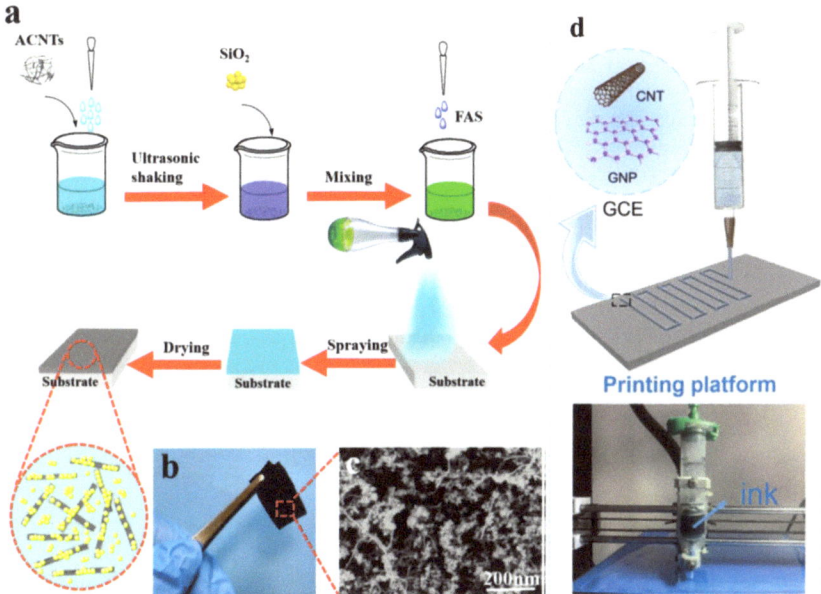

Figure 2. (**a**) Schematic diagram for the preparation of FCNT/SiO$_2$ coating on elastic tape. (**b**) Photograph of one piece of superhydrophobic and conductive elastic tape displaying excellent flexibility. (**c**) SEM image of the FCNT/SiO$_2$ coating [98]. (**d**) Schematic illustration of the direct ink writing (DIW) process of GCE fibers [99].

3.1.2. Fabrication of Graphene Strain Sensors

The development of advanced fabrication approaches for graphene-based strain sensors is in its early stages, with future progress in this direction being crucial. One of the primary goals is achieving highly efficient manufacturing of flexible sensors. Traditional fabrication methods often suffer from high processing costs and complex manufacturing processes, necessitating research into scalable, customizable, and environmentally friendly fabrication techniques suitable for large-scale production. Additive manufacturing and printing technologies show promise in this regard, enabling the creation of intricate 3D structures that expand the potential applications of these sensors. Another essential aspect is ensuring the production of sensors with high repeatability. When these sensors are used as arrays, variations among different sensors can be significant, requiring individual calibration and more complex circuitry. As the number of sensors increases, the tolerance for inter-sensor variability decreases significantly, posing challenges to the sensing operation. In composite strain sensors, inter-sensor variability often arises from the aggregation and uneven dispersion of conductive materials within the elastomeric matrix, especially when the sensor size is reduced. To ensure the reproducibility of fabrication processes, adopting a coupled printing technique may prove beneficial, enabling large-scale production of strain sensors with improved uniformity and reliability. Many researchers used a coating method to prepare graphene-based strain sensors in extreme situations [45,100–102]. Moreover, direct ink writing (DIW) is also a facile way. Zhu et al. developed a DIW assembly approach for fabricating fiber-shaped strain sensors by incorporating graphene nanoplatelet (GNP)/carbon nanotube (CNT) hybrid fillers into the silicone elastomer, as shown in Figure 2d [99].

3.1.3. Fabrication of Nanodiamonds Strain Sensors

Nanodiamonds are typically synthesized through various methods, such as detonation, chemical vapor deposition (CVD) [89,103], or high-pressure high-temperature (HPHT) processes. Detonation synthesis involves detonating explosives in a controlled environment to create high-pressure and high-temperature conditions, leading to the formation of nanodiamonds. CVD involves introducing carbon-containing gases into a reactor at high temperatures to deposit nanodiamonds on a substrate. HPHT utilizes extreme pressure and temperature to convert carbon sources into diamond structures. After synthesis, nanodiamonds may require purification to remove impurities and unwanted by-products. Functionalization can be performed to modify the surface properties of nanodiamonds, enhancing their dispersion and compatibility with the substrate and sensor materials. Choose a suitable substrate material based on the application requirements, such as flexibility, rigidity, and compatibility with nanodiamonds and other sensor components. Common substrates include silicon, glass, flexible polymers, or even textile materials. Nanodiamonds can be deposited on the substrate using techniques like spin coating, dip coating, or inkjet printing. In coating [104], a solution containing nanodiamonds is dispensed onto the substrate and then spun to achieve a uniform coating. In inkjet printing [105], nanodiamonds are dispersed in a liquid solution and printed onto the substrate in a controlled pattern. A flower-like molybdenum disulfide (MoS_2)/nanodiamond (ND) nanocomposite was successfully prepared via an easy hydrothermal synthesis method and fabricated into humidity sensors in Figure 3 [106].

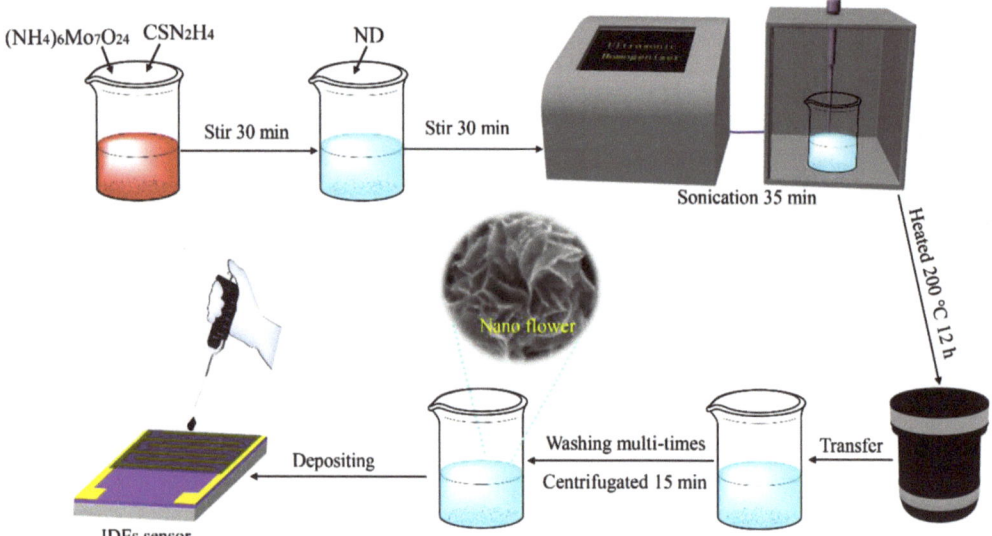

Figure 3. Fabrication route of MoS_2/ND nanocomposite via an easy hydrothermal synthesis method [106].

3.2. Performance of Strain Sensors in Harsh Environment

3.2.1. Ability to Resist Harsh Environment

Strain sensors based on carbon materials possess unique properties that allow them to resist and perform well in harsh environments. These sensors exhibit mechanical robustness, chemical stability, high thermal resistance, and excellent electrical conductivity, making them suitable for monitoring strain in extreme conditions. Carbon materials, such as carbon nanotubes and graphene, are known for their exceptional mechanical strength and toughness. They can withstand high mechanical stress, including tension, compression, and bending, without undergoing significant deformation or failure. This mechanical robustness enables strain sensors to operate reliably in environments with intense mechanical forces, such as aerospace applications, heavy machinery, or offshore structures. The strain sensors used in harsh environments have been summarized in Table 2.

Chemical Stability: Many carbon-based materials, including carbon nanotubes, graphene, and nanodiamond, exhibit outstanding chemical stability. They are resistant to corrosion, oxidation, and chemical reactions, even when exposed to aggressive substances or harsh chemicals. This chemical stability ensures that strain sensors remain functional and accurate in chemically harsh environments, such as chemical processing plants or oil and gas facilities. Especially, Ding and Zhu reported that the strain sensors can resist the strong acid and strong alkaline, pH 2–13 [107] and pH 1–14 [108].

Environment Stability: The sensor's sensitivity can also be affected by environmental factors such as temperature and humidity. Stability under different conditions is essential for accurate strain measurements. Carbon materials are known for their high thermal stability, enabling strain sensors to maintain their structural integrity and electrical conductivity at high and low temperatures and high humidity. Unlike some conventional materials, carbon-based strain sensors do not become brittle or lose their electrical properties under extreme hot or cold conditions. Carbon materials possess excellent thermal conductivity and high thermal stability. Strain sensors based on carbon materials can operate effectively at elevated temperatures, withstanding extreme heat and thermal fluctuations. This thermal resistance is crucial in applications where sensors need to monitor strain in high-temperature environments, such as aerospace propulsion systems or industrial furnaces. Carbon materials generally exhibit low thermal expansion coefficients, meaning they experience minimal dimensional changes in response to temperature variations. This characteristic ensures that strain sensors based on carbon materials can accurately detect mechanical deformations without significant interference from temperature-induced strain. The inherent electrical conductivity of carbon materials, such as carbon nanotubes, graphene, and nanodiamond remains intact at low temperatures. This property allows strain sensors to maintain their piezoresistive effect even in cold environments, ensuring reliable and sensitive strain measurements. Carbon nanotubes-based strain sensors are stable even at high temperatures, 316 °C [109], or at ultralow temperatures, −196 °C [94], and even work under water [99]. For example, a piezoresistive sensor based on nanodiamond proved that the working temperature is 25 °C up to 300 °C [89].

3.2.2. Sensitivity

The sensitivity of a carbon-based strain sensor refers to its ability to detect and respond to small changes in mechanical strain. In other words, it measures how effectively the sensor can convert mechanical deformation into an electrical signal. Sensitivity is a critical parameter for strain sensors as it determines the level of accuracy and precision in measuring strain.

Gauge Factor

The sensitivity of a strain sensor is characterized by the gauge factor (GF), and the GF of a resistive-type strain sensor is expressed as Equation (1).

$$\text{GF} = \frac{(\Delta R/R_0)/R_0}{\varepsilon} = \frac{[(R-R_0)/R_0]/R_0}{\varepsilon} \qquad (1)$$

where R is the actual measured resistance value during sensor deformation, R_0 is the initial resistance value of the sensor, and ε is the strain value that occurs when the sensor is subjected to an external force. The higher GF values indicate that the sensor is more sensitive to the applied strain. The GF values of different CNT-, graphene-, and nanodiamond-based strain sensors in harsh environments are listed in Table 2.

Piezoresistive Effect

The piezoresistive effect is a phenomenon observed in certain materials where their electrical resistance changes in response to applied mechanical stress or strain. In other words, the electrical conductivity of the material is influenced by mechanical deformation. The term "piezoresistive" is derived from "piezo", meaning pressure or stress, and "resistive", referring to electrical resistance. Piezoresistive materials are widely used in the construction of strain sensors and pressure sensors. When these materials are subjected to mechanical stress or strain, their atomic arrangement undergoes changes that alter the movement of charge carriers, such as electrons, within the material. As a result, the electrical resistance of the material changes proportionally to the applied stress [110].

There are two main types of piezoresistive effects:

Positive Piezoresistive Effect: In materials exhibiting a positive piezoresistive effect, the electrical resistance increases with increasing applied stress. This means that as the material undergoes compression or tension, its resistance to the flow of electric current increases.

Negative Piezoresistive Effect: In materials showing a negative piezoresistive effect, the electrical resistance decreases with increasing applied stress. In this case, the material becomes more conductive under mechanical deformation.

The piezoelectric charge coefficient (d) is a material property that relates the electric charge (Q) generated in a material to the applied mechanical stress (σ) or strain (ε). It is a measure of a material's ability to convert mechanical energy into electrical energy and vice versa. The calculation of the piezoelectric charge coefficient depends on the specific measurement setup and units used.

In the direct piezoelectric effect, an applied mechanical stress or strain generates an electric charge in the material, while in the converse piezoelectric effect, an applied electric field induces mechanical strain or stress. The piezoelectric charge coefficient is typically given in units of picocoulombs per newton (pC/N) or coulombs per newton (C/N).

The general formula to calculate the piezoelectric charge coefficient (d) is expressed as Equation (2):

$$d = \frac{Q}{F} \qquad (2)$$

d is the piezoelectric charge coefficient (C/N or pC/N)
Q is the electric charge generated (in coulombs, C, or picocoulombs, pC)
F is the applied force (in newtons, N)

Piezoresistive materials commonly used in strain sensors include silicon and carbon-based materials. These materials are carefully integrated into the sensor's design to measure mechanical strain or pressure accurately and convert it into a corresponding electrical signal. The piezoresistive effect allows for the creation of highly sensitive and responsive sensors, making them valuable in various applications, including structural monitoring, medical devices, robotics, and automotive systems. Piezoelectric charge coefficient of a CNT-based strain sensor is 9.4 pC/N [95].

In some research, different definitions to assess the sensitivity of the strain sensor have been used.

- The sensitivity of the strain sensor is expressed as Equation (3) [94].

$$S = \frac{\Delta I}{I_0} \qquad (3)$$

The relative current change ($\Delta I/I_0$) under a changing tensile load. Here, ΔI defines the current difference between the loaded and unloaded fiber and I_0 is the original current.

- The sensitivity of the composite sensor is expressed as Equation (4) [97].

$$S = \Delta I/I_0 P \ (\Delta R/R_0 P) \qquad (4)$$

where P is the applied pressure, the relative resistance ΔI (ΔR) is defined as $|I(R) - I_0 (R_0)|$, where I_0 (R_0) is the initial electric current (resistance) and I (R) represents the real-time electric current (resistances) under various deformations.

- The sensitivity of the pressure sensor (S) is defined as Equation (5) [111].

$$S = (\Delta R/R_0)/\Delta P \qquad (5)$$

as the derivation of the resistance change rate ($\Delta R/R0$) with pressure.

- The sensor sensitivity (S) is calculated by Equation (6) [106].

$$S = (C_X - C_{11})/(RH_x - RH_{11}), \quad (6)$$

where C_{11} is the sensor capacitance at 11% RH, and C_X represents the capacitance of the sensor in the x% values.

3.2.3. Response and Recovery Time

The response and recovery time of a strain sensor refer to the time it takes for the sensor to detect and register a change in mechanical strain (response time) and the time it takes for the sensor to return to its original state after the strain is removed (recovery time).

The response time of strain sensors is relatively fast, typically in the millisecond to microsecond range, depending on the specific sensor and its application. It is proved that a strain sensor based on rGO has very fast response time, only 22 ms [102]. Most of response time of the strain sensors that we reviewed in this review are all less than 1 s detailed in Table 2. The recovery time may also be relatively quick, but it can vary depending on the material properties and the sensor's design.

It is important to note that the actual response and recovery times for a strain sensor will depend on the specific sensor's characteristics, the materials used, and the intended application. Manufacturers and researchers continuously work to improve the speed and accuracy of strain sensors for various industries and applications, including structural health monitoring, robotics, aerospace, and wearable devices.

3.2.4. Durability

Dynamic durability is the ability of a strain sensor to maintain stable electrical conductivity after a certain cycle or over time. Ideally, the sensor should ensure that the resistance value does not change during the continuous stretch and release cycle. At the same time, the sensor will be able to perform a longer cycle test to ensure that it has a longer service life. To perform durability testing, the sensor is cyclically stretched and released at a certain strain value. Most of the sensors prepared based on CNTs/graphene/nanodiamond in harsh environments reported so far can achieve more than 1000 stretch release cycle tests, as shown in Table 2. The fiber-based sensors can provide better tensile performance while providing excellent dynamic durability of the sensor. A graphene/CNT/silicone composite strain sensor response of the strain sensor possesses excellent stability and durability without an apparent change in the resistance signal, fully proving its stable and repeated recoverability during the 10,000 stretching/releasing cycles [99].

Ensuring reliable and enduring performance, the durability of strain sensors in harsh environments is a paramount concern. Such environments encompass extreme temperatures, corrosive chemicals, high mechanical stress, radiation exposure, or a combination of these factors, presenting formidable challenges to strain sensors. Key factors influencing their durability include material selection, favoring robust and resilient options like carbon nanotubes, graphene, nanodiamonds, ceramics, and piezoelectric materials. To shield sensitive components from the environment, proper encapsulation and packaging techniques are employed, incorporating hermetic sealing and protective coatings. Mechanical durability is addressed through sensor designs tailored to withstand high mechanical stress, shocks, and vibrations, crucial when sensors are integrated into moving or vibrating structures. Washing procedure has a direct impact on electrical conductivity and abrasion resistance is one of the factors that affects the life and wear performance of the fabric. Rehman et al. [104] found that the fabric coated with PANI-ND(polyaniline-nanodiamond) improved significantly along with durability against washing and abrasion, without affecting its breathability. These PANI-ND nanocomposites are still present on the surface of the fabric even after five commercial washing cycles using 20 mL/g liquor-to-fabric ratio at 40 °C for 30 min along with 25 steel balls. Moreover, this nanocomposite helped to resist the degradation of pristine wool fibers during 20,000 abrasion cycles.

Table 2. Summary of fabrication methods and performance of strain sensors in this review.

Carbon Material	Fabrication Method	Property	Sensitivity (GF/S)*	Response Speed	Durability	Ref.
MWCNT	unidirectional freeze-drying method	25, 177, and 316 °C	1.23	stable	high-strain cyclic compression	[109]
SWCNT	one-pot method	−70 to 25 °C	3.76	—	1600 cycles	[96]
MWCNT	a one-pot radical polymerization process	−20 to 80 °C	1.77 at 25 °C	—	capacitance almost the same (1000 cycles)	[36]
Graphene	free radical polymerization	−20 to 70 °C	—	—	88.2% of capacitance retention (6000 cycles)	[112]
CNT	solution mixing	60% RH condition	piezoelectric charge coefficient 9.4 pC/N	—	—	[95]
CNT	freeze drying and thermal imidization technique	250 °C	1.4	response time 50 ms/recovery time 70 ms	1000 cycles	[113]
rGO	the conformal coating	−40 °C/25 °C and 50% RH for 20 days	Pulse rate of 75 beats/min	0.1 s	2000 times (2 kPa)	[45]
CNT	wet spinning method	−196 to 100 °C	$S = 0.12$	—	long life (1000 cycles)	[94]
MWCNT/RGO	dip-coating	−30 and 80 °C	—	—	1000 cycles	[100]
CNT	a convenient solvent replacement strategy	−60 to 60 °C	8.5	200 ms	30 days in normal environment	[114]
rGO/CNT	vacuum filtration process	−40 °C to 200 °C	—	—	90% capacitance (105 cycles)	[115]
Graphene/CNT	DIW	humidity/water exposure	14550.2 (100%)	170 ms	10,000 cycles	[99]
CNT	vacuum-assisted formation of thin CNT networks	RH 35–55%/water exposure	6–10	—	—	[116]
CNT	one-pot	25–70 °C	$S = 20.3$	—	compressive strain (0.1–70%) and loading weight (55–150 g)	[97]
rGO	chemical reduction, thermal reduction	−60 °C	—	—	87.5% capacitance (5000 charge/discharge cycles at −20 °C)	[117]
CNT	CNT suspension was mixed with Ti$_3$C$_2$T$_x$ suspension	−20 to 80 °C	13.3 (60.3%)	—	1000 cycles	[118]

Table 2. Cont.

Carbon Material	Fabrication Method	Property	Sensitivity (GF/S)*	Response Speed	Durability	Ref.
CNT	a direct pyrolysis process	−196 °C	1.65	—	5000 cycles (40%)	[119]
CNT	the film transfer approach	water	2.4	—	—	[93]
CNT	drying mixing method	hot-humidity/−40 to +50 °C	high sensitivity	—	150 cycles (50%)	[120]
SWCNT	dip coating of SWCNTs/chemical polymerization	144.6 °C	9.57	—	500 cycles (20%)	[121]
FCNT	a facile spraying method	moisture, acidic, and other harsh environments	1800	—	—	[98]
CNTs/rGO	coating	water	685.3 (482%)	response/recovery 200 ms	1000 cycles	[92]
MWCNT	spray coating	water, pH 13/2	69.84 (65–80%),	60-80 ms	over 1000 cycles	[107]
CNTs/graphene	freezing-drying	160 °C	high sensitivity (0.25 kPa^{-1})	120 ms	>800 cycles	[111]
graphene–CNT	thin film using a bar-coating technique	−150 °C to +150 °C	118	—	5000 cycles	[101]
GO/RGO	pad dyeing	22.8 °C to 47.3 °C RH 39–71%	2.79 (0–50%)	<60 ms	3000 cycles (10%)	[122]
graphene	one-pot solution-casting process	pH 1–14/humidity	173.17 (100%)	—	2000 cycles (50%)	[108]
rGO	dip-coating	humid, sweaty, underwater, −40 °C	−2.08 (0–60%)	22 ms	4000 cycles	[102]
Nanodiamond	coating/in situ polymerization technique	40 °C for 30 min along with 25 steel balls	1.4 (40–100%)	—	washing/rubbing	[104]
Nanodiamond	hydrothermal synthesized method	11–97% RH	S = 3500, relative humidity 100%	response < 1 s/recovery time~0.9 s	stable one month	[106]
Nanodiamond	vacuum-assisted filtration	100.4 °C	thermal conductivity of 17.43 W/m·K	—	—	[123]

Table 2. Cont.

Carbon Material	Fabrication Method	Property	Sensitivity (GF/S)*	Response Speed	Durability	Ref.
Nanodiamond	self-assessable	thermal stability	thermal conductivity of 30.99 and 6.34 W/m·K	—	1000 cycles	[124]
Ultrananocrystalline diamond	hot filament chemical vapor deposition (HFCVD) and photolithographic and etching processes	25 °C–300 °C	9.54 ± 0.32	—	—	[89]
Nanodiamond	inkjet printing locally	against harsh media	diamond layer does not hamper the stability of the device	—	—	[105]
Nanodiamond	a directed patterned growth of NCD film by microwave plasma-enhanced chemical vapour deposition (CVD)	200 °C	8 ± 0.5	—	—	[103]

4. Carbon Materials Based Strain Sensor on Extreme Temperature Condition

The operating temperature of the strain sensors is crucial to their wide applicability. Some strain sensors are often used at room temperature, but their performance may deteriorate at low and/or high temperatures. Therefore, it is crucial to develop strain sensors that can be used at low and/or high temperatures to expand their range of use.

4.1. CNT-Based Strain Sensor on Extreme Temperature Condition

Anisotropic composite polyimide aerogels (CPIAs) were prepared via the unidirectional freeze-drying method using PI(Polyimide) as the polymer matrix and MWCNTs as the filler [109]. The obtained CPIAs exhibited low density, high porosity, good mechanical behavior, and stable electrical responsiveness across a wide temperature range (from 25 to 316 °C). As a strain sensor, the CPIA also works effectively as a lightweight sensitive strain sensor gauge factor (GF) of 1.23 and stability cyclic testing over 1000 cycles. Due to the good thermal stability of the CPIAs, under high temperature conditions, the CPIAs' responsiveness remained stable, as did their gauge factor values, which were not significantly affected by temperature. There is also other research about strain sensors on high temperature environment. The polypyrrole (PPy)/single-walled carbon nanotube (SWCNT)/polydopamine(PDA)/cotton composite fabric reach showed outstanding heating performance at 144.6 °C [121]. A polyimide (PI)/carbon nanotubes (CNT) composite aerogel with the merits of super elasticity, high porosity, robustness, and high-temperature resistance was successfully prepared, and can be stable even at 250 °C [113]. Li et al. [96] reported a conductive hydrogel consisting of a glycerol (Gly)-water binary solvent and added tannic acid (TA)-coated carboxymethylated cellulose nanofibrils (CMCNFs) to poly (vinyl alcohol) (PVA) as a functional filler to improve the hydrogel's mechanical properties. This frost-resistant strain sensor showed strain sensitivity (GF of 3.76), and cyclic stability (1600 cycles). The differential scanning calorimetry (DSC) curves of the hydrogel with different Gly contents are horizontal straight lines in the temperature range of −70 to 25 °C, indicating that no phase transition occurred in the Gly/water binary solvent system during the heating process in this range. In addition, Huang et al. [94] reported light weight porous aramid nanofibers (ANF) and carbon nanotubes (CNT) aerogel fibers coated with polypyrrole (PPy) layers prepared with low density (56.3 mg/cm^3), conductivity (6.43 S/m), and tensile strength (2.88 MPa), which were used as motion sensors with high sensitivity (0.12) and long life (1000 cycles). They also showed excellent mechanical strength and flexibility, a strong bending property at low temperatures, −196 °C, and high-temperature resistance at 100 °C, which is mainly due to the good network structure of the ANF aerogel. An environment tolerant conductive hydrogel based on nanocomposite polyacrylamide/montmorillonite/carbon nanotube (MMCOHs) was prepared by a convenient solvent replacement strategy, featuring remarkable mechanical properties, anti-freezing ability (−60 °C), long-term environmental stability (>30 days), and anti-drying behavior (60 °C), simultaneously [114]. MMCOH displayed excellent flexibility to twist at will and could be easily bent at −60 °C, as shown in Figure 4a, and had stretchability and conductivity at −60 °C, as shown in Figure 4b,c. The sensor based on MMCOH displays an extremely wide sensing range (0–4196%), high sensitivity (GF = 8.5), rapid response time (200 ms), and excellent durability, benefiting from the inherent stretchability and stability of the organohydrogel. The MMCOH was obtained after directly immersing polyacrylamide/montmorillonite/carbon nanotube hydgrogels (MMCH) into a glycerol solution through a solvent-replacement strategy. Other strain sensors based on CNT are well developed and the temperature resistance is varied (20~80 °C [36,118], 25~70 °C [97], and −40~50 °C [120], and in the evening reaching a temperature of −196 °C [119]).

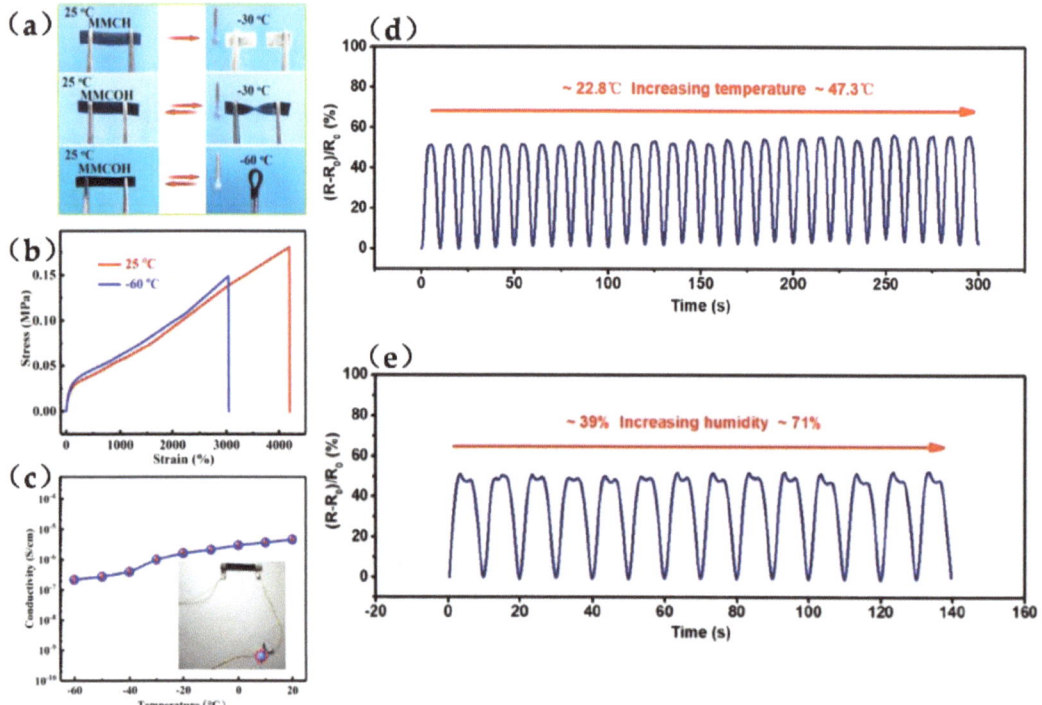

Figure 4. The environmental tolerance of the MMCOH. (**a**) Photographs of the anti-freezing behavior of MMCH and MMCOH. (**b**) Stress–strain curves of MMCOH at 25 and −60 °C. (**c**) Conductivity of MMCOH from −60 to 20 °C [114]. (**d**) Electrical signals of the sensor under a cyclic strain of 15% at temperatures from ≈22.8 °C (room temperature) to ≈47.3 °C, presenting stability of the sensor at different temperatures. %). (**e**) Electrical outputs of the GMF sensor with a recurrent strain of 15% in the process of humidity changing from ≈39 to ≈71%, presenting stability of the sensor in different humidities [122].

4.2. Graphene Based Strain Sensor on Extreme Temperature Condition

A highly breathable graphene-modified fabric (GMF) strain sensor with an insensitive response to changes in temperature and humidity within a certain range was developed, which possesses sensing independence against temperature from 22.8 to 47.3 °C and relative humidity ranges of 39% to 71%. Figure 4d [122] displays the electrical response of the GMF sensor under a cyclic strain of 15% in changeable temperature. The signals maintain stability with highly reproducible signaling patterns when the temperature increases from 22.8 to 47.3 °C. The insensitivity of the GMFs sensor to temperature change is attributed to the rGO on the Calotropis gigantea yarn (CGY) surface exhibiting a decrease in the hopping charge transport number and the slight thermal generation of fewer carriers. The electrical signals of the sensor were measured under a recurrent strain of 15%, and the results indicate that the electrical response remains highly steady and repeatable during the humidity change process (Figure 4e). Moreover, the nanocomposite organohydrogel is prepared from the conformal coating of functionalized rGO network by the hydrogel polymer networks consisting of PVA, phenylboronic acid grafted alginate (Alg-PBA), and polyacrylamide (PAM) in the binary ethylene glycol (EG)/H_2O solvent system (Figure 5a) [45]. The nanocomposite organohydrogel exhibits reliable anti-freezing properties (−40 °C) and self-healing properties without any other external stimuli, and can be stably stored for 20 days (Figure 5b,c). Deng et al. [112] reported that a N-doped hydrogel electrode showed excellent temperature stability in the range of −20 to 70 °C and a stretchability that reached 950%. A low-temperature-resistant organohydrogel electrolyte could be stretched up to 400% strain and still maintain stable mechanical properties up to −60 °C, combining the rGO electrodes with the organohydrogel electrolyte [117]. GO/RGO can be mixed with other carbon materials to form strain sensors that can be stable at low and high temperatures [99–101,115]. Li et al. reported a high performance textile-based stretchable supercapacitor stable over a wide temperature range, from −30 to 80 °C [100]. In this study, a multiwalled carbon nanotubes (MWCNT)/reduced graphene oxide (RGO) nanocomposite was used to fabricate a stretchable electrode. These carbon-based electrode materials are expected to be stable over a wide range of temperatures because the energy storage mechanism of electrical double-layer capacitors relies only on the electrostatic adsorption and desorption of electrolyte ions onto the electrode surface [29,40]. Furthermore, the encapsulated supercapacitors retain the capacitance during being immersed in water for a few days. Similarly, a flexible serpentine-structured graphene–CNT paper strain sensor in variable temperature environments was tested using a cryogenic and hot setup and the device showed robust performances, from −150 °C to 150 °C, which also demonstrated a high sensitivity with a GF of 118 [101]. A flexible rGO/CNT hybrid film can operate at extreme temperatures (down to −40 °C and up to 200 °C) with excellent electrochemical property and high durability [115].

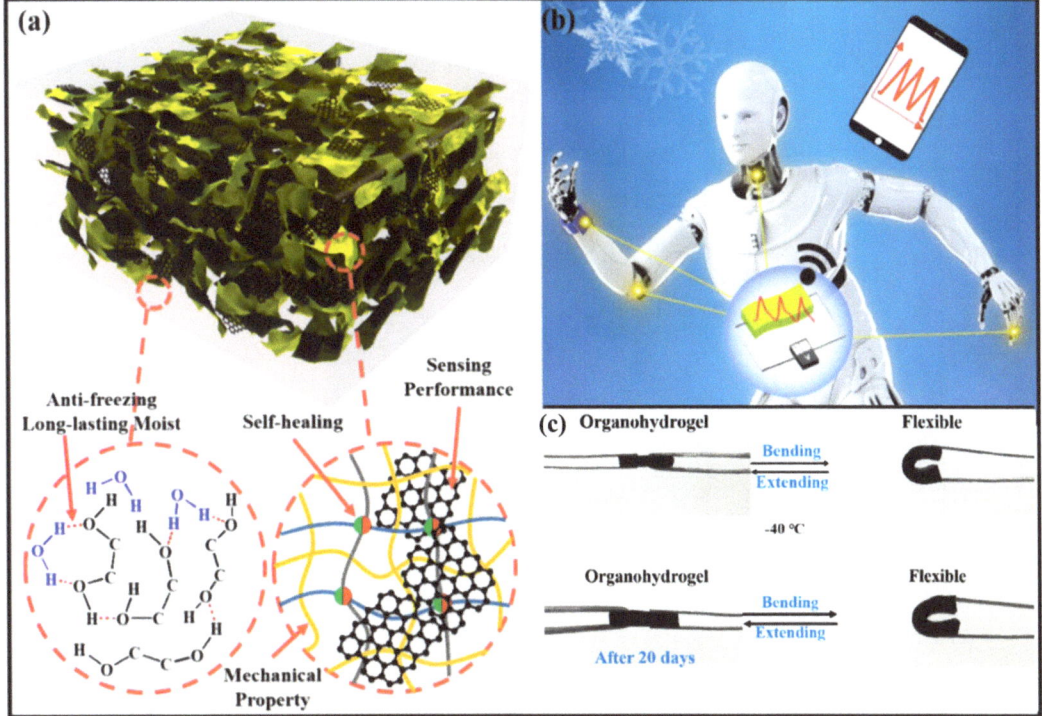

Figure 5. (**a**) Acrylamide (AM), Alg-PBA, and PVA were sequentially added to the rGO-containing EG/H$_2$O solution, and the polymerization was carried out to obtain PAM network and the conductive nanocomposite organohydrogel. (**b**) Conductive nanocomposite organohydrogel could be assembled as human-motion sensors with low temperature tolerance (−40 to 0 °C) for accurate detection of both tiny (swallowing and pulse) and large (finger bending and elbow bending) human activities. (**c**) Organohydrogels were bent after cooling at −40 °C and after 20 days of storage at 25 °C and 50% humidity, the organohydrogels maintained excellent elasticity [45].

4.3. Nanodiamonds Based Strain Sensor on Extreme Temperature Condition

A piezoresistive sensor prototype based on n-type conductive ultrananocrystalline diamond (UNCD) was developed. The respective piezoresistive response of such films was analyzed and the gauge factor was evaluated in both transverse and longitudinal arrangements, also as a function of temperature from 25 °C up to 300 °C [89]. Another piezoresistive sensor was produced. The gauge factor of boron-doped nanocrystalline diamond (NCD) films was around 8, and in the range from room temperature up to 200 °C, the sensor was still stable [103]. Diamond-based film with sensing function has been widely applied and can also be used in wearable sensors in the future. A cellulose nanofibers (CNF)/nanodiamond (ND)/MXene (CNM) composite film introducing ND via vacuum-assisted self-assembled method was produced [123]. Nacre-like structure and strong hydrogen bonding between MXene and CNF endow satisfactory mechanical properties (89.14 ± 3.61 MPa). It is demonstrated that the central surface temperature of LED/CN10M30 rises to 100.4 °C at 30 s, and the temperature difference is 3.7 times that of bare LEDs. Therefore, the above results demonstrate that the CNM composite films have excellent auxiliary heat dissipation capacity, which can greatly improve reliability and extend its service life. Wang et al. [124] prepared poly-(diallyldimethylammonium chloride)-functionalized nanodiamond (ND@PDDA)/aramid nanofiber (ANF) composite films using a self-assembly strategy. Owing to a strong interfacial interaction arising from electrostatic attraction, ND particles attract strongly along the ANF axis to form ANF/ND "core—sheath" arrangements. Thanks to this structure, the incorporation of ND@PDDA can significantly protect the ANF fibrils and improve the thermal stability of the composite film. The heat-resistance index (THRI) of the 50 wt % ND@PDDA/ANF composite film reached 279.33 °C, which was 31.59 °C higher than that of the pure ANF film (247.74 °C). The improved thermal stability enables the ND@PDDA/ANF composite films to operate normally under high temperature conditions without thermal decomposition. Notably, this nanofiber showed outstanding flame retardancy. Flame retardancy is required for applications in high temperature environments. A pure ANF film and a 50 wt % ND@PDDA/ANF composite film were exposed to a flame in air (Figure 6a,d). The pure ANF film ignited and was completely burned out within 3 s (Figure 6b,c). Figure 6 indicated the 50 wt % ND@PDDA/ANF composite film kept its initial shape and exhibited nonflammable behaviors owing to the protection provided by ND@PDDA.

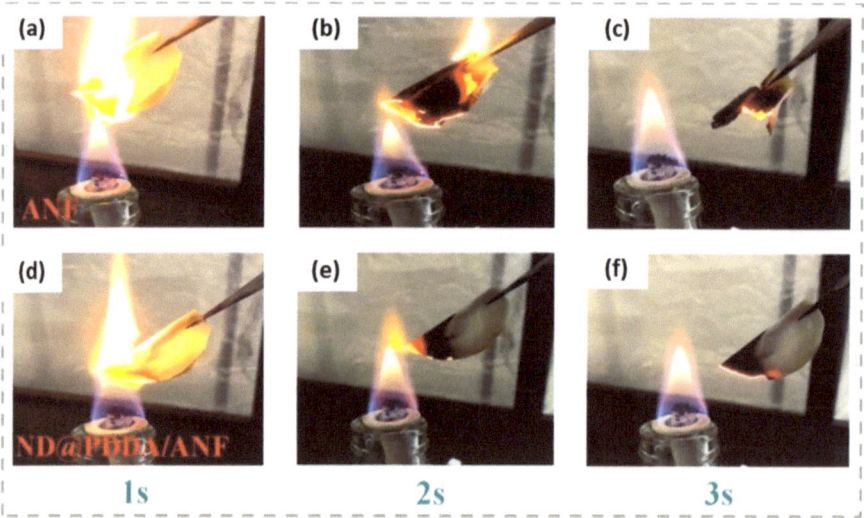

Figure 6. (**a**–**e**) Images showing flame test of the pure ANF film. (**a**–**f**) Images showing flame test of the 50 wt % ND@PDDA/ANF film [124].

5. Carbon Materials Based Strain Sensors on Extreme Humidity or Other Harsh Condition

Current strain sensors still face challenges such as being prone to failure under humid or cold conditions, lack of washing durability and chemical fragility. Exploration of wearable strain sensors in diverse application scenarios is one global requirement for shaping the future of our intelligent community. Apart from extreme temperatures, most strain sensors suffer from the constraint of operation in dry environments due to susceptibility to chemical corrosion or interference in humid circumstances [125,126]. Recently, the fabrication of strain sensors with higher sensitivity and larger detection areas has received considerable attention [55,85,127], but few studies have focused on wearable sensors under extreme conditions, such as intense UV irradiation, large diurnal temperature differences, or extremely cold weather. The exploration of strain sensors with mechanical and chemical robustness, or with the function of water repellency, enables a sensing coverage with a broader application scenario to revolutionize our future [128,129]. For example, a strain sensor with chemical resistivity can be potentially applied for a human machine interaction system as the operation tool even under harsh conditions. A device with waterproof capability also enables underwater applications by transducing the mechanical behaviors, e.g., anger bending, to electrical signals for communications that are difficult to achieve via sound waves [99].

5.1. CNT-Based Strain Sensor on Extreme Humidity or Other Harsh Condition

A flexible film with a sandwich-like structure consisting of thermoplastic elastomer (TPE), multiwalled carbon nanotubes (MWCNTs), and polydimethylsiloxane (PDMS) showed extreme repellency to water, salt, acid, and alkali solutions [107]. In this study, the film was dipped in these solutions and the conductivity was measured every 5 min for 60 min. Figure 7a shows that the resistance of the film sustained around 4.15–4.20 KΩ without obvious change in water, acid, alkali, and salt solutions. This may be due to the air layer trapped on the surface which can inhibit the attack of acid or alkali. After the film was immersed in water, salt, or acid/alkali solutions for 7 days, and then rinsed with water and dried at 60 °C, the contact angle (CA) of the film still remained above 156° (Figure 7b), indicating excellent durability of super-hydrophobicity to acid, alkali, salt, and water solvents corrosion. Moreover, Gao et al. [98] reported a strain sensor that displayed outstanding long-term stability, because CAs and surface resistance of this kind of sensor varied little when dipping it in water, acidic, and alkaline solutions. It is coated by a CNT/SiO_2 that can be used under moist, acidic, or other harsh conditions without sacrificing its conductivity and super-hydrophobicity, displaying outstanding anti-corrosion properties. A highly stretchable and sensitive strain sensor made of CNT–PDMS nanocomposites showed relative stability in high humidity environments [116]. Another water-resilient strain sensor was developed: an SWCNT was encapsulated in a non-fluorinated superhydrophobic coating, providing water resistance during elastic deformation [93,101]. Notably, the water-repellence was maintained in corrosive environments, such as in acidic, alkaline, or saline aqueous solutions, evidencing its applicability in harsh environments. Moreover, some strain sensors showed humidity-resistance, temperature-resistance, and ethane-resistance at the same time. CNT/CB composites show outdoor usage robustness, including excellent salt-spray resistance, hot-humidity stability, and a wide operation temperature range of −40 to +50 °C [120]. It is seen that the modulus of the samples both maintains the constant at the work temperature range between −40 and +50 °C (Figure 7c). It is noticed that the resistance barely changes for the whole temperature range (Figure 7d). The electrical resistance exhibits only a tiny increase of 0.13% when the composites suffer water vapors at the temperature of 60 °C and at a relative humidity of 75% (Figure 7e). Their liquid resistance was identified by the two liquids of water and ethane, common in our daily lives. After they were completely immersed for one day at room temperature, the resistance variations were only 0.5% and 2.6% for water and ethane (Figure 7f). It is really important that CNT/CB composites have desirable ambient environmental resistance to serve their sensor applications [130–133].

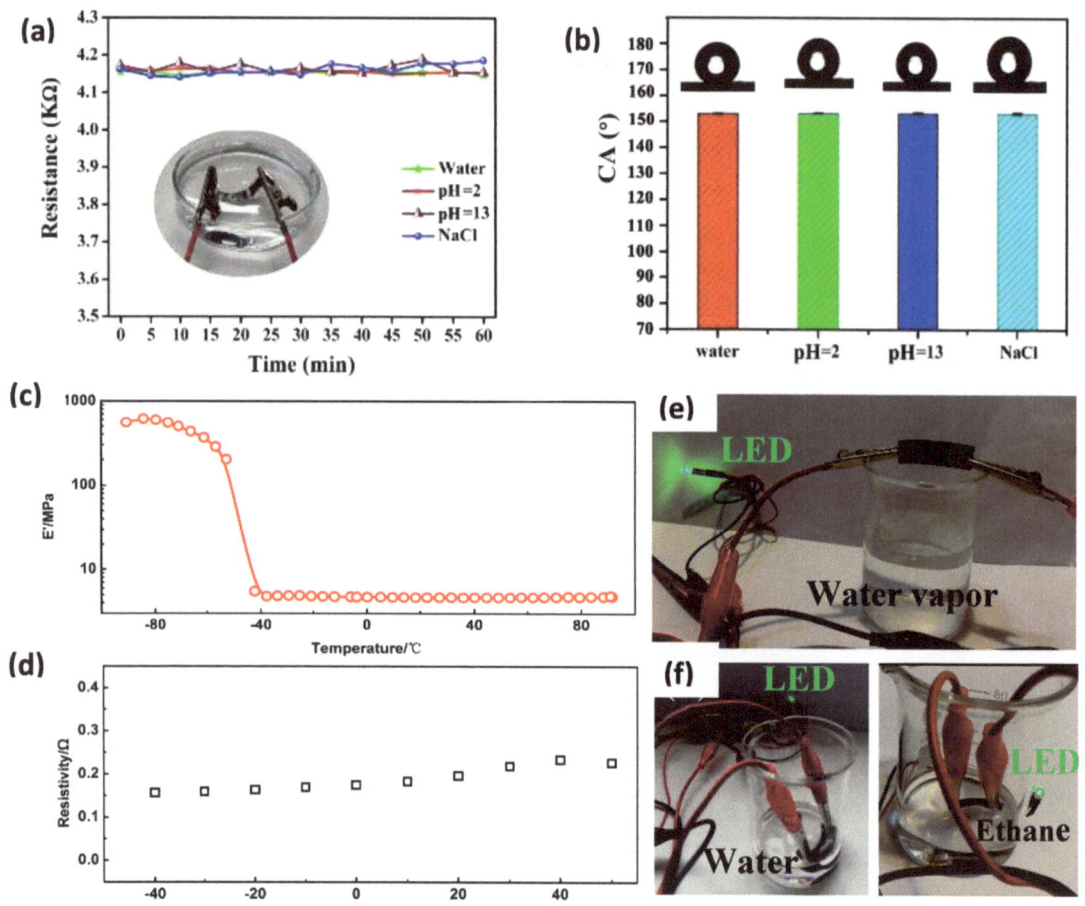

Figure 7. (**a**) Conductivity stability test of the film in water, acidic (pH = 2), alkaline (pH = 13), and salt solutions. Inset of shows the "silver mirror" phenomenon in water and test status of the film. (**b**) CA of the functional film after immersion in different solutions for 7 days [107]. (**c**) The dependence of storage modulus E_0 on the temperatures of −80 to +80 °C. (**d**) Resistivity of the CB/CNT composite in the temperature range of −40 to +50 °C. (**e**,**f**) A self-made device with the composite sample and LED lights is placed in the water vapor with a temperature of +60 °C and a humidity of 75% and is immersed in water and ethane for 1 day. The brightness of the LED light does not change during the two processes [120].

5.2. Graphene Based Strain Sensor on Extreme Humidity or Other Harsh Condition

A reduced graphene oxide (rGO) conductive fabric was first obtained by electrostatic self-assembly of chitosan (CS), as shown in Figure 8 [102]. A polyamide/spandex knitted fabric (PSKF)@rGO/SiO$_2$-PDMS sensor showed that the sensor exhibited a broad strain range of 0 to 60% and a negative GF up to −2.08, with excellent cycling stability and durability during 4000 linear strain cycles. Importantly, this sensor was superhydrophobic and resistant to chemical corrosion, enabling its use in high humidity and low temperature environments and hazardous scenarios. When the fabric sensor was submerged in water, ethanol, toluene, hexane, acetone, hydrochloric acid aqueous solution (pH = 1), and sodium hydroxide aqueous solution (pH = 13) for 24 h, the water contact angles (WCA) of the fabric sensor surface showed only a negligible decrease and remained above 150°, indicating good durability of the fabric sensor against various types of chemical corrosion (Figure 8b). In addition, the laundry endurance of the fabric sensor was further assessed with reference to the national standard (GB/T 12490-2014). Although the WCA of the fabric surface gradually decreased with accumulated washing cycles, the fabric surface still maintained its super-hydrophobicity even after ten washing cycles. Zhu et al. reported an environment-tolerant strain sensor made by incorporating hierarchical cellulose nanocrystal/graphene (MCNC-GN) nanocomplexes into a polydimethylsiloxane (PDMS) matrix [108]. The assembled strain sensor possessed a favorable sensitivity, stretchability, durability, and environment stability. The harsh environment resistance of composite elastomers was evaluated by a series of water resistance and UV irradiation tests. The composite elastomer exhibited a hydrophobicity with a water CA (WCA) of 118° (Figure 8c). Even under more severe conditions, such as dyed acid solution (pH = 1), alkaline solution (pH = 14), and salt solution (pH ≈ 6), the sub-spherical droplets could maintain their original shape and stand on the sample surface with an average CA of 117°. This suggested that the composite elastomer possessed a stable hydrophobicity and resistance to corrosion, although the CA value was not higher due to the oxygen groups on the surface of the elastomer composite. After a seven-day artificial sweat treatment, the water droplet on the elastomer surface still formed a regular sub-sphere shape with a WCA of 109°. The electrical conductivity was almost constant with a slight fluctuation (Figure 8d), indicating the stable conductivity and hydrophobicity of the elastomer in the human sweat environment. Besides, the elastomer exhibited a favorable aging resistance property. The WCA and electrical conductivity could almost keep a constant value after 5 h UV irradiation (Figure 8e). Zhu et al. [99] developed a direct ink writing (DIW) assembly approach for fabricating fiber-shaped strain sensors by incorporating graphene/CNT hybrid fillers into the silicone elastomer. The proposed fiber-shaped GCE strain sensors exhibit a tolerable strain over 300%, high sensitivity (100% strain, GF = 14,550.2), and high durability (10,000 loops), and can maintain excellent electromechanical stability to environmental temperature change (reaching at 60 °C) and show good long-term (30 days) environmental stability and waterproof characteristics. Apart from the temperature-resistance ability, a GMF sensor remains stable at a relative humidity of ≈39% and fluctuates moderately with increasing relative humidity up to ≈71%, which shows that the GMF sensor is insensitive to humidity changes, owing to the reduction in functional groups [122].

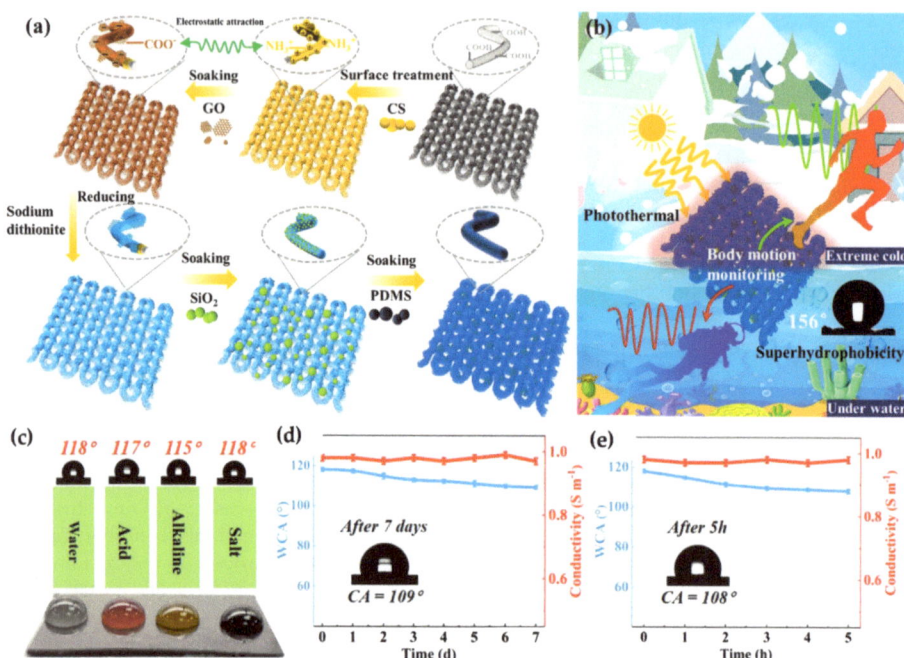

Figure 8. (a) A Schematic illustration of the preparation process of the PSKF@rGO/SiO$_2$-PDMS sensor. (b) Schematic diagram of the application scenarios and functions of the PSKF@rGO/SiO$_2$-PDMS sensor [102]. (c) CAs and photographs of various aqueous droplets on the surface of MCNC-GN/PDMS-2. (d) WCA and conductivity variations of the MCNC-GN/PDMS-2 as a function of the immersion time in artificial sweat. Inset: WCA image after a 7-day immersion. (e) WCA changes of MCNC-GN/PDMS-2 with different light irradiation time. Inset: WCA image after a 5 h UV irradiation.

5.3. Nanodiamonds Based Strain Sensor on Extreme Humidity or Other Harsh Condition

A flower-like molybdenum disulfide (MoS$_2$)/nanodiamond (ND) nanocomposite was successfully prepared via an easy hydrothermal synthesis method and fabricated into humidity sensors to investigate its humidity sensing characteristics, and it could be seen that it was stable at a humidity from 11% RH to 97% RH [106]. A smart multifunctional fabric was produced from the combination of ND with polyaniline (PAIN) with improved mechanical and comfort properties [104]. The nanocomposites based on ND and PAIN are still present on the surface of the fabric even after five commercial washing cycles, which could be due to the bond formation between the nanocomposites and the fabric. The fabrics were washed according to the ISO 105-C08 method. Briefly, the fabrics were washed five times using 20 mL/g liquor-to-fabric ratio at 40 °C for 30 min along with 25 steel balls. ND-PANI showed significant enhancement in abrasion resistance as ND has high abrasion resistance and can increase the abrasion resistance and strengthen the mechanical properties, absorbing the applied energy during abrasion [134,135]. Recently, a pressure sensor device was made of a locally synthesized diamond layer [105]. The rear side of a pressure sensor diaphragm was prepared with an additional diamond layer as protective coating against harsh media, because of the diamond's chemical inertness and resistance.

6. Applications

Carbon material (carbon nanotubes, graphene, and nanodiamond) strain sensors exhibit unique properties and resilience, making them highly versatile in harsh environments. These sensors find diverse applications in healthcare, including human motion sensing, health monitoring, electronic skins (E-skin), and special applications. Integrated into wearable devices, they enable continuous monitoring of vital signs, body movements, and muscle activity, even during rigorous physical activities, thanks to their flexibility and ability to withstand mechanical stresses. In healthcare settings, particularly in monitoring and diagnostics, carbon-based strain sensors, such as those using carbon nanotubes, graphene, and nanodiamond, prove to be well-suited due to their sensitivity, flexibility, and durability. Their integration into various healthcare applications, such as wearable health monitoring devices, orthopedic monitoring, pressure ulcer prevention, smart implants, rehabilitation, cardiac monitoring, prosthetics, and surgical robotics, enables real-time monitoring, precise diagnostics, and customized treatment plans, ultimately enhancing patient care and healthcare outcomes. As research and technology progress, carbon material strain sensors are expected to play an increasingly vital role in healthcare applications, further revolutionizing patient care and healthcare practices.

6.1. Motion Sensing

The wearable sensor serves as a crucial component determining the functionality of current smart devices. However, existing monitoring equipment based on physical indicators like explosive force, respiration, heart rate, blood oxygen saturation, and electromyography lacks wear comfort and sensitivity. Consequently, real-time monitoring of athletes' physical information during free movement is hindered, affecting accurate motion parameter data and ultimately impacting training effectiveness. Thus, there is an urgent demand for the development of flexible wearable strain sensors that offer comfort and high sensitivity, enabling real-time monitoring of athletes' status. To achieve real-time health data during exercise, wearable sensors are strategically placed on various body parts, such as elbows, wrists, fingers, pulses, and knees, facilitating continuous and precise monitoring.

Especially for some outdoor activities, activities take place in an extreme situation, such as at low temperatures or in high humidity. Ma et al. [45] reported that a CNT-based strain sensor exhibits excellent temperature tolerance ($-40\ °C$) and long-lasting moisture (20 days). After being stored at $-40\ °C$ for 24 h, the anti-freezing organohydrogel maintained sensitivity and could be assembled as a wearable device to monitor human biologic activities at low temperatures. In addition, Lu et al. [102] report that a PSKF@rGO/SiO$_2$-PDMS sensor expands the working range. For instance, athletes can use this PSKF@rGO/SiO$_2$-PDMS sensor to detect their finger, wrist, elbow, knee, and ankle movements and different movements (walking and running) of the same joint were also detected (Figure 9a–g). A resistance curve with high repeatability was still observed even when the fabric sensor was submerged in water or placed in a severely cold environment, indicating that the fabric sensor could still maintain good mechanical and electrical properties, as shown in Figure 9h,i. Furthermore, due to its super-hydrophobicity, the fabric sensor maintained normal operation even after water was poured on it and when activity was monitored in sweaty conditions. Based on its multiple superior properties, this sensor could be used as winter sportswear for athletes to track their actions without being impacted by water and as a warmer to ensure the wearer's comfort.

A kind of sensor could detect minor motion. The polypyrrole (PPy)/single-walled carbon nanotube (SWCNT)/polydopamine(PDA)/cotton composite was employed to capture subtle motions originating from the vocal cords during drinking, as depicted in Figure 9 [121]. Remarkably, a unique and repetitive $\Delta R/R_0$ response profile was observed when the volunteer drank 100 mL of orange juice in four sips, highlighting the composite's potential to sense complex and delicate motions of the vocal muscles (Figure 9j). To monitor tiny motions arising from wrist movement during writing, the same composite was applied to the wrist region of the right hand. The sensor's response was examined while writing four different words, "CTN", "PDA", "CNT", and "PPy", with each word showing a distinctive, repetitive, and reproducible $\Delta R/R_0$ response (Figure 9k).

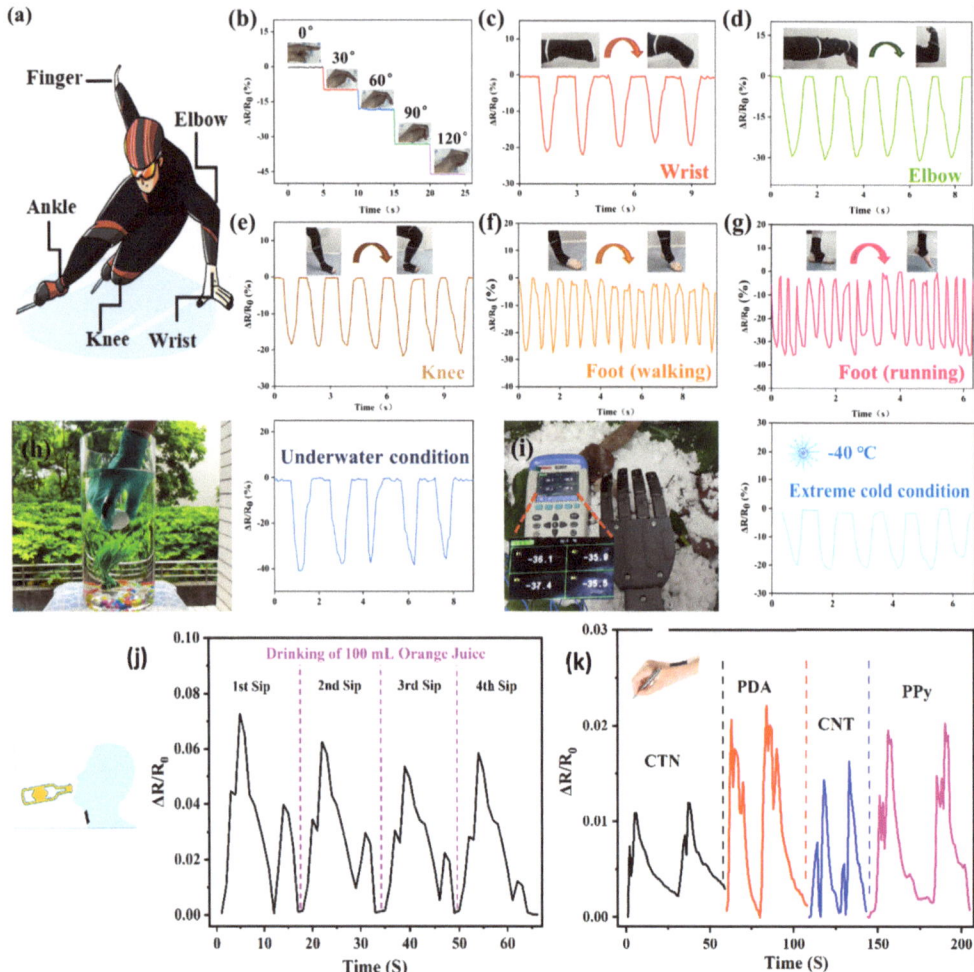

Figure 9. (**a**) Schematic illustration of the PSKF@rGO/SiO$_2$-PDMS sensor mounted on human joints for exercise monitoring and management. (**b**) Relative resistance variation of the PSKF@rGO/SiO$_2$-PDMS sensor based on finger bending at a certain angle. Insets: photos of an index finger at angles of 0°, 30°, 60°, 90°, and 120°. (**c**–**g**) Real-time relative resistance variation of the PSKF@rGO/SiO$_2$-PDMS sensor when monitoring continuous motions of the wrist, elbow, knee, and foot while walking and running. (**h**,**i**) Real-time relative resistance variation of the PSKF@rGO/SiO$_2$-PDMS sensor under extreme conditions (underwater or extremely low temperature (−40 °C)) [102]. (**j**) The strain sensing performance of the PPy/SWCNT/PDA/cotton composite on the index finger; changes of $\Delta R/R_0$ while drinking; (**k**) the sensor attached on the wrist; $\Delta R/R_0$ change profile during writing [121].

6.2. Health Monitor

In the fields of epidemiology and disease control, the use of carbon materials-based flexible strain sensors is vital and essential. In athletic training, real-time recording and analysis of bio-sign signals, including respiration, heart rate, and electromyography, are vital. As a result, they serve as ideal health indicators, playing a crucial role in monitoring public and personal health with precision.

An example of sensor designs for detecting the pulses of the wrist is based on MWCNT [107]. The superhydrophobic sensors were applied to various positions on individuals' skin to facilitate real-time detection. As depicted in Figure 10a, the sensor adhered firmly to the wrist surface, enabling the detection of tiny deformations caused by pulse vibrations and producing a sensitive and stable wrist pulse signal. The amplified image displayed in Figure 10a showcases a typical radial artery pulse waveform with three distinct peaks, affirming the sensor's suitability for human health testing. Moreover, in addition to its capability for precise individual detection, the sensor exhibits high sensitivity in recognizing human joint movements. A flexible and multifunctional temperature/pressure/loading composite foam sensor was developed [97]. Figure 10b,c demonstrate that the PDA-CNT/PDMS sensor exhibits continuous and stable signal peaks with higher intensity compared to the CNT/PDMS sensor when detecting pulse waves, including those from the left superficial temporal artery and the left carotid artery. This strain sensors can also be assembled on a face mask to measure human breathing. To showcase the potential application of the PDA-CNT/PDMS composite sensor, Figure 10d illustrates a flexible composite microarray sewn inside a medical mask to create a smart mask. When an adult breathes with this mask, the inhaled hot air comes into contact with the microarray, leading to a decrease in sensor resistance (Figure 10d). Consequently, the sensor can monitor the human body's breathing rate in relation to exercise and health. Figure 10d displays distinct signals representing normal breathing and shortness of breath, empowering the PDA-CNT/PDMS composite sensor with the capability to predict early diseases in the human body. Moreover, a breath sensor based on rGO-CGY and elastic yarns can monitor speaking, coughing, and breathing [122]. The electrical pattern obtained during gum chewing, as shown in Figure 11a, exhibits reproducibility, revealing its promising application prospects in oral locomotion and training. We affixed the sensor to the throat to discern coughing and speaking states. Figure 11b depicts the resistance change of the sensor during coughing, showing an upward and downward resistance change signal linked to the exertion and release of cough, respectively, consistent with corresponding epidermal vibrations. Furthermore, the electrical response remains highly repeatable with minimal deviations, indicating precise tracking of motion with a reproducible signaling pattern. In Figure 11c, the sensor effectively captures resistance signals generated during speech, exhibiting distinguishable and reproducible patterns when speaking different words ("I love you", "strain sensor", and "wearable"). Additionally, signals of normal human breathing and breathing during running are presented in Figure 11d, showcasing discernible respiratory rates and depths under the two diverse conditions.

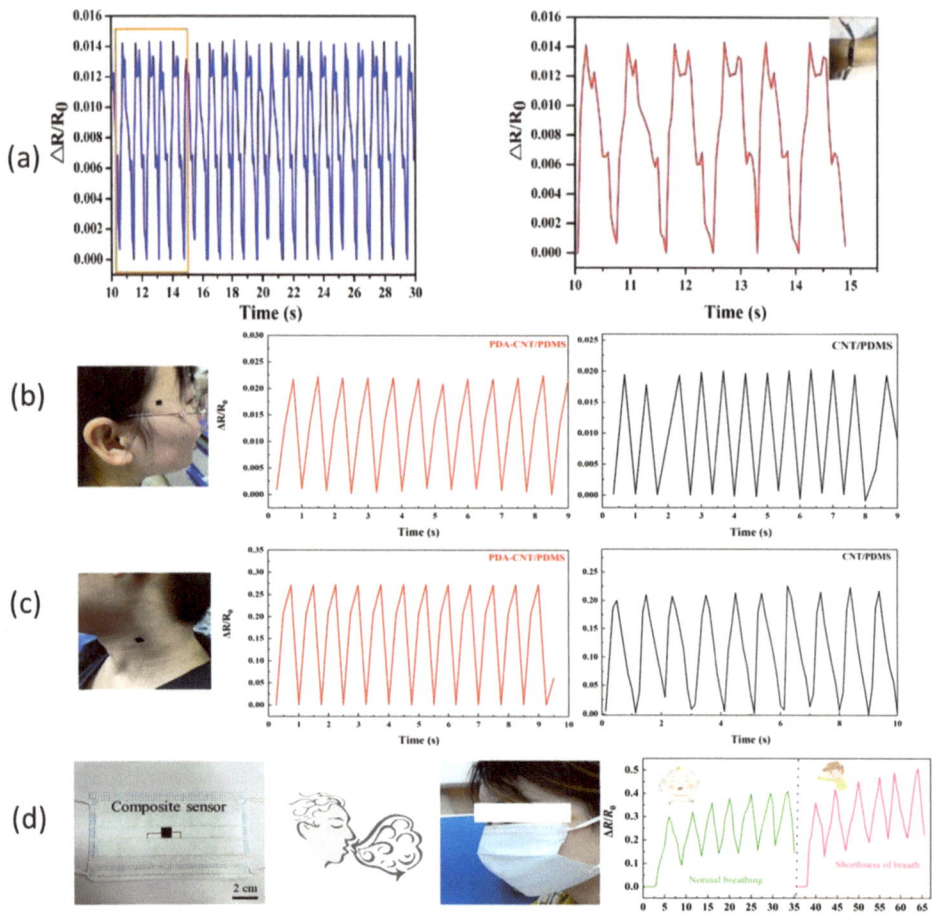

Figure 10. (**a**) Real-time detection of the wrist pulse when the film self-adhered to the wrist surface and the enlarged image of the pulse detection [107]. (**b**,**c**) Comparison of the signals from the CNT/PDMS and PDA-CNT/PDMS composite sensors for recording the left superficial temporal artery and left carotid artery. (**d**) Photograph of a medical mask with a flexible composite microarray inside. Cartoon diagram of breathing. Photograph of breathing with a medical mask embedded with a flexible composite microarray. Comparison of the signals from the medical sensory mask under normal breathing and shortness of breath of the human body [97].

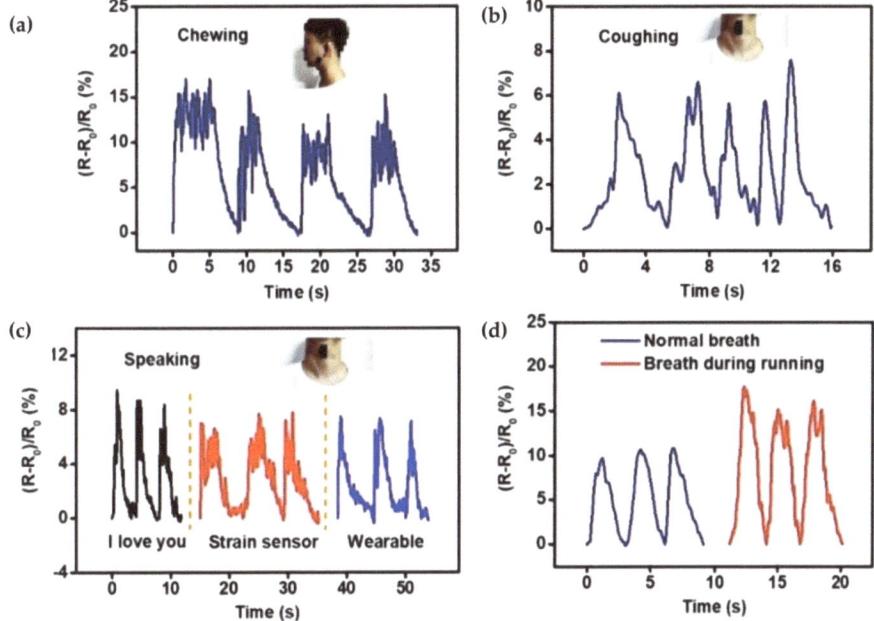

Figure 11. Application demonstration of the GMF sensor. (**a**) Corresponding signals of chewing gum. (**b**) Relative electrical resistance changes of throat vibration during coughing. (**c**) Signals of the throat epidermis vibration while speaking different words. (**d**) Relative electrical resistance changes during normal breathing and breathing during running [122].

6.3. E-Skin

As a tissue directly in contact with the external environment, the skin can perceive various external stimuli simultaneously. Extensive research has been conducted on its intriguing properties, including self-healing capabilities. In the context of artificial intelligent skin, carbon nanotube-, graphene-, and nanodiamond-based flexible strain sensors emerge as excellent candidates, benefiting from the unique strengths of stretchability and conductivity. These sensors are poised to fully replicate the functions of human skin, and several designs with multi-sensing abilities, such as strain and pressure, temperature and strain, have been developed to achieve this goal.

A polyimide (PI)/carbon nanotubes (CNT) composite aerogel was developed as E-skin [113]. Wearable pressure sensors are widely integrated into smart E-skin systems to detect pressure distribution and the location of mechanical stimuli. Figure 12A,B showcase the schematic illustration and top view of the E-skin, consisting of 5 × 5 pixels. As illustrated in Figure 12C, the E-skin accurately recognizes the compressed points by recording resistance variation in a two-dimensional (2D) map based on different applied pressures at various locations (with darker colors representing larger resistance variations) when a finger presses the E-skin affixed to the back of the hand. Moreover, a three-dimensional (3D) pressure distribution mapping is obtained through nonlinear surface fitting (Figure 12F), clearly highlighting significant pressure differences at different points. Specifically, the resistance variation in the pressure center amounts to about 34.2%, equivalent to a pressure of approximately 50 kPa. In addition, when the E-skin is gripped with a stone ball in the hand (Figure 12D,E), it also provides a substantial sensing response, with a 3D pressure distribution mapping shown in Figure 12G,H. Overall, these findings demonstrate the promising capabilities of the E-skin for accurately mapping pressure distribution and location. Moreover, Sun et al. [92] assembled a 4 × 4 electronic fabric using the superhydrophobic conductive RB (based on carbon nanotubes(CNTs)/reduced graphene oxide (rGO) dual conductive layer) to explore its application in the tactile sensing field (Figure 13a). By pressing the fabric with a figure (Figure 13b) or placing a doll (Figure 13c) on it, the deformation of the conductive RB resulted in resistance changes at different cross points, revealing the corresponding spatial pressure distribution. This was effectively captured through a 3D resistance variation mapping (Figure 13e,f), showcasing the immense potential of our superhydrophobic conductive RB for advanced electronic skin applications.

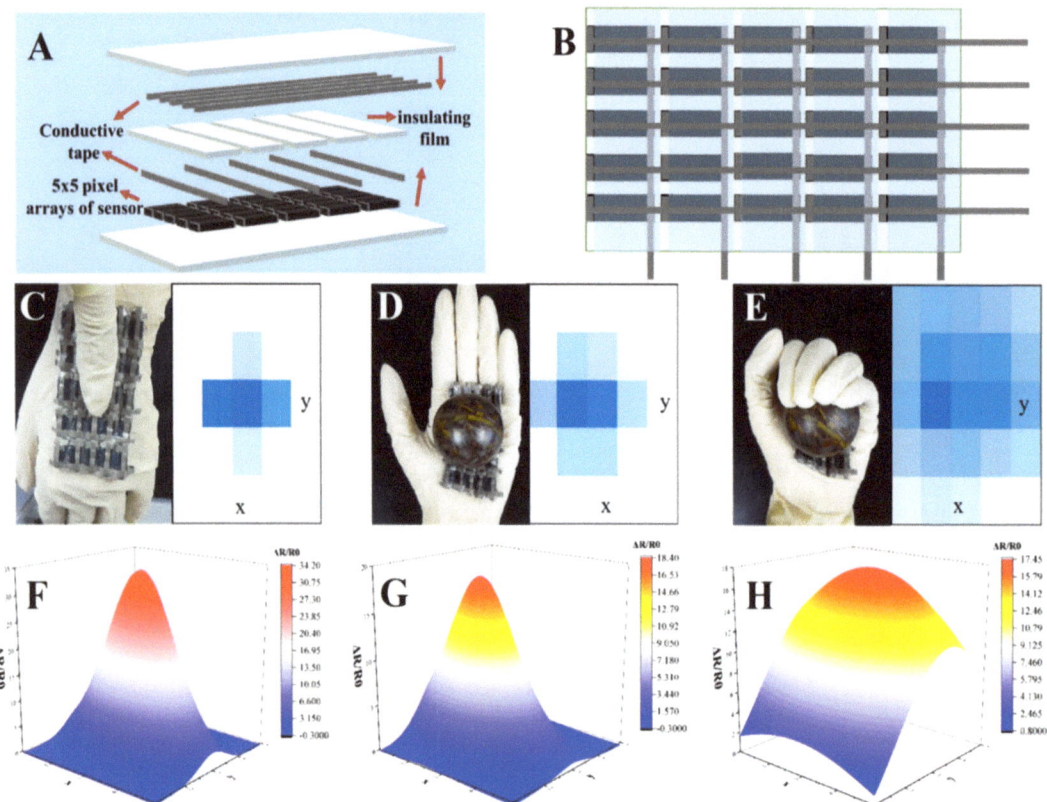

Figure 12. (**A**) Schematic illustration and (**B**) top view of the E-skin assembled from the PI/CNT composite aerogel with a size of 5 × 5 pixels. (**C**) Photograph of a finger pressing the E-skin attached to the back of the hand and the corresponding resistance variation in a 2D map. Photographs of a stone ball (**D**) in the hand and (**E**) being gripped, and their corresponding resistance variation in a 2D map. (**F–H**) Corresponding pressure distribution mappings over a 3D area of (**C–E**) based on the change in resistance, respectively [113].

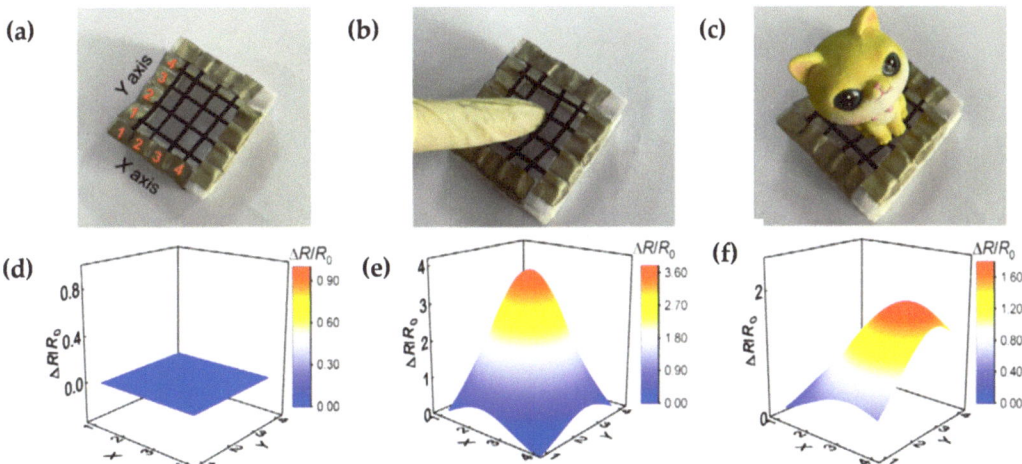

Figure 13. Digital photographs and the corresponding pressure distribution based on the resistance change of (**a**,**d**) electronic fabric assembled from the superhydrophobic conductive RB with 4 × 4 pixels, and (**b**,**e**) a finger and (**c**,**f**) a doll on the electronic fabric.

6.4. Other Application

6.4.1. Intelligent Logistics, Disaster Rescue, and Space Applications

Beside the above-mentioned applications, there are some special applications for strain sensors in harsh environments. For nuclear or space applications, radiation resistance is imperative, allowing sensors to endure without degradation. Ma et al. [109] reported that a MWCNT/polyimide aerogel exhibited desirable EMI SE values (from 99.4 dB to 46.9 dB when CPIA-4 was compressed from 10 mm to 5 mm) with stable electrical responsiveness across a wide temperature range (from 25 to 316 °C). These materials, therefore, showcase great promise as smart high-temperature EMI shielding materials for radar equipment, satellites, or aerobat landing protection equipment. Moreover, the strain/pressure sensing and EMI shielding performance of the hybrid CNT-coated carbonized melamine foams (CNT/CMFs) were systematically investigated, showing an exceptional EMI shielding effectiveness and a high specific shielding effectiveness of 6147.3 dB cm^2/g with an absorption-dominant shielding mechanism due to the highly porous structure [119].

In Shak Sadi's study, to verify the potentiality of the PPy/SWCNT/PDA/Cotton composite as a component of a wearable heating device, the sample was attached to the index finger position of a hand glove while both ends of the composite were connected to an external DC voltage source (2 V) (Figure 14a) [121]. Within 30 s of heating, there was a very uniformly distributed profile of the surface. A component of wearable heating devices is to control the region-specific temperature of the human body as desired by the wearer. Robots with sensors are expected to play a crucial role in extreme environment tasks in the future, such as intelligent logistics, disaster rescue, and aerospace applications [53,54]. The fabric sensor-equipped mechanical arm was carefully maneuvered to approach both a hazardous chemical bottle and a toy doll [102]. The sensor demonstrated its responsiveness, detecting the grasping action regardless of the materials' hardness or softness (see Figure 14b). When the robot successfully grabbed an object, the fabric sensor experienced compression and pressure, leading to a sudden drop in resistance, which promptly activated the signal lamp. Upon releasing the object, the lamp returned to its original state. These results affirm the suitability and reliability of the PSKF@rGO/SiO$_2$-PDMS fabric sensor even in challenging and demanding application conditions [102].

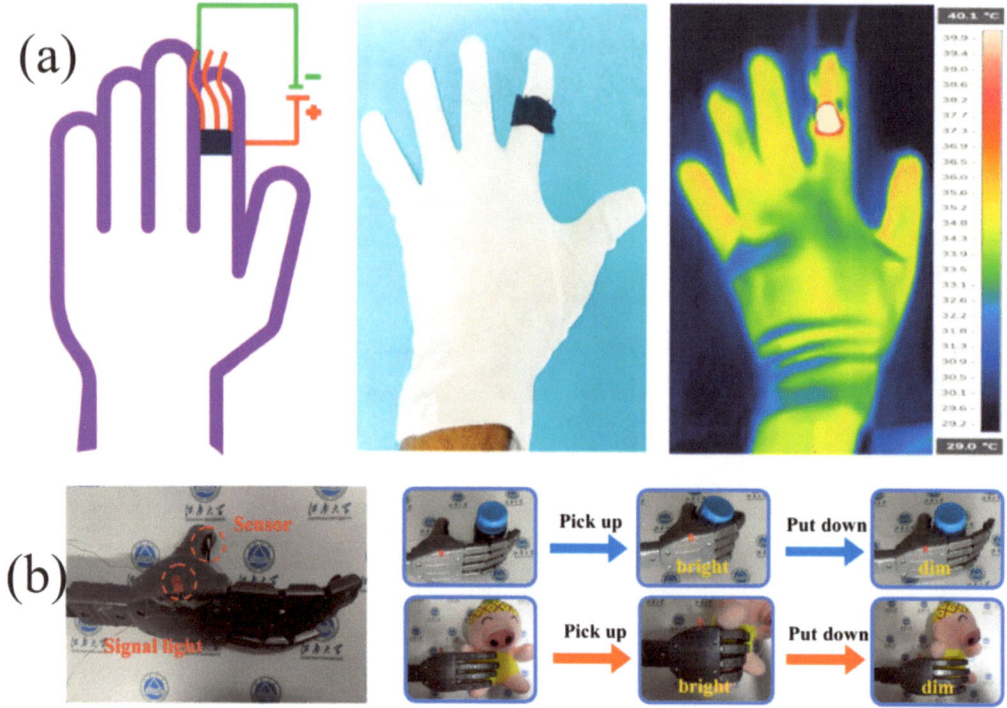

Figure 14. (a) Demonstration of the wearable heating system, attachment of the composite with a hand glove as a wearable heater; temperature response of the heater on the index finger under 2 V [121]. (b) PSKF@rGO/SiO$_2$-PDMS sensor mounted on a robot hand for executing dangerous tasks and intelligent logistics [102]. The background of Figure 14b is the university name for the author in reference [111].

6.4.2. Non-Contact Smart Control

A flower-like nanocomposite of molybdenum disulfide (MoS$_2$) and nanodiamond (ND) was utilized to create humidity sensors [106]. This innovative non-contact sensing control method allows for adjusting the brightness of LED lights by monitoring water evaporation from the skin, presenting a potential and advantageous approach for controlling the brightness of display screens in phones and computers. Leveraging its impressive sensitivity and rapid response, the sensor was applied to monitor water evaporation from the skin. In Figure 15a, the measurement diagram of the sensor during finger water evaporation is displayed, with a micrometer actuator used to control the humidity sensor's distance from the finger. As the distance between the sensor and the finger gradually decreased, the sensor detected the humidity evaporated from sweat glands, leading to an increase in the sensor's response, as depicted in Figure 15a. The humidity values can be determined, as shown in the figure. Additionally, real-time detection of water evaporation occurred when the finger was rapidly brought close to and lifted from the sensor, as demonstrated in Figure 15a. The sensor could clearly and swiftly detect the subtle changes in relative humidity (RH) released from the skin. On the other hand, when uniformly sliding our finger near the top of the sensor, the sensor showed a slow response, indicating a slow adsorption process of RH on the sensor surface. These experimental results underscore the sensor's excellent response sensitivity in monitoring water evaporation from the skin. Furthermore, a non-contact smart sensing control system was constructed by monitoring water evaporation from the skin, as illustrated in Figure 15b. The system primarily consists of an operational amplifier (IC1) and an inverting operational amplifier (IC2). As depicted in Figure 15b, the LED light's brightness exhibited different degrees of variation with decreasing distance between the finger and the sensor. In comparison to other control methods, this non-contact smart control system offers the added advantage of reducing the risk of bacterial transmission.

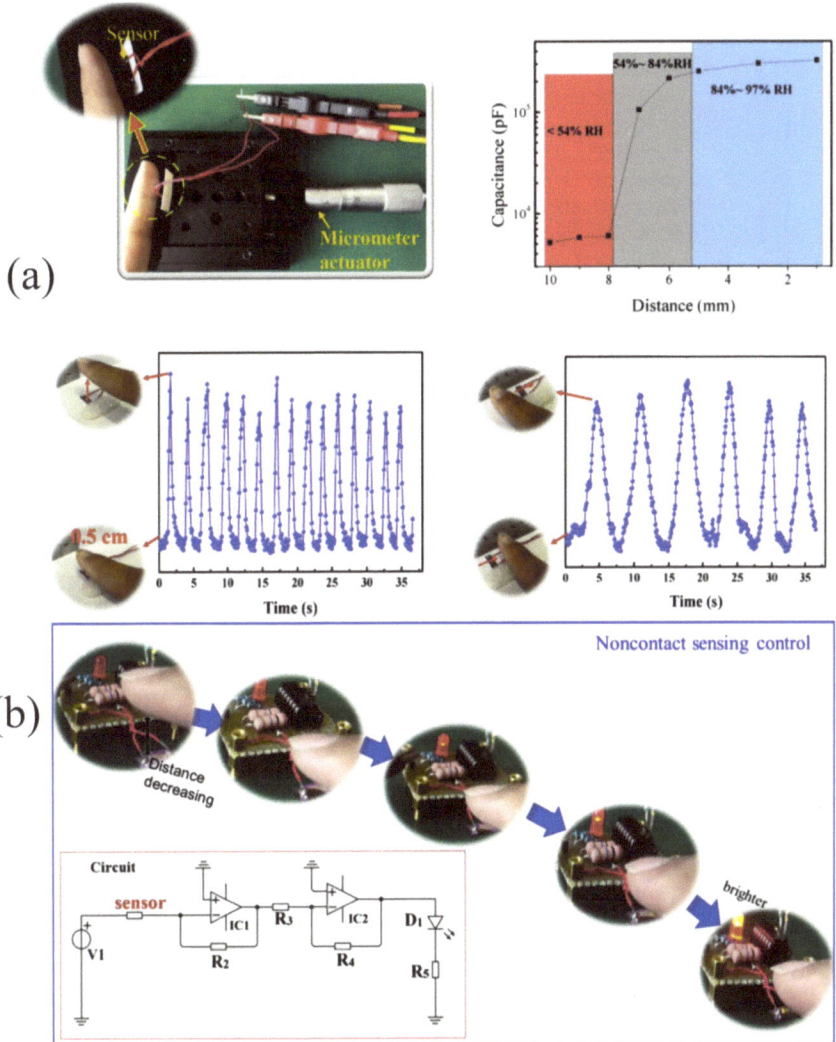

Figure 15. (**a**) The measurement diagram of the sensor on finger water evaporation. The response to a bare finger with different distances. Real-time detection of water evaporation from a finger, moving up and down, uniform sliding. (**b**) Noncontact sensing control. The schematic circuit diagram of noncontact sensing system. The photograph of a finger noncontact humidity control system, displaying the different degrees of brightness of a LED light with the decreasing of distances between the finger and sensor [106].

7. Conclusions and Challenges

A comprehensive evaluation of the carbon-based flexible strain sensor was reviewed and discussed. This paper summarizes the recent breakthroughs in the reasonable fabrication, performance, and application varieties of CNTs, graphene, nanodiamond, and related materials toward wearable electronics with high-performance flexibility. Because of their extraordinary properties (e.g., ultra-translucency, good electrical conductivity, thermal stability, chemical stability, superior mechanical flexibility, installation to be designed into different pliable macroscopic morphologies, and easily functionalized), these carbon material (CNTs, graphene, and nanodiamond), and related materials such as MWCNT, SWCNT, GO, rGO, and ultrananocrystalline diamond, enable a wide range of applications for high-sensitivity strain sensors, which are promising for motion sensing, health indicators, E-skin, and intelligent logistics, disaster rescue, and space applications.

Compared to CNTs and graphene, nanodiamond has a much shorter history. Nanodiamond (ND) is a novel nanocarbon material with favorable mechanical properties, remarkable chemical inertness, low dielectric constant, outstanding thermal stability, and biocompatibility. Owing to these desirable characteristics, it has been applied in the environmental, energy, biological, and drug delivery fields. With the development of nanodiamond in biological and biomedical areas in recent years, a natural question that has come up is that of which one is better for making strain senors. As popular carbon nanomaterials, CNTs, graphene, and nanodiamond share some properties, including excellent mechanic, electronic, and thermal properties. Nanodiamonds have high hardness and wear resistance, but they are not as mechanically flexible as CNTs or graphene. It is obvious that CNTs and graphene have been widely used as strain sensors in harsh environments, while there is not a lot of research about nanodiamond-based strain sensors and most reported research focused on pressure senors. Despite the progress in CNTs, graphene, and nanodiamond-based strain sensors for extreme conditions, challenges remain in terms of sensitivity, stability, and calibration. Researchers continue to explore new carbon materials, sensor designs, and encapsulation methods to enhance their performance and reliability. Future advancements hold the promise of more extensive applications in harsh environments and can revolutionize safety and efficiency in critical industries. Firstly, maintaining high sensitivity and accuracy of carbon material strain sensors in harsh conditions can be challenging. Extreme temperatures, mechanical stresses, and exposure to corrosive substances might affect the sensor's performance, leading to inaccurate readings. Secondly, harsh environments can alter the properties of carbon materials, affecting their calibration and stability over time. Regular calibration is essential to ensure accurate strain measurements, but this process can be complicated and time-consuming in challenging conditions. Thirdly, prolonged exposure to harsh environments, such as extreme temperatures or corrosive substances, can degrade the carbon material's properties. This degradation may lead to reduced sensor lifespan and performance. Fourthly, integrating carbon material strain sensors into existing structures or systems in harsh environments might be difficult due to size constraints, wiring complexities, and compatibility issues. Harsh environments can introduce various sources of signal interference, such as electromagnetic noise, which can impact the sensor's ability to transmit data accurately. Finally, carbon material strain sensors, particularly those designed to withstand extreme conditions, may be more expensive compared to conventional strain sensors, posing a challenge for their widespread adoption.

Strain sensors are used to measure the deformation of an object under stress. They are widely used in various fields such as aerospace, civil engineering, and medical engineering. The prospects of strain sensors in harsh environments are promising. In recent years, there has been a growing demand for sensors that can withstand harsh environments such as high temperatures, high radiation, high shock, and chemically corrosive environments. TE Connectivity has developed a range of industrial sensors that can withstand harsh environments. These sensors are designed to operate in demanding conditions and provide accurate and reliable data. The latest sensor developments are being influenced by IoT trends including miniaturization, digitization, sensor fusion, lower power, and the integration of wireless communication technologies. The commercialization of strain sensors for harsh environments is already underway. However, the requirements for sensors are becoming more demanding as factories are becoming more autonomous and IoT (Internet of

Things) ready. The development of sensors and actuators to operate in harsh environmental conditions has been gaining momentum in recent years.

Overall, the future prospects of carbon nanotubes, graphene, and nanodiamond as strain sensors in harsh environments are exciting and dynamic. Ongoing research and development efforts are expected to lead to significant advancements in materials, fabrication techniques, and sensor design, making these carbon-based sensors increasingly reliable, versatile, and applicable in a wide range of extreme conditions, including aerospace, automotive, oil and gas, structural monitoring, and healthcare applications. As research continues to push the boundaries of materials science and sensor technology, carbon-based strain sensors are expected to play an increasingly vital role in shaping the future of wearable and flexible electronics, leading to improved healthcare monitoring, industrial applications, and the realization of a smarter, interconnected community.

Author Contributions: All authors contributed equally to this work. All authors have read and agreed to the published version of the manuscript.

Funding: This research was funded by the Jiangsu Science and Technology Programme–Young Scholar (BK20200251) and the XJTLU AI University Research Centre, Jiangsu Province Engineering Research Centre of Data Science and Cognitive Computation at XJTLU and SIP AI innovation platform (YZCXPT2022103).

Data Availability Statement: Data are contained within the article.

Acknowledgments: The authors also acknowledge the support from the State Key Laboratory for Manufacturing Systems Engineering (Xi'an Jiaotong University).

Conflicts of Interest: The authors declare no conflict of interest.

References

1. Gao, S.; Zhao, X.; Fu, Q.; Zhang, T.; Zhu, J.; Hou, F.; Ni, J.; Zhu, C.; Li, T.; Wang, Y. Highly transmitted silver nanowires-SWCNTs conductive flexible film by nested density structure and aluminum-doped zinc oxide capping layer for flexible amorphous silicon solar cells. *J. Mater. Sci. Technol.* **2022**, *126*, 152–160. [CrossRef]
2. Li, G.; Wang, L.; Lei, X.; Peng, Z.; Wan, T.; Maganti, S.; Huang, M.; Murugadoss, V.; Seok, I.; Jiang, Q. Flexible, yet robust polyaniline coated foamed polylactic acid composite electrodes for high-performance supercapacitors. *Adv. Compos. Hybrid Mater.* **2022**, *5*, 853–863. [CrossRef]
3. Wang, Y.; Yang, D.; Hessien, M.M.; Du, K.; Ibrahim, M.M.; Su, Y.; Mersal, G.A.; Ma, R.; El-Bahy, S.M.; Huang, M. Flexible barium titanate@ polydopamine/polyvinylidene fluoride/polymethyl methacrylate nanocomposite films with high performance energy storage. *Adv. Compos. Hybrid Mater.* **2022**, *5*, 2106–2115. [CrossRef]
4. Das, H.S.; Roymahapatra, G.; Kumar, P.; Das, R. Study the effect of ZnO/Cu/ZnO multilayer structure by RF magnetron sputtering for flexible display applications. *ES Energy Environ.* **2021**, *13*, 50–56. [CrossRef]
5. Lai, C.; Wang, Y.; Fu, L.; Song, H.; Liu, B.; Pan, D.; Guo, Z.; Seok, I.; Li, K.; Zhang, H. Aqueous flexible all-solid-state NiCo-Zn batteries with high capacity based on advanced ion-buffering reservoirs of $NiCo_2O_4$. *Adv. Compos. Hybrid Mater.* **2022**, *5*, 536–546. [CrossRef]
6. Wang, R.; Sun, L.; Zhu, X.; Ge, W.; Li, H.; Li, Z.; Zhang, H.; Huang, Y.; Li, Z.; Zhang, Y.F. Carbon nanotube-based strain sensors: Structures, fabrication, and applications. *Adv. Mater. Technol.* **2023**, *8*, 2200855. [CrossRef]
7. Li, Z.; Li, H.; Zhu, X.; Peng, Z.; Zhang, G.; Yang, J.; Wang, F.; Zhang, Y.F.; Sun, L.; Wang, R. Directly printed embedded metal mesh for flexible transparent electrode via liquid substrate electric-field-driven jet. *Adv. Sci.* **2022**, *9*, 2105331. [CrossRef]
8. Shintake, J.; Piskarev, Y.; Jeong, S.H.; Floreano, D. Ultrastretchable strain sensors using carbon black-filled elastomer composites and comparison of capacitive versus resistive sensors. *Adv. Mater. Technol.* **2018**, *3*, 1700284. [CrossRef]
9. Li, X.; Yang, T.; Yang, Y.; Zhu, J.; Li, L.; Alam, F.E.; Li, X.; Wang, K.; Cheng, H.; Lin, C.T. Large-area ultrathin graphene films by single-step marangoni self-assembly for highly sensitive strain sensing application. *Adv. Funct. Mater.* **2016**, *26*, 1322–1329. [CrossRef]
10. Eom, J.; Jaisutti, R.; Lee, H.; Lee, W.; Heo, J.-S.; Lee, J.-Y.; Park, S.K.; Kim, Y.-H. Highly sensitive textile strain sensors and wireless user-interface devices using all-polymeric conducting fibers. *ACS Appl. Mater. Interfaces* **2017**, *9*, 10190–10197. [CrossRef]
11. Ramalingame, R.; Lakshmanan, A.; Müller, F.; Thomas, U.; Kanoun, O. Highly sensitive capacitive pressure sensors for robotic applications based on carbon nanotubes and PDMS polymer nanocomposite. *J. Sens. Sens. Syst.* **2019**, *8*, 87–94. [CrossRef]
12. You, I.; Kim, B.; Park, J.; Koh, K.; Shin, S.; Jung, S.; Jeong, U. Stretchable E-skin apexcardiogram sensor. *Adv. Mater.* **2016**, *28*, 6359–6364. [CrossRef] [PubMed]

13. Gong, S.; Lai, D.T.; Wang, Y.; Yap, L.W.; Si, K.J.; Shi, Q.; Jason, N.N.; Sridhar, T.; Uddin, H.; Cheng, W. Tattoolike polyaniline microparticle-doped gold nanowire patches as highly durable wearable sensors. *ACS Appl. Mater. Interfaces* **2015**, *7*, 19700–19708. [CrossRef]
14. Yu, X.; Xie, Z.; Yu, Y.; Lee, J.; Vazquez-Guardado, A.; Luan, H.; Ruban, J.; Ning, X.; Akhtar, A.; Li, D. Skin-integrated wireless haptic interfaces for virtual and augmented reality. *Nature* **2019**, *575*, 473–479. [CrossRef] [PubMed]
15. Zheng, M.; Li, W.; Xu, M.; Xu, N.; Chen, P.; Han, M.; Xie, B. Strain sensors based on chromium nanoparticle arrays. *Nanoscale* **2014**, *6*, 3930–3933. [CrossRef] [PubMed]
16. Ho, M.D.; Ling, Y.; Yap, L.W.; Wang, Y.; Dong, D.; Zhao, Y.; Cheng, W. Percolating network of ultrathin gold nanowires and silver nanowires toward "invisible" wearable sensors for detecting emotional expression and apexcardiogram. *Adv. Funct. Mater.* **2017**, *27*, 1700845. [CrossRef]
17. Duan, S.; Wang, Z.; Zhang, L.; Liu, J.; Li, C. A highly stretchable, sensitive, and transparent strain sensor based on binary hybrid network consisting of hierarchical multiscale metal nanowires. *Adv. Mater. Technol.* **2018**, *3*, 1800020. [CrossRef]
18. Narongthong, J.; Das, A.; Le, H.H.; Wießner, S.; Sirisinha, C. An efficient highly flexible strain sensor: Enhanced electrical conductivity, piezoresistivity and flexibility of a strongly piezoresistive composite based on conductive carbon black and an ionic liquid. *Compos. Part A Appl. Sci. Manuf.* **2018**, *113*, 330–338. [CrossRef]
19. Seong, M.; Hwang, I.; Lee, J.; Jeong, H.E. A pressure-insensitive self-attachable flexible strain sensor with bioinspired adhesive and active CNT layers. *Sensors* **2020**, *20*, 6965. [CrossRef]
20. Chen, S.; Wu, R.; Li, P.; Li, Q.; Gao, Y.; Qian, B.; Xuan, F. Acid-interface engineering of carbon nanotube/elastomers with enhanced sensitivity for stretchable strain sensors. *ACS Appl. Mater. Interfaces* **2018**, *10*, 37760–37766. [CrossRef]
21. Liu, H.; Gao, H.; Hu, G. Highly sensitive natural rubber/pristine graphene strain sensor prepared by a simple method. *Compos. Part B Eng.* **2019**, *171*, 138–145. [CrossRef]
22. Yang, Y.-F.; Tao, L.-Q.; Pang, Y.; Tian, H.; Ju, Z.-Y.; Wu, X.-M.; Yang, Y.; Ren, T.-L. An ultrasensitive strain sensor with a wide strain range based on graphene armour scales. *Nanoscale* **2018**, *10*, 11524–11530. [CrossRef] [PubMed]
23. Rycewicz, M.; Ficek, M.; Gajewski, K.; Kunuku, S.; Karczewski, J.; Gotszalk, T.; Wlasny, I.; Wysmołek, A.; Bogdanowicz, R. Low-strain sensor based on the flexible boron-doped diamond-polymer structures. *Carbon* **2021**, *173*, 832–841. [CrossRef]
24. Lee, H.; Seong, B.; Moon, H.; Byun, D. Directly printed stretchable strain sensor based on ring and diamond shaped silver nanowire electrodes. *Rsc Adv.* **2015**, *5*, 28379–28384. [CrossRef]
25. Liu, J.; Zhang, H.; Gong, H.; Zhang, X.; Wang, Y.; Jin, X. Polyethylene/polypropylene bicomponent spunbond air filtration materials containing magnesium stearate for efficient fine particle capture. *ACS Appl. Mater. Interfaces* **2019**, *11*, 40592–40601. [CrossRef] [PubMed]
26. Zhang, R.; Ying, C.; Gao, H.; Liu, Q.; Fu, X.; Hu, S. Highly flexible strain sensors based on polydimethylsiloxane/carbon nanotubes (CNTs) prepared by a swelling/permeating method and enhanced sensitivity by CNTs surface modification. *Compos. Sci. Technol.* **2019**, *171*, 218–225. [CrossRef]
27. Kim, H.-J.; Thukral, A.; Yu, C. Highly sensitive and very stretchable strain sensor based on a rubbery semiconductor. *ACS Appl. Mater. Interfaces* **2018**, *10*, 5000–5006. [CrossRef]
28. Xu, W.; Hu, S.; Zhao, Y.; Zhai, W.; Chen, Y.; Zheng, G.; Dai, K.; Liu, C.; Shen, C. Nacre-inspired tunable strain sensor with synergistic interfacial interaction for sign language interpretation. *Nano Energy* **2021**, *90*, 106606. [CrossRef]
29. Wrbanek, J.; Fralick, G.; Gonzalez, J. Developing multilayer thin film strain sensors with high thermal stability. In Proceedings of the 42nd AIAA/ASME/SAE/ASEE Joint Propulsion Conference & Exhibit, Sacramento, CA, USA, 9–12 July 2006; p. 4580.
30. Read, I.; Foote, P. Sea and flight trials of optical fibre Bragg grating strain sensing systems. *Smart Mater. Struct.* **2001**, *10*, 1085. [CrossRef]
31. Zrelli, A.; Ezzedine, T. Design of optical and wireless sensors for underground mining monitoring system. *Optik* **2018**, *170*, 376–383. [CrossRef]
32. Mattmann, C.; Clemens, F.; Tröster, G. Sensor for measuring strain in textile. *Sensors* **2008**, *8*, 3719–3732. [CrossRef] [PubMed]
33. Wang, M.; Mu, L.; Zhang, H.; Ma, S.; Liang, Y.; Ren, L. Flexible strain sensor with ridge-like microstructures for wearable applications. *Polym. Adv. Technol.* **2022**, *33*, 96–103. [CrossRef]
34. Liu, X.; Wei, Y.; Qiu, Y. Advanced flexible skin-like pressure and strain sensors for human health monitoring. *Micromachines* **2021**, *12*, 695. [CrossRef]
35. Wang, J.; Dai, T.; Zhou, Y.; Mohamed, A.; Yuan, G.; Jia, H. Adhesive and high-sensitivity modified Ti_3C_2TX (MXene)-based organohydrogels with wide work temperature range for wearable sensors. *J. Colloid Interface Sci.* **2022**, *613*, 94–102. [CrossRef] [PubMed]
36. Jung, G.; Lee, H.; Park, H.; Kim, J.; Kim, J.W.; Kim, D.S.; Keum, K.; Lee, Y.H.; Ha, J.S. Temperature-tolerant flexible supercapacitor integrated with a strain sensor using an organohydrogel for wearable electronics. *Chem. Eng. J.* **2022**, *450*, 138379. [CrossRef]
37. Qin, Z.; Sun, X.; Zhang, H.; Yu, Q.; Wang, X.; He, S.; Yao, F.; Li, J. A transparent, ultrastretchable and fully recyclable gelatin organohydrogel based electronic sensor with broad operating temperature. *J. Mater. Chem. A* **2020**, *8*, 4447–4456. [CrossRef]
38. Yang, J.; Xu, Z.; Wang, J.; Gai, L.; Ji, X.; Jiang, H.; Liu, L. Antifreezing zwitterionic hydrogel electrolyte with high conductivity of 12.6 mS cm^{-1} at $-40°$C through hydrated lithium ion hopping migration. *Adv. Funct. Mater.* **2021**, *31*, 2009438. [CrossRef]
39. Liu, Z.; Zhang, J.; Liu, J.; Long, Y.; Fang, L.; Wang, Q.; Liu, T. Highly compressible and superior low temperature tolerant supercapacitors based on dual chemically crosslinked PVA hydrogel electrolytes. *J. Mater. Chem. A* **2020**, *8*, 6219–6228. [CrossRef]

40. Yang, Y.; Wang, K.-P.; Zang, Q.; Shi, Q.; Wang, Y.; Xiao, Z.; Zhang, Q.; Wang, L. Anionic organo-hydrogel electrolyte with enhanced ionic conductivity and balanced mechanical properties for flexible supercapacitors. *J. Mater. Chem. A* **2022**, *10*, 11277–11287. [CrossRef]
41. Rong, Q.; Lei, W.; Huang, J.; Liu, M. Low temperature tolerant organohydrogel electrolytes for flexible solid-state supercapacitors. *Adv. Energy Mater.* **2018**, *8*, 1801967. [CrossRef]
42. Feng, E.; Li, J.; Zheng, G.; Yan, Z.; Li, X.; Gao, W.; Ma, X.; Yang, Z. Long-term anti-freezing active organohydrogel based superior flexible supercapacitor and strain sensor. *ACS Sustain. Chem. Eng.* **2021**, *9*, 7267–7276. [CrossRef]
43. Wang, P.; Wei, W.; Li, Z.; Duan, W.; Han, H.; Xie, Q. A superhydrophobic fluorinated PDMS composite as a wearable strain sensor with excellent mechanical robustness and liquid impalement resistance. *J. Mater. Chem. A* **2020**, *8*, 3509–3516. [CrossRef]
44. Huang, J.; Peng, S.; Gu, J.; Chen, G.; Gao, J.; Zhang, J.; Hou, L.; Yang, X.; Jiang, X.; Guan, L. Self-powered integrated system of a strain sensor and flexible all-solid-state supercapacitor by using a high performance ionic organohydrogel. *Mater. Horiz.* **2020**, *7*, 2085–2096. [CrossRef]
45. Ma, D.; Wu, X.; Wang, Y.; Liao, H.; Wan, P.; Zhang, L. Wearable, antifreezing, and healable epidermal sensor assembled from long-lasting moist conductive nanocomposite organohydrogel. *ACS Appl. Mater. Interfaces* **2019**, *11*, 41701–41709. [CrossRef] [PubMed]
46. Wang, X.; Wang, X.; Yin, J.; Li, N.; Zhang, Z.; Xu, Y.; Zhang, L.; Qin, Z.; Jiao, T. Mechanically robust, degradable and conductive MXene-composited gelatin organohydrogel with environmental stability and self-adhesiveness for multifunctional sensor. *Compos. Part B Eng.* **2022**, *241*, 110052. [CrossRef]
47. Yang, Z.; Huang, T.; Cao, P.; Cui, Y.; Nie, J.; Chen, T.; Yang, H.; Wang, F.; Sun, L. Carbonized silk nanofibers in biodegradable, flexible temperature sensors for extracellular environments. *ACS Appl. Mater. Interfaces* **2022**, *14*, 18110–18119. [CrossRef]
48. Lee, J.-H.; Chen, H.; Kim, E.; Zhang, H.; Wu, K.; Zhang, H.; Shen, X.; Zheng, Q.; Yang, J.; Jeon, S. Flexible temperature sensors made of aligned electrospun carbon nanofiber films with outstanding sensitivity and selectivity towards temperature. *Mater. Horiz.* **2021**, *8*, 1488–1498. [CrossRef] [PubMed]
49. Dai, Z.; Ding, S.; Lei, M.; Li, S.; Xu, Y.; Zhou, Y.; Zhou, B. A superhydrophobic and anti-corrosion strain sensor for robust underwater applications. *J. Mater. Chem. A* **2021**, *9*, 15282–15293. [CrossRef]
50. Hu, X.; Yang, F.; Wu, M.; Sui, Y.; Guo, D.; Li, M.; Kang, Z.; Sun, J.; Liu, J. A Super-Stretchable and Highly Sensitive Carbon Nanotube Capacitive Strain Sensor for Wearable Applications and Soft Robotics. *Adv. Mater. Technol.* **2022**, *7*, 2100769. [CrossRef]
51. Tsang, A.C.H.; Hui, K.N.; Hui, K.S.; Liu, B.; Yu, L.; Shi, B.; Huang, H. Ink-printed metal/graphene aerogel for glucose electro-oxidation. *Battery Energy* **2022**, *1*, 20220004. [CrossRef]
52. Shao, C.; Qiu, S.; Wu, G.; Cui, B.; Chu, H.; Zou, Y.; Xiang, C.; Xu, F.; Sun, L. Rambutan-like hierarchically porous carbon microsphere as electrode material for high-performance supercapacitors. *Carbon Energy* **2021**, *3*, 361–374. [CrossRef]
53. Zhang, Q.; Deng, C.; Huang, Z.; Zhang, Q.; Chai, X.; Yi, D.; Fang, Y.; Wu, M.; Wang, X.; Tang, Y. Dual-Silica Template-Mediated Synthesis of Nitrogen-Doped Mesoporous Carbon Nanotubes for Supercapacitor Applications. *Small* **2023**, *19*, 2205725. [CrossRef]
54. Wang, H.; Liu, C.; Li, B.; Liu, J.; Shen, Y.; Zhang, M.; Ji, K.; Mao, X.; Sun, R.; Zhou, F. Advances in Carbon-Based Resistance Strain Sensors. *ACS Appl. Electron. Mater.* **2023**, *5*, 674–689. [CrossRef]
55. Afroze, J.D.; Tong, L.; Abden, M.J.; Chen, Y. Multifunctional hierarchical graphene-carbon fiber hybrid aerogels for strain sensing and energy storage. *Adv. Compos. Hybrid Mater.* **2023**, *6*, 18. [CrossRef]
56. Bariya, M.; Nyein, H.Y.Y.; Javey, A. Wearable sweat sensors. *Nat. Electron.* **2018**, *1*, 160–171. [CrossRef]
57. Kim, S.; Xiao, X.; Chen, J. Advances in Photoplethysmography for Personalized Cardiovascular Monitoring. *Biosensors* **2022**, *12*, 863. [CrossRef] [PubMed]
58. Zhang, S.; Bick, M.; Xiao, X.; Chen, G.; Nashalian, A.; Chen, J. Leveraging triboelectric nanogenerators for bioengineering. *Matter* **2021**, *4*, 845–887. [CrossRef]
59. Zhou, Y.; Xiao, X.; Chen, G.; Zhao, X.; Chen, J. Self-powered sensing technologies for human Metaverse interfacing. *Joule* **2022**, *6*, 1381–1389. [CrossRef]
60. Rim, Y.S.; Bae, S.H.; Chen, H.; De Marco, N.; Yang, Y. Recent progress in materials and devices toward printable and flexible sensors. *Adv. Mater.* **2016**, *28*, 4415–4440. [CrossRef] [PubMed]
61. Araromi, O.A.; Graule, M.A.; Dorsey, K.L.; Castellanos, S.; Foster, J.R.; Hsu, W.-H.; Passy, A.E.; Vlassak, J.J.; Weaver, J.C.; Walsh, C.J. Ultra-sensitive and resilient compliant strain gauges for soft machines. *Nature* **2020**, *587*, 219–224. [CrossRef]
62. Chao, M.; Wang, Y.; Ma, D.; Wu, X.; Zhang, W.; Zhang, L.; Wan, P. Wearable MXene nanocomposites-based strain sensor with tile-like stacked hierarchical microstructure for broad-range ultrasensitive sensing. *Nano Energy* **2020**, *78*, 105187. [CrossRef]
63. Yi, J.; Xianyu, Y. Gold Nanomaterials-Implemented Wearable Sensors for Healthcare Applications. *Adv. Funct. Mater.* **2022**, *32*, 2113012. [CrossRef]
64. Peng, Z.; Shi, J.; Xiao, X.; Hong, Y.; Li, X.; Zhang, W.; Cheng, Y.; Wang, Z.; Li, W.J.; Chen, J. Self-charging electrostatic face masks leveraging triboelectrification for prolonged air filtration. *Nat. Commun.* **2022**, *13*, 7835. [CrossRef] [PubMed]
65. Yan, T.; Wu, Y.; Yi, W.; Pan, Z. Recent progress on fabrication of carbon nanotube-based flexible conductive networks for resistive-type strain sensors. *Sens. Actuators A Phys.* **2021**, *327*, 112755. [CrossRef]
66. He, S.; Hong, Y.; Liao, M.; Li, Y.; Qiu, L.; Peng, H. Flexible sensors based on assembled carbon nanotubes. *Aggregate* **2021**, *2*, e143. [CrossRef]

67. Wang, C.; Xia, K.; Wang, H.; Liang, X.; Yin, Z.; Zhang, Y. Advanced carbon for flexible and wearable electronics. *Adv. Mater.* **2019**, *31*, 1801072. [CrossRef]
68. Zhang, Y.; Xiao, Q.; Wang, Q.; Zhang, Y.; Wang, P.; Li, Y. A review of wearable carbon-based sensors for strain detection: Fabrication methods, properties, and mechanisms. *Text. Res. J.* **2023**, *93*, 2918–2940. [CrossRef]
69. Liu, F.; Xie, D.; Lv, F.; Shen, L.; Tian, Z.; Zhao, J. Additive Manufacturing of Stretchable Polyurethane/Graphene/Multiwalled Carbon Nanotube-Based Conducting Polymers for Strain Sensing. *ACS Appl. Nano Mater.* **2023**, *6*, 4522–4531. [CrossRef]
70. Chen, H.; Zhuo, F.; Zhou, J.; Liu, Y.; Zhang, J.; Dong, S.; Liu, X.; Elmarakbi, A.; Duan, H.; Fu, Y. Advances in graphene-based flexible and wearable strain sensors. *Chem. Eng. J.* **2023**, *464*, 142576. [CrossRef]
71. O'connell, M.J. *Carbon Nanotubes: Properties and Applications*; CRC Press: Boca Raton, FL, USA, 2018.
72. Maniecki, T.; Shtyka, O.; Mierczynski, P.; Ciesielski, R.; Czylkowska, A.; Leyko, J.; Mitukiewicz, G.; Dubkov, S.; Gromov, D. Carbon nanotubes: Properties, synthesis, and application. *Fibre Chem.* **2018**, *50*, 297–300. [CrossRef]
73. Popov, V.N. Carbon nanotubes: Properties and application. *Mater. Sci. Eng. R Rep.* **2004**, *43*, 61–102. [CrossRef]
74. Neto, A.C.; Guinea, F.; Peres, N.M.; Novoselov, K.S.; Geim, A.K. The electronic properties of graphene. *Rev. Mod. Phys.* **2009**, *81*, 109. [CrossRef]
75. Elapolu, M.S.; Tabarraei, A. Mechanical and fracture properties of polycrystalline graphene with hydrogenated grain boundaries. *J. Phys. Chem. C* **2021**, *125*, 11147–11158. [CrossRef]
76. Abergel, D.; Apalkov, V.; Berashevich, J.; Ziegler, K.; Chakraborty, T. Properties of graphene: A theoretical perspective. *Adv. Phys.* **2010**, *59*, 261–482. [CrossRef]
77. Kumar, S.; Nehra, M.; Kedia, D.; Dilbaghi, N.; Tankeshwar, K.; Kim, K.-H. Nanodiamonds: Emerging face of future nanotechnology. *Carbon* **2019**, *143*, 678–699. [CrossRef]
78. Mochalin, V.; Shenderova, O.; Ho, D.; Gogotsi, Y. The properties and applications of nanodiamonds. In *Nano-Enabled Medical Applications*; Jenny Stanford Publishing: New York, NY, USA, 2020; pp. 313–350.
79. Hilding, J.; Grulke, E.A.; George Zhang, Z.; Lockwood, F. Dispersion of carbon nanotubes in liquids. *J. Dispers. Sci. Technol.* **2003**, *24*, 1–41. [CrossRef]
80. Szroeder, P.; Sagalianov, I.Y.; Radchenko, T.M.; Tatarenko, V.A.; Prylutskyy, Y.I.; Strupiński, W. Effect of uniaxial stress on the electrochemical properties of graphene with point defects. *Appl. Surf. Sci.* **2018**, *442*, 185–188. [CrossRef]
81. Balog, R.; Jørgensen, B.; Nilsson, L.; Andersen, M.; Rienks, E.; Bianchi, M.; Fanetti, M.; Lægsgaard, E.; Baraldi, A.; Lizzit, S. Bandgap opening in graphene induced by patterned hydrogen adsorption. *Nat. Mater.* **2010**, *9*, 315–319. [CrossRef]
82. Zeiger, M.; Jäckel, N.; Mochalin, V.N.; Presser, V. Carbon onions for electrochemical energy storage. *J. Mater. Chem. A* **2016**, *4*, 3172–3196. [CrossRef]
83. Mohr, M.; Caron, A.; Herbeck-Engel, P.; Bennewitz, R.; Gluche, P.; Brühne, K.; Fecht, H.-J. Young's modulus, fracture strength, and Poisson's ratio of nanocrystalline diamond films. *J. Appl. Phys.* **2014**, *116*, 124308. [CrossRef]
84. Zhu, Y.; Murali, S.; Cai, W.; Li, X.; Suk, J.W.; Potts, J.R.; Ruoff, R.S. Graphene and graphene oxide: Synthesis, properties, and applications. *Adv. Mater.* **2010**, *22*, 3906–3924. [CrossRef]
85. Zhang, Y.; Lin, H.; Zhang, L.; Peng, S.; Weng, Z.; Wang, J.; Wu, L.; Zheng, L. Mechanical exfoliation assisted with carbon nanospheres to prepare a few-layer graphene for flexible strain sensor. *Appl. Surf. Sci.* **2023**, *611*, 155649. [CrossRef]
86. De Marchi, L.; Pretti, C.; Gabriel, B.; Marques, P.A.; Freitas, R.; Neto, V. An overview of graphene materials: Properties, applications and toxicity on aquatic environments. *Sci. Total Environ.* **2018**, *631*, 1440–1456. [CrossRef] [PubMed]
87. Turcheniuk, K.; Mochalin, V.N. Biomedical applications of nanodiamond. *Nanotechnology* **2017**, *28*, 252001. [CrossRef] [PubMed]
88. Petrák, V.; Živcová, Z.V.; Krýsová, H.; Frank, O.; Zukal, L.; Klimša, L.; Kopeček, J.; Taylor, A.; Kavan, L.; Mortet, V. Fabrication of porous boron-doped diamond on SiO2 fiber templates. *Carbon* **2017**, *114*, 457–464. [CrossRef]
89. Wiora, N.; Mertens, M.; Mohr, M.; Bruehne, K.; Fecht, H.-J. Piezoresistivity of n-type conductive ultrananocrystalline diamond. *Diam. Relat. Mater.* **2016**, *70*, 145–150. [CrossRef]
90. Chauhan, S.; Jain, N.; Nagaich, U. Nanodiamonds with powerful ability for drug delivery and biomedical applications: Recent updates on in vivo study and patents. *J. Pharm. Anal.* **2020**, *10*, 1–12. [CrossRef]
91. Rifai, A.; Pirogova, E.; Fox, K. Diamond, carbon nanotubes and graphene for biomedical applications. In *Reference Module in Biomedical Sciences*; Elsevier: Amsterdam, The Netherlands, 2018; pp. 1–11.
92. Sun, H.; Bu, Y.; Liu, H.; Wang, J.; Yang, W.; Li, Q.; Guo, Z.; Liu, C.; Shen, C. Superhydrophobic conductive rubber band with synergistic dual conductive layer for wide-range sensitive strain sensor. *Sci. Bull.* **2022**, *67*, 1669–1678. [CrossRef]
93. Ahuja, P.; Akiyama, S.; Ujjain, S.K.; Kukobat, R.; Vallejos-Burgos, F.; Futamura, R.; Hayashi, T.; Kimura, M.; Tomanek, D.; Kaneko, K. A water-resilient carbon nanotube based strain sensor for monitoring structural integrity. *J. Mater. Chem. A* **2019**, *7*, 19996–20005. [CrossRef]
94. Huang, J.; Li, J.; Xu, X.; Hua, L.; Lu, Z. In situ loading of polypyrrole onto aramid nanofiber and carbon nanotube aerogel fibers as physiology and motion sensors. *ACS Nano* **2022**, *16*, 8161–8171. [CrossRef]
95. Badatya, S.; Bharti, D.K.; Sathish, N.; Srivastava, A.K.; Gupta, M.K. Humidity sustainable hydrophobic poly(vinylidene fluoride)-carbon nanotubes foam based piezoelectric nanogenerator. *ACS Appl. Mater. Interfaces* **2021**, *13*, 27245–27254. [CrossRef] [PubMed]

96. Li, H.; Yang, Y.; Li, M.; Zhu, Y.; Zhang, C.; Zhang, R.; Song, Y. Frost-resistant and ultrasensitive strain sensor based on a tannic acid-nanocellulose/sulfonated carbon nanotube-reinforced polyvinyl alcohol hydrogel. *Int. J. Biol. Macromol.* **2022**, *219*, 199–212. [CrossRef] [PubMed]
97. Zhang, C.; Song, S.; Li, Q.; Wang, J.; Liu, Z.; Zhang, S.; Zhang, Y. One-pot facile fabrication of covalently cross-linked carbon nanotube/PDMS composite foam as a pressure/temperature sensor with high sensitivity and stability. *J. Mater. Chem. C* **2021**, *9*, 15337–15345. [CrossRef]
98. Gao, J.; Wu, L.; Guo, Z.; Li, J.; Xu, C.; Xue, H. A hierarchical carbon nanotube/SiO$_2$ nanoparticle network induced superhydrophobic and conductive coating for wearable strain sensors with superior sensitivity and ultra-low detection limit. *J. Mater. Chem. C* **2019**, *7*, 4199–4209. [CrossRef]
99. Zhu, W.-B.; Xue, S.-S.; Zhang, H.; Wang, Y.-Y.; Huang, P.; Tang, Z.-H.; Li, Y.-Q.; Fu, S.-Y. Direct ink writing of a graphene/CNT/silicone composite strain sensor with a near-zero temperature coefficient of resistance. *J. Mater. Chem. C* **2022**, *10*, 8226–8233. [CrossRef]
100. Lee, H.; Jung, G.; Keum, K.; Kim, J.W.; Jeong, H.; Lee, Y.H.; Kim, D.S.; Ha, J.S. A textile-based temperature-tolerant stretchable supercapacitor for wearable electronics. *Adv. Funct. Mater.* **2021**, *31*, 2106491. [CrossRef]
101. Singh, K.; Gupta, M.; Tripathi, C. Fabrication of flexible and sensitive laser-patterned serpentine-structured graphene–CNT paper for strain sensor applications. *Appl. Phys. A* **2022**, *128*, 1131. [CrossRef]
102. Lu, D.; Liao, S.; Chu, Y.; Cai, Y.; Wei, Q.; Chen, K.; Wang, Q. Highly durable and fast response fabric strain sensor for movement monitoring under extreme conditions. *Adv. Fiber Mater.* **2023**, *5*, 223–234. [CrossRef]
103. Kulha, P.; Kromka, A.; Babchenko, O.; Vanecek, M.; Husak, M.; Williams, O.A.; Haenen, K. Nanocrystalline diamond piezoresistive sensor. *Vacuum* **2009**, *84*, 53–56. [CrossRef]
104. Rehman, S.; Houshyar, S.; Reineck, P.; Padhye, R.; Wang, X. Multifunctional smart fabrics through nanodiamond-polyaniline nanocomposites. *ACS Appl. Polym. Mater.* **2020**, *2*, 4848–4855. [CrossRef]
105. Bähr, M.; Käpplinger, I.; Pobedinskas, P.; Frank, T.; Grün, A.; Haenen, K.; Ortlepp, T. Pressure Sensor Devices Featuring a Chemical Passivation Made of a Locally Synthesized Diamond Layer. *Phys. Status Solidi (a)* **2023**, *220*, 2200309. [CrossRef]
106. Yu, X.; Chen, X.; Ding, X.; Yu, X.; Zhao, X.; Chen, X. Facile fabrication of flower-like MoS2/nanodiamond nanocomposite toward high-performance humidity detection. *Sens. Actuators B Chem.* **2020**, *317*, 128168. [CrossRef]
107. Ding, Y.-R.; Xue, C.-H.; Fan, Q.-Q.; Zhao, L.-L.; Tian, Q.-Q.; Guo, X.-J.; Zhang, J.; Jia, S.-T.; An, Q.-F. Fabrication of superhydrophobic conductive film at air/water interface for flexible and wearable sensors. *Chem. Eng. J.* **2021**, *404*, 126489. [CrossRef]
108. Zhu, S.; Lu, Y.; Wang, S.; Sun, H.; Yue, Y.; Xu, X.; Mei, C.; Xiao, H.; Fu, Q.; Han, J. Interface design of stretchable and environment-tolerant strain sensors with hierarchical nanocellulose-supported graphene nanocomplexes. *Compos. Part A Appl. Sci. Manuf.* **2023**, *164*, 107313. [CrossRef]
109. Ma, S.; Jia, T.; Wang, C.; Xu, H.; Zhou, H.; Zhao, X.; Chen, C.; Wang, D.; Liu, C.; Qu, C. Anisotropic MWCNT/polyimide aerogels with multifunctional EMI shielding and strain sensing capabilities. *Compos. Part A Appl. Sci. Manuf.* **2022**, *163*, 107208. [CrossRef]
110. Barlian, A.A.; Park, W.-T.; Mallon, J.R.; Rastegar, A.J.; Pruitt, B.L. Semiconductor piezoresistance for microsystems. *Proc. IEEE* **2009**, *97*, 513–552. [CrossRef] [PubMed]
111. Zhai, J.; Zhang, Y.; Cui, C.; Li, A.; Wang, W.; Guo, R.; Qin, W.; Ren, E.; Xiao, H.; Zhou, M. Flexible waterborne polyurethane/cellulose nanocrystal composite aerogels by integrating graphene and carbon nanotubes for a highly sensitive pressure sensor. *ACS Sustain. Chem. Eng.* **2021**, *9*, 14029–14039. [CrossRef]
112. Deng, Y.; Wang, H.; Zhang, K.; Shao, J.; Qiu, J.; Wu, J.; Wu, Y.; Yan, L. A high-voltage quasi-solid-state flexible supercapacitor with a wide operational temperature range based on a low-cost "water-in-salt" hydrogel electrolyte. *Nanoscale* **2021**, *13*, 3010–3018. [CrossRef]
113. Chen, X.; Liu, H.; Zheng, Y.; Zhai, Y.; Liu, X.; Liu, C.; Mi, L.; Guo, Z.; Shen, C. Highly compressible and robust polyimide/carbon nanotube composite aerogel for high-performance wearable pressure sensor. *ACS Appl. Mater. Interfaces* **2019**, *11*, 42594–42606. [CrossRef]
114. Sun, H.; Zhao, Y.; Jiao, S.; Wang, C.; Jia, Y.; Dai, K.; Zheng, G.; Liu, C.; Wan, P.; Shen, C. Environment tolerant conductive nanocomposite organohydrogels as flexible strain sensors and power sources for sustainable electronics. *Adv. Funct. Mater.* **2021**, *31*, 2101696. [CrossRef]
115. Zang, X.; Zhang, R.; Zhen, Z.; Lai, W.; Yang, C.; Kang, F.; Zhu, H. Flexible, temperature-tolerant supercapacitor based on hybrid carbon film electrodes. *Nano Energy* **2017**, *40*, 224–232. [CrossRef]
116. Nankali, M.; Nouri, N.M.; Navidbakhsh, M.; Malek, N.G.; Amindehghan, M.A.; Shahtoori, A.M.; Karimi, M.; Amjadi, M. Highly stretchable and sensitive strain sensors based on carbon nanotube–elastomer nanocomposites: The effect of environmental factors on strain sensing performance. *J. Mater. Chem. C* **2020**, *8*, 6185–6195. [CrossRef]
117. Hou, X.; Zhang, Q.; Wang, L.; Gao, G.; Lu, W. Low-temperature-resistant flexible solid supercapacitors based on organohydrogel electrolytes and microvoid-incorporated reduced graphene oxide electrodes. *ACS Appl. Mater. Interfaces* **2021**, *13*, 12432–12441. [CrossRef]
118. Xu, X.; Chen, Y.; He, P.; Wang, S.; Ling, K.; Liu, L.; Lei, P.; Huang, X.; Zhao, H.; Cao, J. Wearable CNT/Ti$_3$C$_2$T$_x$ MXene/PDMS composite strain sensor with enhanced stability for real-time human healthcare monitoring. *Nano Res.* **2021**, *14*, 2875–2883. [CrossRef]

119. Zhu, S.; Peng, S.; Qiang, Z.; Ye, C.; Zhu, M. Cryogenic-environment resistant, highly elastic hybrid carbon foams for pressure sensing and electromagnetic interference shielding. *Carbon* **2022**, *193*, 258–271. [CrossRef]
120. Zhang, J.; Liu, X.; Feng, H.; Wang, C.; Rong, Q.; Liu, M. Conductive, sensing stable and mechanical robust silicone rubber composites for large-strain sensors. *Polym. Compos.* **2021**, *42*, 6394–6402. [CrossRef]
121. Sadi, M.S.; Kumpikaitė, E. Highly conductive composites using polypyrrole and carbon nanotubes on polydopamine functionalized cotton fabric for wearable sensing and heating applications. *Cellulose* **2023**, *30*, 7981–7999. [CrossRef]
122. Zhang, J.; Liu, J.; Zhao, Z.; Sun, W.; Zhao, G.; Liu, S.; Xu, J.; Li, Y.; Liu, Z.; Li, Y. Calotropis gigantea Fiber-Based Sensitivity-Tunable Strain Sensors with Insensitive Response to Wearable Microclimate Changes. *Adv. Fiber Mater.* **2023**, *5*, 1378–1391. [CrossRef]
123. Jiao, E.; Wu, K.; Liu, Y.; Zhang, H.; Zheng, H.; Xu, C.-a.; Shi, J.; Lu, M. Nacre-like robust cellulose nanofibers/MXene films with high thermal conductivity and improved electrical insulation by nanodiamond. *J. Mater. Sci.* **2022**, *57*, 2584–2596. [CrossRef]
124. Wang, X.; Cao, W.; Su, Z.; Zhao, K.; Dai, B.; Gao, G.; Zhao, J.; Zhao, K.; Wang, Z.; Sun, T. Fabrication of High Thermal Conductivity Nanodiamond/Aramid Nanofiber Composite Films with Superior Multifunctional Properties. *ACS Appl. Mater. Interfaces* **2023**, *15*, 27130–27143. [CrossRef]
125. Li, L.; Bai, Y.; Li, L.; Wang, S.; Zhang, T. A superhydrophobic smart coating for flexible and wearable sensing electronics. *Adv. Mater.* **2017**, *29*, 1702517. [CrossRef]
126. Wang, L.; Chen, Y.; Lin, L.; Wang, H.; Huang, X.; Xue, H.; Gao, J. Highly stretchable, anti-corrosive and wearable strain sensors based on the PDMS/CNTs decorated elastomer nanofiber composite. *Chem. Eng. J.* **2019**, *362*, 89–98. [CrossRef]
127. Liu, H.; Xu, T.; Cai, C.; Liu, K.; Liu, W.; Zhang, M.; Du, H.; Si, C.; Zhang, K. Multifunctional superelastic, superhydrophilic, and ultralight nanocellulose-based composite carbon aerogels for compressive supercapacitor and strain sensor. *Adv. Funct. Mater.* **2022**, *32*, 2113082. [CrossRef]
128. Chen, Y.; Wang, L.; Wu, Z.; Luo, J.; Li, B.; Huang, X.; Xue, H.; Gao, J. Super-hydrophobic, durable and cost-effective carbon black/rubber composites for high performance strain sensors. *Compos. Part B Eng.* **2019**, *176*, 107358. [CrossRef]
129. Seyedin, S.; Zhang, P.; Naebe, M.; Qin, S.; Chen, J.; Wang, X.; Razal, J.M. Textile strain sensors: A review of the fabrication technologies, performance evaluation and applications. *Mater. Horiz.* **2019**, *6*, 219–249. [CrossRef]
130. Qu, R.; Zhang, W.; Liu, N.; Zhang, Q.; Liu, Y.; Li, X.; Wei, Y.; Feng, L. Antioil Ag_3PO_4 nanoparticle/polydopamine/Al_2O_3 sandwich structure for complex wastewater treatment: Dynamic catalysis under natural light. *ACS Sustain. Chem. Eng.* **2018**, *6*, 8019–8028. [CrossRef]
131. Wu, H.; Liu, Q.; Du, W.; Li, C.; Shi, G. Transparent polymeric strain sensors for monitoring vital signs and beyond. *ACS Appl. Mater. Interfaces* **2018**, *10*, 3895–3901. [CrossRef] [PubMed]
132. Liu, X.; Lu, C.; Wu, X.; Zhang, X. Self-healing strain sensors based on nanostructured supramolecular conductive elastomers. *J. Mater. Chem. A* **2017**, *5*, 9824–9832. [CrossRef]
133. Pang, Y.; Yang, Z.; Han, X.; Jian, J.; Li, Y.; Wang, X.; Qiao, Y.; Yang, Y.; Ren, T.-L. Multifunctional mechanical sensors for versatile physiological signal detection. *ACS Appl. Mater. Interfaces* **2018**, *10*, 44173–44182. [CrossRef]
134. Houshyar, S.; Nayak, R.; Padhye, R.; Shanks, R.A. Fabrication and characterization of nanodiamond coated cotton fabric for improved functionality. *Cellulose* **2019**, *26*, 5797–5806. [CrossRef]
135. Houshyar, S.; Padhye, R.; Shanks, R.A.; Nayak, R. Nanodiamond fabrication of superhydrophilic wool fabrics. *Langmuir* **2019**, *35*, 7105–7111. [CrossRef] [PubMed]

Disclaimer/Publisher's Note: The statements, opinions and data contained in all publications are solely those of the individual author(s) and contributor(s) and not of MDPI and/or the editor(s). MDPI and/or the editor(s) disclaim responsibility for any injury to people or property resulting from any ideas, methods, instructions or products referred to in the content.

Article

Synthesis and Characterization of Ni–Co–O Nanosheets on Silicon Carbide Microspheres/Graphite Composite for Supercapacitor Applications

Han-Wei Chang [1,2,*], Zong-Ying Tsai [3], Jia-Jun Ye [1], Kuo-Chuang Chiu [4], Tzu-Yu Liu [4] and Yu-Chen Tsai [3,*]

1. Department of Chemical Engineering, National United University, Miaoli 360302, Taiwan; u0914010@o365.nuu.edu.tw
2. Pesticide Analysis Center, National United University, Miaoli 360302, Taiwan
3. Department of Chemical Engineering, National Chung Hsing University, Taichung 402202, Taiwan; g111065070@mail.nchu.edu.tw
4. Material and Chemical Research Laboratories, Industrial Technology Research Institute, Hsinchu 310401, Taiwan; ckc@itri.org.tw (K.-C.C.); jill.t.y.liu@itri.org.tw (T.-Y.L.)
* Correspondence: hwchang@nuu.edu.tw (H.-W.C.); yctsai@dragon.nchu.edu.tw (Y.-C.T.); Tel.: +886-37-382216 (H.-W.C.); +886-4-22857257 (Y.-C.T.)

Abstract: The well-interconnected ternary Ni–Co–O nanosheets were grown on silicon carbide microspheres/graphite composite (gra@SiC/Ni–Co–O) by optimizing the electrodeposition method. Silicon carbide microspheres/graphite composite (gra@SiC) serves as a conductive template for the growth of Ni–Co–O nanosheets to form a binder-free 3D well-designed hierarchical interconnected network between the Ni–Co–O nanosheets and SiC microspheres. The obtained gra@SiC/Ni–Co–O is proposed as a great capacitance performance for supercapacitors. Field emission scanning electron microscopy (FESEM), Raman spectroscopy, high-resolution transmission electron microscopy (HRTEM) with selected area electron diffraction (SAED) and energy dispersive X-ray spectroscopy (EDS), X-ray photoelectron spectroscopy, and electrochemical analysis were employed to investigate the morphology and structural and electrochemical characteristics. The synergistic effects of EDLC (SiC microspheres) and pseudo-capacitance (Ni–Co–O nanosheets) can effectively improve the supercapacitive performance. It is also worth mentioning that after electrochemical testing, the redox reaction of Ni–Co–O nanosheets greatly promoted the faradic pseudo-capacitance contribution, and silicon carbide microspheres/graphite composite contributed to the formation of a 3D interconnected network, improving the cycling stability during the charging/discharging processes.

Keywords: ternary Ni–Co–O nanosheets; silicon carbide microspheres/graphite composite; supercapacitor; EDLC and pseudo-capacitance

Citation: Chang, H.-W.; Tsai, Z.-Y.; Ye, J.-J.; Chiu, K.-C.; Liu, T.-Y.; Tsai, Y.-C. Synthesis and Characterization of Ni–Co–O Nanosheets on Silicon Carbide Microspheres/Graphite Composite for Supercapacitor Applications. *C* **2023**, *9*, 101. https://doi.org/10.3390/c9040101

Academic Editors: Gil Gonçalves and Olena Okhay

Received: 3 October 2023
Revised: 22 October 2023
Accepted: 27 October 2023
Published: 29 October 2023

Copyright: © 2023 by the authors. Licensee MDPI, Basel, Switzerland. This article is an open access article distributed under the terms and conditions of the Creative Commons Attribution (CC BY) license (https:// creativecommons.org/licenses/by/ 4.0/).

1. Introduction

Resource depletion, extreme climate change, and environmental damage are the major driving forces for the development of renewable and sustainable energy technologies (e.g., supercapacitors, fuel cells, hydrogen generation, and rechargeable batteries) [1–3]. As promising energy storage devices, supercapacitors have attracted significant research interest among researchers owing to their high power density, fast charge/discharge rate, excellent reversibility, and high durability. Additionally, they can be integrated into hybrid energy storage systems to meet the demand for renewable energy sources that can address the aforementioned issues. The charge storage mechanisms of supercapacitors are mainly divided into electrochemical double-layer capacitance (EDLC) and pseudo-capacitance. The EDLC accumulates the charges at the interface between the electrode and electrolyte through non-Faradaic physical processes. Carbon, semiconductor, and cermet materials are widely used as electrode materials in EDLC supercapacitors [4–6]. The performance of pseudo-capacitance is derived from the pseudocapacitive materials through

Faradaic processes involving surface or near-surface redox reactions. Transition metal-based compounds are considered the most ideal electrode materials for pseudocapacitance. Generally, binary and ternary transition metal compounds as electrode materials have been widely investigated due to their rich redox chemistry and valence state transitions, promising pseudocapacitive characteristics [7–9].

Particularly, the unique tuneable bandgaps of silicon carbide (SiC) EDLC-type semiconductor electrode materials enable high electron mobility, good mechanical performance, and excellent temperature stability, indicating that it may be a good candidate for supercapacitor electrode materials. However, the relatively low energy density of SiC-based materials for EDLC limits their applicability in supercapacitors. Hybrid supercapacitors combine the intrinsic properties of both EDLC and pseudo-capacitance to solve the existing limitation of EDLC to significantly improve the supercapacitor performance. Significant efforts have been devoted to combining SiC-based EDLC and transition metal-based pseudo-capacitance-type materials as electrode materials for supercapacitors. A previous study reported that the combination of microspherical SiC as EDLC materials and birnessite-type MnO_2 as pseudo-capacitance materials provides an effective method to achieve high supercapacitor performance compared to the performance of microspherical SiC used in EDLC. Typically, SiC should possess a high surface area to enable its use as support structures for the growth of birnessite-type MnO_2 with intimate interfacial contact, significantly facilitating the electrons and ions transportation in the electrode, resulting in excellent capacitive performance [10]. Another study reported that the obtained ferroferric oxide (Fe_3O_4) grown on SiC flakes (Fe_3O_4/SiC) is a feasible configuration for supercapacitors, as the introduction of Fe_3O_4 into the composite significantly enhanced the capacitive performance of the SiC, which can be attributed to the additional pseudo-capacitance contributed by the Fe_3O_4 [11]. Further, another study synthesized $MgCo_2O_4$/SiC composite as an electrode material for supercapacitors by growing ternary transition metal oxides (spinel cobaltites ($MgCo_2O_4$)) on SiC flakes. The synergistic effects of the EDLC behavior of SiC and the pseudo-capacitance behavior of $MgCo_2O_4$ enhanced the supercapacitive performance of the composite. In particular, they observed that SiC provided a more accessible surface area for charge storage through ion adsorption to further enhance the capacitance of the EDLC, and $MgCo_2O_4$ simultaneously exhibited pseudo-capacitive energy storage efficiency through the multiple oxidation states/structures of Mg and Co ions [12]. The findings of the aforementioned studies indicated that hybrid supercapacitors (EDLC and pseudo-capacitance) can be fabricated by incorporating SiC with transition metal-based compounds, thus presenting a feasible strategy to enhance the capacitance of supercapacitors.

In this work, well-interconnected ternary Ni–Co–O nanosheets were directly grown on silicon carbide microspheres (gra@SiC/Ni–Co–O) by optimizing the electrodeposition method. Silicon carbide (SiC) possesses excellent surface characteristics, electron mobility, mechanical performance, and temperature stability to exhibit better EDLC properties. A simple and low-cost electrodeposition method is suitable for the preparation of binder-free electrode materials without adding conductive agents and binders. Observations revealed that the Ni–Co–O nanosheets were uniformly grown on the surface of silicon carbide microspheres using a binder-free electrodeposition process, which enabled the maximal exposure of active sites, thus ensuring sufficient charge transport kinetics at the electrode/electrolyte interface. Simultaneously, the synergistic effect between the Ni–Co–O nanosheets and silicon carbide microspheres further endowed the EDLC of silicon carbide microspheres with the additional pseudocapacitance of Ni–Co–O nanosheets. This study demonstrated the promising potential of the designed gra@SiC/Ni–Co–O and provided guidelines for enhancing the capacitance of SiC-based supercapacitors.

2. Materials and Methods

2.1. Reagents

Nickel(II) chloride hexahydrate ($NiCl_2 \cdot 6H_2O$), nickel(II) nitrate hexahydrate ($Ni(NO_3)_2 \cdot 6H_2O$), cobalt(II) nitrate hexahydrate ($Co(NO_3)_2 \cdot 6H_2O$), and potassium hydroxide (KOH) were purchased from Sigma-Aldrich (St. Louis, MO, USA). Silicon carbide (SiC) was provided by the Industrial Technology Research Institute (ITRI) (Chutung, Hsinchu, Taiwan). All chemicals in this work were obtained as analytical grade and directly used without further purification. All aqueous solutions were prepared using a Milli-Q water purification system (Millipore, Milford, MA, USA).

2.2. Preparation of gra@SiC/Ni–Co–O

First, 10 mg of SiC and graphite conductive additive are uniformly dispersed via ultra-sonication for 1 h to make well-dispersed and homogenous suspensions. A graphite electrode with a 3 mm diameter was used for surface modification. Then, 10 µL of SiC homogeneous suspension with a concentration of 3 mg mL^{-1} was modified on the graphite electrode surface by a drop coating method and dried the SiC modified electrode under 60 °C for 30 min. Graphite electrodes without and with SiC modification were designated as gra and gra@SiC, respectively. Subsequently, the gra/Ni–Co–O and gra@SiC/Ni–Co–O were fabricated through the electrodeposition method by applying a constant potential at room temperature. The electrodeposition method was performed potentiostatically with an electrochemical analyzer using a three-electrode system. The three-electrode system was used consisting of gra and gra@SiC as working electrode, a platinum wire as counter electrode, and an Ag/AgCl (3 M KCl) as reference electrode (the volume of precursor electrolyte for electrodeposition was 20 mL). The electrodeposition precursor electrolyte of 20 mL was a mixture of $Ni(NO_3)_2 \cdot 6H_2O$ (6 mM) and $Co(NO_3)_2 \cdot 6H_2O$ (12 mM) and the mixture solution was continuously stirred at room temperature for 10 min to form a homogeneous solution. And, the precursor electrolyte was kept at room temperature for the subsequent Ni-Co hydroxide precursor electrodeposition. Then, gra and gra@SiC working electrodes were immersed in the precursor electrolyte. The electrodeposition of Ni-Co hydroxide precursor at a constant potential of −0.6 V for 240 s, and washed with DI water trice for 15 min each, and then dried in an oven to remove the remaining reagents and collected for subsequent characterization. The resulting modified electrode was designated as gra/Ni–Co–O and gra@SiC/Ni–Co–O, respectively.

2.3. Characterization

The morphology was characterized using a Field emission scanning electron microscopic (FESEM, JSM-7800F, JEOL, Akishima, Japan) and high-resolution transmission electron microscopy (HRTEM, JEM-2010, JEOL, Japan) with selected area electron diffraction (SAED) and energy dispersive X-ray spectroscopy (EDS). Raman spectra were characterized using an automated Raman spectrometer equipped with an argon laser excitation wavelength of 532 nm (Unidron, CL Technology Co., Ltd., New Taipei city, Taiwan). The chemical structure and composition were determined by X-ray photoelectron spectroscopy (XPS, PHI-5000 Versaprobe, ULVAC-PHI, Chigasaki, Kanagawa, Japan). Electrochemical measurements were performed using a three-electrode system by an electrochemical analyzer (Autolab, model PGSTAT30, Eco Chemie, Utrecht, The Netherlands). The supercapacitor with a conventional three-electrode system comprised gra/Ni–Co–O and gra@SiC/Ni–Co–O working electrode, a platinum wire counter electrode, and an Ag/AgCl (3 M KCl) reference electrode in 1 M KOH electrolyte (pH 13.8) (the volume of KOH electrolyte for electrochemical measurements was 20 mL). The electrochemical measurements of gra/Ni–Co–O and gra@SiC/Ni–Co–O were examined by cyclic voltammetry (CV) and galvanostatic charge–discharge (GCD).

3. Results

The surface morphologies of gra, gra@SiC, gra/Ni–Co–O, and gra@SiC/Ni–Co–O were characterized using field emission scanning electron microscopy (FESEM). Compared to that of gra (Figure 1a), a notable apophysis was observed on the surface of the coating material after the addition of SiC (Figure 1b); marked in the region of the orange dashed circle), and SiC was observed to exhibit a spherical structure with a size of few tens of micrometers. Next, gra and gra@SiC were used as growth templates for the electrodeposition growth of ternary Ni–Co–O nanosheets. The SEM image confirmed the dense and even coverage of the entire surface of gra and gra@SiC by the ternary Ni–Co–O nanosheets (Figure 1c,d). Further, high-magnification FESEM images (Figure 1e,f) revealed that the ternary Ni–Co–O nanosheets formed a uniformly interconnected network with gra and gra@SiC (enlarged view of the yellow framed region in Figure 1c,d). This well-defined interconnected network structure formed a conductive network for the transport of electrons, indicating its significant potential in supercapacitors. Additionally, the SEM images revealed the aggregation tendency of the ternary Ni–Co–O nanosheets on the surface of gra in gra/Ni–Co–O, and the average diameter of the Ni–Co–O nanosheet aggregates was approximately 1 μm. In contrast, the ternary Ni–Co–O nanosheets in gra@SiC/Ni–Co–O were well-dispersed on the surface of the SiC without notable aggregation, resulting in increased surface area. Consequently, this is expected to further accelerate the electron/ion transport, thus enhancing their electrochemical performances. Further, gra/Ni–Co–O and gra@SiC/Ni–Co–O were further characterized using Raman, high-resolution transmission electron microscopy (HRTEM), and X-ray photoelectron spectroscopy (XPS).

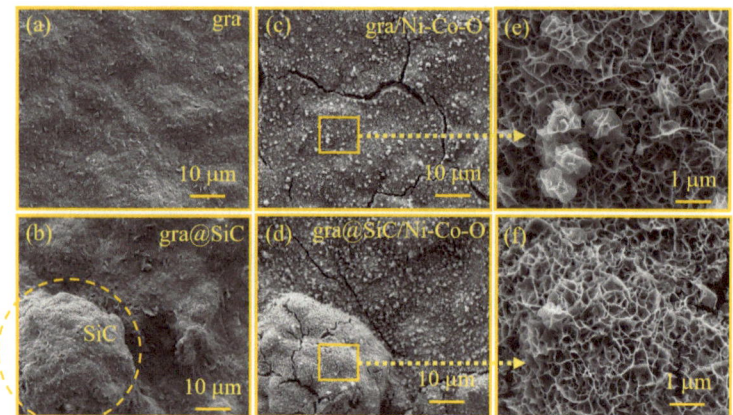

Figure 1. FESEM images of (**a**) gra, (**b**) gra@SiC, (**c**) gra/Ni–Co–O, and (**d**) gra@SiC/Ni–Co–O; (**e**,**f**) the enlarged area of the marked place in (**c**,**d**).

The phase composition and structural features of gra, gra@SiC, gra/Ni–Co–O, and gra@SiC/Ni–Co–O were characterized using Raman spectroscopy in the Raman shift range of 100–2000 cm^{-1}. Two characteristic peaks were observed at approximately 1567 (G band) and 1347 cm^{-1} (D band) in the Raman spectra of all samples and can be attributed to the presence of graphite conductive additive (gra) during the electrode slurry preparation (Figure 2). The D and G bands correspond to disordered carbon and graphitic sp^2 carbon, respectively. Additionally, a notable characteristic peak was observed in the Raman spectrum of gra@SiC at approximately 785 cm^{-1}, indicating the presence of hexagonal (mainly, 2H-SiC) polytype form phase within the gra@SiC [13], implying the successful coordination of SiC with graphite conductive additive to form gra@SiC. During the electrodeposition process, Ni–Co–O nanosheets uniformly covered the surface of gra and gra@SiC to form an interconnected structure. It is well known that the characteristic peaks observed in the low Raman shift region of the Raman spectrum are associated with Ni–Co compounds. The characteristic peak of 2H-SiC was observed to disappear in the Raman spectra of gra/Ni–Co–O and gra@SiC/Ni–Co–O (Figure 2), and four newly formed characteristic peaks were observed at approximately 196, 468, 522, and 666 cm^{-1}, corresponding to the E_g mode, O–M–O bending, M–O A_g vibrations, and A_g modes of Ni–Co binary hydroxides, respectively. Additionally, a weak and broad band observed at approximately 1070 cm^{-1} was linked with the presence of residual nitrate ions from the nitrate precursor [14,15]. These results confirmed the successful synthesis of Ni–Co compounds with Ni–Co binary hydroxides structure by the electrodeposition process and the complete coverage of gra and gra@SiC by Ni–Co binary hydroxides.

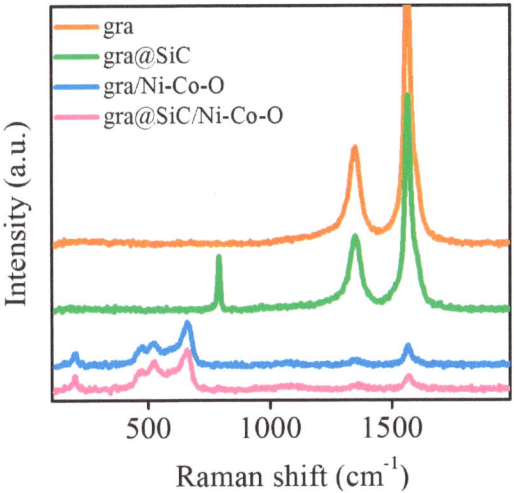

Figure 2. Raman spectra of gra, gra@SiC, gra/Ni–Co–O, and gra@SiC/Ni–Co–O.

To evaluate the Ni–Co–O nanosheet that grew along the outer wall of gra@SiC, the morphologies and relevant element compositions of gra@SiC/Ni–Co–O (the interior and exterior surfaces of the Ni–Co–O nanosheet shell, designated Pt1 and Pt2, respectively) were characterized using HRTEM, and corresponding selected area electron diffraction (SAED) pattern and energy dispersive X-ray spectroscopy (EDS), (Figure 3). A thin layer of Ni–Co–O nanosheet was conformally grown along the outer wall of gra@SiC (Figure 3a,b). SAED (Figure 3c) showed well-defined polymorphic rings confirming the polycrystalline nature of Ni–Co–O nanosheet shell. The broadening of the diffraction SAED rings suggested that the Ni–Co–O nanosheet shell was comprised of nanocrystalline or had a relatively high degree of crystallinity. The lattice planes of (100), (101), and (110) were indexed from SAED rings, which confirmed that the Ni–Co–O nanosheet shell was in the phases of Ni–Co binary hydroxides, in agreement with Raman results (Figure 2) [16,17]. In the EDS results (Figure 3d), the two selected regions of the interior and exterior surfaces within the Ni–Co–O nanosheet shell enclosed by the HRTEM image indicated that the interior surfaces of the Ni–Co–O nanosheet (Pt2) exhibited significantly higher amount of Si element than the exterior surfaces (Pt1). Additionally, it revealed the presence of Ni, Co, and O elements in the two selected regions where the formation of a uniform interconnected network structure was confirmed, indicating the strong coupling between Ni–Co–O nanosheet and the SiC. The EDS results were consistent the aforementioned results. Indeed, the inset in Figure 3d showed the atomic ratio of Ni and Co elements in the two selected regions (Pt1 and Pt2). The Ni:Co atomic ratio at Pt1 is 1.0:2.0 higher than that of the atomic ratio at Pt2 (1.0:3.5). The result further revealed that the dissimilar growth rates of the Ni–Co binary hydroxides observed in the interior and exterior surfaces of the Ni–Co–O nanosheet shell.

It can be concluded that the contact interface between the SiC core/Ni–Co–O nanosheet shell, and the surface characteristics of SiC were decisive factors influencing the growth rates of the Ni–Co binary hydroxides.

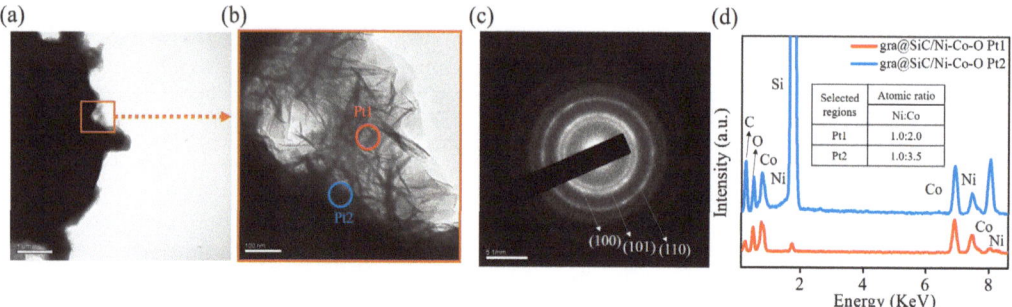

Figure 3. HRTEM images of (**a**) gra@SiC/Ni–Co–O, (**b**) the enlarged area of the marked place in (**a**). (**c**) SAED pattern. (**d**) EDS of the two selected regions (Pt1 and Pt2) enclosed in (**b**) and the inset in (**d**) represents the atomic ratio of Ni and Co elements.

The surface elemental composition and valance states of gra/Ni–Co–O and gra@SiC/Ni–Co–O were characterized using XPS, and the XPS results are shown in Figure 4. Figure 4a,b shows the high-resolution Ni 2p XPS profiles of gra/Ni–Co–O and gra@SiC/Ni–Co–O. Two spin–orbit-split doublets (Ni $2p_{1/2}$ and Ni $2p_{3/2}$) were observed in the Ni 2p XPS profiles. To identify the specific Ni species, Ni $2p_{1/2}$ (Ni $2p_{3/2}$) doublets were assigned to two fitted peaks located at approximately 872.7 (854.7 eV) and 873.8 eV (856.1 eV), which correspond to Ni^{2+} and Ni^{3+}, respectively, and their shake-up satellite (Sat.) was located at approximately 880.0 eV (862.2 eV). The Ni 2p XPS quantitative analysis results are summarized in Table 1. The intensity ratio of Ni^{2+}/Ni^{3+} for gra/Ni–Co–O and gra@SiC/Ni–Co–O was 0.46 and 0.33 (mainly in the oxidation state of Ni^{3+}), respectively, indicating that the oxidation state of Ni in gra@SiC/Ni–Co–O is higher than that in gra/Ni–Co–O [18,19]. Figure 4c,d shows the high-resolution Co 2p XPS profiles of gra/Ni–Co–O and gra@SiC/Ni–Co–O. Two spin–orbit-split doublets were observed in the Co 2p XPS profiles at binding energies of 780.8 (Co $2p_{1/2}$) and 796.7 eV (Co $2p_{3/2}$). The spin-doublet Co $2p_{1/2}$ (Co $2p_{3/2}$) region was further fitted into Co^{2+} (located at about 796.7 (781.0) eV) and Co^{3+} (located at about 794.9 (779.8) eV). The results revealed that Co^{2+} was the main oxidation state of Co on both gra/Ni–Co–O and gra@SiC/Ni–Co–O. The Co 2p XPS quantitative analysis results (Table 2) indicate the presence of more Co^{3+} species on the gra@SiC/Ni–Co–O sample [20,21]. The XPS results confirmed the formation of more Ni^{3+} and Co^{3+} active species on the surface gra@SiC/Ni–Co–O compared to gra/Ni–Co–O. Figure 4e,f shows the high-resolution O 1s XPS profiles of gra/Ni–Co–O and gra@SiC/Ni–Co–O. The O 1s XPS profile was deconvoluted into four peaks located at binding energies of peaks at 532.5, 531.3, 530.5, and 529.1 eV, respectively, which are assigned to physical/chemical adsorbed water (H–O–H), oxygen defective site (O-defect site), hydroxyl groups (M–O–H) and metal–oxygen bonds (M–O–M) on the surface of Ni–Co–O nanosheet. The O 1s XPS quantitative analysis results (Table 3) revealed reveal that both gra/Ni–Co–O and gra@SiC/Ni–Co–O possessed abundant and accessible oxygen vacancies, indicating the existence of low coordinated metal oxygen structures, which are very helpful for enhanced supercapacitive performance [22,23].

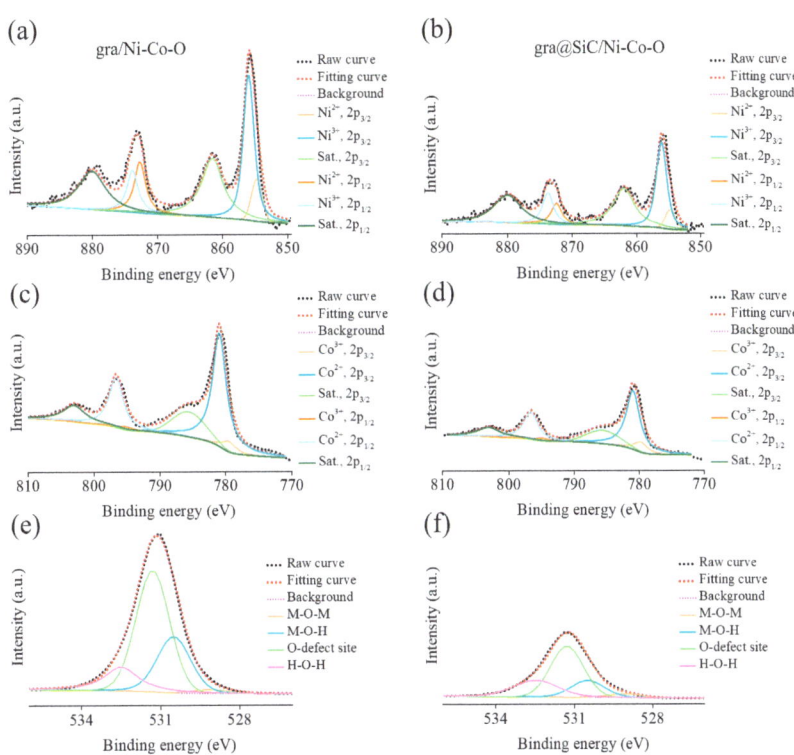

Figure 4. (**a**) Ni 2p, (**c**) Co 2p, and (**e**) O 1s XPS spectra of gra/Ni–Co–O. (**b**) Ni 2p, (**d**) Co 2p, and (**f**) O 1s XPS spectra of gra@SiC/Ni–Co–O.

Table 1. Fitted Ni 2p XPS spectra results of gra/Ni–Co–O and gra@SiC/Ni–Co–O.

Sample	Fitted Results of Ni (2p) XPS Spectra						
	Ni^{2+} $2p_{3/2}$ (%)	Ni^{3+} $2p_{3/2}$ (%)	Sat. $2p_{3/2}$ (%)	Ni^{2+} $2p_{3/2}$ (%)	Ni^{3+} $2p_{3/2}$ (%)	Sat. $2p_{3/2}$ (%)	Ni^{2+}/Ni^{3+}
gra/Ni–Co–O	5.9	27.9	25.3	11.2	9.4	20.3	0.46
gra@SiC/Ni–Co–O	4.8	26.9	26.5	7.7	11.3	22.8	0.33

Table 2. Fitted Co 2p XPS spectra results of gra/Ni–Co–O and gra@SiC/Ni–Co–O.

Sample	Fitted Results of Co (2p) XPS Spectra						
	Co^{3+} $2p_{3/2}$ (%)	Co^{2+} $2p_{3/2}$ (%)	Sat. $2p_{3/2}$ (%)	Co^{3+} $2p_{3/2}$ (%)	Co^{2+} $2p_{3/2}$ (%)	Sat. $2p_{3/2}$ (%)	Co^{2+}/Co^{3+}
gra/Ni–Co–O	4.2	46.6	16.8	1.3	20.6	10.5	12.27
gra@SiC/Ni–Co–O	7.4	43.4	16.3	2.1	20.1	10.7	6.74

Table 3. Fitted O 1s XPS spectra results of gra/Ni–Co–O and gra@SiC/Ni–Co–O.

Sample	Fitted Results of O (1s) XPS Spectra			
	M-O-M (%)	M-O-H (%)	O-Defect Site (%)	H-O-H (%)
gra/Ni–Co–O	2.0	27.0	55.4	15.6
gra@SiC/Ni–Co–O	4.0	18.7	52.2	25.1

To evaluate the capacitive performance of gra/Ni–Co–O and gra@SiC/Ni–Co–O, cyclic voltammetry (CV) and galvanostatic charges/discharge (GCD) measurements were performed from 0.0 to 0.6 V versus Ag/AgCl in 1 M KOH solution. The CV curves of all Ni–Co–O@3D Ni-x samples at a scan rate of 50 mV s^{-1} exhibited a distorted semi-rectangular shape, indicating pseudocapacitance characteristics, which could be attributed to the reversible redox couple of Ni–Co–O nanosheet (Figure 5a). Moreover, the CV curve area of gra@SiC/Ni–Co–O (Figure 5a) was larger than that of gra/Ni–Co–O, indicating the excellent electrochemical capacitance performance of gra@SiC/Ni–Co–O, which can be attributed to the synergistic effects between the Ni–Co–O nanosheets and silicon carbide microspheres. The exposed Ni–Co–O nanosheet interfaces offer a large specific surface area, which could enhance the electrocatalytic active sites to accelerate electron/ion transfer characteristics. Figure 5b shows the CV curve of gra@SiC/Ni–Co–O at different scan rates. At a higher scan rate of 100 mV s^{-1}, the CV curve exhibited a slightly distorted shape, but still maintained its semi-rectangular CV shape, demonstrating its good capacitance behavior. Figure 5c,d displays the GCD curve of gra/Ni–Co–O and gra@SiC/Ni–Co–O at different current densities. The areal capacitance can be calculated from the above GCD results by the equation C = (I Δt)/ΔV, where C is the areal capacitance (mF cm^{-2}); I is the current density (mA cm^{-2}); Δt is the discharge time (s); and ΔV is the voltage change during discharge (V). Rather than a triangular distribution, the GCD curves of both samples exhibited a quasi-triangular symmetrical distribution with plateaus, indicating the synergistic effect between the pseudocapacitance (Ni–Co–O nanosheets) and EDLC (SiC microspheres) behaviors. Additionally, the maximum areal capacitance of gra@SiC/Ni–Co–O (521 mF cm^{-2}) was higher than that of gra/Ni–Co–O (322 mF cm^{-2}) at a low current density of 1.4 mA cm^{-2}. In particular, the capacitive performance of gra@SiC/Ni–Co–O was higher or comparable to those of other previously reported results on combining semiconductor materials to obtain composite electrode materials (Table 4) [24–27]. This excellent performance of gra@SiC/Ni–Co–O may be attributed to the following reasons: (1) gra@SiC functioned as a conductive template for the subsequent growth of Ni–Co–O nanosheets to form a binder-free 3D well-designed hierarchical interconnected network between the Ni–Co–O nanosheets and SiC microspheres, leading to improved electrochemical performance [28]. (2) The contact interface between the SiC core/Ni–Co–O nanosheet shell, and the surface characteristics of SiC were decisive factors influencing the growth rates of the Ni–Co binary hydroxides. The Ni:Co atomic ratio at the interior surface of the Ni–Co–O nanosheet shell was higher than that at the exterior surface of the Ni–Co–O nanosheet shell, indicating that a relatively large amount of Ni hydroxide was present within gra@SiC/Ni–Co–O compared with gra/Ni–Co–O. This difference may affect supercapacitor performance [29]. (3) gra@SiC/Ni–Co–O possessed abundant oxygen vacancies and further exposed rich electroactive sites, providing dense diffusion channels for energy storage [30]. (4) The synergistic effects of EDLC (SiC microspheres) and pseudo-capacitance (Ni–Co–O nanosheets) effectively enhanced the supercapacitive performance. This study revealed that the use of SiC as the growth template prevented the aggregation of Ni–Co–O nanosheets, which can largely improve the specific surface area. Ni–Co–O nanosheets simultaneously possess pseudo-capacitive energy storage efficiency through the multiple oxidation states/structures of Ni and Co atoms, resulting in improved supercapacitive performance [31].

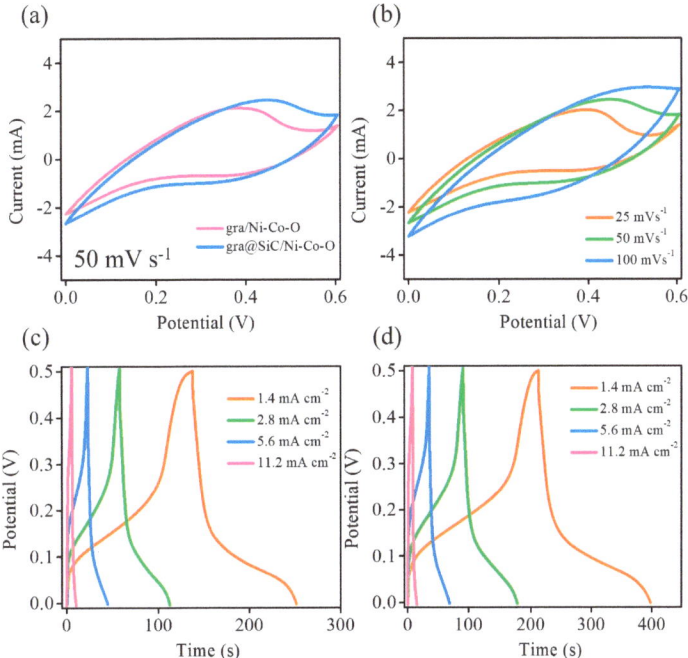

Figure 5. (**a**) Cyclic voltammetry of gra/Ni–Co–O, and gra@SiC/Ni–Co–O. (**b**) Cyclic voltammetry of gra@SiC/Ni–Co–O at different scan rates. Galvanostatic charge/discharge of (**c**) gra/Ni–Co–O and (**d**) gra@SiC/Ni–Co–O at different current densities.

Table 4. Comparing the performance of semiconductor materials-based supercapacitors.

Samples	Electrolyte	Current Density (mA cm^{-2})	Areal Capacitance (mF cm^{-2})	Reference
gra@SiC/Ni–Co–O	KOH (1 M)	1.4	521	This work
SiC@PANI	H$_2$SO$_4$ (1 M)	1	352	[24]
SiC@SiO$_2$/MnO$_2$	Na$_2$SO$_4$ (1 M)	0.2	271	[25]
SiC/HG/MnO$_2$	LiCl (3 M)	7.2	1000	[26]
SiNWs/NC@NiO	KOH (6 M)	1	110	[27]

To further understand the stability and the charge storage mechanisms of gra@SiC/Ni–Co–O after electrochemical testing, the electrochemical behavior of gra@SiC/Ni–Co–O before and after electrochemical testing was evaluated using FESEM and XPS (Figure 6). Figure 6a shows the FESEM images of gra@SiC/Ni–Co–O after electrochemical testing. The morphology of gra@SiC/Ni–Co–O after electrochemical testing indicated its structural stability, as the electrochemical testing had no effect on its morphology. The gra@SiC/Ni–Co–O possessed a 3D interconnected network with high mechanical strength, which makes the electrode more stable with electrochemical cycling. Figure 6b,c further revealed that gra@SiC/Ni–Co–O promotes the formation of a larger amount of Ni^{2+} and Co^{3+} after electrochemical testing. Interestingly, it was apparent that the opposite tendency was observed in both Ni and Co oxidation states near the surface. The Ni oxidation states were partially reduced to Ni^{2+}, and the Co oxidation states were partially oxidized to Co^{3+} (see XPS results in Figures 4 and 6). This phenomenon provided strong evidence that gra@SiC/Ni–Co–O underwent an electrochemically driven phase transition, and, simultaneously, reversible redox couples of Ni^{2+}/Ni^{3+} and Co^{2+}/Co^{3+} occurred on or near the surface of gra@SiC/Ni–Co–O to contribute remarkable pseudocapacitance [32,33]. The tests mentioned above can aid the understanding of charge-storage mechanisms in gra@SiC/Ni–Co–O. Gra@SiC/Ni–Co–O exhibited excellent synergistic effects of EDLC (SiC microspheres) and pseudo-capacitance (Ni–Co–O nanosheets) to synergistically enhance the capacitive performance. SiC microspheres contributed to the formation of the 3D interconnected network between Ni–Co–O nanosheets and SiC microspheres, which enlarged the large accessible surface area for charge absorption and accumulation to exhibit excellent EDLC performance. Faraday redox reaction contributed by Ni–Co–O nanosheets. The coexistence of Ni^{2+}/Ni^{3+} and Co^{2+}/Co^{3+} redox couples in interconnected 3D Ni–Co–O nanosheets network on SiC microspheres enabled excellent multiple redox reactions. Due to the unique synergistic effect of SiC microspheres and Ni–Co–O nanosheets, gra@SiC/Ni–Co–O significantly improved its capacitive performance and cycling stability during the charging/discharging processes.

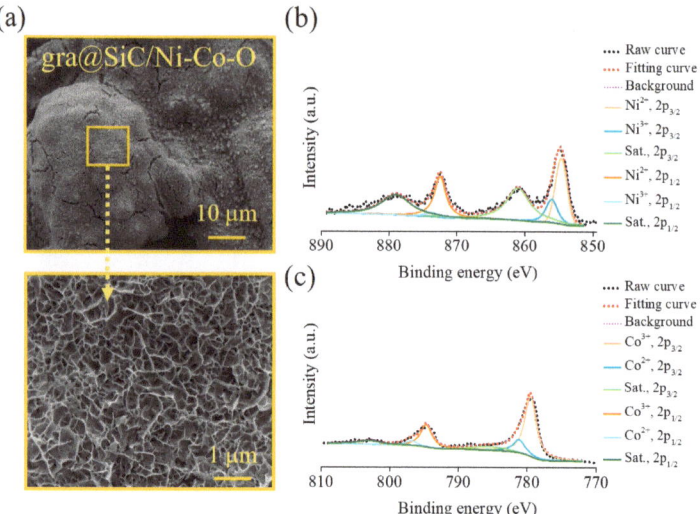

Figure 6. (a) FESEM images, (b) Ni 2p XPS spectra, and (c) Co 2p XPS spectra of gra@SiC/Ni–Co–O after electrochemical testing.

4. Conclusions

In this study, ternary Ni–Co–O nanosheets are directly grown on silicon carbide microspheres/graphite composite (gra@SiC/Ni–Co–O) by optimizing the electrodeposition method. The surface characteristics endowed SiC with the capability as growth templates for the electrodeposition growth of ternary Ni–Co–O nanosheets to construct a binder-free 3D well-designed hierarchical interconnected network structure of gra@SiC/Ni–Co–O. The gra@SiC/Ni–Co–O electrode materials own several advantages including: (1) 3D interconnected network enables enhanced mechanical properties. (2) The surface characteristics of SiC possess the difference in the growth rate of the Ni–Co binary hydroxides. A relatively large amount of Ni hydroxide was present within gra@SiC/Ni–Co–O compared with gra/Ni–Co–O. (3) The abundant and accessible oxygen vacancies lead to low coordinated metal oxygen structures. (4) The synergistic effects between two components (Ni–Co–O nanosheets and silicon carbide microspheres) contribute to faradaic redox reactions and EDLC properties. Based on the above advantages, the gra@SiC/Ni–Co–O could exhibit excellent supercapacitive performance. Further, FESEM and XPS results were used to evaluate the electrochemical behavior of gra@SiC/Ni–Co–O before and after electrochemical testing. It demonstrated that SiC microspheres contribute to the formation of the 3D interconnected network between Ni–Co–O nanosheets and SiC microspheres, improving the cycling stability during the charging/discharging processes. And, the pseudo-capacitive charge storage mechanism in gra@SiC/Ni–Co–O involves electrochemically driven phase transition and reversible redox couples of Ni^{2+}/Ni^{3+} and Co^{2+}/Co^{3+} occurred on or near the surface of gra@SiC/Ni–Co–O. The as-fabricated gra@SiC/Ni–Co–O exhibited excellent capacitative performance and is a potential candidate for future electrochemical energy storage applications (lithium batteries, metal-air batteries, fuel cells, and supercapacitors), which can meet key expectations to integrate renewable energy into electrical grid systems.

Author Contributions: Conceptualization, H.-W.C., K.-C.C., T.-Y.L. and Y.-C.T.; methodology, H.-W.C., K.-C.C. and T.-Y.L.; software, Z.-Y.T. and J.-J.Y.; formal analysis, H.-W.C., Z.-Y.T. and J.-J.Y.; investigation, H.-W.C., Z.-Y.T., J.-J.Y., K.-C.C., T.-Y.L. and Y.-C.T.; data curation, H.-W.C., Z.-Y.T. and J.-J.Y.; writing—original draft preparation, H.-W.C. and Y.-C.T.; writing—review and editing, H.-W.C. and Y.-C.T.; visualization, H.-W.C.; supervision, H.-W.C.; project administration, H.-W.C. and Y.-C.T. All authors have read and agreed to the published version of the manuscript.

Funding: This research was funded by the National Science and Technology Council (NSTC), the National United University (NUU), the Ministry of Economic Affairs (MEA), and the Industrial Technology Research Institute/Material and Chemical Laboratories (ITRI/MCL), Taiwan (NSTC 112-2221-E-239-001-MY3, NSTC 112-2221-E-005-007-MY3, SE112002, and 112-EC-17-A-24-1771).

Data Availability Statement: Not applicable.

Acknowledgments: The authors are grateful to the NSTC, NUU, MEA, and ITRI/MCL for the financial assistance granted in support of this work. For instrumentation support, we thank the NSTC and the Instrument Center of National Chung Hsing University, Taiwan, for help with FESEM, HRTEM, and XPS measurements (NSTC 112-2740-M-005-001).

Conflicts of Interest: The authors declare no conflict of interest.

References

1. Minakshi, M.; Wickramaarachchi, K. Electrochemical aspects of supercapacitors in perspective: From electrochemical configurations to electrode materials processing. *Prog. Solid State Chem.* **2023**, *69*, 100390. [CrossRef]
2. Cai, D.; Xiao, S.; Wang, D.; Liu, B.; Wang, L.; Liu, Y.; Li, H.; Wang, Y.; Li, Q.; Wang, T. Morphology controlled synthesis of $NiCo_2O_4$ nanosheet array nanostructures on nickel foam and their application for pseudocapacitors. *Electrochim. Acta* **2014**, *142*, 118–124. [CrossRef]
3. Chen, J.; Zhang, Y.; Hou, X.; Su, L.; Fan, H.; Chou, K.-C. Fabrication and characterization of ultra light SiC whiskers decorated by RuO_2 nanoparticles as hybrid supercapacitors. *RSC Adv.* **2016**, *6*, 19626–19631. [CrossRef]
4. Gu, L.; Wang, Y.; Fang, Y.; Lu, R.; Sha, J. Performance characteristics of supercapacitor electrodes made of silicon carbide nanowires grown on carbon fabric. *J. Power Sources* **2013**, *243*, 648–653. [CrossRef]
5. Kim, M.; Kim, J. Development of high power and energy density microsphere silicon carbide–MnO_2 nanoneedles and thermally oxidized activated carbon asymmetric electrochemical supercapacitors. *Phys. Chem. Chem. Phys.* **2014**, *16*, 11323–11336. [CrossRef]
6. Oh, I.; Kim, M.; Kim, J. Fe_3O_4/carbon coated silicon ternary hybrid composite as supercapacitor electrodes. *Appl. Surf. Sci.* **2015**, *328*, 222–228. [CrossRef]
7. Pramitha, A.; Raviprakash, Y. Recent developments and viable approaches for high-performance supercapacitors using transition metal-based electrode materials. *J. Energy Storage* **2022**, *49*, 104120. [CrossRef]
8. Dai, M.; Zhao, D.; Wu, X. Research progress on transition metal oxide based electrode materials for asymmetric hybrid capacitors. *Chin. Chem. Lett.* **2020**, *31*, 2177–2188. [CrossRef]
9. Yadav, S.; Sharma, A. Importance and challenges of hydrothermal technique for synthesis of transition metal oxides and composites as supercapacitor electrode materials. *J. Energy Storage* **2021**, *44*, 103295. [CrossRef]
10. Kim, M.; Kim, J. Redox deposition of birnessite-type manganese oxide on silicon carbide microspheres for use as supercapacitor electrodes. *ACS Appl. Mater. Interfaces* **2014**, *6*, 9036–9045. [CrossRef]
11. Kim, M.; Kim, J. Synergistic interaction between pseudocapacitive Fe_3O_4 nanoparticles and highly porous silicon carbide for high-performance electrodes as electrochemical supercapacitors. *Nanotechnology* **2017**, *28*, 195401. [CrossRef]
12. Kim, M.; Yoo, J.; Kim, J. Fast and reversible redox reaction of $MgCo_2O_4$ nanoneedles on porous β-polytype silicon carbide as high performance electrodes for electrochemical supercapacitors. *J. Alloys Compd.* **2017**, *710*, 528–538. [CrossRef]
13. Popov, O.; Vishnyakov, V.; Poperenko, L.; Yurgelevych, I.; Avramenko, T.; Ovcharenko, A. Reactively sintered TiB_2-based heteromodulus UHT ceramics with in-situ formed graphene for machinable concentrated solar light absorbers. *Ceram. Int.* **2022**, *48*, 17828–17836. [CrossRef]
14. Liang, J.; Xiang, C.; Zou, Y.; Hu, X.; Chu, H.; Qiu, S.; Xu, F.; Sun, L. Spacing graphene and Ni-Co layered double hydroxides with polypyrrole for high-performance supercapacitors. *J. Mater. Sci. Technol.* **2020**, *55*, 190–197. [CrossRef]

15. Adán-Más, A.; Duarte, R.G.; Silva, T.M.; Guerlou-Demourgues, L.; Montemor, M.F.G. Enhancement of the Ni-Co hydroxide response as energy storage material by electrochemically reduced graphene oxide. *Electrochim. Acta* **2017**, *240*, 323–340. [CrossRef]
16. Zhou, X.-C.; Yang, X.-Y.; Fu, Z.-B.; Yang, Q.; Yang, X.; Tang, Y.-J.; Wang, C.-Y.; Yi, Y. Single-crystalline ultrathin nanofilms of Ni aerogel with Ni(OH)$_2$ hybrid nanoparticles towards enhanced catalytic performance for ethanol electro-oxidation. *Appl. Surf. Sci.* **2019**, *492*, 756–764. [CrossRef]
17. Zhou, Q.; Bian, Q.; Liao, L.; Yu, F.; Li, D.; Tang, D.; Zhou, H. In situ electrochemical dehydrogenation of ultrathin Co(OH)$_2$ nanosheets for enhanced hydrogen evolution. *Chin. Chem. Lett.* **2023**, *34*, 107248. [CrossRef]
18. Salarizadeh, P.; Askari, M.B.; Seifi, M.; Rozati, S.M.; Eisazadeh, S.S. Pristine NiCo$_2$O$_4$ nanorods loaded rGO electrode as a remarkable electrode material for asymmetric supercapacitors. *Mater. Sci. Semicond. Process.* **2020**, *114*, 105078. [CrossRef]
19. Pore, O.; Fulari, A.; Chavare, C.; Sawant, D.; Patil, S.; Shejwal, R.; Fulari, V.; Lohar, G. Synthesis of NiCo$_2$O$_4$ microflowers by facile hydrothermal method: Effect of precursor concentration. *Chem. Phys. Lett.* **2023**, *824*, 140551. [CrossRef]
20. Xiong, D.; Du, Z.; Li, H.; Xu, J.; Li, J.; Zhao, X.; Liu, L. Polyvinylpyrrolidone-assisted hydrothermal synthesis of CuCoO$_2$ nanoplates with enhanced oxygen evolution reaction performance. *ACS Sustain. Chem. Eng.* **2018**, *7*, 1493–1501. [CrossRef]
21. Cai, P.; Zhao, J.; Zhang, X.; Zhang, T.; Yin, G.; Chen, S.; Dong, C.-L.; Huang, Y.-C.; Sun, Y.; Yang, D.; et al. Synergy between cobalt and nickel on NiCo$_2$O$_4$ nanosheets promotes peroxymonosulfate activation for efficient norfloxacin degradation. *Appl. Catal. B Environ.* **2022**, *306*, 121091. [CrossRef]
22. Pathak, M.; Mutadak, P.; Mane, P.; More, M.A.; Chakraborty, B.; Late, D.J.; Rout, C.S. Enrichment of the field emission properties of NiCo$_2$O$_4$ nanostructures by UV/ozone treatment. *Mater. Adv.* **2021**, *2*, 2658–2666. [CrossRef]
23. Yan, D.; Wang, W.; Luo, X.; Chen, C.; Zeng, Y.; Zhu, Z. NiCo$_2$O$_4$ with oxygen vacancies as better performance electrode material for supercapacitor. *Chem. Eng. J.* **2018**, *334*, 864–872. [CrossRef]
24. Wang, R.; Li, W.; Jiang, L.; Liu, Q.; Wang, L.; Tang, B.; Yang, W. Rationally designed hierarchical SiC@PANI core/shell nanowire arrays: Toward high-performance supercapacitors with high-rate performance and robust stability. *Electrochim. Acta* **2022**, *406*, 139867. [CrossRef]
25. Zhang, Y.; Chen, J.; Fan, H.; Chou, K.-C.; Hou, X. Characterization of modified SiC@SiO$_2$ nanocables/MnO$_2$ and their potential application as hybrid electrodes for supercapacitors. *Dalton Trans.* **2015**, *44*, 19974–19982. [CrossRef] [PubMed]
26. Chen, Y.; Zhang, X.; Xue, W.; Xie, Z. Three-dimensional SiC/holey-graphene/holey-MnO$_2$ architectures for flexible energy storage with superior power and energy densities. *ACS Appl. Mater. Interfaces* **2020**, *12*, 32514–32525. [CrossRef] [PubMed]
27. Zhou, Q.; Bao, M.; Ni, X. A novel surface modification of silicon nanowires by polydopamine to prepare SiNWs/NC@NiO electrode for high-performance supercapacitor. *Surf. Coat. Technol.* **2021**, *406*, 126660. [CrossRef]
28. Yu, M.; Chen, J.; Liu, J.; Li, S.; Ma, Y.; Zhang, J.; An, J. Mesoporous NiCo$_2$O$_4$ nanoneedles grown on 3D graphene-nickel foam for supercapacitor and methanol electro-oxidation. *Electrochim. Acta* **2015**, *151*, 99–108. [CrossRef]
29. Chen, J.-C.; Hsu, C.-T.; Hu, C.-C. Superior capacitive performances of binary nickel–cobalt hydroxide nanonetwork prepared by cathodic deposition. *J. Power Sources* **2014**, *253*, 205–213. [CrossRef]
30. Wei, S.; Wan, C.; Zhang, L.; Liu, X.; Tian, W.; Su, J.; Cheng, W.; Wu, Y. N-doped and oxygen vacancy-rich NiCo$_2$O$_4$ nanograss for supercapacitor electrode. *Chem. Eng. J.* **2022**, *429*, 132242. [CrossRef]
31. Bai, Y.; Wang, R.; Lu, X.; Sun, J.; Gao, L. Template method to controllable synthesis 3D porous NiCo$_2$O$_4$ with enhanced capacitance and stability for supercapacitors. *J. Colloid Interface Sci.* **2016**, *468*, 1–9. [CrossRef]
32. Pan, C.; Liu, Z.; Li, W.; Zhuang, Y.; Wang, Q.; Chen, S. NiCo$_2$O$_4$@polyaniline nanotubes heterostructure anchored on carbon textiles with enhanced electrochemical performance for supercapacitor application. *J. Phys. Chem. C* **2019**, *123*, 25549–25558. [CrossRef]
33. Zhang, H.; Sun, C.; Xie, X. Understanding electrochemical structural degradation of NiCo$_2$S$_4$ nanoparticles in aqueous alkaline electrolyte. *Electrochim. Acta* **2023**, *466*, 143057. [CrossRef]

Disclaimer/Publisher's Note: The statements, opinions and data contained in all publications are solely those of the individual author(s) and contributor(s) and not of MDPI and/or the editor(s). MDPI and/or the editor(s) disclaim responsibility for any injury to people or property resulting from any ideas, methods, instructions or products referred to in the content.

Article

Chemical Activation of Apricot Pit-Derived Carbon Sorbents for the Effective Removal of Dyes in Environmental Remediation

Vitalii Vashchynskyi [1,*], Olena Okhay [2,3,*] and Tetiana Boychuk [4]

1. The Department of General Physics, Lviv Polytechnic National University, 12 S. Bandery Str., 79013 Lviv, Ukraine
2. Centre for Mechanical Technology and Automation (TEMA), University of Aveiro, 3810-193 Aveiro, Portugal
3. LASI-Intelligent Systems Associate Laboratory, 4800-058 Guimaraes, Portugal
4. Institut de Ciència dels Materials (ICMUV), Universitat de València, Catedrático José Beltrán 2, 46980 Paterna, Spain
* Correspondence: v.vashchynskyi@gmail.com (V.V.); olena@ua.pt (O.O.)

Abstract: The aim of this work is to study the properties of carbon materials prepared from apricot stones by carbonization at 300–900 °C and chemical activation by KOH with different ratios between components. It was found that increasing the carbonization temperature to 800–900 °C leads to the degradation of narrow micropores and the carbon matrix. The adsorbent materials were characterized with FTIR and SEM, and a specific surface area was calculated. Moreover, additional activation by HNO_3 and annealing at 450 °C led to an increase in surface area up to 1300 m^2/g. The obtained N-enriched sorbents show adsorption activities of 190–235 mg/g for methylene blue and 210–260 mg/g for methyl orange. The results of this study can be useful for future scale-up using the apricot material as a low-cost adsorbent for the removal of dyes in environmental remediation production.

Keywords: activated carbon; sorbent; chemical activation; carbonization; sorption; N-doping

Citation: Vashchynskyi, V.; Okhay, O.; Boychuk, T. Chemical Activation of Apricot Pit-Derived Carbon Sorbents for the Effective Removal of Dyes in Environmental Remediation. C 2023, 9, 93. https://doi.org/10.3390/c9040093

Academic Editor: Jorge Bedia

Received: 11 September 2023
Revised: 22 September 2023
Accepted: 27 September 2023
Published: 29 September 2023

Copyright: © 2023 by the authors. Licensee MDPI, Basel, Switzerland. This article is an open access article distributed under the terms and conditions of the Creative Commons Attribution (CC BY) license (https://creativecommons.org/licenses/by/4.0/).

1. Introduction

In the modern world, one of the key problems that concern humanity is environmental pollution, which has become a growing problem and is of serious negative effects. In the modern world, one of the key problems that concern humanity is the reduction in environmental pollution, especially of the hydrosphere because of rapidly growing industrialization and mining. Pollutants enter our waters, soils, and atmosphere due to the rapid development of agriculture and metallurgical industry and the improper disposal of waste, fertilizers, and pesticides [1]. One of the most common pollutants of the water environment are dyes used in various industries such as textile, food, and cosmetics. Industrial wastewater contains colored compounds that can be dangerous due to toxicity, low biodegradability, etc. Dyes in surface water significantly affect nature/ecosystem, e.g., affect the process of photosynthesis and negatively affect living organisms. Metals such as lead, cadmium, and mercury are particularly dangerous because they are practically not removed from biological objects [2]. In addition, coloring compounds can be harmful to the human body and lead to various diseases [3]. Thus, effective treatment of wastewater containing dyes is required, and chemical oxidation, precipitation and coagulation, biological treatment, membrane methods, and adsorption on activated carbon can be used [4,5]. Reagent treatment, ion exchange, and membrane purification methods are usually used to remove metal ions from water environments. On the contrary, sorption purification methods are simple, less expensive, available, and effective. Carbon sorbents, natural and synthetic sorbents, clay rocks, zeolites, etc., are used as filter materials. However, the ideal materials/technologies—that are cheap and effective—have yet to be found. In addition, porous carbon materials can be used as catalysts/electrode components for energy storage in supercapacitors [6,7], batteries [8], etc.

The main factors when choosing a sorbent for targeted separation are surface area, pore volume, porous structure, and physicochemical properties. In addition, due to environmental friendliness, low cost, and good surface characteristics, biomass-based adsorbents have attracted attention and are widely studied [9]. Carbon sorbent (CS) is a material used to remove both inorganic and organic impurities from the environment. It is characterized by its high specific surface area, well-developed porous structure, and functional groups on the surface of the sorbent. The presence of these groups within the carbon structure, along with stable radicals and fragments with donor–acceptor properties, determines the reaction nature of the carbon interaction with chemical reagents [10]. In addition, CS derived from secondary vegetable raw materials find extensive use across various industries where they are utilized as catalysts and absorbers. These materials can also be employed in hemosorption systems, which perform targeted purification of blood and other physiological fluids from various toxicants, as well as in absorbed preparations such as probiotics. When choosing a raw material for use as a sorbent, attention should also be paid to its availability, cost, and environmental impact. Correspondingly, examples of suitable materials include wood in the form of sawdust, charcoal, peat, fossil coal, peat coke, oil refining products [11], and products of plant origin, such as cellulose, fruit pits, soybean meal, sunflower and coffee husks, nutshells, and empty pods of agricultural crops [12,13]. The preparation of activated carbon from different types of shells was studied for a long time and includes tests with coconut [14], pistachio nut [15], almond [16–19], and many others. Among the wide range of raw materials for obtaining porous carbon materials (PCM), preference was given to plant-based raw materials, specifically apricot pits, which are essentially waste from the industrial production of large fruit companies. In addition to availability, cost effectiveness, and safety in handling, fruit pits are characterized by natural micro- and mesoporosity. The material obtained by carbonizing apricot pits has a high carbon content (up to 90%). Reported activated carbon obtained from apricot stones with a surface area of 566 m^2/g was studied as an adsorbent to remove heavy metal ions Ni(II), Co(II), Cd(II), Cu(II), Pb(II), Cr(III), and Cr(VI) ions from aqueous solutions [20]. Methylene cyanine adsorption by activated carbon with a surface area of 866 m^2/g obtained from the apricot shell was tested by Zhu et al. [21]. The activated carbon from apricot stones was used by Abbas as an adsorbent to remove malachite green from the aqueous solution [22]. Abd Ali prepared activated carbon surface areas of 1115 m^2/g from apricot stones and evaluated them as an adsorbent for the basic dye (Reactive Blue 222) from aqueous solutions [23].

Currently, chemical activation methods are commonly used to produce sorption materials from natural raw materials, leading to the creation of highly porous activated carbon (AC). The interaction mechanisms between carbon and inorganic substances are determined by the presence of functional groups, radicals, and fragments with catalytic and electrolytic properties in the organic mass. The authors note [24] that the preliminary treatment of carbon with aqueous solutions of acids and alkalis activates the decomposition process of the organic mass, resulting in changes in the mass yield and the composition of liquid and gaseous decomposition products. Therefore, the main goal of the research is to investigate the influence of the chemical activator nature and the reagent/carbon ratio on the yield, structure, and specific surface of carbon, with the aim of obtaining carbon sorption materials.

2. Materials and Methods

2.1. Preparation of Samples

The current study involves three distinct steps: (1) the initial sample obtained by carbonization of raw material of plant origin (RMPO), specifically apricot pits, purchased from the company "T.B. Fruit", Ukraine (Figure 1); (2) the subsequent activation of derived material with potassium hydroxide (KOH) (purchased from SFERA SIM Ltd., 1 Lviv, Ukraine) [25], resulting in the formation of porous carbon material (PCM); and (3) nitrogen (N)-enriched samples of PCM in concentrated HNO_3 (purchased from SFERA SIM Ltd., 1 Lviv, Ukraine) with the following thermal treatment in an argon (Ar) stream [26].

Figure 1. Raw material of plant origin (apricot pits).

The RMPO was carbonized in a closed furnace in the temperature range of 300–900 °C, and the first digit in the sample number indicates the temperature. The carbonization process aims to remove volatile components from the carbon matrix, maximize the specific carbon content, and obtain a porous material with a high surface area and developed porosity. However, achieving the optimal pore size distribution during carbonization is practically impossible. Therefore, additional processing of the material is required, such as activation. The obtained carbonized carbon was mechanically crushed and mixed with KOH in different mass ratios, stirred for 1–2 h, heated in an Ar atmosphere to 900 °C, and subjected to isothermal exposure. Correspondingly, mineral impurities and ash were chemically washed from the samples using hydrochloric acid (HCl) (purchased from SFERA SIM Ltd., 1 Lviv, Ukraine). Then, the material was rinsed in distilled water until it reached a neutral pH and dried in an oven until a constant mass was achieved. The resulting samples were designated as Cxy, where x = 3, 4, 5, 6, 7, 8, and 9 and corresponds to the used temperature 300, 400, 500, 600, 700, 800, and 900 °C; y = 1, 2, 3, and 4 and corresponds to the ratio of KOH and C as 1:1, 1:2, 1:3, and 1:4. For example, sample C31 was prepared at 300 °C with an equal amount of KOH and C (see Table 1).

Table 1. List of the prepared activated carbon samples after the activation process by KOH.

Designation	Carbonization Temperature, °C	Ratio of KOH and C
C31, C41, C51, C61, C71, C81, C91	300, 400, 500, 600, 700, 800, 900, respectively	1:1 in all samples
C32, C42, C52, C62, C72, C82, C92	300, 400, 500, 600, 700, 800, 900, respectively	1:2 in all samples
C33, C43, C53, C63, C73, C83, C93	300, 400, 500, 600, 700, 800, 900, respectively	1:3 in all samples
C34, C44, C54, C64, C74, C84, C94	300, 400, 500, 600, 700, 800, 900, respectively	1:4 in all samples

Sample C41, prepared at 400 °C and with the ratio KOH and C as 1:1 was used to investigate the formation of surface N-containing functional groups, which enhance reactivity. The proposed process was carried out as follows: 160 mL of 65% HNO_3 solution (purchased from SFERA SIM Ltd., 1 Lviv, Ukraine) was added to 12 g of activated PCM (samples C41). The resulting suspension was thoroughly mixed with a magnetic stirrer at room temperature for 3 h, and then washed with distilled water to a neutral pH and dried in air at a temperature of 65 °C for a day. The obtained finely dispersed material exhibited low chemical stability, necessitating additional thermal activation. This activation of the obtained N-containing PCM was carried out in a vertical tubular furnace at temperatures in the range of 250 ÷ 650 °C in an Ar stream for 1 h. The prepared N-enriched samples were labeled as N-C41-t, where t corresponds to the temperature of the heat treatment (from 250 to 650 °C). For example, N-C41-250 represents PCM obtained by chemical activation of C41 material in concentrated HNO_3 followed by thermal treatment in Ar at 250 °C.

For a better understanding, all N-enriched carbon samples are organized in Table 2.

Table 2. List of N-doped carbon materials prepared based on sample C41.

Designation	Carbonization Temperature, °C	Ratio KOH:C	Activation by HNO_3	Annealing in Ar, °C
N-C41-0	400	1:1	Yes	-
N-C41-250	400	1:1	Yes	250
N-C41-450	400	1:1	Yes	450
N-C41-650	400	1:1	Yes	650

2.2. Characterization of the Samples

The characteristics of the PCM structure including specific surface area (SSA) and total pore volume were determined based on the analysis of nitrogen adsorption/desorption isotherms at its boiling temperature (77 K) using a Quantachrome NOVA2200e Analyzer (Quantachrome Instruments, Boynton Beach, FL, USA). Before measurements, the samples were degassed at 180 °C for 18 h. The specific surface area was calculated using the Brunauer–Emmett–Teller (BET) method (S_{BET}, m^2/g) in the region of the isotherm limited by the range of relative pressure $P/P_0 = 0.050 - 0.035$. The total pore volume (V_{total}, cm^3/g) was determined based on the amount of nitrogen adsorbed at $P/P \sim 1.0$. Additionally, the micropore volume (V_{micro}, cm^3/g), as well as the specific surface areas of micro-(S_{micro}, m^2/g) and mesopores (S_{mezo}, m^2/g) were found using the t-method, the Barrett–Joyner–Halenda (BJH) method, and the Density Functional Theory (DFT) [27].

The structural study was conducted using a scanning electron microscope JSM-6700 F (JEOL, Peabody, MA, USA) equipped with an energy system JED-2300.

The infrared (IR) spectra of PCM samples were obtained using a Thermo Nicolet FT-IR spectrometer in reflectance mode. The test sample was mixed with KBr in a ratio of 1:100.

To evaluate the adsorption properties, methylene blue (A_{MB}, mg/g), methyl orange (A_{MO}, mg/g), and iodine (A_{Iodine}, mg/g) were used [28]. The optical density of these solutions was measured using a spectrophotometer ULAB 102 UV with a blue light filter and a wavelength of 400 nm in cuvettes with an optical path length of 10 mm.

The carbon adsorption activity by the dye, measured in milligrams per 1 g of the product was calculated as

$$A = \frac{(C_1 - C_2 K) \times 0.025}{m} \quad (1)$$

where C_1 is the mass concentration of the initial indicator solution (mg/dm^3); C_2 is the mass concentration of the solution after contact with activated carbon (AC) (obtained after activating PCM with KOH) (mg/dm^3); K is the dilution factor of the solution; 0.025 is the volume of the indicator solution taken for illumination (dm^3); and m is the mass of AC (g).

The residual mass concentration of the indicator was determined in comparison with the previously obtained graduation graph [29].

3. Results and Discussion

3.1. Porous Carbon Material (PCM) and Its Activation

In this context, the question of the state of the particle surface becomes important. Functional groups affect the surface hydrophilicity with an electrolyte solution. Since the micropores of carbon materials are not completely wetted due to the presence of graphite crystallites, functional compounds can improve the hydrophilic properties of their surface. They locally change the electrostatic field of the pores (which strengthens the interaction with polarized water molecules [30,31], and thereby increase the carbon sorption activity owing to the improvement in the availability of chemical reagent ions to the pores and, accordingly, the involvement of the additional surface where sorption is possible. In order to confirm the statement about the presence of surface functional groups on the porous carbon material (PCM) surface, the structural analysis of the chemically modified PCM was carried out using IR spectroscopy (Figure 2). The IR spectra of carbon materials are characterized by absorption bands in the entire working range, which is within the so-called fundamental infrared region (500 cm^{-1}–4000 cm^{-1}), mostly without clearly defined absorption peaks. This indicates the presence of various types of surface functionalities. Each type of functional group is characterized by its own set of absorption bands corresponding to the vibrational modes of the chemical bonds of certain atoms. It is worth noting that the absorption band can simultaneously indicate different types of chemical bonds or result from overlapping spectra of various groups [32]. Therefore, IR spectra are analyzed by dividing them into sections that correspond to the vibration modes of certain chemical bonds. The type of surface functional groups is identified by the combination of these bonds.

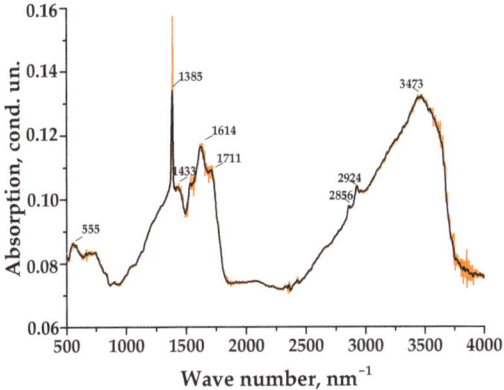

Figure 2. IR spectrum of surface of activated carbon material C41 prepared at 400 °C and chemically modified by KOH (KOH:C as 1:1).

The IR spectrum of the initial material shows several characteristic modes. Vibrations at 2856 cm^{-1} and 2924 cm^{-1} are attributed to asymmetric and symmetric C-H vibrations in CH$_2$ groups. A vibration band at 1385 cm^{-1} indicates the presence of alcohol compounds on the PCM surface and corresponds to the deformation O-H vibrations. The presence of alkenes is confirmed by the modes at 1433 cm^{-1} and 1614 cm^{-1}, which characterize the valence vibrations of C-H bonds in CH$_2$/CH$_3$ groups and C=C aromatic rings, respectively. In addition, the starting material spectrum contains absorption bands in the vicinity of 1711 cm^{-1} and 2924 cm^{-1}, which are attributed to the valence vibrations of C=O and C-H bonds in ketones [33]. Furthermore, there are intense absorption bands with maxima at 1385 and 1614 cm^{-1} in the spectrum, which may indicate the presence of carboxylic acids in the composition of potassium salts. According to Marsh et al. [34], the activation of carbon material with KOH is accompanied by the flow of gasification reactions, where carbon is oxidized to CO and CO$_2$, and this contributes to the development of a porous structure with the formation of secondary products (K$_2$CO$_3$). It is also worth noting that at high temperatures (550–900 °C), the formation of metallic potassium is possible, which is introduced between graphite layers as a result of the reduction of K$_2$O by carbon. The destruction of the chemically treated sample to a powder state at the carbonization stage and the subsequent porosity formation after washing with water are connected precisely with the emergence of these salts. They do not ensure the consolidation of polymer complexes during the heat treatment stage and are removed by washing with water, thereby revealing the internal structure.

Thus, when exposed to a chemical activator such as KOH, significant structural changes occur in the carbon material with the formation of carboxylic acid salts. This transformation includes the removal of oxygen functional groups from the surface of the carbon material due to chemical modification and heat treatment. These changes affect the chemical composition of the surface of carbon materials, thereby determining their catalytic and sorption properties.

The methods applied to examine the adsorption activity of carbon by the indicator are static and involve determining adsorption isotherms. The adsorption capacity of materials was measured by the interaction of the dye solution (methylene blue (MB)) with the solvent containing a certain amount of adsorbent, which, in our case, is AC (PCM after KOH treatment). After the dye monolayer formation, the solvent was almost completely displaced from the surface of the adsorbent.

Porous carbon samples denoted Cxy, where x = 3, 4, 5, 6, 7, 8, and 9 (corresponds to the used temperature 300, 400, 500, 600, 700, 800, and 900 °C) and y = 2, 3, and 4 (represents to the ratio of KOH and C 1:2, 1:3, and 1:4) were used in the reported measurements. In order to conduct experiments to determine the MB adsorption activity, a calibration graph was previously constructed and is provided in Figure A1 in Appendix A following the main text.

To determine the adsorption capacity, pre-dried and ground PCM weighing 0.1 g was placed in a 50 mL flask, and an aqueous solution of the adsorbent (25 mL of the MB solution with a concentration of 1.5 g/L) was added and stirred for 20 min. Since the solute diffusion rate is much lower compared to gaseous agents, slow adsorption processes were observed without the need for mechanical mixing. Next, the mixed samples were placed in a centrifuge, and after complete precipitation of the adsorbent, the solution was separated. Afterward, it was carefully sampled with a pipette, and, using a spectrophotometer ULAB 102 UV, the value of the optical density was determined according to a previously constructed calibration curve (see Figure A1 in Appendix A). If the optical density exceeded the experimental limits (according to [28], it should not exceed 0.8 optical units), then this solution was diluted by adding 5 mL of it into a 25 mL volumetric flask and topping it up with distilled water to reach the mark.

During the alkaline treatment phase, KOH interacts with oxygen-containing functional groups of carbon, and this leads to the formation of potassium phenolates and carboxylates, as well as the splitting of complex ester groups. These reactions are the driving force for the introduction of K^+ and OH^- ions, KOH molecules, and their incorporation into the three-dimensional structure of carbon. This can occur in the form of individual ions, molecules, or aggregates of molecules [35,36]. The carbonization of the carbon material activated by KOH contributes to the growth of the specific volume of pores in the structure of the obtained PCM.

To study the influence of the alkali amount on the structural and adsorption properties of the studied samples, the chemical activation with potassium hydroxide was carried out. Figure 3 presents microscopic images of the samples C4y prepared at 400 °C at different KOH:C ratios. These images reveal that the introduction of alkali contributes to the formation of numerous pores and the development of a spongy carbon structure [37]. At lower alkali concentrations, the primary carbonation almost does not destroy the morphology of the raw material. However, an increase in the potassium hydroxide concentration leads to the degradation of cellulose and its migration from the particle volume to the surface. This, in turn, affects the porous structure of the carbon material. While closer examining the inner surface, it becomes evident that the material is permeated with a multitude of pores.

Figure 3 demonstrates that the C4y materials have a more destroyed surface, i.e., with an increase in the KOH concentration, significant changes occur in the surface structure. These findings suggest that the presence of "free" alkali within the framework of carbon material leads to the surface development via reactions with framework fragments: heterolysis of C-O and polarized C-C bonds [38], and the cleavage of O- and S-containing heterocycles [39]. These processes release hydrogen and methane and contribute to the intensive development of the microporous structure thanks to an increase in the number and volume of pores. The use of potassium hydroxide as an activating agent in the process of chemical activation of raw materials of plant origin allows for obtaining a material with a distinctive porous structure and the lace morphology of texture. The sufficient number of pores on the surface serves as an excellent transport system, facilitating the entry of sorbed substances into the micro-or mesopores from either liquid or gaseous phases.

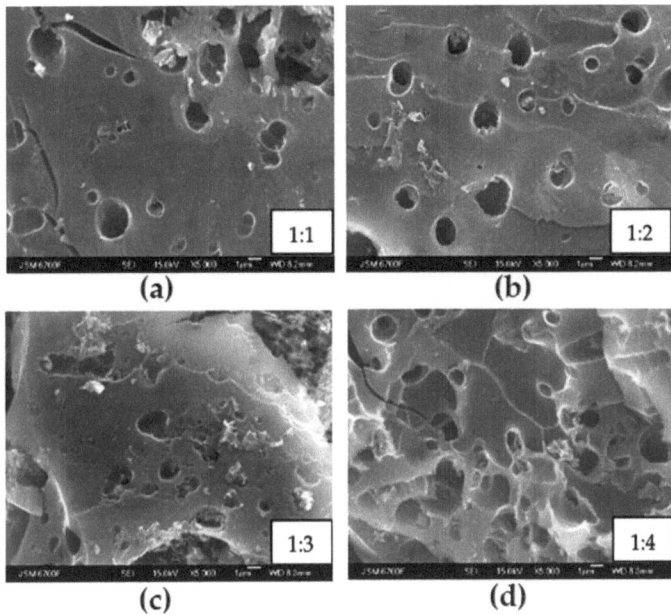

Figure 3. SEM surface images of the C4y sample surface at different ratios of KOH and C: (a) 1:1 for C41, (b) 1:2 for C42, (c) 1:3 for C44, and (d) 1:4 for C44.

The nature of the change in the adsorption activity depending on the KOH:C ratio is quite complex, and its dependency on the temperature is unclear, although it decreases after 400 °C for all studied samples using MB (Figure 4). The amount of adsorption activity by MB makes it possible to conclude the content of micropores in the adsorbent since the molecule size of MB is quite large (1.97 nm) [28] and their diffusion into micropores (with a diameter of up to 2 nm) is challenging. The analysis of experimental data showed that chemical activation improves the adsorption capacity of the resulting carbons. The most significant increase in activity occurs at a ratio of KOH to C as 1:3, reaching 320 mg/g. The nature of the change in the adsorption activity of the sorbents due to variations in the KOH concentration is similar for most samples. However, for samples C3y, the sorption amount increases with the addition of alkali in the ratio KOH:C = 1:1, but at higher concentrations, the activity decline is 31%. For samples of the C4y series, the maximum capacitance when absorbing methylene blue reaches 280–320 mg/g.

The adsorption by methylene blue (A_{MB}) characterizes the ability of AC to sorb large molecules of organic substances from aqueous solutions. It was established that the carbonation temperature and the KOH:C ratio are interconnected and significantly affect the sorption properties of the studied samples [40]. Thus, there is a positive correlation between the capacitance growth and increasing the KOH dosage. Samples C43 and C44 demonstrated the maximum adsorption activity (Figure 3, carbonization temperature for both samples was 400 °C). In the same KOH:C ratio (1:1), the maximum activity by MB (290 mg/g) was observed for sorbents obtained at T = 400 °C. A further increase in temperature leads to a decrease in sorption activity to 150 mg/g. This deterioration is because, at temperatures of 600–900 °C, volatile substances do not form condensation products within carbon pores, which leads to the burnout of the pore walls.

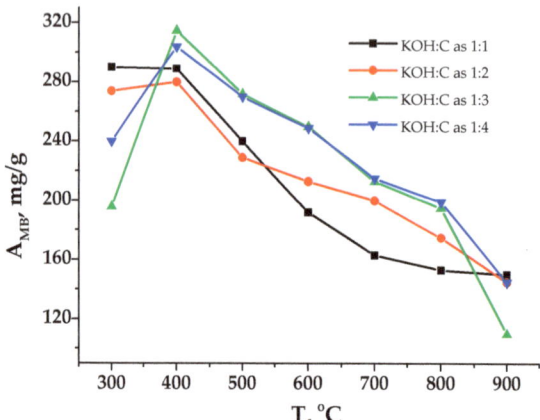

Figure 4. Dependence of adsorption capacity by MB on carbonization temperature.

The fact of the matter is that the chemical nature of the PCM surface determines the degree of the adsorption activity by the indicator [41]. When carbon has a nearly perfect porous surface with a high concentration of delocalized electrons, it enables the specific sorption of the dye complex ions due to the interaction of these ions with the central atom of the complex ion. However, as the surface is filled with chemisorbed oxygen, the effective concentration reduces and, accordingly, sorption decreases. Consequently, the adsorption activity declines, and sample C93 exhibits the poorest absorption properties, with an adsorption capacity of only 110 mg/g.

It is worth noting that the presence of certain pore sizes in adsorbents can be inferred from the adsorption activity of the dyes. While the MB adsorption method is the most effective for examining large-porous materials with pore radii exceeding 100 nm, it would be advisable to use another method for studying the adsorption activity of iodine. This method makes it possible to investigate the micropores with effective diameters of 0.5–1.7 nm in the adsorbent.

Likely, the carbonization temperature of the raw material is also an essential factor that affects the sorption properties of carbon. The selection of carbonization and activation temperatures of the raw material is guided by the fact that the main processes of decomposition (pyrolysis) and the initial carbonization of the original solid-phase substance primarily take place in the temperature range of 300–900 °C [42].

Figure 5 presents scanning electron microscopy (SEM) photographs at various magnifications of carbon materials prepared with the same KOH:C ratio as 1:1 but at different carbonized temperatures: C31 at 300 °C (Figure 5a), C51 at 500 °C (Figure 5b), C71 at 700 °C (Figure 5c), which have a meso- and mostly microporous structure, obtained using the method of scanning electron microscopy. The images clearly show surface microcracks and the presence of round or oval micro-sized transport pores in all studied samples. White inclusions associated with ash residues and products of the interaction of potassium hydroxide with carbonaceous material are observed over the entire surface. The structure of the material also contains needle-shaped and fiber inclusions that have the appearance of blurred rings [43].

It can be assumed that the increase in carbonization temperature initiates the process of carbon material graphitization and leads to the formation of graphite-like clusters [44]. In such a case, at temperatures of 700 °C and higher, the formation of the porous structure slows down as the degeneration of narrow micropores occurs via their merging, resulting in a reduction in both pore volume and their SSA.

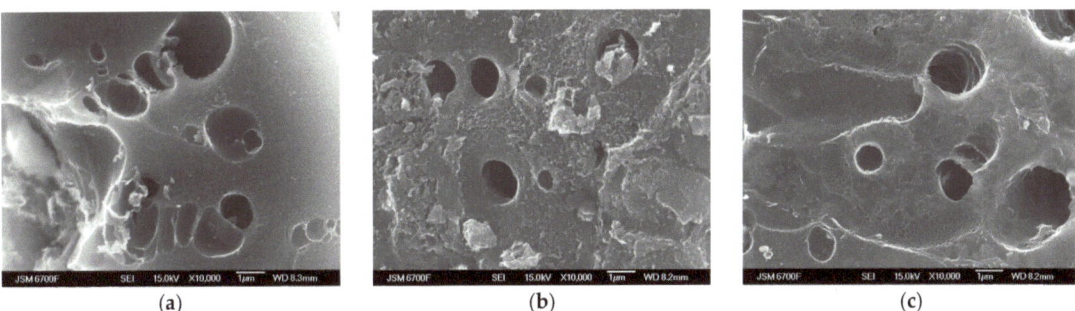

Figure 5. SEM surface images of porous carbon materials prepared with the same KOH:C as 1:1 but at different carbonized temperatures: (**a**) C31 at 300 °C, (**b**) C51 at 500 °C, and (**c**) C71 at 700 °C.

Figure 6 presents typical nitrogen adsorption/desorption isotherms for the studied carbon materials. They belong to isotherms that are characteristic of multilayer adsorption in micro- and mesopores of organic materials. Notably, a hysteresis loop of the Type H4 according to the IUPAC classification [45] is observed for all samples and is associated with sorption processes within narrow pores.

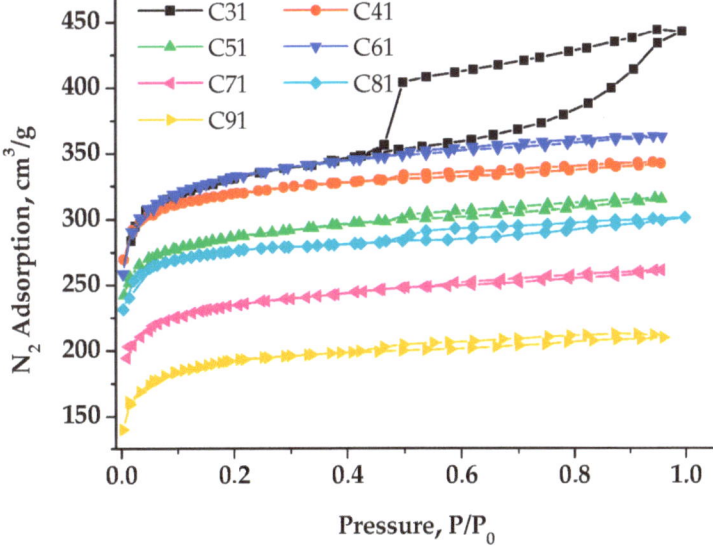

Figure 6. Nitrogen adsorption isotherms at 77 K of carbon samples.

The sorption isotherms obtained for samples C41–C91 are characteristic of typical microporous materials. This type of isotherm is associated with sorption processes occurring mainly in narrow micropores. A significant hysteresis associated with the process of nitrogen sorption in mesopores is observed for sample C31. The histogram (Figure 7) presents the parameters of the porous structure (S_{BET}, V_{total}) for samples C31–C91 with the contribution of pores of various sizes. The C31 material is characterized by the mesopore area that is three times larger than that of other carbons activated by KOH (with the ratio KOH and C as 1:1).

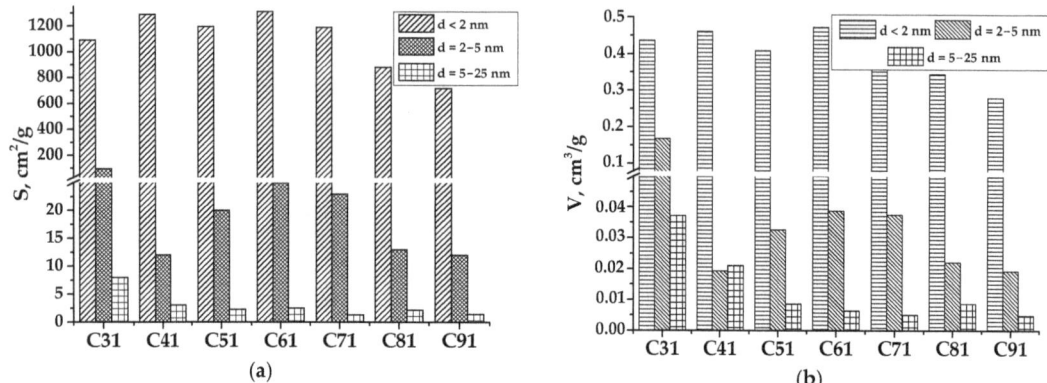

Figure 7. Dependence of specific surface area (**a**) and pore volume (**b**) on pore diameter of PCM prepared at different temperatures with the same ratio between KOH and C (1:1).

The analysis of sorption isotherms made it possible to determine the parameters of the porous surface of carbon materials after chemical modification and temperature treatment (Table 3). Comparing the results obtained by the Brunauer–Emmett–Teller (BET) method, two maxima of the active area growth of the material at temperatures of 300 °C (sample C31) and 600 °C (sample 91) are visible. A further increase in the carbonization temperature leads to a decrease in the specific surface, and for sample C91, it is only 721 m^2/g. The nature of the change in the total pore volume (V_{total}) is also different. The largest V_{total} was calculated for sample C31, equal to 0.684 cm^3/g, and it was decreased with the heat treatment temperature increased, carbon materials C51 and C71 have almost the same total pore volume. Sample C61 drops out from this group, with the average values of the total pore volume V_{total}~ 0.555 cm^3/g, which can be an advantage in microvolume over all other samples.

Table 3. Main characteristics of porous structure of carbon material.

Sample	S_{BET}, m^2/g	S_{DFT}, m^2/g	S_{micro}, m^2/g	S_{meso}, m^2/g	V_{micro}, cm^3/g	V_{total}, cm^3/g
C31	1313	1196	1067	246	0.438	0.684
C41	1188	1303	1148	40	0.470	0.521
C51	1068	1216	1018	50	0.416	0.470
C61	1213	1332	1160	53	0.477	0.555
C71	1042	1214	984	58	0.398	0.466
C81	837	894	804	33	0.330	0.404
C91	721	731	692	29	0.273	0.317

As already mentioned, the mechanisms of interaction of carbon with inorganic substances are determined by the presence of functional groups, radicals, and fragments with catalytic and electrolytic properties in the organic mass. The preliminary treatment of carbon with aqueous solutions of acids and alkalis activates the process of decomposition of the material, organic mass, leading to changes in the mass yield and composition of liquid and gaseous decomposition products. Therefore, it becomes essential to investigate how these structural changes affect the sorption properties of the samples under study.

The adsorption activity of PCM was determined according to standard methods involving methyl orange (MO) and elemental iodine.

It was observed that the carbonation temperature and the KOH:C ratio are interrelated and have a significant impact on the sorption properties of the studied samples. Specifically, there is a positive trend of capacitance growth with increasing the KOH dosage. Materials carbonized at a temperature of 400 °C have the maximum adsorption activity by methyl orange (MO) (Figure 8a). A characteristic feature of the C4y series is high sorption values across all three methods. When using iodine, the value of A_{iodine} is the highest for sample C4y (Figure 8b) since it has a large total pore volume V_{total} = 0.521 cm^3/g (Table 3), primarily attributed to mesopores with sizes from 2 to 25 nm (Figure 7a).

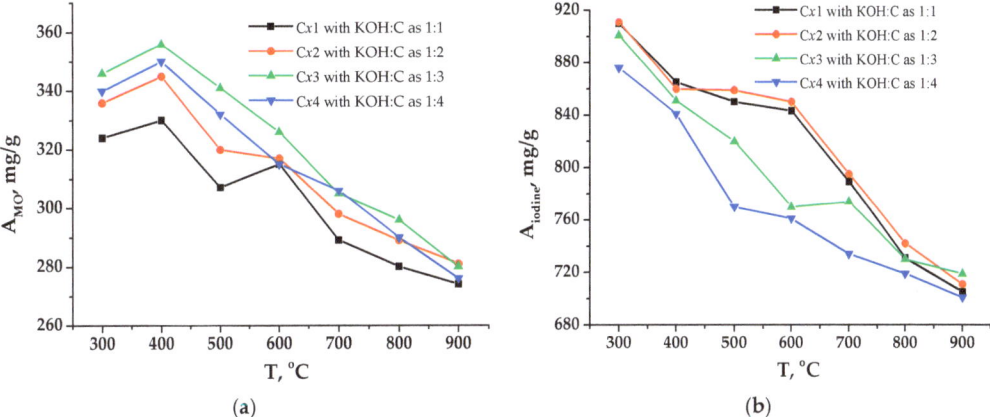

Figure 8. Dependence of adsorption capacity on carbonization temperature: (**a**) methyl orange; (**b**) iodine.

Thus, it could be assumed that as the processing temperature of the carbon material increases, the development of its porous structure slows down, and therefore the sorption properties of these materials also deteriorate. This is due to the formation of predominantly small micropores that are inaccessible to methylene blue and methyl orange. Carbon with a well-developed porous surface (S_{meso} = 246 m^2/g for C31) and a wide pore size distribution (Figure 7b) shows the highest sorption capacitance.

3.2. Nitrogen (N)-Enriched Samples of PCM

The surface of AC is highly heterogeneous both in terms of geometry and electronics. Carbon atoms on the surface are in slightly different electronic states than atoms of the bulk phase, especially in areas with defects in the crystal lattice and on the faces, vertices, and edges of the crystallites. These surface carbon atoms, due to their free valency, readily engage in chemical and sorption interactions, which are more accessible and require less energy. Chemical modification of AC can be achieved using various reagents, and this process involves multiple chemical reactions, including the catalytic decomposition of the reagent, redox interactions with the formation of surface and phase oxides, and partial structural degradation. In this study, nitrogen (N)-enriched samples of PCM obtained at different temperatures were prepared and studied. Considering the previous research results, namely the developed porous structure of the material and high adsorption activity for dyes, C41 was chosen as the starting sample due to the highest adsorption activity by MO (see Figure 8a). The original PCM was obtained via carbonization of RMPO and subsequent activation with KOH at a ratio of KOH and C of 1:1 (sample C41). Subsequently, N-C41-250, N-C41-450, and N-C41-650 were produced by subjecting the C41 material to chemical activation using concentrated HNO$_3$ followed by thermal modification in an argon stream at temperatures of 250, 450, and 650 °C, respectively. It is important to note that thermal modification was not performed for the sample CN-0.

Adsorption/desorption isotherms for nitrogen-containing PCM are depicted in Figure 9. Notably, the isotherm shape remained consistent across all chemically activated materials. However, a slight decrease in the volume of sorbed nitrogen can be observed for sample N-C41-0 compared to porous carbon material that underwent annealing at 400 °C (sample C41 presented in Figure 6). This reduction is likely due to pore blocking caused by the introduction of nitrogen heteroatoms, which probably appeared during the chemical modification of C41 using HNO_3. As the temperature rises, an increase in the volume of sorbed nitrogen is evident for materials N-C41-250 and N-C41-250, but at 650 °C, it decreases (sample N-C41-650).

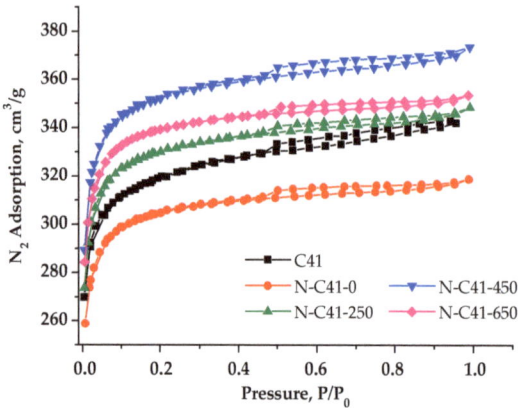

Figure 9. Nitrogen adsorption/desorption isotherms at 77 K for samples.

The measured isotherms are typical for Type 1 according to the IUPAC classification [27], and they are characterized by loops of the capillary condensation hysteresis of the H4 category at a relative pressure of ~0.5. The analysis of sorption isotherms allowed for the determination of the porous structure parameters of the samples. According to the presented data in Table 4, the nature of the chemical activator significantly affects the change in the porous structure and surface characteristics of PCM. Sample N-C41-450 has the largest SSA with the most developed microporous structure, which makes up 97% of the total surface area.

Table 4. Structural and adsorption characteristics of PCM before (sample C41) and after modification by nitrogen.

Sample	S_{total}, m²/g	S_{micro}, m²/g	S_{meso}, m²/g	V_{total}, cm³/g	V_{micro}, cm³/g
C41	1188	1148	40	0.521	0.469
N-C41-0	1158	1130	27	0.493	0.453
N-C41-250	1251	1219	31	0.539	0.491
N-C41-450	1339	1303	36	0.577	0.523
N-C41-650	1292	1261	31	0.547	0.504

The carbon adsorption activity by the indicator (MO and MB) was calculated in milligrams per 1 g of product (Figure 10) according to standard methods. The final result was determined by averaging two independent calculations, ensuring that the absolute deviation between them did not exceed 10 mg per 1 g of product.

Samples N-C41-0 and N-C41-250 are characterized by a microporous structure (Table 4) and have low adsorption activity by MB compared to the C41 material. This is explained by the fact that when the C41 sample is oxidized with nitric acid, the SSA and the pore volume N-C41-0) decrease. Such changes are due to the adsorption of ions and molecules of reactive substances that can occupy a certain volume of pores, as well as an increase in the number of acidic oxygen-containing groups and the formation of new surface heteroatoms that can firmly settle at the entrance and the pore walls available for the N_2 adsorption. Heat treatment at T $\leq$ 450 °C causes an increase in the SSA (N-C41-250), which is thanks to the water desorption, by-products of synthesis, and the removal of functional groups from the material surface and an increase in the adsorption activity by MO and MB, which is associated with an increase in the volume of micropores compared to N-C41-0.

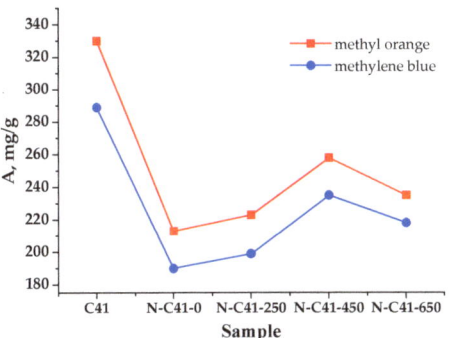

Figure 10. Sorptive activity of nitrogen-containing carbon materials activated at different temperatures.

Sample N-C41-450 has the highest specific surface area (1339 m^2/g), primarily attributed to its micropores, accounting for 97% of the total surface area. The SSA obtained in this study (1339 m^2/g) for KOH-activated carbon prepared at a temperature ≤450 °C is much higher than that reported by Foo et al. (807 m^2/g) for KOH-activated carbons prepared by carbonizing oil palm empty fruit bunch at a high temperature of 700 °C and treated in the microwave [46]. Song et al. also reported a slightly lower SSA (1135 m^2/g) for activated carbons from coconut shells after carbonization at 500 °C and subsequent heat treatment at 800 °C following KOH-activation [47].

At the same time, Li et al. reported a slightly lower adsorption capacity of 210.97 mg/g for methylene blue obtained using loofah sponge (LSC)-based porous carbons as starting materials by carbonization in an argon atmosphere at 800 °C [48] in contrast to 235 mg/g for N-C41-450 °C obtained in the current work at much lower temperature.

As can be seen in Table 5, the combination of two acids (KOH-HNO$_3$) and two-step heat treatment (400 °C for the carbonization process and 450 °C for annealing), the specific surface area of 1339 m^2/g was obtained, is the highest value in comparison with activated carbon materials prepared from similar starting apricot but at different temperatures, time and acids.

Table 5. Details of the preparation of porous carbon materials in already published papers and results of ongoing work (ordered by processing temperature).

Starting Material	Processing Temperature, °C/ Time, Hours	Used Acid for Activation Process	Surface Area, m^2/g	Adsorption Capacity, mg/g/Type of Dye	Refs.
apricot kernel shells	105/24	KOH	359	33.67/methylene blue	[49]
apricot stone	200/24	H$_2$SO$_4$	642	metal ions 27.21/Ni(II) 30.07/Co(II) 33.57/Cd(II) 24.21/Cu(II) 22.85/Pb(II) 29.47/Cr(III) 7.86/Cr(VI)	[20]

Table 5. Cont.

Starting Material	Processing Temperature, °C/ Time, Hours	Used Acid for Activation Process	Surface Area, m^2/g	Adsorption Capacity, mg/g/Type of Dye	Refs.
apricot stones	250/4	H_3PO_4	88	23.94/malachite green	[22]
apricot stones	600–700/2–3	H_3PO_4	1115	<80/reactive blue	[23]
apricot stones	700/1	H_3PO_4-HNO_3	359	98/methylene blue 81/methyl orange	[50]
apricot shell	700–900/<1	-(by moisture and CO_2)	866	-/methylene blue	[21]
apricot pits (sample N-C41-450)	400–450/1	KOH-HNO_3	1339	235/methylene blue 260/methyl orange	current work

A distinctive feature is the increase in pore volume, which provides high values of the adsorption activity of 235 mg/g for MB and ~260 mg/g for MO. However, for the N-C41-650 sample, carbon material burnout occurs with the participation of oxygen heteroatom material, and as a result, the microporous surface decreases. The sample obtained at temperatures above 650 °C as a sorbent exhibits lower adsorption capacity concerning the absorption of the indicator from the aqueous solution.

4. Conclusions

In conclusion, this research demonstrated a scientific approach to the development of highly effective sorbents based on carbon materials. The methods of scanning electron microscopy and adsorption porosimetry proved that carbon sorbents possess diverse morphologies and porous structures depending on the conditions and modes of production. The study highlights that carbonization of raw materials of plant origin at a temperature of 300 °C, followed by chemical modification with potassium hydroxide, results in PCM with well-developed surfaces and high porosity. Notably, in the case of samples treated at 300 °C (series C3y), mesopores contribute to approximately 20% of the total specific surface area. The analysis of adsorption isotherms showed that by increasing the carbonization temperature of raw material, the pore size distribution in the resulting nanoporous carbon can be controlled. Specifically, at 400–600 °C, micropores dominate, whereas carbonization temperatures exceeding 600 °C lead to a transformation in the carbon matrix, along with the burnout of pore walls and a subsequent reduction in the total pore volume.

It was established that the use of potassium hydroxide (KOH) as a chemical activator influences the activity of surface centers of the sorbent due to the removal of surface functional oxygen compounds from the surface of carbon materials. As the percentage ratio of KOH:C increases, the value of sorption activity for the dyes (methylene blue and methyl orange) also increases. The peak sorption is observed in PCM synthesized by the carbonization of the initial raw material at 400 °C, followed by alkaline activation with a KOH:C ratio of 3:1. In this case, the maximum capacitance was achieved for samples of the C4y series, reaching values of 280–320 mg/g when absorbing methylene blue. However, the highest SSA was found in the KOH-activated carbons prepared at the lowest carbonized temperature and vice versa.

Moreover, the additional chemical modification by HNO_3 of the prepared AC showed little increase in the specific surface area. Nevertheless, when the HNO_3 and heat treatment were followed by annealing at 450 °C, the SSA increased from 1188 up to 1339 m^2/g. This treatment results in a 97% predominance of micropores for sample N-C41-450. The observed decrease in adsorption activity in N-doped carbon materials is likely attributed to the formation of surface nitrogen and oxygen heterostructures, as well as the disordered distribution of pores.

Author Contributions: Conceptualization, resources, formal analysis, V.V.; investigation, V.V. and T.B.; methodology, validation, data curation, V.V., O.O. and T.B.; V.V. writing—original draft preparation, V.V. and O.O.; writing—review and editing, V.V., O.O. and T.B. All authors have read and agreed to the published version of the manuscript.

Funding: V.V.'s work was supported by Project 0121U108649 of the Ministry of Education and Science of Ukraine. O.O. thanks FCT (Fundacao para a Ciencia e a Tecnologia) in the scope of the framework contract foreseen in numbers 4, 5, and 6 of article 23 of the Decree Law 57/2016, of 29 August, UIDB/00481/2020 and UIDP/00481/2020—FCT; and CENTRO-01-0145- FEDER-022083—Centro Portugal Regional Operational Programme (Centro2020) under the PORTUGAL 2020 Partnership Agreement via the European Regional Development Fund.

Data Availability Statement: Not applicable.

Conflicts of Interest: The authors declare no conflict of interest.

Appendix A

In order to conduct experiments for determining the MB adsorption activity, a calibration graph was previously constructed. Accordingly, 0.5; 1.0; 1.5; 2.0; 3.0; 4.0; 5.0; 6.0; 7.0; and 8.0 mL of the MB solution with a concentration of 1500 mg/L were injected into 10 dry volumetric flasks with a capacity of 50 mL each. Respectively, the resulting solutions contain 15; 30; 45; 60; 90; 120; 150; 180; 210; and 240 mg/L of the indicator in 1 L [21]. Then, the optical density for these solutions was determined using a spectrophotometer equipped with a blue light filter and a wavelength of 400 nm in cuvettes with a thickness of the absorbing layer of 10 mm. Distilled water was used as a control solution. Based on the received data, the calibration curve was constructed, showing the dependence of optical density on the concentration of calibration solutions (Figure 2).

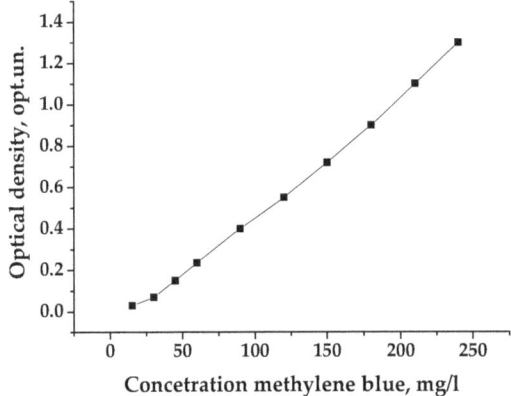

Figure A1. Calibration curve for measuring concentration by MB.

References

1. Pyrzynska, K. Application of carbon sorbents for the concentration and separation of metal ions. *Anal. Sci.* **2007**, *23*, 631–637. [CrossRef] [PubMed]
2. Seitkhan, A. Synthesis of carbonized nano mesoporous sorbents based on vegetable raw materials. *J. Nanosci. Nanoeng.* **2003**, *1*, 41–44. [CrossRef]
3. Briffa, J.; Sinagra, E.; Blundell, R. Heavy metal pollution in the environment and their toxicological effects on humans. *Heliyon* **2020**, *6*, 04691. [CrossRef] [PubMed]
4. Nidheesh, P.V.; Zhou, M.; Oturan, M.A. An overview on the removal of synthetic dyes from water by electrochemical advanced oxidation processes. *Chemosphere* **2018**, *197*, 210–227. [CrossRef]
5. Aragaw, T.A.; Bogale, F.M. Biomass-based adsorbents for removal of dyes from wastewater: A review. *Front. Environ. Sci.* **2021**, *9*, 558. [CrossRef]

6. Zhai, Z.; Zhang, L.; Du, T.; Ren, B.; Xu, Y.; Wang, S.; Miao, J.; Liu, Z. A review of carbon materials for supercapacitors. *Mater. Des.* **2022**, *221*, 111017. [CrossRef]
7. Boychuk, T.Y.; Budzulyak, I.M.; Ivanichok, N.Y.; Lisovskiy, R.P.; Rachiy, B.I. Electrochemical properties of hybrid supercapacitors formed from nanosized spinel LiMn$_{1.5}$Fe$_{0.5}$O$_4$. *J. Nano Electron. Phys.* **2015**, *7*, 1019.
8. Yuan, S.; Lai, Q.; Duan, X.; Wang, Q. Carbon-based materials as anode materials for lithium-ion batteries and lithium-ion capacitors: A review. *J. Energy Storage* **2023**, *6*, 106716. [CrossRef]
9. Ketabchi, M.R.; Babamohammadi, S.; Davies, W.G.; Gorbounov, M.; Soltani, S.M. Latest advances and challenges in carbon capture using bio-based sorbents: A state-of-the-art review. *Carbon. Capture Sci. Technol.* **2023**, *6*, 100087. [CrossRef]
10. Ostafiychuk, B.K.; Budzulyak, I.M.; Mandzyuk, V.I.; Lisovskyy, R.P. Electrochemical characteristics of capacitor systems formed on chemically modified carbon base. *Nanosistemi Nanomater. Nanotehnologii* **2008**, *6*, 1207–1217.
11. Wang, H.; Xu, J.; Liu, X.; Sheng, L. Preparation of straw activated carbon and its application in wastewater treatment: A review. *J. Clean. Prod.* **2021**, *283*, 124671. [CrossRef]
12. Lisovska, S.A.; Ilnytskyy, R.V.; Lisovskyy, R.P.; Ivanichok, N.Y.; Bandura, K.V.; Rachiy, B.I. Structural and sorption properties of nanoporous carbon materials obtained from walnut shells. *Phys. Chem. Solid. State* **2023**, *24*, 348–353. [CrossRef]
13. Sklepova, S.V.; Gasyuk, I.M.; Ivanichok, N.Y.; Kolkovskyi, P.I.; Kotsyubynsky, V.O.; Rachiy, B.I. The porous structure of activated carbon-based on waste coffee grounds. *Phys. Chem. Solid State* **2022**, *23*, 484–490. [CrossRef]
14. Laine, J.; Calafat, A.; Labady, M. Preparation and characterization of activated carbons from coconut shell impregnated with phosphoric acid. *Carbon* **1989**, *27*, 191–195. [CrossRef]
15. Yang, T.; Lua, A.C. Characteristics of activated carbons prepared from pistachio-nut shells by physical activation. *J. Colloid Interface Sci.* **2003**, *267*, 408–417. [CrossRef] [PubMed]
16. Linares-Solano, A.; Gonzalez, L.J.d.D.; Sabio, M.M. Active carbons from almond shells as adsorbents in gas and liquid phases. *J. Chem. Technol. Biotech.* **1980**, *30*, 65–72. [CrossRef]
17. Bevla, F.R.; Rico, D.P.; Gomis, A.F.M. Activated carbon from almond shells. Chemical activation. 2. Zinc chloride activation temperature influence. *Ind. Eng. Chem. Prod. Res. Dev.* **1984**, *23*, 269–271.
18. Balcı, S.; Doğu, T.; Yücel, H. Characterization of activated carbon produced from almond shell and hazelnut shell. *J. Chem. Technol. Biotechnol.* **1994**, *60*, 419–426. [CrossRef]
19. Toles, C.A.; Marshall, W.E.; Johns, M.M.; Wartelle, L.H.; McAloon, A. Acid-activated carbons from almond shells: Physical, chemical and adsorptive properties and estimated cost of production. *Bioresour. Technol.* **2000**, *71*, 87–92. [CrossRef]
20. Kobya, M.; Demirbas, E.; Senturk, E.; Ince, M. Adsorption of heavy metal ions from aqueous solutions by activated carbon prepared from apricot stone. *Bioresour. Technol.* **2005**, *96*, 1518–1521. [CrossRef]
21. Zhu, G.; Duan, J.; Zhao, H.; Liu, M.; Li, F. Apricot shell: A potential high-quality raw material for activated carbon. *Adv. Mater. Res.* **2013**, *798*, 3–7. [CrossRef]
22. Abbas, M. Experimental investigation of activated carbon prepared from apricot stones material (ASM) adsorbent for removal of malachite green (MG) from aqueous solution. *Adsorpt. Sci. Technol.* **2020**, *38*, 24–45. [CrossRef]
23. Abd Ali, K.M. Synthesis of activated carbon by chemical activation of apricot stone with adsorption kinetics. *J. Mater. Environ. Sci.* **2021**, *12*, 887–898.
24. Tan, H.; Tall, O.E.; Liu, Z.; Wei, N.; Yapici, T.; Zhan, T.; Han, Y. Selective oxidation of glycerol to glyceric acid in base-free aqueous solution at room temperature catalyzed by platinum supported on carbon activated with potassium hydroxide. *ChemCatChem* **2016**, *8*, 1699–1707. [CrossRef]
25. Ostafiychuk, B.K.; Budzulyak, I.M.; Rachiy, B.I.; Vashchynsky, V.M.; Mandzyuk, V.I.; Lisovsky, R.P.; Shyyko, L.O. Thermochemical activated carbon as an electrode material for supercapacitors. *Nanoscale Res. Lett.* **2015**, *10*, 65. [CrossRef] [PubMed]
26. Deng, C.; Zhu, M. New type nitrogen-doped carbon material applied to deep adsorption desulfurization. *Energy Fuels* **2020**, *34*, 9320–9327. [CrossRef]
27. Sing, K.S. Adsorption methods for the characterization of porous materials. *Adv. Colloid Interface Sci.* **1998**, *76*, 3–11. [CrossRef]
28. Bedin, K.C.; Martins, A.C.; Cazetta, A.L.; Pezoti, O.; Almeida, V.C. KOH-activated carbon prepared from sucrose spherical carbon: Adsorption equilibrium, kinetic and thermodynamic studies for Methylene Blue removal. *Chem. Eng. J.* **2016**, *286*, 476–484. [CrossRef]
29. Aktaş, Ö.; Çeçen, F. Effect of type of carbon activation on adsorption and its reversibility. *J. Chem. Technol. Biotechnol.* **2006**, *81*, 94–101. [CrossRef]
30. Lang, J.W.; Yan, X.B.; Liu, W.W.; Wang, R.T.; Xue, Q.J. Influence of nitric acid modification of ordered mesoporous carbon materials on their capacitive performances in different aqueous electrolyte. *J. Power Sources* **2012**, *24*, 220–229. [CrossRef]
31. Su, F.; Poh, C.K.; Chen, J.S.; Xu, G.; Wang, D.; Li, Q.; Lin, J.; Lou, X.W. Nitrogen-containing microporous carbon nanospheres with improved capacitive properties. *Energy Environ. Sci.* **2011**, *4*, 717–724. [CrossRef]
32. Zhu, M.; Weber, C.J.; Yang, Y.; Konuta, M.; Starke, U.; Kern, K.; Bittner, A.M. Chemical and electrochemical ageing of carbon materials used in supercapacitor electrodes. *Carbon* **2008**, *46*, 1829–1840. [CrossRef]
33. Huck, C.W. Advances of infrared spectroscopy in natural product research. *Phytochem. Lett.* **2015**, *11*, 384–393. [CrossRef]
34. Marsh, H.; Yan, D.S.; O'Grady, T.M.; Wennerberg, A. Formation of active carbons from cokes using potassium hydroxide. *Carbon* **1984**, *22*, 603–611. [CrossRef]

35. Kucherenko, V.A.; Tamarkina, Y.V.; Raenko, G.F.; Chernyshova, M.I. Thermolysis of brown coal in the presence of alkali metal hydroxides. *Solid Fuel Chem.* **2017**, *51*, 147–154. [CrossRef]
36. Rachiy, B.I.; Budzulyak, I.M.; Vashchynsky, V.M.; Ivanichok, N.Y.; Nykoliuk, M.O. Electrochemical properties of nanoporous carbon material in aqueous electrolytes. *Nanoscale Res. Lett.* **2016**, *11*, 18. [CrossRef] [PubMed]
37. Vashchynskyi, V.M.; Semkiv, I.V.; Kashuba, A.I.; Petrus', R.Y. Influence of carbonization conditions on porous structure of carbon materials. *Khimiya Fiz. Tekhnologiya Poverhni* **2022**, *13*, 349–357. [CrossRef]
38. Zhou, M.; Cai, T.; Pu, F.; Chen, H.; Wang, Z.; Zhang, H.; Guan, S. Graphene/carbon-coated Si nanoparticle hybrids as high-performance anode materials for Li-ion batteries. *ACS Appl. Mater. Interfaces* **2013**, *5*, 3449–3455. [CrossRef]
39. Mandzyuk, V.I.; Nagirna, N.I.; Strelchuk, V.V.; Budzulyak, S.I.; Budzulyak, I.M.; Rachiy, B.I. Electrical and optical properties of porous carbon material. *Phys. Chem. Solid State* **2012**, *13*, 94–101.
40. Zubrik, A.; Matik, M.; Hredzák, S.; Lovás, M.; Danková, Z.; Kováčová, M.; Briančin, J. Preparation of chemically activated carbon from waste biomass by single-stage and two-stage pyrolysis. *J. Clean. Prod.* **2017**, *143*, 643–653. [CrossRef]
41. Budzulyak, I.M.; Vashchynsky, V.M.; Rachiy, B.I. Adsorption properties of porous carbon materials obtained by chemical activation. *JSPE* **2015**, *13*, 84–90.
42. Şentorun-Shalaby, Ç.; Uçak-Astarlıoglu, M.G.; Artok, L.; Sarıcı, Ç. Preparation and characterization of activated carbons by one-step steam pyrolysis/activation from apricot stones. *Microporous Mesoporous Mater.* **2006**, *88*, 126–134. [CrossRef]
43. Ostafiychuk, B.K.; Budzulyak, I.M.; Rachiy, B.I.; Kuzyshyn, M.M.; Vashchynskyi, V.M.; Mykyteichuk, P.M.; Merena, R.I. Adsorption properties of carbon activated with orthophosphoric acid. *Khimiya Fiz. Tekhnologiya Poverhni* **2014**, *5*, 204–209.
44. Hassan, M.M.; Carr, C.M. Biomass-derived porous carbonaceous materials and their composites as adsorbents for cationic and anionic dyes: A review. *Chemosphere* **2021**, *265*, 129087. [CrossRef] [PubMed]
45. Thommes, M.; Kaneko, K.; Neimark, A.V.; Olivier, J.P.; Rodriguez-Reinoso, F.; Rouquerol, J.; Sing, K.S. Physisorption of gases, with special reference to the evaluation of surface area and pore size distribution (IUPAC Technical Report). *Pure Appl. Chem* **2015**, *87*, 1051–1069. [CrossRef]
46. Foo, K.Y.; Hameed, B.H. Preparation of oil palm (Elaeis) empty fruit bunch activated carbon by microwave-assisted KOH activation for the adsorption of methylene blue. *Desalination* **2011**, *275*, 302–305. [CrossRef]
47. Song, C.; Wu, S.; Cheng, M.; Tao, P.; Shao, M.; Gao, G. Adsorption studies of coconut shell carbons prepared by KOH activation for removal of lead (II) from aqueous solutions. *Sustainability* **2014**, *6*, 86–98. [CrossRef]
48. Li, Z.; Wang, G.; Zhai, K.; He, C.; Li, Q.; Guo, P. Methylene blue adsorption from aqueous solution by loofah sponge-based porous carbons. *Colloids Surf. A Physicochem. Eng. Asp.* **2018**, *538*, 28–35. [CrossRef]
49. Namal, O.O.; Kalipci, E. Adsorption kinetics of methylene blue removal from aqueous solutions using potassium hydroxide (KOH) modified apricot kernel shells. *Int. J. Environ. Anal. Chem.* **2020**, *100*, 1549–1565. [CrossRef]
50. Djilani, C.; Zaghdoudi, R.; Djazi, F.; Bouchekima, B.; Lallam, A.; Modarressi, A.; Rogalski, M. Adsorption of dyes on activated carbon prepared from apricot stones and commercial activated carbon. *J. Taiwan Inst. Chem. Eng.* **2015**, *53*, 112–121. [CrossRef]

Disclaimer/Publisher's Note: The statements, opinions and data contained in all publications are solely those of the individual author(s) and contributor(s) and not of MDPI and/or the editor(s). MDPI and/or the editor(s) disclaim responsibility for any injury to people or property resulting from any ideas, methods, instructions or products referred to in the content.

Article

Impacts of Mn Content and Mass Loading on the Performance of Flexible Asymmetric Solid-State Supercapacitors Using Mixed-Phase MnO₂/N-Containing Graphene Composites as Cathode Materials

Hsin-Ya Chiu and Chun-Pei Cho *

Department of Applied Materials and Optoelectronic Engineering, National Chi Nan University, Nantou 54561, Taiwan
* Correspondence: cpcho@ncnu.edu.tw

Abstract: MnO_2/nitrogen-containing graphene (x-NGM) composites with varying contents of Mn were used as the electrode materials for flexible asymmetric solid-state supercapacitors. The MnO_2 was a two-phase mixture of γ- and α-MnO_2. The combination of nitrogen-containing graphene and MnO_2 improved reversible Faraday reactions and charge transfer. However, excessive MnO_2 reduced conductivity, hindering ion diffusion and charge transfer. Overloading the electrode with active materials also negatively affected conductivity. Both the mass loading and MnO_2 content were crucial to electrochemical performance. x-NGM composites served as cathode materials, while graphene acted as the anode material. Operating by two charge storage mechanisms enabled a synergistic effect, resulting in better charge storage purposes. Among the supercapacitors, the 3-NGM1//G1 exhibited the highest conductivity, efficient charge transfer, and superior capacitive characteristics. It showed a superior specific capacitance of 579 $F·g^{-1}$, leading to high energy density and power density. Flexible solid-state supercapacitors using x-NGM composites demonstrated good cycle stability, with a high capacitance retention rate of 86.7% after 2000 bending cycles.

Keywords: nitrogen-containing graphene; manganese dioxide; composite; asymmetric solid-state supercapacitor; ASSC

Citation: Chiu, H.-Y.; Cho, C.-P. Impacts of Mn Content and Mass Loading on the Performance of Flexible Asymmetric Solid-State Supercapacitors Using Mixed-Phase MnO_2/N-Containing Graphene Composites as Cathode Materials. C **2023**, 9, 88. https://doi.org/10.3390/c9030088

Academic Editors: Olena Okhay and Gil Gonçalves

Received: 8 July 2023
Revised: 5 September 2023
Accepted: 6 September 2023
Published: 10 September 2023

Copyright: © 2023 by the authors. Licensee MDPI, Basel, Switzerland. This article is an open access article distributed under the terms and conditions of the Creative Commons Attribution (CC BY) license (https://creativecommons.org/licenses/by/4.0/).

1. Introduction

The fast evolution of portable and wearable consumer electronics has triggered a demand for high-performance power storage devices. The escalating concerns about energy and the environment have further underscored the significance of developing stable energy storage systems [1–8]. Over the past two decades, supercapacitors have garnered considerable attention because of their exceptional attributes, including high charge-discharge rates, high power density, excellent cycle stability, and cost-effective manufacturing and maintenance [1–8]. Regarded as one of the foremost electrochemical energy storage devices, supercapacitors depend on the electrostatic interactions among electrolyte ions to store and discharge electrical energy [9–11]. They are categorized into two primary types based on the storage mechanisms: double-layer capacitors and pseudocapacitors. The former entails energy storage between two electrodes, while the latter involves electrochemical adsorption on the electrode surface through redox reactions. Optimal electrode materials for supercapacitors should possess a large surface area, favorable wettability, and conductivity. However, the existing electrode materials currently exhibit relatively low specific capacitance (ranging from 80 $F·g^{-1}$ to 120 $F·g^{-1}$), preventing supercapacitors from meeting the requirements of commercial energy storage [12,13].

The energy (E) stored in a capacitor is determined by its capacitance (C) and operation potential window (V), as shown by the formula $E = CV^2/2$ [14–17]. Electrode materials play an essential role in determining the capacitance and energy density of supercapacitors. Extensive efforts have been dedicated to developing active electrode materials with high specific capacitances. Materials with excellent electrochemical performance for positive electrodes can be matched with appropriate counterparties for negative electrodes. By leveraging the electrolyte decomposition overpotential at both electrodes [18–21], the battery-like Faraday electrode acts as the energy source, while the capacitor electrode functions as the power source [22]. Thus, the operating potential range can be extended by the complementary potentials on both sides, and the energy density can be further enhanced for more diverse applications [23]. The selection and design of a material system that combines the benefits of each component is a strategic approach to asymmetric supercapacitors [24]. Using solid electrolytes when assembling asymmetric supercapacitors can expand the operation potential window. Compared to conventional liquid electrolyte supercapacitors, solid-state supercapacitors (SSCs) are anticipated to offer higher energy density, rate capability, safety, and cycle stability [11,25–29].

Active electrode materials can be chosen from various materials such as carbon fibers, activated carbon, carbon nanotubes, graphene (G), transition metal oxides, conductive polymers, and their derivatives/composites [30]. Among them, manganese dioxide (MnO_2) has gained significant attention due to its low toxicity, affordability, ease of preparation, and high theoretical specific capacitance (approximately 1300 $F \cdot g^{-1}$). Consequently, MnO_2 has been widely utilized as a positive electrode material [31]. The pseudocapacitance behavior of MnO_2 can be attributed to the single-electron transfer in the Mn^{4+}/Mn^{3+} redox system [32]. However, it is limited by some inherent drawbacks, such as poor conductivity, a smaller specific surface area, larger volume changes, and high ion/charge transport resistances, leading to rapid capacitance decay during the charge/discharge processes. This makes it difficult to fully discharge and obtain high capacitive performance, which hinders practical application. To address these issues, incorporating conductive materials with MnO_2 has proven to be an effective solution. By doping or formation of composites, different morphologies and crystalline structures can be obtained to tailor the electrochemical characteristics [18,33,34]. The combination of MnO_2 nanostructures and modified carbonaceous materials may result in higher specific capacitances in the composites.

Liu et al. developed a simple and cost-effective method to synthesize nitrogen (N)-doped G/MnO_2 nanosheet composites with good electrochemical properties. The N-doped G was utilized as the template to grow layered δ-MnO_2 nanosheets. At a scan rate of 5 $mV \cdot s^{-1}$, the highest specific capacitance was 305 $F \cdot g^{-1}$. The composites were then used as the cathode material in flexible asymmetric solid-state supercapacitors (ASSCs) with activated carbon as the anode material. The ASSCs could work reversibly over the potential range of 0 V–1.8 V and demonstrate high energy density and good cycle stability. The maximum energy density reached 3.5 $mWh \cdot cm^{-3}$ at a power density of 0.019 $W \cdot cm^{-3}$ [30]. Zhu et al. used N-doped carbon nanotubes (N-CNTs) derived from polypyrrole as a support material for MnO_2 nanoparticles in the preparation of N-CNTs/MnO_2 composites. By optimizing the mass ratio between N-CNTs and MnO_2, an electrode with superior electrochemical performance was achieved. The capacitance reached 366.5 $F \cdot g^{-1}$ at a current density of 0.5 $A \cdot g^{-1}$ and remained at 245.5 $F \cdot g^{-1}$ as the current density increased to 25 $A \cdot g^{-1}$. The asymmetric supercapacitors using N-CNTs in the negative electrode and N-CNTs/MnO_2 composites in the positive electrode had a high energy density of 20.9 $Wh \cdot kg^{-1}$ when operating within a potential window of 1.8 V. After 5000 cycles, the capacitance retention rate was 91.6% [13]. Du et al. employed yeast cells as a template to prepare N-doped porous hollow carbon spheres (HCS) and in-situ deposited MnO_2 nanowires. Varied morphologies and electrochemical properties were achieved by controlling reaction parameters. A high specific capacitance of 255 $F \cdot g^{-1}$ was obtained at a current density of 1 $A \cdot g^{-1}$ utilizing a liquid electrolyte. The asymmetric supercapacitor was constructed using the MnO_2/HCS composite in the positive electrode and HCS in the negative electrode. Operating at a power density of 500 $W \cdot kg^{-1}$ within a potential window of 2.0 V, the maximum energy density reached 41.4 $Wh \cdot kg^{-1}$. Even at a higher power density of 7901 W·kg^{-1}, the energy density remained at 23.0 Wh·kg^{-1}. After 5000 cycles, the capacitance retention rate was 93.9%. This work showcased an environmentally friendly approach that combines biomass and energy for the preparation of electrode materials [35].

Despite the progress made in MnO$_2$-based asymmetric supercapacitors, their energy density still lags behind lead-acid batteries. The power density would inevitably be reduced due to the differences in kinetics and specific capacitances between the two electrodes. Thus, maintaining high power density while increasing specific capacitance and energy density is a key challenge for MnO$_2$-based supercapacitors. One approach to improve conductivity is the incorporation of conductive carbonaceous materials into MnO$_2$ composites. In this study, we address the challenge by utilizing G as the support for growing MnO$_2$ nanostructures, aiming to elevate the electrochemical performance of supercapacitors. As demonstrated in previous studies [36–38], N doping can increase wettability and active sites, promoting the accessibility of electrolyte ions and carrier density by enriching free charges on the electrode surface. To exploit this, we fabricated positive electrodes based on MnO$_2$/N-containing G composites, with G serving as the active material for the negative electrode, to produce flexible ASSCs. N-containing G enhanced conductivity and reduced interfacial impedance. Equilibrium conditions where both electrodes exhibited optimal capacitance were achieved by using identical mass loading on both sides. Mixed-phase MnO$_2$ has been demonstrated to exhibit superior performance over δ- MnO$_2$ [30]. By optimizing the MnO$_2$ content and mass loading in the electrode, a high specific capacitance of 110 F·g^{-1} was achieved at a current density of 1.0 A·g^{-1}. Operating at a power density of 4400 W·kg^{-1} over a wide potential window of 3.9 V, a peak energy density of 73.6 Wh·kg^{-1} was achieved. After 2000 bending cycles, the capacitance retention rate was 86.7% at the current density of 1.0 A·g^{-1}. The results have revealed that our materials processing strategy improved not only the capacitive behavior of individual electrodes but also the overall performance of ASSCs. The synergistic effect arising from the combination of N-containing G and MnO$_2$ nanostructures led to higher specific capacitances and good capacitance retention at high charge/discharge rates.

2. Experimental

2.1. Preparation of G

Graphite oxide (GO) was synthesized using the modified Hummers' method with graphite powder [39]. Prior to GO synthesis, a pre-oxidation procedure was carried out [40,41]. Four grams of graphite powder (UniRegion Bio-Tech, New Taipei City, Taiwan, <20 μm) was added to a solution composed of 2 g of potassium persulfate (K$_2$S$_2$O$_8$, J. T. Baker, 99%), 2 g of phosphorus pentoxide (P$_2$O$_5$, J. T. Baker, 99%), and 30 mL of concentrated sulfuric acid (H$_2$SO$_4$, Sigma Aldrich, 95 vol%–97 vol%). The mixture was heated to 80 °C and stirred continuously for 6 h. After cooling to room temperature, it was rinsed repeatedly with deionized (DI) water through centrifugation until a neutral pH level was reached. Subsequently, 4 g of pre-oxidized graphite powder was added to 100 mL of concentrated H$_2$SO$_4$ solution in an ice bath. Then, 12 g of potassium permanganate (KMnO$_4$, J. T. Baker, 99%) was slowly added at 35 °C, and the stirring continued for 2 h until the color turned dark brown. Following this, a solution containing 200 mL of DI water and 40 mL of hydrogen peroxide (H$_2$O$_2$, 30 vol% in water) was slowly added, resulting in a vigorous chemical reaction. Once the reaction was complete, a yellow-brown intermediate was produced and placed in a dilute aqueous hydrochloric acid (HCl, Showa, 36 vol%) solution to remove metal ions. After 1 h of ultrasonication, it was repeatedly rinsed with DI water through centrifugation until a neutral pH level was reached, resulting in the obtained GO powder.

To prepare the GO solution, 20 mg of GO powder was added to 100 mL of DI water. The mixture underwent ultrasonication for 2 h to ensure proper dispersion. Subsequently, the suspension was transferred to an autoclave and subjected to a 200 °C hydrothermal process for 2 h in a furnace. After cooling to room temperature, the product was collected by filtration and dried at 80 °C for 12 h. Finally, the dried product was ground to obtain G powder [42].

2.2. Preparation of N-Doped G (NG) Composites

To prepare the NG powder, 55 mg of G was combined with 8.6 mL of ammonia hydroxide solution (28 vol%–30 vol%) in 70 mL of DI water. The mixture underwent ultrasonication for 2 h to ensure proper dispersion. Subsequently, the suspension was transferred to an autoclave and subjected to a 140 °C hydrothermal process for 6 h in a furnace. After cooling to room temperature, the product was rinsed repeatedly with DI water until a neutral pH level was reached and then collected by centrifugation. After being dried at 80 °C for 12 h, the collected product was ground to obtain NG powder [30].

2.3. Preparation of NG/MnO_2 (NGM) Composites

G (55 mg) was mixed with 8.6 mL of ammonia hydroxide solution (28 vol%–30 vol%) in 70 mL of DI water. After ultrasonication for 2 h to ensure a better G dispersion, the suspension was transferred to an autoclave placed in a furnace for a 140 °C hydrothermal process for 6 h. Upon cooling to room temperature, it was divided into five portions. Different weights of $KMnO_4$ were added to each portion to create solutions with concentrations of 8.9 mM, 17.8 mM, 26.7 mM, 35.6 mM, and 44.5 mM of $KMnO_4$, respectively.

Each mixture underwent ultrasonication for 30 min and was then transferred to the autoclave for another hydrothermal process at 160 °C for 2 h. After cooling to room temperature, each was rinsed repeatedly with DI water until the neutral pH level was reached and then collected by centrifugation. The collected products were dried at 80 °C for 12 h, followed by grinding. This resulted in five NGM composites with varying amounts of Mn [30], which were named x-NGM, where x is 1, 2, 3, 4, and 5, respectively, corresponding to the abovementioned five $KMnO_4$ concentrations.

2.4. Fabrication of Electrodes

G, NG, and x-NGM composites (100 mg) were mixed with 12.5 mg of carbon black (UniRegion Bio-Tech, New Taipei City, Taiwan) in 2 mL of absolute ethanol (Shimakyu, Samut Sakhon, Thailand, 99.5 vol%), respectively. After ultrasonication for 10 min to improve dispersion, 0.5 g of ethyl cellulose (Aldrich, St. Louis, MI, USA) and 1 mL of terpineol (Aldrich, St. Louis, MI, USA) were added to each suspension. Another ultrasonication for 10 min ensured thorough mixing. Subsequently, stirring for 10 min was performed to allow ethanol to evaporate and achieve the desired consistencies of the G, NG, and x-NGM slurries for electrode fabrication.

A 3.5 cm × 2.5 cm polyimide (PI) tape was affixed to graphite paper to create a flexible PI/graphite substrate (Figure 1a,b). Meanwhile, a square hole measuring 1.6 cm in length was created at the center of a transparency sheet, which was placed above the flexible substrate and secured with 3M tape (Figure 1c). The transparency adhered closely to the substrate, defining the effective area. Next, an even layer of the G, NG, and x-NGM slurries was uniformly coated within the square hole on the flexible substrate using the doctor-blade method (Figure 1c). After allowing it to stand overnight at room temperature, the transparency was removed, leaving behind the electrodes (Figure 1d). They were then subjected to calcination at 200 °C for 1 h to remove organic residues. Three different mass loadings were employed for coating the active materials. The resulting electrodes were named Gy, NGy, and x-NGMy, where y represents the mass loadings of 1 mg, 2 mg, and 3 mg, respectively.

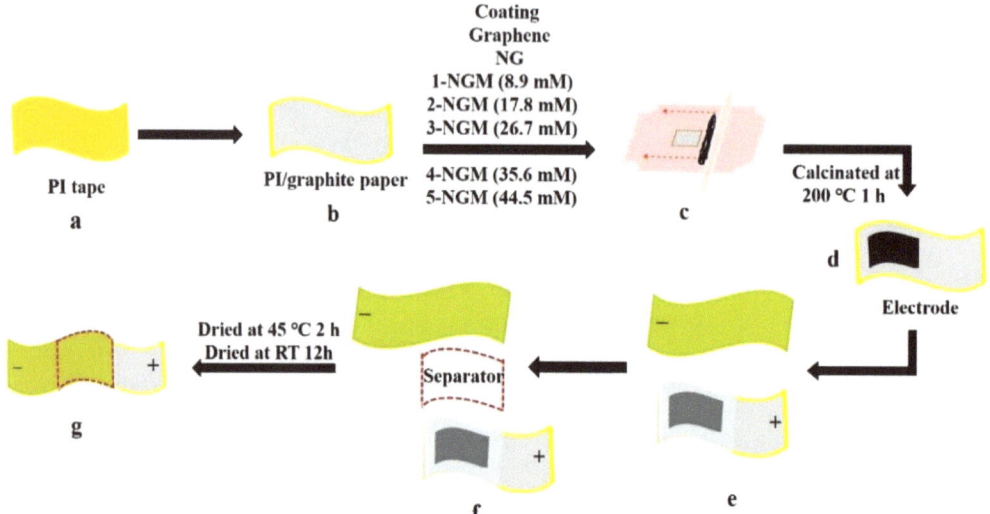

Figure 1. Fabrication flow chart of the ASSCs: (a, b) creation of a flexible PI/graphite substrate; (c) defining the effective area and spreading electrode materials; (d) formation of positive electrodes; (e) formation of the negative electrode; (f) placement of the separator between two electrodes; (g) assembling flexible ASSCs with a sandwich architecture.

2.5. Fabrication of ASSCs

A 10.6 g portion of lithium chloride (LiCl, Alfa Aesar, 99%) was mixed with 5.0 g of polyvinyl alcohol (PVA, Aldrich, 87–89%) in 50 mL of DI water to prepare a 5 M PVA/LiCl solution, which was stirred at 85 °C for 3 h until it transformed into a clear gel. After cooling to room temperature, a piece of filter paper cut to the desired size was immersed in the gel for 15 min. After removal from the gel, the PVA/LiCl gel membrane serving as the solid electrolyte was obtained [30].

The positive electrodes were fabricated using the three groups of active materials: Gy, NGy, and x-NGMy (Figure 1d). Only G1 was used to fabricate the negative electrode (Figure 1e). As a result, the ASSCs were designated as Gy//G1, NGy//G1, and x-NGMy//G1. A total of 21 flexible devices were obtained. The PVA/LiCl gel membrane also served as the separator between two electrodes (Figure 1f). To secure and position each component, two slide glasses were placed on the top and bottom of the devices, respectively. After baking at 45 °C for 2 h and air drying for 12 h, the slide glasses were removed, resulting in lightweight and flexible ASSCs with a sandwich architecture (Figure 1g) [39,43–45].

2.6. Characterization

The surface morphologies of the active materials were examined using field-emission gun scanning electron microscopes (SEM, JEOL JSM-7610F, Tokyo, Japan). The microstructures and lattice fringes were investigated using a high-resolution transmission electron microscope (HRTEM, JEOL JEM-ARM200F, Tokyo, Japan). Elemental mappings were obtained through energy-dispersive X-ray spectroscopy (EDS). The chemical compositions were examined using an X-ray photoelectron spectrometer (XPS, Thermo VG-Scientific, Sigma Probe, Waltham, MA, USA). The chemical states of the elements were determined based on the binding energies of emitted photoelectrons. Raman spectroscopy (Horiba Jobin Yvon, iHR320, East Kilbride, UK) was employed to identify molecular vibrational modes in the range of 400 cm^{-1} to 2000 cm^{-1}, allowing for determination of the chemical compositions, bonding configurations, and molecular structures of the active materials.

2.7. Electrochemical Measurements

The electrochemical properties were characterized by cyclic voltammetry (CV), galvanostatic charge/discharge (GCD), and electrochemical impedance spectroscopy (EIS) using a potentiostat/galvanostat (CH Instruments, CHI627C, Austin, TX, USA). For measurements on electrodes coated with active materials, a 5 M LiCl solution was used as the electrolyte, and a three-electrode configuration was employed with a platinum (Pt) wire auxiliary electrode and a silver chloride (Ag/AgCl) reference electrode. For the corresponding ASSCs, a 5 M PVA/LiCl gel electrolyte was utilized, and a two-electrode configuration was adopted, with the G1 counter electrode serving as both the reference and auxiliary electrodes.

The capacitance characteristics of the electrodes and ASSCs were determined by the areas inside the CV curves at different scan rates and the symmetry of the GCD curves at different current densities. The gravimetric specific capacitances, C_{CV} and $C_{C\text{-}DC}$ of individual electrodes, and ASSCs were calculated from the CV and GCD curves using Equations (1) and (2), respectively [46]:

$$C_{CV} = k \frac{\int i}{m \cdot s} \qquad (1)$$

$$C_{C-DC} = k \frac{i \cdot \Delta t}{\Delta V \cdot m} \qquad (2)$$

$$E_{EL}(Wh/kg) = (\frac{1}{4} \times C_{CV} \times V^2)/3.6 \qquad (3)$$

$$E_{Cell}(Wh/kg) = (\frac{1}{2} \times C_{CV} \times V^2)/3.6 \qquad (4)$$

$$E_{EL}(Wh/kg) = (\frac{1}{4} \times C_{C-DC} \times \Delta V^2)/3.6 \qquad (5)$$

$$E_{Cell}(Wh/kg) = (\frac{1}{2} \times C_{C-DC} \times \Delta V^2)/3.6 \qquad (6)$$

$$P_{EL}(W/kg) = E_{EL}/(\Delta t) \qquad (7)$$

$$P_{Cell}(W/kg) = E_{Cell}/(\Delta t) \qquad (8)$$

where k represents the electrode constant, typically 2 for a single electrode and 4 for a pair of electrodes. The variables i, $\int i$, m, s, Δt, and ΔV correspond to the discharging current, integral area of a CV curve, mass of electrode materials, scan rate (100 mV·s^{-1} in this study), discharging time, and potential window subtracting the initial potential drop, respectively. For individual electrodes and ASSCs in this study, the potential windows were set as −1.9 V to 1.0 V and −2.9 V to 1.0 V, respectively. By substituting C_{CV} into Equations (3) and (4) and $C_{C\text{-}DC}$ into Equations (5) and (6), the energy densities of electrodes (E_{EL}) and corresponding capacitors (E_{Cell}) were calculated. Subsequently, by substituting E_{EL} into Equation (7) and E_{Cell} into Equation (8), the power densities of electrodes (P_{EL}) and corresponding capacitors (P_{Cell}) were obtained [46].

EIS was used to investigate electronic and ionic transports at the interface of an electrode-active material. The frequency range for EIS was 10^{-2} Hz to 10^5 Hz, with an AC amplitude of 10 mV applied between two electrodes. To demonstrate the flexibility of our ASSCs, bending tests were conducted on 3-NGM1//G1, which exhibited superior electrochemical performance. The device was cycled between diameters of 10 mm, 5 mm, and 2 mm before returning to its original unbent state. A total of 2000 cycles were completed. The capacitance parameters were recorded every 50 cycles using the GCD method, with a current density of 1 A·g^{-1}, allowing for comparing the cycle stability of the ASSCs.

3. Results and Discussion

Ultrathin sheet-like structures were discovered in graphite oxide, G, and NG. Fine needle-like structures were observed on the surface of NG. The morphology and structure of x-NGM composites were significantly influenced by the Mn content [47]. Higher concentrations of $KMnO_4$ during preparation resulted in the formation of larger needle-like structures. To demonstrate the presence and uniform distribution of carbon (C), oxygen (O), N, and Mn elements in x-NGM, elemental mappings were again conducted, as displayed in Figure 2. This serves as one of the pieces of evidence for the successful N doping and preparation of x-NGM composites. NG played a dual role in enhancing conductivity and acting as a template for inducing the growth of MnO_2 nanostructures [47]. In comparison to the literature [48], the NG template acted as a stronger reducing agent, reacting more readily with $KMnO_4$ and facilitating MnO_2 growth. The sparser distribution of the N element, as revealed in Figure 2f, is attributed to its relatively lower proportion in the composite. TEM microstructures used to confirm further the deposition of MnO_2 nanostructures on the surface of NG have been presented elsewhere [47]. High-resolution images revealed that the MnO_2 in x-NGM composites was a mixture of γ- and α-MnO_2 [47]. The XRD patterns presented in Figure 3 further substantiate the presence of both phases. Specifically, the diffraction peaks at 12.5°, 18.1°, 26.1°, 28.9°, 37.2°, 42.8°, 51.2°, 56.4°, 60.1°, and 65.6° correspond to the (110), (200), (220), (310), (211), (301), (411), (600), (521), and (002) planes of α-MnO_2, respectively [49]. Additionally, the diffraction peaks at 21.2°, 32.4°, 36.1°, 38.1°, 44.6°, 55.1°, 58.6°, and 64.7° can be attributed to the (120), (031), (131), (230), (300), (160), (401), and (421) planes of γ-MnO_2, respectively [50]. Notably, the peak intensities of both phases in 4-NGM and 5-NGM are reduced. The reduction is a consequence of the excess MnO_2 content, which adversely affects crystallinity. The combined results from XRD, Raman, and XPS have validated the successful preparation of the x-NGM composites containing NG and MnO_2 using the hydrothermal method [47].

Based on the EIS results of the 21 different active materials on the PI/graphite flexible substrate, it was concluded that the optimal mass loading was 1 mg. Given the stable potential windows and the high specific capacitances exhibited by x-NGM composites, they served as excellent cathode candidates for ASSCs. By pairing G as the anode material with x-NGM as the cathode material, the operating voltage of the capacitor was extended from 2.9 V to 3.9 V. Figure 4 depicts the Nyquist plots of the supercapacitors. The equivalent circuit used for EIS analysis is similar to that reported earlier and consists of the following components [47]: (1) charge transfer impedance at the electrode/electrolyte interface in the high-frequency region (R_{CT}), (2) solution resistance (R_S) indicating the contact series resistance between the substrate and current collector, (3) Warburg impedance (W) representing the diffusion resistance of ions in the electrolyte and related to the line tail slope in the low-frequency region, (4) electric double-layer capacitance (C_1), and (5) Faraday pseudocapacitance (C_2) [51–53]. The R_{CT} values obtained by simulation using the equivalent circuit are listed in Table 1.

Table 1. R_{CT} values of the 21 supercapacitors with Gy, NGy, and x-NGMy composites obtained by EIS simulation.

Capacitor	R_{CT} (Ω)	Capacitor	R_{CT} (Ω)	Capacitor	R_{CT} (Ω)
G1//G1	3.46	1-NGM2//G1	2.59	3-NGM3//G1	1.89
G2//G1	3.74	1-NGM3//G1	8.97	4-NGM1//G1	9.14
G3//G1	15.40	2-NGM1//G1	1.43	4-NGM2//G1	9.43
NG1//G1	2.68	2-NGM2//G1	1.74	4-NGM3//G1	9.74
NG2//G1	3.41	2-NGM3//G1	2.24	5-NGM1//G1	9.87
NG3//G1	9.46	3-NGM1//G1	1.24	5-NGM2//G1	10.24
1-NGM1//G1	2.35	3-NGM2//G1	1.32	5-NGM3//G1	13.60

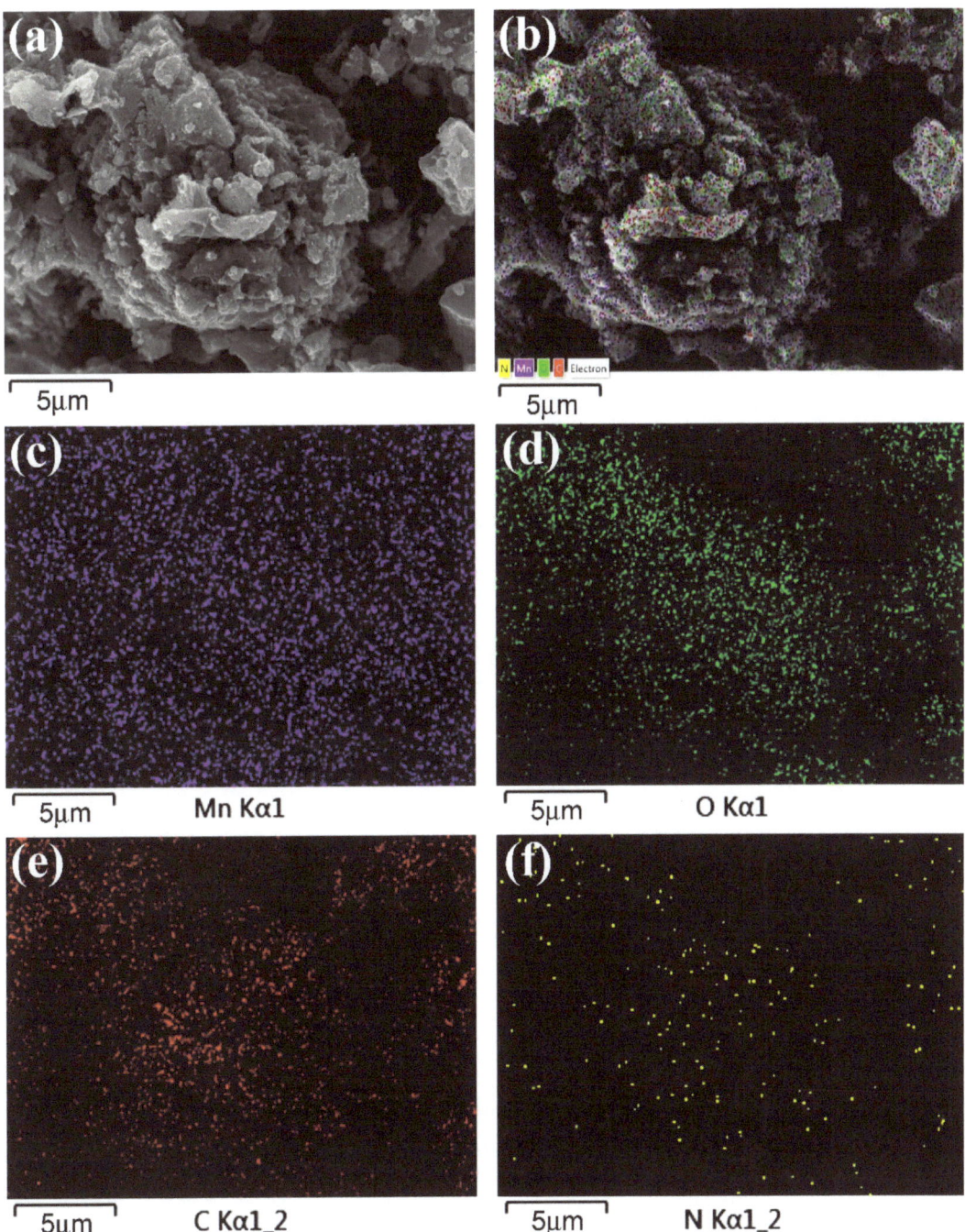

Figure 2. (**a**) SEM and (**b**) EDS layered images of 3-NGM. Elemental mappings of 3-NGM: (**c**) Mn, (**d**) O, (**e**) C, and (**f**) N.

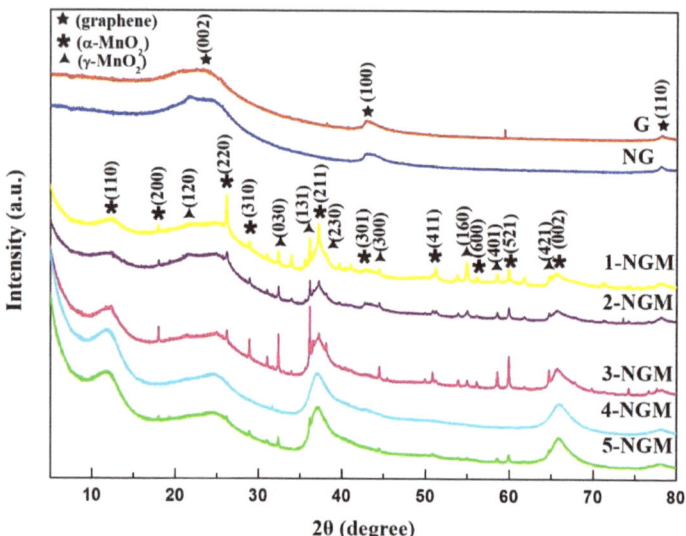

Figure 3. XRD patterns of G, NG, and x-NGM composites.

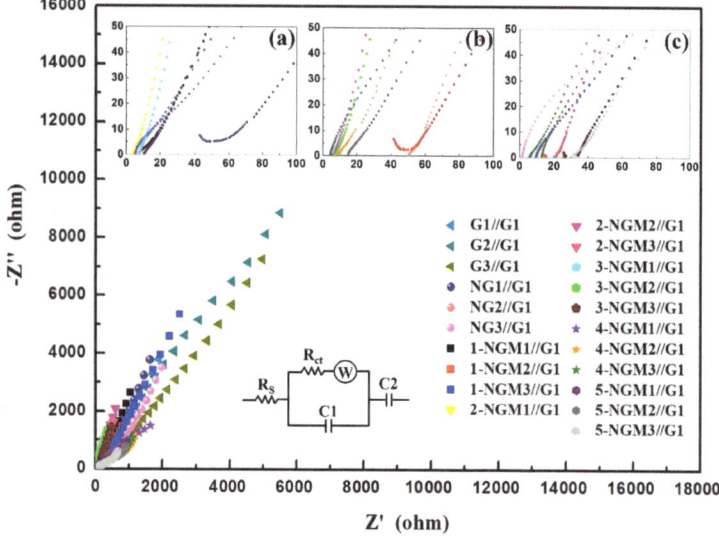

Figure 4. Nyquist plots of the 21 supercapacitors with Gy, NGy, and x-NGMy composites. Insets are enlargements when the mass loading is (**a**) 1 mg, (**b**) 2 mg, and (**c**) 3 mg.

As depicted in the inset (a) of Figure 4, when the mass loading of active materials is 1 mg, the Nyquist plots of the G1//G1, NG1//G1, 1-NGM1//G1, 2-NGM1//G1, 3-NGM1//G1, 4-NGM1//G1, and 5-NGM1//G1 devices exhibit smaller semicircles in the high-frequency region, resulting in the following R_{CT} values: 3.46 Ω, 2.68 Ω, 2.35 Ω, 1.43 Ω, 1.24 Ω, 9.14 Ω, and 9.87 Ω, respectively. Table 1 demonstrates that the addition of N to G decreases the R_{CT} from 3.46 Ω to 2.68 Ω. A further reduction to 2.35 Ω is obtained with the addition of MnO_2. The combination of NG and MnO_2 enhances the reversible Faraday reactions on the surface of active materials and improves charge transfer capability, leading to a more linear response in the low-frequency region of the Nyquist plot. A steeper slope indicates lower diffusion resistance, faster ion transport, and better capacitive properties [51,54]. Among the 21 supercapacitors, the 3-NGM1//G1 device exhibits the smallest R_{CT}, implying the best conductivity. It also shows the steepest slope in the Nyquist plot's low-frequency region, suggesting rapid charge transfer and excellent capacitive characteristics. However, the R_{CT} value significantly increases in the 4-NGM1//G1 and 5-NGM1//G1 devices with higher MnO_2 content. Consistent with the results observed in active material electrodes [47], an excess of MnO_2 in the 4-NGM1 and 5-NGM1 electrodes led to larger R_{CT} and poorer conductivity due to fewer contacts. Herein, excess MnO_2 is observed to result in a smaller slope in the Nyquist plot's low-frequency region, which hampers ion diffusion and Faraday charge transfer.

As displayed in the inset (b) of Figure 4, when the mass loading of active materials is 2 mg, the Nyquist plots of the G2//G1, NG2//G1, 1-NGM2//G1, 2-NGM2//G1, 3-NGM2//G1, 4-NGM2//G1, and 5-NGM2//G1 devices exhibit larger semicircles in the high-frequency region, resulting in the corresponding R_{CT} values of 3.74 Ω, 3.41 Ω, 2.59 Ω, 1.74 Ω, 1.32 Ω, 9.43 Ω, and 10.24 Ω. Both insets (a) and (b) demonstrate a decrease in electrochemical impedance due to N doping. Moreover, it is observed that the R_{CT} increases with higher mass loading. Similarly, as shown in the inset (c) of Figure 4, when the mass loading of active materials is 3 mg, the Nyquist plots of the G3//G1, NG3//G1, 1-NGM3//G1, 2-NGM3//G1, 3-NGM3//G1, 4-NGM3//G1, and 5-NGM3//G1 devices exhibit even larger semicircles in the high-frequency region, resulting in larger corresponding R_{CT} values of 15.40 Ω, 9.46 Ω, 8.97 Ω, 2.24 Ω, 1.89 Ω, 9.74 Ω, and 13.60 Ω. This suggests that an excessive mass of active materials on flexible electrodes is unfavorable for conductivity improvement. As proven, 1 mg is the most appropriate mass loading. Amounts of 2 mg and 3 mg are excessive and cause reduced specific capacitance, energy density, and power density. The results once again demonstrate that both the mass loading of active materials and the content of MnO_2 impact the conductivity in the ASSCs. Consistent with the EIS results, better charge transfer efficiency can only be achieved when the most appropriate MnO_2 content is used in an x-NGM composite. Therefore, in this study, x-NGM composites and G were utilized as the active materials for cathodes and anodes, respectively, to manufacture ASSCs operating based on both (pseudocapacitance and electric double-layer capacitor) mechanisms simultaneously. The synergistic effect of the two types of materials enabled improved charge storage.

Table 2 presents the capacitor parameters obtained from the CV results for the 21 flexible ASSCs using Gy, NGy, and x-NGMy composites. Among the nine capacitors using Gy, NGy, and 1-NGMy composites, the 1-NGM1//G1 had the largest integral area within its CV curve loop, resulting in a high specific capacitance of 209 $F \cdot g^{-1}$. Its energy density and power density were 441.9 $Wh \cdot kg^{-1}$ and 1744.3 $W \cdot kg^{-1}$, respectively. The capacitive characteristics of the 12 x-NGMy//G1 capacitors with varying MnO_2 content and three mass loadings of active materials for each MnO_2 content were also explored and compared. Although their CV curves deviated from rectangles due to the synergistic effect of the two charge storage mechanisms, they remained symmetric, indicating a faster and reversible charge/discharge process. Consistent with the findings reported in the literature [30,35], the specific capacitance decreased significantly with increasing mass loading. Among the 15 capacitors using x-NGMy composites, the 3-NGM1//G1 had the largest CV curve loop, implying the highest conductivity. The charge transfer path was shortened, facilitating electrolyte ions to reach the electrode surface and enabling faster electron transfer [32]. It exhibited the highest specific capacitance of 579 $F \cdot g^{-1}$, resulting in corresponding high energy density and power density of 1223.3 $Wh \cdot kg^{-1}$ and 73,153.6 $W \cdot kg^{-1}$, respectively, as listed in Table 2. Previous results have shown that the 3-NGM1 electrode exhibited excellent capacitance performance [47]. Therefore, x = 3 and y = 1 were identified as the optimal parameters for x-NGMy materials. The utilization of the 3-NGM1 composite in ASSCs facilitated reversible redox (pseudocapacitive) reactions on the electrode surface. When combined with the fast charge/discharge property of the electric double-layer capacitor material, the overall performance of the 3-NGM1//G1 capacitor was enhanced by the synergistic effect of the two charge storage mechanisms. However, when x = 4 and 5, there was an excess of MnO_2 in the active materials, leading to the growth of nanorod structures that impeded ion diffusion in the electrolyte and the functioning of the pseudocapacitive storage mechanism. As a result, the synergistic effect could not be effectively achieved. The excessive mass loading of active materials on electrodes resulted in a reduction in specific capacitance, energy density, and power density. The CV results of ASSCs were consistent with the findings for the electrodes [47]. Figure 5a shows the Ragone plot of the supercapacitors obtained by plotting energy density vs. power density calculated using Equations (4) and (8). Again, it is evident that the 3-NGM1//G1 capacitor exhibits the best capacitance characteristic among the ASSCs.

Figure 5b shows the plots of specific capacitance vs. current density obtained from the GCD curves of three Gy//G1 symmetric capacitors and 18 NGy//G1 and x-NGMy//G1 asymmetric capacitors. Evidently, the Gy//G1, NGy//G1, 1-NGMy//G1, 4-NGMy//G1, and 5-NGMy//G1 devices are unable to sustain current densities exceeding $1\ A \cdot g^{-1}$. The 2-NGM2//G1 capacitor can only withstand current densities of $1\ A \cdot g^{-1}$, $3\ A \cdot g^{-1}$, and $5\ A \cdot g^{-1}$. In contrast, the 3-NGM2//G1 capacitor can endure current densities of $1\ A \cdot g^{-1}$, $3\ A \cdot g^{-1}$, $5\ A \cdot g^{-1}$, and $7\ A \cdot g^{-1}$, while the 2-NGM1//G1 and 3-NGM1//G1 capacitors can handle all current densities. Among them, the 3-NGM1//G1 capacitor stands out for its exceptional endurance and sustainability, indicating that an optimal Mn content in an NGM composite leads to superior performance. Table 3 shows the specific capacitance of the 3-NGM1//G1 capacitor under a current density of $1\ A \cdot g^{-1}$, which is $110\ F \cdot g^{-1}$. This corresponds to the energy density and power density of $73.6\ Wh \cdot kg^{-1}$ and $4400.0\ W \cdot kg^{-1}$, respectively, across a wide potential window of 3.9 V. When compared to the results in Ref. [30], it can be inferred that mixed-phase MnO_2 nanostructures may offer better overall capacitive characteristics than δ-MnO_2. Additionally, the GCD results indicate an inverse relationship between charge/discharge time and current density. Supercapacitors operate at higher current density charges and discharge faster but have lower specific capacitance due to limited time for redox reactions [55]. Conversely, those operated at lower current density charges discharge more slowly but have higher specific capacitance due to sufficient time for redox reactions. This is consistent with the observation that electrodes performed more efficiently at low current densities. The GCD results of the ASSCs reveal a similar trend to those of the electrodes [47]. Combining the pseudocapacitive material 3-NGM1 with the fast charge/discharge electric double-layer capacitor material G1 produced a synergistic effect, enhancing the performance of the 3-NGM1//G1 capacitor.

Bending tests were conducted to evaluate the flexibility and cycle stability of the solid-state ASSCs. The devices were bent to diameters of 10 mm, 5 mm, and 2 mm, and then returned to their nonbending state to complete a cycle. Figure 6a illustrates the CV curves of the 3-NGM1//G1 capacitor obtained from the bending test at a scan rate of $100\ mV \cdot s^{-1}$. After one cycle, the CV curve closely resembles the original curve. Figure 6b presents the GCD curves of the same capacitor obtained from the bending test at the current density of $1\ A \cdot g^{-1}$. The curves exhibit slight distortion from the ideal triangular shape due to the pseudocapacitive contribution from MnO_2. However, after one cycle, the GCD curve closely matches the original curve, confirming the excellent flexibility and high charge/discharge stability of the solid-state ASSCs utilizing x-NGMy active materials. Figure 6c displays the GCD curves obtained after subjecting the 3-NGM1//G1 capacitor to 2000 bending cycles at the current density of $1\ A \cdot g^{-1}$. Data were recorded every 50 cycles. The capacitance parameters (specific capacitance, energy density, and power density) after 50, 100, 200, 400, 800, 1600, and 2000 cycles, as listed in Table 4, are compared with the results from the first cycle. With an increasing cycle count, the shape and position of the GCD plot exhibit no significant deviations compared to the first cycle. As depicted in Figure 6d, the capacitance retention rate after 2000 bending cycles is 86.7%, confirming the excellent cycle stability of the flexible ASSCs using x-NGMy composites. Figure 6e shows the Ragone plot, comparing the energy density of the 3-NGM1//G1 capacitor calculated from the GCD results at different current densities with those in the literature [45,56–67]. The performance achieved in this study surpasses that of most results reported elsewhere, highlighting the potential of x-NGMy composites as electrode-active materials for flexible supercapacitors.

Table 2. Capacitance parameters obtained from the CV results of the 21 supercapacitors with Gy, NGy, and x-NGMy composites.

Capacitor	Scan Rate ($mV \cdot s^{-1}$)	Specific Capacitance ($F \cdot g^{-1}$)	Energy Density ($Wh \cdot kg^{-1}$)	Power Density ($W \cdot kg^{-1}$)
G1//G1	100	72	144.6	1177.9
G2//G1	100	12	25.8	4837.2
G3//G1	100	4	7.5	2713.2
NG1//G1	100	73	154.5	1185.6
NG2//G1	100	26	50.2	2730.2
NG3//G1	100	6	12.2	1034.8
1-NGM1//G1	100	209	441.9	1744.3
1-NGM2//G1	100	95	191.1	1346.1
1-NGM3//G1	100	7	13.2	305.7
2-NGM1//G1	100	480	1014.1	82,407.0
2-NGM2//G1	100	263	556.4	116,454.6
2-NGM3//G1	100	196	413.6	160,087.4
3-NGM1//G1	100	579	1223.3	73,153.6
3-NGM2//G1	100	334	705.9	123,958.1
3-NGM3//G1	100	243	513.4	150,269.1
4-NGM1//G1	100	13	27.7	6726.1
4-NGM2//G1	100	3	5.4	12,221.9
4-NGM3//G1	100	1	2.9	17,744.4
5-NGM1//G1	100	12	25.6	11,496.8
5-NGM2//G1	100	3	6.5	25,867.5
5-NGM3//G1	100	2	5.1	22,750.9

Table 3. Capacitance parameters obtained from the GCD results of the supercapacitors with Gy, NGy, and x-NGMy composites.

Capacitor	Current Density ($A \cdot g^{-1}$)	Specific Capacitance ($F \cdot g^{-1}$)	Energy Density ($Wh \cdot kg^{-1}$)	Power Density ($W \cdot kg^{-1}$)
2-NGM1//G1	1	81	54.1	4400.0
2-NGM2//G1	1	31	21.0	4400.0
2-NGM3//G1	1	17	11.4	4400.0
3-NGM1//G1	1	110	73.6	4400.0
3-NGM2//G1	1	37	25.0	4400.0
3-NGM3//G1	1	22	15.0	4400.0
2-NGM1//G1	3	33	22.0	13,200.0
2-NGM2//G1	3	16	10.6	13,200.0
2-NGM3//G1	3	8	5.1	13,200.0
3-NGM1//G1	3	50	33.4	13,200.0
3-NGM2//G1	3	16	11.0	13,200.0
3-NGM3//G1	3	9	6.2	13,200.0
2-NGM1//G1	5	21	14.1	22,000.0
2-NGM2//G1	5	10	6.7	22,000.0
3-NGM1//G1	5	42	28.1	22,000.0
3-NGM2//G1	5	13	8.6	22,000.0
2-NGM1//G1	7	17	11.1	30,800.0
3-NGM1//G1	7	36	24.0	30,800.0
3-NGM2//G1	7	10	6.8	30,800.0
2-NGM1//G1	9	13	8.8	39,600.0
3-NGM1//G1	9	31	20.9	39,600.0

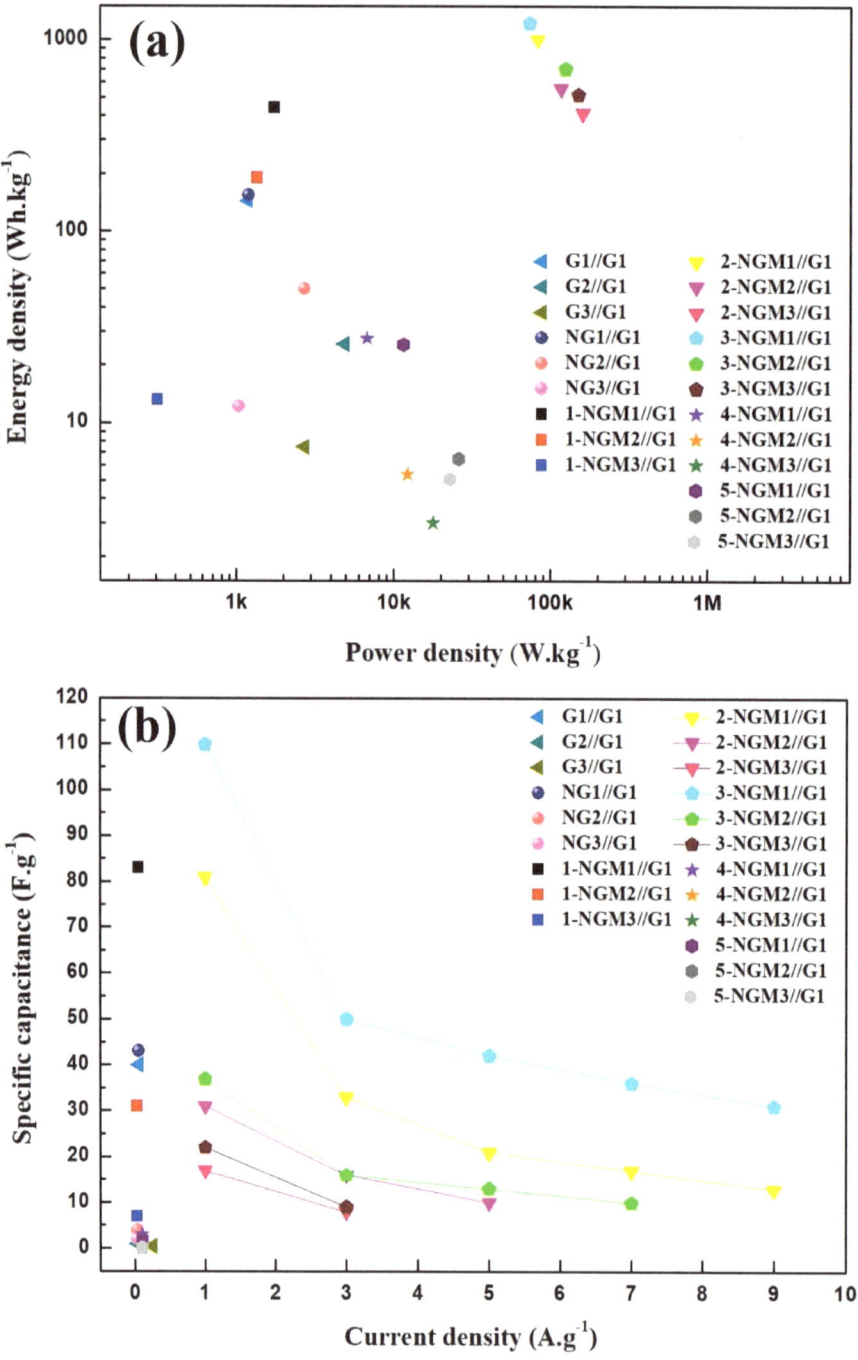

Figure 5. (a) Ragone plot obtained from the CV results. (b) Plots of specific capacitance vs. current density were obtained from the GCD results of the Gy//G1, NGy//G1, and x-NGMy//G1 supercapacitors.

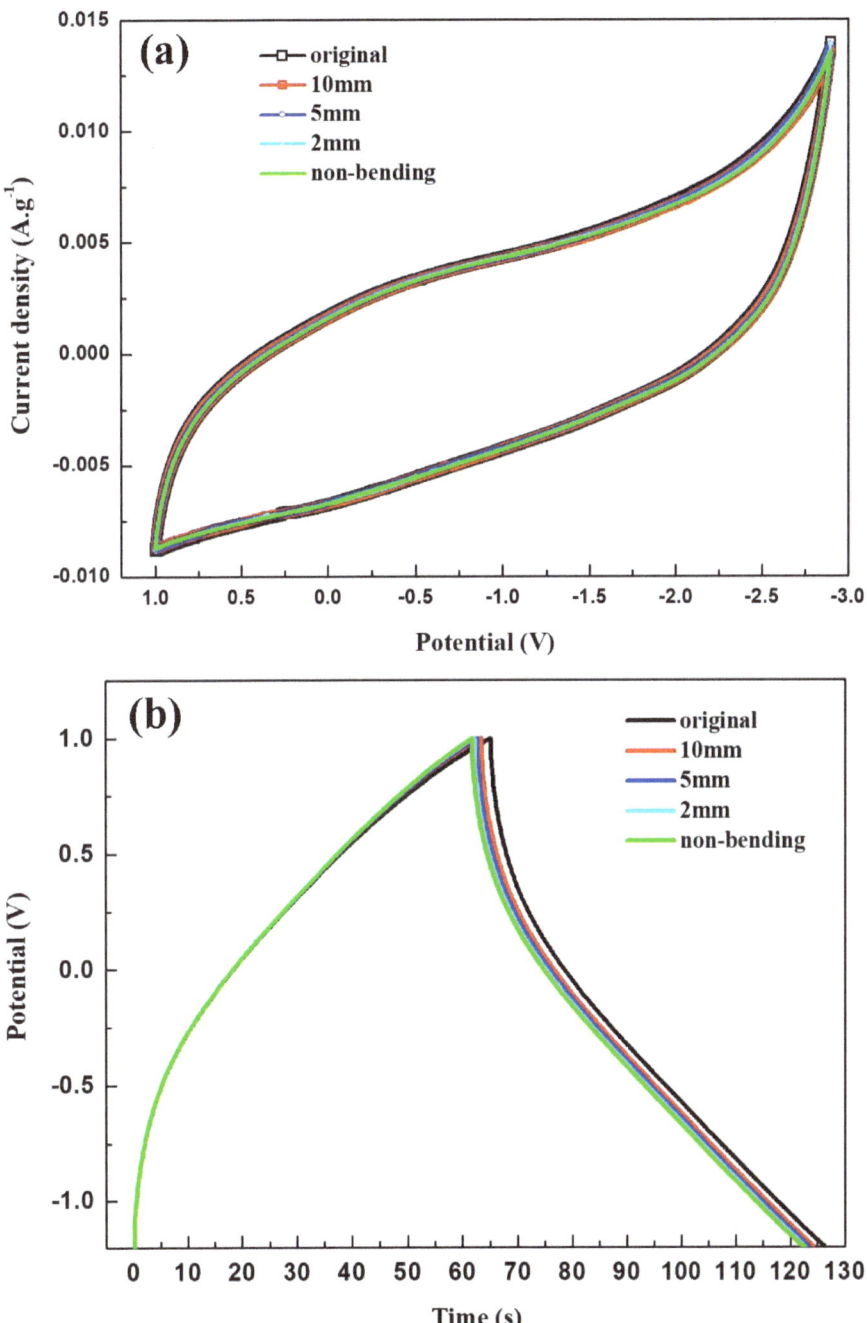

Figure 6. *Cont.*

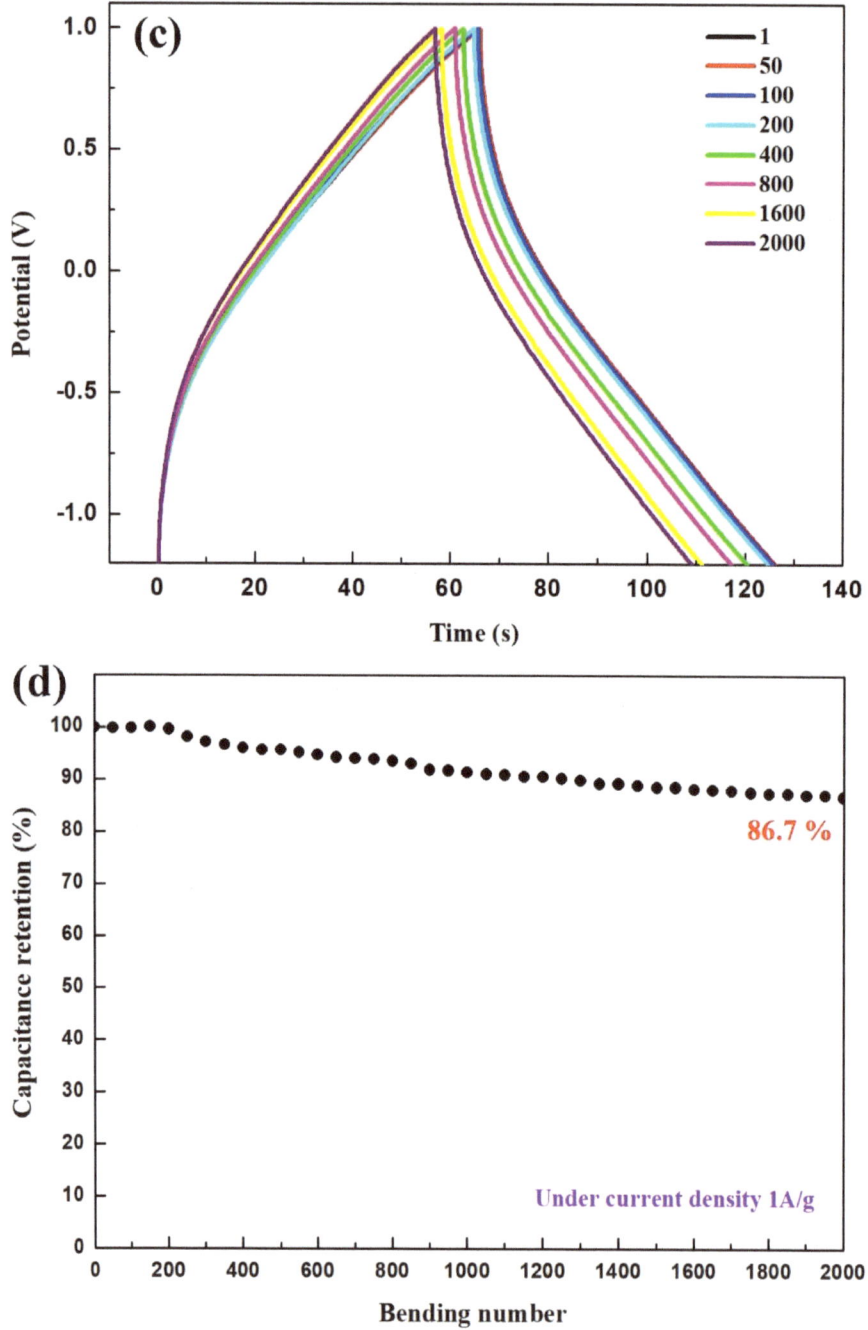

Figure 6. *Cont.*

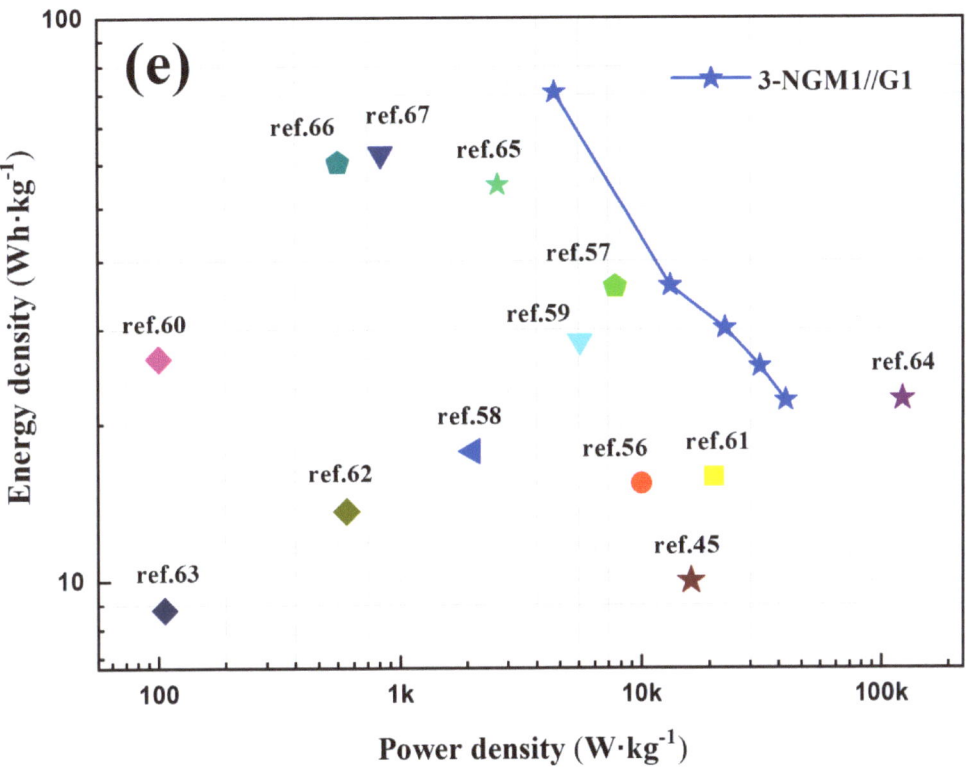

Figure 6. 3-NGM1//G1 supercapacitor: (**a**) CV and (**b**) GCD curves obtained by one bending cycle; (**c**) GCD curves and (**d**) retention rate of specific capacitance obtained by 2000 bending cycles; (**e**) Ragone plot, which compares the energy density calculated by GCD results with those reported in the literature [45,56–67].

Table 4. Capacitance parameters obtained from the GCD results of the 3-NGM1//G1 capacitor after 50, 100, 200, 400, 800, 1600, and 2000 bending cycles.

Bending Cycle	Current Density (A·g^{-1})	Specific Capacitance (F·g^{-1})	Energy Density (Wh·kg^{-1})	Power Density (W·kg^{-1})
1	1	110	73.6	4400.0
50	1	109	73.5	4400.0
100	1	109	73.5	4400.0
200	1	109	73.3	4400.0
400	1	105	70.6	4400.0
800	1	102	68.8	4400.0
1600	1	97	64.9	4400.0
2000	1	95	63.8	4400.0

4. Conclusions

Hydrothermal synthesis produced x-NGM composites comprising NG and MnO$_2$ with varying Mn content for use in ASSCs as active electrode materials. SEM, EDS mappings, XPS, and Raman results confirmed the presence of NG and MnO$_2$, as well as the successful preparation of the composites. TEM analysis revealed a two-phase mixture of γ- and α-MnO$_2$ in x-NGM composites. The combination of NG and MnO$_2$ enhanced reversible Faraday reactions and charge transfer capability. Excessive MnO$_2$ reduced conductivity and the Nyquist plot's slope in the low-frequency region, hindering ion diffusion and charge transfer. Optimum charge transfer efficiency requires the appropriate MnO_2 content. Overloading the flexible electrode with active materials also showed a negative impact on conductivity. A mass loading of 1 mg was the most suitable, while 2 mg and 3 mg reduced specific capacitance, energy density, and power density. Both the mass loading and the MnO_2 content played crucial roles in determining capacitor performance.

By utilizing x-NGM composites as cathode materials and G as the anode material, the simultaneous operation of two charge-storage mechanisms was enabled, leading to improved charge storage. Among the supercapacitors, the 3-NGM1//G1 device showed superior conductivity and the steepest slope in the Nyquist plot's low-frequency region, indicating its efficient charge transfer and best capacitive properties. It demonstrated a high specific capacitance of 579 $F \cdot g^{-1}$, giving rise to the high energy density (1223.3 $Wh \cdot kg^{-1}$) and power density (73,153.6 $W \cdot kg^{-1}$). Furthermore, the flexible ASSCs using x-NGMy composites exhibited good cycle stability. The capacitance retention rate was 86.7% after 2000 bending cycles. The exceptional performance in this study surpassed most literature results, highlighting the potential of x-NGMy composites as promising electrode materials for supercapacitors.

Author Contributions: Conceptualization, H.-Y.C. and C.-P.C.; methodology, H.-Y.C. and C.-P.C.; software, H.-Y.C. and C.-P.C.; validation, H.-Y.C. and C.-P.C.; formal analysis, H.-Y.C. and C.-P.C.; investigation, H.-Y.C. and C.-P.C.; resources, C.-P.C.; data curation, H.-Y.C. and C.-P.C.; writing—original draft preparation, C.-P.C.; writing—review and editing, C.-P.C.; visualization, H.-Y.C. and C.-P.C.; supervision, C.-P.C.; project administration, C.-P.C.; funding acquisition, C.-P.C. All authors have read and agreed to the published version of the manuscript.

Funding: This research was funded by National Science and Technology Council, Taiwan under the grant number of MOST 109-2221-E-260-009-. The APC was funded by National Chi Nan University.

Data Availability Statement: Not applicable.

Acknowledgments: Supports from the National Science and Technology Council, Taiwan and National Chi Nan University are gratefully appreciated.

Conflicts of Interest: The authors declare no conflict of interest.

References

1. Zuo, W.; Li, R.; Zhou, C.; Li, Y.; Xia, J.; Liu, J. Battery-supercapacitor hybrid devices: Recent progress and future prospects. *Adv. Sci.* **2017**, *4*, 1600539. [CrossRef] [PubMed]
2. El-Kady, M.F.; Shao, Y.; Kaner, R.B. Graphene for batteries, supercapacitors and beyond. *Nat. Rev. Mater.* **2016**, *1*, 16033. [CrossRef]
3. Wang, F.; Wu, X.; Yuan, X.; Liu, Z.; Zhang, Y.; Fu, L.; Zhu, Y.; Zhou, Q.; Wu, Y.; Huang, W. Latest advances in supercapacitors: From new electrode materials to novel device designs. *Chem. Soc. Rev.* **2017**, *46*, 6816. [CrossRef] [PubMed]
4. Wang, J.; Li, X.; Du, X.; Wang, J.; Ma, H.; Jing, X. Polypyrrole composites with carbon materials for supercapacitors. *Chem. Pap.* **2017**, *71*, 293. [CrossRef]
5. González, A.; Goikolea, E.; Barrena, J.A.; Mysyk, R. Review on supercapacitors: Technologies and materials. *Renew. Sustain. Energy Rev.* **2016**, *58*, 1189. [CrossRef]
6. Lakshmi, K.C.S.; Vedhanarayanan, B. High-performance supercapacitors: A comprehensive review on paradigm shift of conventional energy storage devices. *Batteries* **2023**, *9*, 202. [CrossRef]
7. Sajedi-Moghaddam, A.; Gholami, M.; Naseri, N. Inkjet printing of MnO_2 nanoflowers on surface-modified A4 paper for flexible all-solid-state microsupercapacitors. *ACS Appl. Mater. Interfaces* **2023**, *15*, 3894. [CrossRef]
8. Zhang, Y.; Zhou, C.G.; Yan, X.H.; Cao, Y.; Gao, H.L.; Luo, H.W.; Gao, K.Z.; Xue, S.C.; Jing, X. Recent advances and perspectives on graphene-based gels for superior flexible all-solid-state supercapacitors. *J. Power Sources* **2023**, *565*, 232916. [CrossRef]
9. Arico, A.S.; Bruce, P.; Scrosati, B.; Tarascon, J.M.; Schalkwijk, W.V. Nanostructured materials for advanced energy conversion and storage devices. *Nat. Mater.* **2005**, *4*, 366. [CrossRef]
10. Zhang, L.L.; Zhao, X.S. Carbon-based materials as supercapacitor electrodes. *Chem. Soc. Rev.* **2009**, *38*, 2520. [CrossRef]
11. Simon, P.; Gogotsi, Y. Materials for electrochemical capacitors. *Nat. Mater.* **2008**, *7*, 845. [CrossRef] [PubMed]
12. Béguin, F.; Presser, V.; Balducci, A.; Frackowiak, E. Carbons and electrolytes for advanced supercapacitors. *Adv. Mater.* **2014**, *26*, 2219. [CrossRef]

13. Zhu, J.; Xu, Y.; Hu, J.; Wei, L.; Liu, J.; Zheng, M. Facile synthesis of MnO_2 grown on nitrogen-doped carbon nanotubes for asymmetric supercapacitors with enhanced electrochemical performance. *J. Power Sources* **2018**, *393*, 135. [CrossRef]
14. Liu, J.; Zhang, L.; Wu, H.B.; Lin, J.; Shen, Z.; Lou, X.W. High-performance flexible asymmetric supercapacitors based on a new graphene foam/carbon nanotube hybrid film. *Energy Environ. Sci.* **2014**, *7*, 3709. [CrossRef]
15. Ning, P.; Duan, X.; Ju, X.; Lin, X.; Tong, X.; Pan, X.; Wang, T.; Li, Q. Facile synthesis of carbon nanofibers/MnO_2 nanosheets as high-performance electrodes for asymmetric supercapacitors. *Electrochim. Acta* **2016**, *210*, 754. [CrossRef]
16. Ghosh, K.; Yue, C.Y.; Sk, M.M.; Jena, R.K. Development of 3D urchin-shaped coaxial manganese dioxide@polyaniline (MnO_2@PANI) composite and self-assembled 3D pillared graphene foam for asymmetric all-solid-state flexible supercapacitor application. *ACS Appl. Mater. Interfaces* **2017**, *9*, 15350. [CrossRef]
17. Kong, S.; Cheng, K.; Ouyang, T.; Gao, Y.; Ye, K.; Wang, G.; Cao, D. Facile dip coating processed 3D MnO_2-graphene nanosheets/MWNT-Ni foam composites for electrochemical supercapacitors. *Electrochim. Acta* **2017**, *226*, 29. [CrossRef]
18. Fan, Z.; Yan, J.; Wei, T.; Zhi, L.; Ning, G.; Li, T.; Wei, F. Asymmetric supercapacitors based on graphene/MnO_2 and activated carbon nanofiber electrodes with high power and energy density. *Adv. Funct. Mater.* **2011**, *21*, 2366. [CrossRef]
19. Wang, X.; Liu, W.S.; Lu, X.; Lee, P.S. Dodecyl sulfate-induced fast faradic process in nickel cobalt oxide-reduced graphite oxide composite material and its application for asymmetric supercapacitor device. *J. Mater. Chem.* **2012**, *22*, 23114. [CrossRef]
20. Wang, H.; Liang, Y.; Mirfakhrai, T.; Chen, Z.; Casalongue, H.S.; Dai, H. Advanced asymmetrical supercapacitors based on graphene hybrid materials. *Nano Res.* **2011**, *4*, 729. [CrossRef]
21. Liu, Y.; He, D.; Wu, H.; Duan, J.; Zhang, Y. Hydrothermal self-assembly of manganese dioxide/manganese carbonate/reduced graphene oxide aerogel for asymmetric supercapacitors. *Electrochim. Acta* **2015**, *164*, 154. [CrossRef]
22. Lu, X.; Yu, M.; Wang, G.; Zhai, T.; Xie, S.; Ling, Y.; Tong, Y.; Li, Y. H-TiO_2@MnO_2//H-TiO_2@C core-shell nanowires for high performance and flexible asymmetric supercapacitors. *Adv. Mater.* **2013**, *25*, 267. [CrossRef] [PubMed]
23. Miniach, E.; Śliwak, A.; Moyseowicz, A.; Fernández-Garcia, L.; González, Z.; Granda, M.; Menendez, R.; Gryglewicz, G. MnO_2/thermally reduced graphene oxide composites for high-voltage asymmetric supercapacitors. *Electrochim. Acta* **2017**, *240*, 53. [CrossRef]
24. Tseng, L.H.; Hsiao, C.H.; Nguyen, D.D.; Hsieh, P.Y.; Lee, C.Y.; Tai, N.H. Activated carbon sandwiched manganese dioxide/graphene ternary composites for supercapacitor electrodes. *Electrochim. Acta* **2018**, *266*, 284. [CrossRef]
25. Niu, Z.; Dong, H.; Zhu, B.; Li, J.; Hng, H.H.; Zhou, W.; Chen, X.; Xie, S. Highly stretchable, integrated supercapacitors based on single-walled carbon nanotube films with continuous reticulate architecture. *Adv. Mater.* **2013**, *25*, 1058. [CrossRef]
26. Lu, X.; Yu, M.; Zhai, T.; Wang, G.; Xie, S.; Liu, T.; Liang, C.; Tong, Y.; Li, Y. High energy density asymmetric quasi-solid-state supercapacitor based on porous vanadium nitride nanowire anode. *Nano Lett.* **2013**, *13*, 2628. [CrossRef] [PubMed]
27. Xiao, X.; Ding, T.; Yuan, L.; Shen, Y.; Zhong, Q.; Zhang, X.; Cao, Y.; Hu, B.; Zhai, T.; Gong, L. WO_{3-x}/MoO_{3-x} core/shell nanowires on carbon fabric as an anode for all-solid-state asymmetric supercapacitors. *Adv. Energy Mater.* **2012**, *2*, 1328. [CrossRef]
28. Hu, C.C.; Chang, K.H.; Lin, M.C.; Wu, Y.T. Design and tailoring of the nanotubular arrayed architecture of hydrous RuO_2 for next generation supercapacitors. *Nano Lett.* **2006**, *6*, 2690. [CrossRef]
29. Meng, C.; Liu, C.; Chen, L.; Hu, C.; Fan, S. Highly flexible and all-solid-state paper like polymer supercapacitors. *Nano Lett.* **2010**, *10*, 4025. [CrossRef]
30. Liu, Y.; Miao, X.; Fang, J.; Zhang, X.; Chen, S.; Li, W.; Feng, W.; Chen, Y.; Wang, W.; Zhang, Y. Layered-MnO_2 nanosheet grown on nitrogen-doped graphene template as a composite cathode for flexible solid-state asymmetric supercapacitor. *ACS Appl. Mater. Interfaces* **2016**, *8*, 5251. [CrossRef]
31. Liu, M.; Gan, L.; Xiong, W.; Xu, Z.; Zhu, D.; Chen, L. Development of MnO_2/porous carbon microspheres with a partially graphitic structure for high performance supercapacitor electrodes. *J. Mater. Chem.* **2014**, *2*, 2555. [CrossRef]
32. Mu, B.; Zhang, W.; Xu, W.; Wang, A. Hollowed-out tubular carbon@MnO_2 hybrid composites with controlled morphology derived from kapok fibers for supercapacitor electrode materials. *Electrochim. Acta* **2015**, *178*, 709. [CrossRef]
33. He, Y.; Chen, W.; Li, X.; Zhang, Z.; Fu, J.; Zhao, C.; Xie, E. Freestanding three-dimensional graphene/MnO_2 composite networks as ultralight and flexible supercapacitor electrodes. *ACS Nano* **2013**, *7*, 174. [CrossRef] [PubMed]
34. Chen, S.; Zhu, J.; Wu, X.; Han, Q.; Wang, X. Graphene oxide-MnO_2 nanocomposites for supercapacitors. *ACS Nano* **2010**, *4*, 2822. [CrossRef]
35. Du, W.; Wang, X.; Zhan, J.; Sun, X.; Kang, L.; Jiang, F.; Zhang, X.; Shao, Q.; Dong, M.; Liu, H.; et al. Biological cell template synthesis of nitrogen-doped porous hollow carbon spheres/MnO_2 composites for high-performance asymmetric supercapacitors. *Electrochim. Acta* **2019**, *296*, 907. [CrossRef]
36. Wen, Z.; Wang, X.; Mao, S.; Bo, Z.; Kim, H.; Cui, S.; Lu, G.; Feng, X.; Chen, J. Crumpled nitrogen-doped graphene nanosheets with ultrahigh pore volume for high-performance supercapacitor. *Adv. Mater.* **2012**, *24*, 5610. [CrossRef]
37. Deng, Y.; Xie, Y.; Zou, K.; Ji, X. Review on recent advances in nitrogen-doped carbons: Preparations and applications in supercapacitors. *J. Mater. Chem. A* **2016**, *4*, 1144. [CrossRef]
38. Zhang, W.; Xu, C.; Ma, C.; Li, G.; Wang, Y.; Zhang, K.; Li, F.; Liu, C.; Cheng, H.M.; Du, Y.; et al. Nitrogen-superdoped 3D graphene networks for high-performance supercapacitors. *Adv. Mater.* **2017**, *29*, 1701677. [CrossRef]

39. Hummers, W.S.; Offeman, R.E. Preparation of graphitic oxide. *J. Am. Chem. Soc.* **1958**, *80*, 1339. [CrossRef]
40. Kovtyukhova, N.I.; Ollivier, P.J.; Martin, B.R.; Mallouk, T.E.; Chizhik, S.A.; Buzaneva, E.V.; Gorchinskiy, A.D. Layer-by-layer assembly of ultrathin composite films from micron-sized graphite oxide sheets and polycations. *Chem. Mater.* **1999**, *11*, 771. [CrossRef]
41. Wen, Y.; Ding, H.; Shan, Y. Preparation and visible light photocatalytic activity of Ag/TiO_2/graphene nanocomposite. *Nanoscale* **2011**, *3*, 4411. [CrossRef]
42. Zhang, Z.; Xiao, F.; Guo, Y.; Wang, S.; Liu, Y. One-pot self-assembled three-dimensional TiO_2-graphene hydrogel with improved adsorption capacities and photocatalytic and electrochemical activities. *ACS Appl. Mater. Interfaces* **2013**, *5*, 2227. [CrossRef]
43. Zhu, Y.Q.; Cao, C.B.; Tao, S.; Chu, W.S.; Wu, Z.Y.; Li, Y.D. Ultrathin nickel hydroxide and oxide nanosheets: Synthesis, characterizations and excellent supercapacitor performances. *Sci. Rep.* **2015**, *4*, 5787. [CrossRef]
44. Zeiger, M.; Jackel, N.; Mochalin, V.N.; Presser, V. Review: Carbon onions for electrochemical energy storage. *J. Mater. Chem. A* **2016**, *4*, 3172. [CrossRef]
45. Brousse, T.; Taberna, P.L.; Crosnier, O.; Dugas, R.; Guillemet, P.; Scudeller, Y.; Zhou, Y.; Favier, F.; Bélanger, D.; Simon, P. Long-term cycling behavior of asymmetric activated carbon/MnO_2 aqueous electrochemical supercapacitor. *J. Power Sources* **2007**, *173*, 633. [CrossRef]
46. Chee, W.K.; Lim, H.N.; Zainal, Z.; Huang, N.M.; Harrison, I.; Andou, Y. Flexible graphene-based supercapacitors: A review. *J. Phys. Chem. C* **2016**, *120*, 4153. [CrossRef]
47. Chiu, H.Y.; Cho, C.P. Mixed-phase MnO_2/N-Containing graphene composites applied as electrode active materials for flexible asymmetric solid-state supercapacitors. *Nanomaterials* **2018**, *8*, 924. [CrossRef]
48. Liu, L.; Su, L.; Lang, J.; Hu, B.; Xu, S.; Yan, X. Controllable synthesis of Mn_3O_4 nanodots@nitrogen-doped graphene and its application for high energy density supercapacitors. *J. Mater. Chem. A* **2017**, *5*, 5523. [CrossRef]
49. Li, Y.; Zhao, N.; Shi, C.; Liu, E.; He, C. Improve the supercapacity performance of MnO_2-decorated graphene by controlling the oxidization extent of graphene. *J. Phys. Chem. C* **2012**, *116*, 25226. [CrossRef]
50. Devaraj, S.; Munichandraiah, N. Effect of crystallographic structure of MnO_2 on its electrochemical capacitance properties. *J. Phys. Chem. C* **2008**, *112*, 4406. [CrossRef]
51. Li, Z.; Mi, Y.; Liu, X.; Liu, S.; Yang, S.; Wang, J. Flexible graphene/MnO_2 composite papers for supercapacitor electrodes. *J. Mater. Chem.* **2011**, *21*, 14706. [CrossRef]
52. Unnikrishnan, B.; Wu, C.W.; Chen, I.W.P.; Chang, H.T.; Lin, C.H.; Huang, C.C. Carbon dot-mediated synthesis of manganese oxide decorated graphene nanosheets for supercapacitor application. *ACS Sustain. Chem. Eng.* **2016**, *4*, 3008. [CrossRef]
53. Du, D.; Li, P.; Ouyang, J. Nitrogen-doped reduced graphene oxide prepared by simultaneous thermal reduction and nitrogen doping of graphene oxide in air and its application as an electrocatalyst. *ACS Appl. Mater. Interfaces* **2015**, *7*, 26952. [CrossRef]
54. Yang, Y.; Zeng, B.; Liu, J.; Long, Y.; Li, N.; Wen, Z.; Jiang, Y. Graphene/MnO_2 composite prepared by a simple method for high performance supercapacitor. *Mater. Res. Innov.* **2016**, *20*, 92. [CrossRef]
55. Xiao, W.; Zhou, W.; Yu, H.; Pu, Y.; Zhang, Y.; Hu, C. Template synthesis of hierarchical mesoporous δ-MnO_2 hollow microspheres as electrode material for high-performance symmetric supercapacitor. *Electrochim. Acta* **2018**, *264*, 1. [CrossRef]
56. Gao, H.; Xiao, F.; Ching, C.B.; Duan, H. High-performance asymmetric supercapacitor based on graphene hydrogel and nanostructured MnO_2. *ACS Appl. Mater. Interfaces* **2012**, *4*, 2801. [CrossRef]
57. Yang, M.; Kim, D.S.; Hong, S.B.; Sim, J.W.; Kim, J.; Kim, S.S.; Choi, B.G. MnO_2 nanowire/biomass-derived carbon from hemp stem for high-performance supercapacitors. *Langmuir* **2017**, *33*, 5140. [CrossRef]
58. Qu, Q.; Zhang, P.; Wang, B.; Chen, Y.; Tian, S.; Wu, Y.; Holze, R. Electrochemical performance of MnO_2 nanorods in neutral aqueous electrolytes as a cathode for asymmetric supercapacitors. *J. Phys. Chem. C* **2009**, *113*, 14020. [CrossRef]
59. Cheng, H.; Duon, H.M. Three dimensional manganese oxide on carbon nanotube hydrogels for asymmetric supercapacitors. *RSC Adv.* **2016**, *6*, 36954. [CrossRef]
60. Śliwak, A.; Gryglewicz, G. High-voltage asymmetric supercapacitors based on carbon and manganese oxide/oxidized carbon nanofiber composite electrodes. *Energy Technol.* **2014**, *2*, 819. [CrossRef]
61. Chou, T.C.; Doong, R.A.; Hu, C.C.; Zhang, B.; Su, D.S. Hierarchically porous carbon with manganese oxides as highly efficient electrode for asymmetric supercapacitor. *ChemSusChem* **2014**, *7*, 841. [CrossRef]
62. Li, L.; Hu, Z.A.; An, N.; Yang, Y.Y.; Li, Z.M.; Wu, H.Y. Facile synthesis of MnO_2/CNTs composite for supercapacitor electrodes with long cycle stability. *J. Phys. Chem. C* **2014**, *118*, 22865. [CrossRef]
63. Cheng, Y.; Lu, S.; Zhang, H.; Varanasi, C.V.; Liu, J. Synergistic effects from graphene and carbon nanotubes enable flexible and robust electrodes for high-performance supercapacitors. *Nano Lett.* **2012**, *12*, 4206. [CrossRef]
64. Khomenko, V.; Pinero, E.R.; Beguin, F. Optimisation of an asymmetric manganese oxide/activated carbon capacitor working at 2V in aqueous medium. *J. Power Sources* **2006**, *153*, 183. [CrossRef]
65. Zhang, M.; Chen, Y.; Yang, D.; Li, J. High performance MnO_2 supercapacitor material prepared by modified electrodeposition method with different electrodeposition voltages. *J. Energy Storage* **2020**, *29*, 101363. [CrossRef]

66. Wu, J.; Raza, W.; Wang, P.; Hussain, A.; Ding, Y.; Yu, J.; Wu, Y.; Zhao, J. Zn-doped MnO_2 ultrathin nanosheets with rich defects for high performance aqueous supercapacitors. *Electrochim. Acta* **2022**, *418*, 140339. [CrossRef]
67. Li, M.; Zhu, K.; Zhao, H.; Meng, Z.; Wang, C.; Chu, P.K. Construction of α-MnO_2 on carbon fibers modified with carbon nanotubes for ultrafast flexible supercapacitors in ionic liquid electrolytes with wide voltage windows. *Nanomaterials* **2022**, *12*, 2020. [CrossRef]

Disclaimer/Publisher's Note: The statements, opinions and data contained in all publications are solely those of the individual author(s) and contributor(s) and not of MDPI and/or the editor(s). MDPI and/or the editor(s) disclaim responsibility for any injury to people or property resulting from any ideas, methods, instructions or products referred to in the content.

Article

Phosphorus/Sulfur-Enriched Reduced Graphene Oxide Papers Obtained from Recycled Graphite: Solid-State NMR Characterization and Electrochemical Performance for Energy Storage

Mariana A. Vieira [1,*], Tainara L. G. Costa [1], Gustavo R. Gonçalves [1], Daniel F. Cipriano [1], Miguel A. Schettino, Jr. [1], Elen L. da Silva [2], Andrés Cuña [2] and Jair C. C. Freitas [1,*]

[1] Laboratory of Carbon and Ceramic Materials, Department of Physics, Federal University of Espírito Santo, Vitória 29075-910, Brazil; guerratainara@gmail.com (T.L.G.C.); gustavo.rgoncalves@hotmail.com (G.R.G.); danielcipriano.fisica@gmail.com (D.F.C.); miguel.ufes@gmail.com (M.A.S.J.)

[2] Área Fisicoquímica, DETEMA, Facultad de Química, Universidad de la República, CC 1157, Montevideo 11800, Uruguay; elenlealdasilva@gmail.com (E.L.d.S.); acunasuarez@gmail.com (A.C.)

* Correspondence: m.arpinivieira@gmail.com (M.A.V.); jairccfreitas@yahoo.com.br (J.C.C.F.)

Abstract: The reduction of graphene oxide (GO) by means of thermal and/or chemical treatments leads to the production of reduced graphene oxide (rGO)—a material with improved electrical conductivity and considered a viable and low-cost alternative to pure graphene in several applications, including the production of supercapacitor electrodes. In the present work, GO was prepared by the oxidation of graphite recycled from spent Li-ion batteries using mixtures of sulfuric and phosphoric acids (with different H_2SO_4/H_3PO_4 ratios), leading to the production of materials with significant S and P contents. These materials were then thermally reduced, resulting in rGO papers that were investigated by solid-state ^{13}C and ^{31}P nuclear magnetic resonance, along with other methods. The electrochemical properties of the produced rGO papers were evaluated, including the recording of cyclic voltammetry and galvanostatic charge–discharge curves, besides electrochemical impedance spectroscopy analyses. The samples obtained by thermal reduction at 150 °C exhibited good rate capability at high current density and high capacitance retention after a large number of charge–discharge cycles. The results evidenced a strong relationship between the electrochemical properties of the produced materials and their chemical and structural features, especially for the samples containing both S and P elements. The methods described in this work represent, then, a facile and low-cost alternative for the production of rGO papers using graphite recycled from spent batteries, with promising applications as supercapacitor electrodes.

Keywords: graphene oxide; reduced graphene oxide; solid-state NMR; electrochemical properties

1. Introduction

Since 2004, when Novoselov et al. [1] reported a method for obtaining graphene layers by mechanical exfoliation of graphite, the search for alternative routes to produce graphene-based materials has grown remarkably. The oxidation of graphite to produce graphite oxide, first reported almost 160 years ago [2], has attracted considerable interest as a potential method for the low-cost and mass production of graphene-based materials [3,4]. Graphite oxide has a structure somewhat similar to graphite, but containing oxygenated functions such as hydroxyls, epoxides, carboxylic acids, ketones, and esters linked to its basal planes, which allows the intercalation of water molecules between them, causing a significant increase in the interlayer spacing. As a result, these layers can be easily exfoliated in water, giving rise to a material made up of single layers (or groups of a few stacked layers) called graphene oxide (GO) [5,6]. GO can then be reduced to form sheets similar to graphene by removing oxygenated functions and recovering most of the

sp^2 bonds in the basal planes, leading to the formation of a material known as reduced graphene oxide (rGO) which exhibits good electrical conductivity and, despite the presence of structural defects and chemical heterogeneity, can be a viable, low-cost alternative to pristine graphene in several applications [6]. The reduction of GO can be accomplished through various approaches, such as chemical reduction with hydrazine [7,8], hydrogen plasma [9], pulsed laser irradiation [10], gamma ray irradiation [11], and thermal treatments [12,13]. Different reduction methods lead to the production of rGO samples with different degrees of reduction which may therefore contain some remaining oxygen functionalities, directly affecting the properties and performance of the material depending on the area in which it is to be applied. Some methods, such as reduction with hydrazine and thermal treatments, can also be used together to obtain samples of rGO with a minimum content of oxygenated functions and with maximum restoration of the sp^2 layers (and, consequently, of the electrical conductivity) [7,14]. In cases where there is no need for a high degree of structural order or excellent electrical conductivity, thermal annealing at low temperatures (typically 90 to 300 °C) represents a simple and low-cost approach for GO reduction, allowing the obtention of rGO samples with good properties for applications in gas sensing or electrochemical devices [12,15].

Graphene-based materials obtained by liquid-phase exfoliation, such as GO and rGO, offer the possibility of changing the chemical and surface properties through the insertion of heteroatoms and/or functional groups, which can also contribute to many sensing and electrochemical device applications [16]. For instance, the methods used in most previous studies for the preparation of P-functionalized GO or rGO-based materials involve reactions conducted at several steps and using varied phosphorus sources such as phytic acid [17], diammonium hydrogen phosphate [18], red phosphorus [18], phosphorus trichloride [19], triphenylphosphine [20] and phosphoric acid [21]. A substantial simplification of the process of the incorporation of phosphorus into the structure of GO and rGO can be achieved by using H_3PO_4 directly in the reaction of graphite oxidation. In fact, the so-called Marcano–Tour method [22] involves a modification of the original graphite oxidation method proposed by Hummers and Offeman [23], replacing H_2SO_4 with a 9:1 (v/v ratio) H_2SO_4/H_3PO_4 mixture. Previous reports have shown that GO samples prepared by the Marcano–Tour method exhibit a high degree of oxidation, resulting in general in O/C atomic ratios larger than those reached by using the Hummers method [22–25].

In this work, an investigation was conducted aiming to better understand the nature of the S- and P-containing groups present in GO samples prepared from the oxidation of graphite in acid mixtures with varied H_2SO_4/H_3PO_4 proportions and in rGO samples produced by subsequent thermal treatments; solid-state ^{13}C and ^{31}P nuclear magnetic resonance (NMR), Fourier-transform infrared spectroscopy (FTIR), and X-ray diffraction (XRD) were the main tools used to probe the chemical and structural features of the GO and rGO samples. Finally, the electrochemical behavior of the rGO samples was investigated, showing an intimate connection between the electrochemical performance of the produced materials and their chemical and structural features which were defined by the conditions used in the GO synthesis.

2. Experimental Methods

2.1. Samples Preparation

Materials

The precursor used for the production of GO was recycled graphite derived from spent lithium-ion batteries, as reported elsewhere [25]. The graphite present in the anode of the spent batteries was separated from the copper foil by dissolution in concentrated HNO_3, followed by filtration, washing with abundant distilled water and drying at 100 °C for 3 h. The other reactants were $KMnO_4$, H_3PO_4 (85%) and H_2SO_4 (95%) from Vetec (Duque de Caxias, Brazil), $NaNO_3$ and H_2O_2 (30 vol.%) from Cromoline (Diadema, Brazil).

Preparation of phosphorus and sulfur-containing graphene oxide (P/S-GO) and reduced graphene oxide papers (P/S-rGO):

The P/S-GO samples were prepared using the Marcano–Tour method [22], varying the ratio of H_2SO_4/H_3PO_4 used. For the conventional Marcano–Tour method, 1 g of graphite, 6 g of $KMnO_4$, and 135 mL of a 9:1 (v/v ratio) H_2SO_4/H_3PO_4 solution were mixed. The mixture was kept in a water bath at 50 °C and stirred for 12 h. After this time, the mixture was poured into 400 mL of crushed ice, and 3 mL of a 30% v/v H_2O_2 were added, leading to a change in the color of the mixture from dark brown to intense yellow. The mixture was allowed to stand for 24 h to decant the product. To remove the excess acid, the precipitate was washed with distilled water and centrifuged until the supernatant reached a pH between 5 and 6. In order to obtain the GO samples with varying P and S contents, reactions were also performed using 1:1 and 1:2 (v/v ratios) H_2SO_4/H_3PO_4 solutions, following the same protocol as in the conventional method. The GO samples containing P and S (generally named P/S-GO samples) were then dried at 80 °C and were called GOTx, in which x indicates the H_2SO_4/H_3PO_4 ratio: x = 91, 11, or 12 for 9:1, 1:1, and 1:2 v/v ratios, respectively.

After preliminary studies (as stated in the Supplementary Information), samples of thermally reduced graphene oxide paper were produced by the thermal treatment of the GO samples that were synthesized in the previous step. The P/S-GO suspensions obtained at the end of each reaction and washing process were divided into 2 mL aliquots, which were directly submitted to two different treatments in a lab oven, under ambient atmosphere, using two treatment temperatures (100 and 150 °C) for 24 h. At the end of the process, rGO papers were obtained, and these samples were named GOTx_n, where x indicates the H_2SO_4/H_3PO_4 volume ratio used in the synthesis (as described above) and n indicates the heat treatment temperature (100 or 150 °C).

2.2. Characterization

2.2.1. X-ray Diffraction (XRD)

The X-ray diffractograms of the precursor graphite, GO and rGO papers were recorded on a Shimadzu (Columbia, MD, USA) XRD-6000 diffractometer. The samples were ground and the experiments were conducted at room temperature using Cu-Kα radiation (λ = 1.5418 Å) and with the angle of diffraction (2θ) varying from 5 to 50° in steps of 0.04°.

2.2.2. Solid-State ^{13}C and ^{31}P NMR

Solid-state NMR experiments were performed at room temperature on a Varian/Agilent (Palo Alto, CA, USA) commercial spectrometer operating in a 9.4 T magnetic field, which corresponds to a frequency of 100.5 MHz for ^{13}C and 161.8 MHz for ^{31}P. Magic angle spinning (MAS) at 14 kHz was employed in all experiments, with the powdered samples packed into a 4 mm diameter zirconia rotor. The ^{13}C NMR experiments were conducted using a pulse sequence specially designed to avoid background signals from contributions external to the probe (which are common in ^{13}C NMR spectra obtained by direct excitation of the ^{13}C nuclei), with a $\pi/2$ pulse (4.3 µs) immediately followed by a pair of π pulses (8.6 µs) and subsequent detection of the free induction decay (FID) [26]. In all the ^{13}C NMR experiments, the recycle delay was 15 s and the spectral width was 250 kHz. All spectra were obtained by the Fourier transform of the FID after the accumulation of ca. 4000 transients. The ^{13}C chemical shifts were referenced to tetramethylsilane (TMS) using hexamethylbenzene as a secondary reference. The ^{31}P NMR experiments were conducted using single-pulse excitation (SPE) with a $\pi/2$ pulse with duration of 4.0 µs and a recycle delay of 60 s. The spectra were obtained by the Fourier transform of the FID after the accumulation of ca. 500 transients, and the chemical shifts were referenced to an 85 wt.% aqueous H_3PO_4 solution using $NH_4H_2PO_4$ (δ = 0.9 ppm) as a secondary reference.

2.2.3. Fourier-Transform Infrared (FTIR) Spectroscopy

The infrared absorption spectroscopy analyses were conducted directly using pulverized GO and rGO samples on a Spectrum 100 (Perkin Elmer, Waltham, MA, USA) spectrometer in transmittance mode, equipped with an attenuated total reflection (ATR) attachment with a diamond crystal. The spectra were acquired using 64 scans and a 2 cm^{-1} resolution. The pulverized samples were directly deposited on the top of the ATR crystal, without the need of any additional sample preparation steps.

2.2.4. X-ray Fluorescence (XRF)

The XRF analyses to determine the P and S contents of the GO and rGO samples were performed on an EDX700 (Shimadzu, (Columbia, MD, USA) spectrometer equipped with a Rhodium (Rh) X-ray tube. This instrument operated at 15 kV, with collimation of 10 mm of the incident beam for 4 min per sample. The Sodium–Scandium (Na-Sc) energy channel was used and the analyzed energy interval was from 0.00 to 20.48 keV, with a scanning step of 0.01 keV. To minimize matrix effects and enhance the accuracy of the quantification, a carbonaceous sample with negligible P and S contents was selected as the matrix for the construction of the calibration curves. The calibration curves for phosphorus (P) and sulfur (S) were constructed by adding known concentrations of $NH_4H_2PO_4$ (corresponding to P contents of 0.1, 0.5, 1.0, 2.0, 3.0, and 5.0 wt.%) and Na_2SO_4 (corresponding to S contents of 1.0, 5.0, 10.0, 15.0 and 20.0 wt.%), respectively, to the carbon powder (with a mass of ca. 1 g). The intensities of the XRF lines corresponding to P and S were recorded for each calibration sample. The calibration curves were obtained through linear regression; high values of the coefficients of determination (~0.99) were obtained in the full range of P and S contents used to construct the calibration curves, allowing the determination of the P contents between 0.1 and 5.0 wt.% and of S contents between 1.0 and 20.0 wt.%. To ensure representative sampling and homogeneity, the samples were pulverized and sieved prior to the analyses. GO and rGO samples with approximately 300 mg were consistently used for all XRF measurements. The samples were submitted to the analyses without any additional treatment or manipulation, and 3–5 independent analyses were performed for each sample. The standard deviations obtained from these measurements were assumed to represent the uncertainties of the average values.

2.2.5. Electrochemical Characterization

For the electrochemical analysis, a symmetric two-electrode Swagelok® (Buenos Aires, Argentina)-type cell with two tantalum rods as current collectors was used. The cell electrodes were prepared by cutting an rGO paper (GO treated at 150 °C for 24 h) sample as a rectangular electrode with a cross-section area of 0.32 cm^2 and weighing ca. 1.0 mg (Figure 1). A sulfuric acid aqueous solution (2 mol L^{-1}) was used as the electrolyte, and a glassy microfiber paper (Whatman 934 AH—Maidstone, UK) was chosen as the electrode separator. Electrochemical characterization included the recording of galvanostatic charge–discharge curves (in a current density range of 1–8 A g^{-1}), cyclic voltammograms (at 5, 10, 20 and 50 mV s^{-1}) obtained in a voltage range of 0–1 V, and electrochemical impedance spectroscopy (EIS) measurement in the frequency range of 5.0×10^{-4}–2.5×10^5 Hz. The gravimetric specific capacitance (C_s) was determined from galvanostatic charge–discharge measurements at each current according to equation

$$C_s = 2 \cdot I \cdot t_d / E_2 \cdot m_e. \tag{1}$$

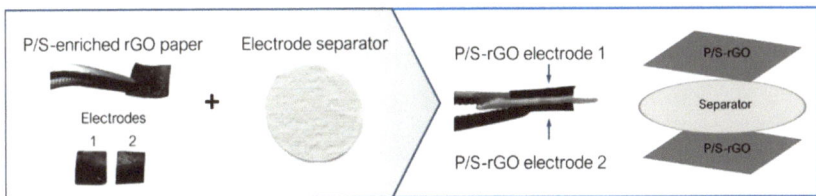

Figure 1. Schematic diagram showing the assembly of the laboratory supercapacitor for electrochemical analyses.

In this expression, I is the applied current, t_d is the discharge time, E_2 is the voltage range during the discharge and m_e is the mass of one electrode. All measurements were carried out at room temperature using a potentiostat/galvanostat/FRA Vertex (Ivium Technologies, De Lismortel, Netherlands).

3. Results and Discussion

A preliminary study (described in the Supplementary Information section) showed that the thermal treatment of a GO sample—subjecting it to the temperature of 150 °C for 24 h in a lab oven—proved to be efficient for obtaining an rGO paper with few remaining oxygen functionalities (mostly carboxylic acids, as verified by solid-state ^{13}C NMR spectroscopy). Based on that, the GO suspension obtained at the end of each reaction of the employed Marcano–Tour method [22] (using three H_2SO_4/H_3PO_4 mixtures at 9:1, 1:1 and 1:2 v/v ratios) was submitted to two different thermal treatments in the same lab oven at ambient atmosphere and at temperatures of 100 and 150 °C for 24 h, which provided several P/S-GO and P/S-rGO samples in the form of papers [27], with thicknesses ranging from ca. 10 to 20 μm as estimated by scanning electron microscopy (SEM); details are provided in the Supplementary Information section (Figure S7).

The sulfur and phosphorus contents of the P/S-GO and P/S-rGO papers were evaluated by XRF analysis; the results (given in Table 1) confirmed that by changing the H_2SO_4/H_3PO_4 ratio, one can vary the S and P contents in the synthesized GO samples, with S contents varying from 1.1 to 3.6 wt.% (GOT11 and GOT91, respectively), and P contents varying from 0.26 to 2.8 wt.% (for GOT91 and GOT12, respectively). It is worth noting that these S contents are somewhat inferior to the value corresponding to a similar GO sample prepared by a modified Hummers method (with no addition of H_3PO_4), as described in the Supplementary Information section (Table S2). The thermal treatment at 150 °C did not cause the complete removal of whatever S and P functionalities present in the synthesized GO samples; thus, the method presented here can be considered effective for the production S- and P-containing rGO samples, with S contents varying from 2.1 to 6.8 wt.% (GOT12_150 and GOT91_150, respectively), and P contents varying from 0.27 to 5.4 wt.% (for GOT91_150 and GOT12_150, respectively). Given that most synthetic approaches involve the use of concentrated sulfuric acid, such as the methods of Hummers [23] and Marcano–Tour [22], the presence of sulfur in GO can be expected. Previous elemental analysis results showed that the concentration of S can reach 6 wt.% depending on the chosen synthetic method, and this concentration occurs predominantly due to the presence of organosulfate-type groups [28]. Due to their low hydrolysis rate, these sulfonated groups tend to remain in the GO structure even after a thorough washing and, although often disregarded, these S concentrations can directly affect the GO properties [6,29]. Similarly, phosphorus can also be present in low concentrations in GO (<0.1 wt.%), especially in samples prepared by methods that follow the Marcano–Tour route [22], in which the oxidation reaction takes place in the presence of concentrated phosphoric acid. The hydrolysis of phosphorus functionalities also occurs slowly, and therefore phosphate derivatives are expected to occur in the structure of the GO samples synthesized in the presence of H_3PO_4 [22,25,30].

Table 1. XRF results for P and S contents of the P/S-GO samples synthesized by the Marcano–Tour method and the corresponding reduced samples.

Sample	P Content (wt.%)	S Content (wt.%)
GOT91	0.26 ± 0.01	3.6 ± 0.2
GOT91_150	0.27 ± 0.01	6.8 ± 0.3
GOT11	1.4 ± 0.1	1.1 ± 0.1
GOT11_150	2.3 ± 0.1	2.7 ± 0.1
GOT12	2.8 ± 0.1	1.4 ± 0.1
GOT12_150	5.4 ± 0.3	2.1 ± 0.1

The XRD results (Figure 2) of the synthesized GOs clearly demonstrated the efficiency of the reactions following both conventional and modified Marcano–Tour methods [22], indicating the formation of the structure expected for GOs by the appearance of an intense diffraction peak at 2θ ≅ 8–10° for all samples [31,32]. It is possible to note that treating samples at 100 °C did not cause significant changes in any of the corresponding XRD patterns. A small displacement of the main diffraction peak from 2θ ≅ 8–10° to higher angles was observed in the case of the samples thermally treated at 100 °C when compared to the case of the XRD patterns of the GO parent materials. In addition, a broadening of this peak and an attenuation of its intensity were observed, accompanied by the appearance of a broad peak at 2θ ≅ 18–22°; this new diffraction peak indicated the formation of a new phase with a drastic decrease in its interlayer spacing to ca. 4 Å, which is a value much smaller than the corresponding interlayer spacing of 9 Å found for the GO samples (GOT91, GOT11 and GOT12). The coexistence of these two peaks shows that the reduction process is not abrupt (i.e., there are regions with intermediate interplanar spacing formed during the thermal reduction) and does not happen at the same time throughout the whole sample, in agreement with previous reports [12,15,33].

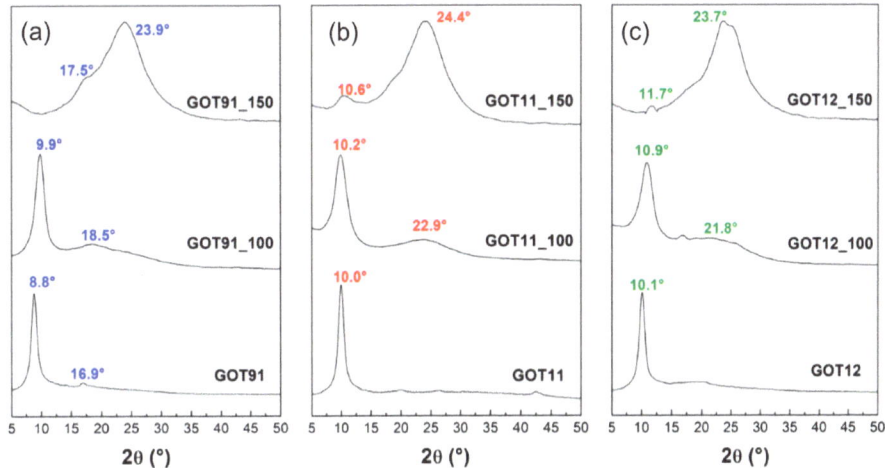

Figure 2. X-ray diffractograms of the (**a**) GOT91, (**b**) GOT11 and (**c**) GOT12 samples as prepared and after the thermal reduction treatments performed at 100 and 150 °C.

On the other hand, for samples treated at 150 °C, it is possible to notice in the X-ray diffractograms of all P/S-rGO samples that the peak near 2θ ≅ 8–10° (for the original GO samples) is almost completely eliminated. In addition, the presence of a broad and intense diffraction peak at 2θ ≅ 24° indicates the formation of a turbostratic-like structure; in this arrangement, the basal planes are stacked in an approximately parallel way, with some degree of curvature and defects, but randomly oriented in relation to each other, with an interlayer spacing (~3.7 Å) somewhat larger than the value corresponding to graphite (3.35 Å) [5,33]. Nevertheless, the XRD patterns of all samples treated at 150 °C still exhibit broad and low-intensity peaks at lower angles, indicating the presence of a phase with nanostructured arrangement of defective graphene planes and possibly still containing residues of oxygenated functional groups [12,34]. Therefore, the samples thermally reduced for 24 h at 150 °C have characteristics compatible with those of the material known as reduced graphene oxide [12,34].

The solid-state ^{13}C NMR results (Figure 3) confirm the successful synthesis of GO in all cases, as evidenced by the presence of the typical resonances expected for the GO structure, such as those at 180–190 ppm (carbonyl groups in ketones), 160–170 ppm (carbonyl groups in esters or carboxylic acids), 60 ppm (epoxide carbons), 70 pm (carbons bonded to hydroxyl groups), 100 ppm (lactol carbons) and 120–140 ppm (aromatic carbons) [25,35,36]. The degree of oxidation of the GOs can be evaluated by ^{13}C NMR by integrating the spectral regions associated with C-OH/C-O-C groups and aromatic sp^2 carbon atoms, indicated as regions A and B, respectively, in Figure 3; the higher the A/B ratio, the higher the degree of oxidation [37,38]. Accordingly, the results shown in Figure 3 demonstrate that the GOT11 sample has an oxidation degree comparable to that of GOT91 (A/B ratios of 2.5 and 2.7, respectively) even after changing the H_2SO_4/H_3PO_4 ratio used in the conventional Marcano–Tour method [22] from 9:1 to 1:1 (v/v ratio). However, when the amount of H_3PO_4 is increased even more by changing the acid ratio to 1:2 v/v, the GOT12 sample shows a reduced oxidation degree (with an A/B ratio of 1.6), despite the remaining significant presence of oxygenated functionalities and the highest P content (2.8 wt.%) among all as-prepared GO samples analyzed here (see Table 1). The decrease in the size of the basal planes caused by the addition of oxygenated functionalities during the oxidation reaction causes changes in the chemical shift of the peak associated with sp^2 carbons. This occurs due to the reduction in the magnitude of the effects related to the electrical conductivity and diamagnetic susceptibility of the graphene layers, consequently leading to a reduced shielding of the ^{13}C nuclei and, therefore, a higher chemical shift of the peak associated with aromatic sp^2 carbon atoms when compared to the chemical shift of this same peak in non-oxidized graphite samples [37,38]. In light of this information, it is possible to note in the NMR spectra shown in the ^{13}C NMR spectra demonstrated in Figure 3 that the sp^2 carbon peak of the synthesized GOs occurs at slightly different chemical shifts (134 ppm for GOT11 and 129 ppm for GOT12 sample) and has different linewidths depending on the degree of oxidation of each sample. This effect follows the trend predicted by the A/B ratio: GOs with the highest A/B ratios (which indicate a higher degree of oxidation) generally present the highest chemical shifts for the peak associated with aromatic sp^2 carbon atoms.

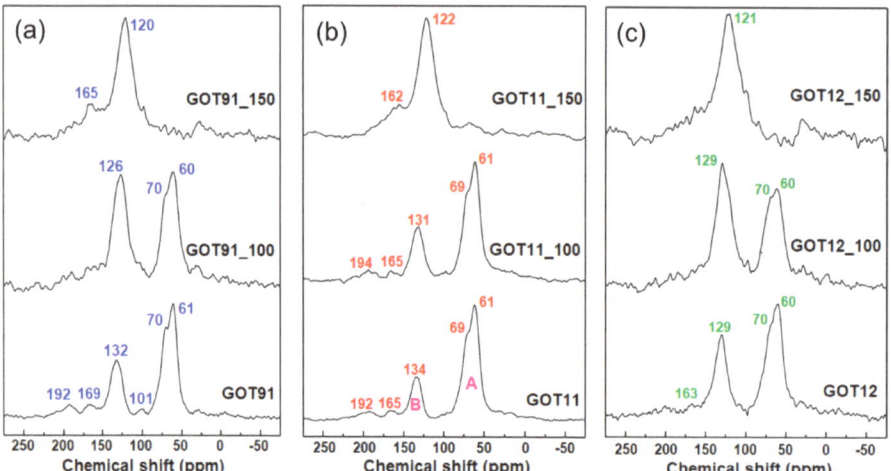

Figure 3. Solid-state ^{13}C NMR spectra of the (**a**) GOT91, (**b**) GOT11 and (**c**) GOT12 samples as prepared and after the thermal reduction treatments performed at 100 and 150 °C.

Analyzing the ^{13}C NMR spectra of the reduced samples, it is possible to note that the peak associated with aromatic sp^2 carbon atoms (chemical shift $\cong$ 130 ppm) is dominant in all samples after the thermal treatment at 150 °C. However, the spectra obtained for samples treated at 100 °C still show significant contribution of the peaks at 60 and 70 ppm, associated with C-O-C and C-OH carbons, respectively [35,38]. Furthermore, the ^{13}C NMR spectra of all thermally treated samples exhibit peaks near 165 and 190 ppm, associated with carbonyl groups; the permanence of groups such as carboxylic acids in these samples is plausible, since these groups decompose typically above 300 °C [3,15,38]. Another notable aspect in these spectra is the difference in chemical shifts of the peaks relative to aromatic sp^2 carbon atoms. The displacement of these peaks to lower frequencies in the ^{13}C NMR spectra recorded for the thermally treated samples (more notably in the case of the samples treated at 150 °C) is indicative of the reestablishment of the effects related to the electrical conductivity and diamagnetic susceptibility of the graphene layers in the rGO samples [37,38].

The use of solid-state ^{31}P NMR spectroscopy is essential for the characterization of the GO samples synthesized by the conventional and modified Marcano–Tour methods [22] as it allows a detailed understanding of the possible interaction between the GO matrix and phosphorus-containing groups present in the material [25,30]. As previously discussed, GO has a large number of functional groups containing O and H, which are capable of interactions with phosphate groups (with P in a tetrahedral coordination). In solid-state ^{31}P NMR, tetrahedral orthophosphates, in which the central phosphorus atom is pentavalent and bonded to four oxygen atoms, are usually referred to as Q units. The Q units can bridge through the oxygen atoms, forming chains or networks, wherein the amount of bridging oxygen ranges from 0 to 3 depending on the degree of polymerization. In this way, four different Q^n units have been described, where n corresponds to the number of bridging oxygens, that is, of other phosphate groups to which the main unit is linked. The ^{31}P chemical shift associated with such groups is sensitive to the chemical environment of the oxygen atoms directly connected to the phosphorus atom [39–41].

Figure 4 shows the ^{31}P NMR spectra of the GO samples synthesized by conventional and modified Marcano–Tour methods [22]. It is possible to readily observe the differences between the spectra of the three samples, starting with the weak signal-to-noise ratio in the GOT91 spectrum. As all spectra were recorded under similar conditions (including nearly the same number of scans), such difference in signal-to-noise ratio indicates that the P content in the GOT91 sample is considerably lower than that in the GOT11 and GOT12 samples, as confirmed by the XRF results (Table 1). Nevertheless, the detected peak near −9 ppm present in the GOT91 spectrum can be associated with Q^1 units, or pyrophosphoric acid [25,39].

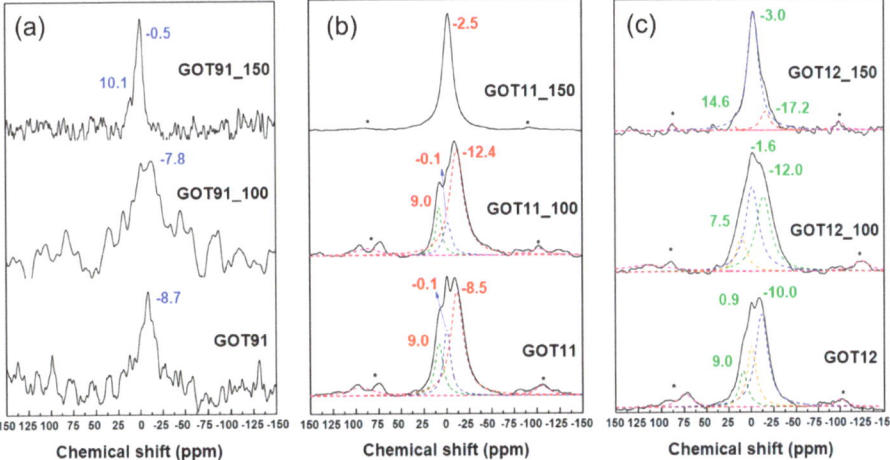

Figure 4. Solid-state ^{31}P NMR spectra of the (**a**) GOT91, (**b**) GOT11 and (**c**) GOT12 samples as prepared and after the thermal reduction treatments performed at 100 and 150 °C. Spinning sidebands are indicated by stars (*).

The ^{31}P NMR spectra of the GOT11 and GOT12 samples show similar profiles, with two peaks centered around 0 and −10 ppm, associated with Q^0 and Q^1 units (or orthophosphoric and pyrophosphoric acids, respectively). The signal intensity in these spectra is higher when compared to that of the spectrum of the GOT91 sample, indicating the increase in the phosphorus content due to the increase in the amount of phosphoric acid used in the synthesis. Thus, through solid-state ^{31}P NMR spectroscopy, it was possible to verify the presence of phosphorus in the samples synthesized by both conventional and modified Marcano–Tour methods [22] using different H_2SO_4/H_3PO_4 volume ratios in a one-step synthesis.

As mentioned above, the peak near −9 ppm present in the ^{31}P NMR spectrum of the GOT91 sample is commonly associated with Q^1 units, or pyrophosphoric acid. After thermally treating this sample at 100 °C (GOT91_100), it is possible to detect the presence of a peak of the same nature, probably due to the fact that, at this treatment temperature, the final structure of the product does not differ much from the structure of the conventional sample dried at 80 °C (GOT91), as also verified by ^{13}C NMR and XRD. On the other hand, the ^{31}P NMR spectrum of the GOT91_150 sample presents a different behavior, with a peak at −0.5 ppm associated with the presence of orthophosphoric acid (Q^0 units), suggesting a rearrangement of the phosphate groups into isolated H_3PO_4 units, and another peak around 10 ppm, which may be associated with the formation of P-O-C and/or P-C bonds [25,39].

The ^{31}P NMR spectra of the GOT11 and GOT12 samples treated at 100 °C exhibit peaks in the region from 7 to 10 ppm, associated with the formation of P-O-C and/or P-C bonds; these bonds are known to occur in reactions in which the amount of phosphorus used is much higher than that of carbon, as used in this work [30,40]. Peaks in the region from −10 to −20 ppm are also detected, which are attributed to the condensation of the phosphate groups leading to the formation of polyphosphoric acids (Q^2 and Q^3 units, corresponding to the final and average groups in phosphate linear chains, respectively) [41].

It is worth observing that the most significant changes of the phosphate species only occur after the treatment at 150 °C. The first notable aspect of the ^{31}P NMR spectra of the thermally reduced samples is the difference in the signal-to-noise ratio of the spectrum of the GOT91_150 sample compared to the GOT11_150 and GOT12_150 samples, which is due to the much larger P content of these two latter samples, as indicated by the XRF results (Table 1). Therefore, it is possible to note the similarity between the profiles of the ^{31}P NMR spectra of these samples (GOT11_150 and GOT12_150), exhibiting a main peak around −3.0 ppm, associated with orthophosphoric acid. It is interesting to note the tendency of a slight shift of the peaks closer to 0 ppm to lower chemical shifts with increasing thermal treatment temperature, which can be attributed once more to the increase in the effects associated with the electrical conductivity and the diamagnetic susceptibility of the graphene layers in these samples, as already reported in studies involving chemically activated carbons [41].

Figure 5 shows the FTIR spectra of the prepared GOs and rGOs. The GO samples synthesized with different proportions of H_2SO_4/H_3PO_4 (GOT91, GOT11 and GOT12) exhibit all typical GO vibration bands around 1050, 1618 and 1720 cm^{-1} commonly related to the stretching vibrations of C-O, C=C and C=O (of carboxylic acids) bonds, respectively. A broad band can also be observed at ~3200 cm^{-1}, associated with the stretching vibrations of -OH groups. According to previous studies on sulfur species in the chemical structure of GO [28,42,43], bands near 1215 and 1416 cm^{-1}, such as the ones observed in the spectra of the samples of the GOT91 group, can be related to S=O stretching vibrations of sulfates. The same set of bands can be observed in the spectra of the samples of the GOT11 group (at 1225 and 1390 cm^{-1}) but, knowing the considerable phosphate content of these samples (as verified by XRF and ^{31}P NMR), these bands can be related to P=O stretching vibration as well [44,45].

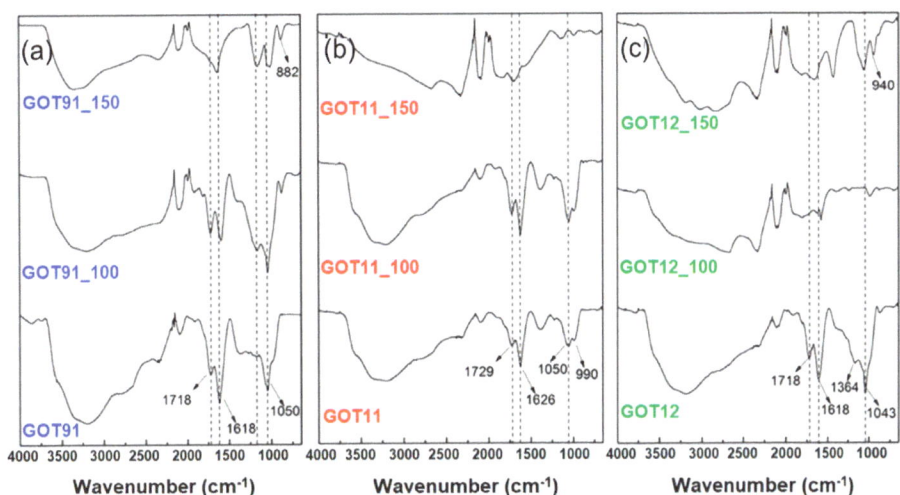

Figure 5. FTIR spectra of the (**a**) GOT91, (**b**) GOT11 and (**c**) GOT12 samples as prepared and after the thermal reduction treatments performed at 100 and 150 °C.

Phosphate groups are known to assemble into a polymeric arrangement and this network is dominated by linkages between PO_4 tetrahedra. As discussed previously, ^{31}P NMR results suggest that most phosphates are present in GO samples in the form of Q^0 and Q^1 units (or orthophosphoric and pyrophosphoric acids, respectively). After the thermal treatment of the GO samples at 150 °C, the ^{31}P NMR results suggested that there was a rearrangement of the phosphate groups into H_3PO_4 clusters (Q^0 units), probably interacting with the water molecules present in the GO structure [25]. In FTIR spectra, vibration bands of Q^0 and Q^1 units often appear around 890–990 cm^{-1} and 1080–1100 cm^{-1}, respectively [19,20,46]. As shown in Figure 5, it is possible to observe bands in these ranges in the FTIR spectra of GO and rGO samples (with different sulfur and phosphorus contents), mainly at 990 cm^{-1}, which is associated with H_3PO_4 molecules (Q^0 units). The band associated with Q^1 units is likely to be overlapped by the C-O vibration band at 1050 cm^{-1} of GO. However, after treatment at 150 °C, a band at this range can still be observed in the spectra of the rGO samples, suggesting the presence of ortho- and pyrophosphoric acids, in good agreement with the ^{31}P NMR results. In addition, after thermal treatment at 150 °C, broad bands around 2300 and 3000 cm^{-1} are observed, associated with vibrations of the -OH group in P-OH bonds. The stretching vibration of these P-OH bonds was previously associated with bands at 2430 and 2870 cm^{-1} and bending modes at 766 and 1256 cm^{-1} [47]. Water stretching vibrations were found at 3050 and 3350 cm^{-1} [21]. These bands were similarly observed in the spectra recorded for the samples investigated here. It is also interesting to note that even after GO thermal reduction the band related to water stretching vibrations can still be observed, which suggests the presence of adsorbed water molecules closely interacting with orthophosphoric acid.

Finally, selected P/S-rGO paper samples (prepared at 150 °C) had their electrochemical behavior evaluated. Electrodes containing the selected samples were built as shown in Figure 1. Aiming to evaluate the influence of phosphorus-containing groups on the electrochemical behavior of the studied samples, the electrochemical analyses were also conducted using an rGO sample (treated at 150 °C) prepared from a GO sample that had been synthesized by a modified Hummers method. This latter sample, labeled as GOH_150, was prepared in the same way as the other rGO samples, but without the addition of phosphoric acid during the synthesis (see Supplementary Information).

Figure 6 shows the cyclic voltammograms of the P/S-rGO and GOH_150 samples obtained at 10 mV s^{-1} (Figure 6a) and 50 mV s^{-1} (Figure 6b). All samples exhibited a quasi-rectangular shape voltammogram, even at a high scan rate, demonstrating a good capacitive behavior of the samples. In addition, a very good electrochemical stability of the electrolyte at high potentials was observed, with no current increment related to oxygen evolution being observed even up to 1 V. The P/S-rGO samples showed higher current densities than the GOH_150 sample, which points to a higher C_s for the phosphorous-containing samples. In the voltammograms obtained at low scan rate (Figure 6a), it can be seen that the P/S-rGO hybrid samples exhibited a broad peak between 0.1 and 0.4 V, suggesting a pseudocapacitive contribution reaction that could be linked to the oxygenated and/or phosphorus-rich surface functional groups present in these samples [17].

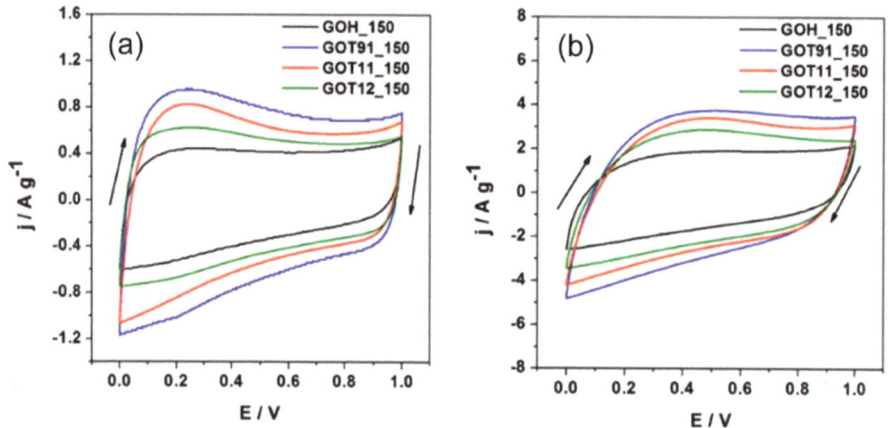

Figure 6. Cyclic voltammograms recorded at (**a**) 10 mV s^{-1} and (**b**) 50 mV s^{-1}.

The C_s values were determined from the galvanostatic charge–discharge curves according to Equation (1); their dependence on the current density is shown in Figure 7a. For all current density ranges, the P/S-rGO samples presented higher C_s values than the GOH_150 sample. The best behavior was obtained for the GOT91_150 sample, which reached values up to 155 and 110 F g^{-1} at 1 and 8 A g^{-1}, respectively. These values are similar to those reported in the literature for other carbon-based materials [48–51].

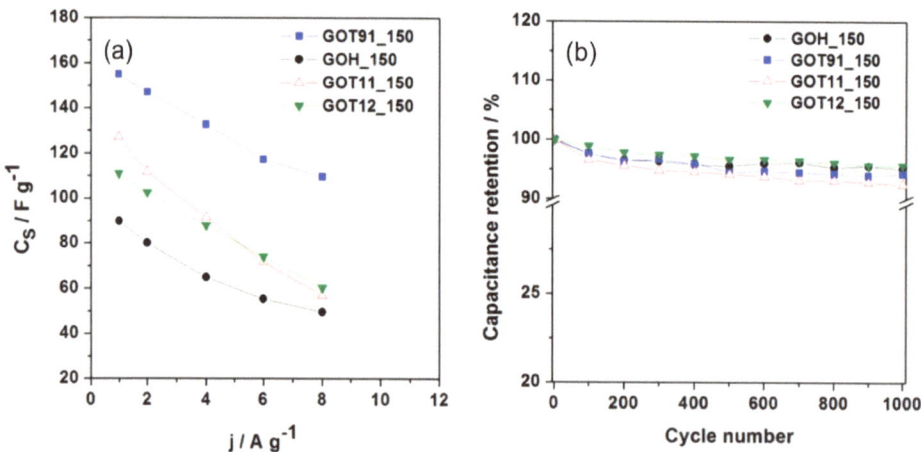

Figure 7. (**a**) Specific capacitance (C_s) vs. gravimetric current density (j); (**b**) Cyclic performance performed at 4 A g^{-1} during 1000 charge–discharge cycles.

Taking into account that the specific surface area values (see Table S3, Supplementary Information) of the rGO samples are small (less than 20 m^2 g^{-1}), the analyzed samples show a significant capacitance per unit of surface area. This value is above ca. 5 F m^{-2} for the GOT91_150 sample, which is substantially higher than the values generally reported for carbon materials (0.1–0.2 F m^{-2}) [50]. These results suggest the occurrence of a high pseudocapacitive contribution through the oxygenated and/or phosphorus-rich surface functional groups. Regarding this, it is observed that, on the one hand, the samples simultaneously containing phosphorus and sulfur (P/S-rGO samples) show a higher C_s value when compared to the sample not containing phosphorus (GOH_150). On the other hand, among the P/S-rGO samples, the sample with lower phosphorus content (GOT91_150) shows the highest C_s value. Therefore, the highest C_S value determined for the GOT91_150 sample could be explained by the simultaneous presence of sulfur and phosphorus functional groups linked to the pseudocapacitive phenomena, as observed in previous works on S and P co-doped porous carbons [52,53]. Still regarding the analysis of the C_s vs. j plot (Figure 7a), the GOT91_150 sample presents the best rate capability at 8 A g^{-1} (70% with respect to the C_s determined at 1 A g^{-1}), whereas the lowest rate capability (45%) is found for the GOT11_150 sample. These findings can be explained in terms of the differences in the equivalent series resistance (ESR) values of these samples (see Figure S7, in the Supplementary Information), as detailed below. Long-term charge–discharge cycling tests were performed at 4 A g^{-1} within the potential range of 0–1.0 V. The capacitance retention was calculated from the C_s value determined for each cycle divided by the C_s value of the first cycle. Figure 7b shows the capacitance retention as function of the cycle number for each sample. All the rGO samples showed good capacitance retention (≈95%) after 1000 cycles. This demonstrates a low degradation of the samples after successive charge–discharge cycles.

The Nyquist plots of the rGO samples obtained from the EIS measurements are shown in Figure 8. It is possible to observe that all samples show a typical spectrum of a non-ideal electrochemical capacitor, with straight lines with a nearly 45° phase angle at high frequencies and more vertical lines at the low-frequency portions of the diagrams [54]. At low frequencies, the GOT11_150 sample shows the most vertical curve, suggesting a better capacitive behavior [55]. From the intercept of the curve with the real axis at high frequency, it is possible to determine the series resistance (R_s), which includes the electrolyte solution resistance, separator resistance and electrode resistance [54]. Sample GOT11_150 shows the lowest R_s value (ca. 1 Ω), which indicates its good electrical conductivity. On the other hand, the GOH_150 sample shows the lowest charge transfer resistance (R_{ct}), which is associated with the diameter of the semi-circle at high frequency [55]. The sum of $R_s + R_{ct}$ is in agreement with the values of ESR determined from the galvanostatic measurements (see Figure S7, in the Supplementary Information section), with the GOT11_150 sample having the highest $R_s + R_{ct}$ value and the GOH_150 sample having the lowest one. Taking into account that, in general, a higher content of oxygenated functional groups can determine a drop in the electrical conductivity of carbon materials [51,54,56], these EIS results suggest a higher content of oxygenated functionalities in the rGO samples obtained by the modified Marcano–Tour method [22] than in the reduced sample obtained by the modified Hummers method [23].

The specific energy density (W_s) and the specific power density (P_s) were calculated considering the total mass of the two electrodes in the supercapacitor, as follows:

$$W_s \left(\text{Wh·kg}^{-1} \right) = \frac{C_S \cdot E_2^2}{28.8}, \quad (2)$$

$$P_s \left(\text{W·kg}^{-1} \right) = \frac{W_S}{t_d}. \quad (3)$$

In these expressions, C_s is the specific capacitance of one electrode (F g^{-1}), E_2 is the voltage range during the galvanostatic discharge (V), and t_d is the discharge time (s). Figure 9 shows the Ragone plots obtained for the rGO samples. At low P_s values, the highest W_s value was achieved for the GOT91_150 sample (4.9 Wh kg^{-1} at 238.7 W), while the lowest one corresponded to the GOH_150 sample, which is in accordance with the C_s values obtained for these samples (i.e., the sample with the higher C_s reached the higher W_s value). On the other hand, for all the samples, as P_s increases, the W_s values decrease. This behavior is particularly noticeable for the GOT11_150 sample, which is consistent with its higher ESR (Figure S7, Supplementary Material) and relatively low rate capability (Figure 8a).

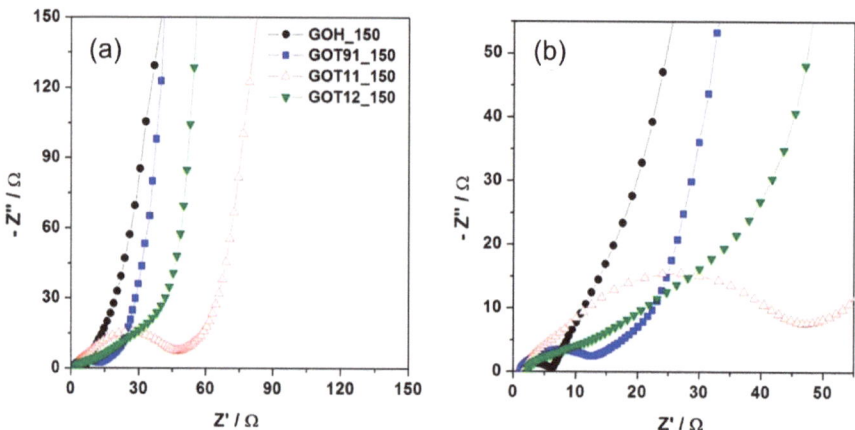

Figure 8. (**a**) Nyquist plots obtained for the rGO samples. (**b**) Zoom of a selected area of the plots exhibited in (**a**).

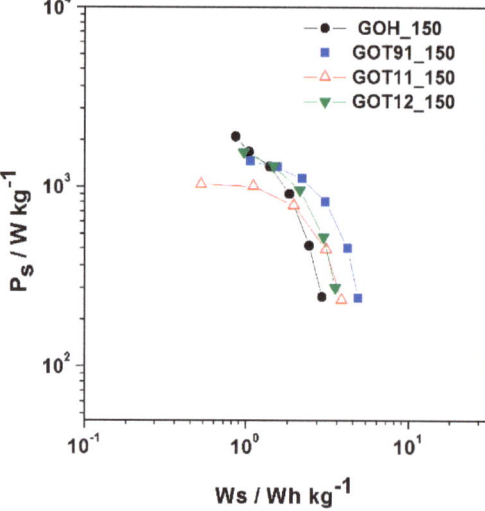

Figure 9. Specific power vs. specific energy of the electrochemical analyzed samples.

It is interesting to note that the GOH_150 sample exhibited a particularly high P_s value for the lowest W_s value (0.9 Wh kg^{-1} at 1902.5 W). If we take into account the C_s values determined at higher currents (i.e., high power), this may seem inconsistent at first, since this sample does not have a high C_s value among the rGO samples analyzed here. However, its lower ESR enables a higher voltage range (E_2) during the charge–discharge experiment (see Figure S7 in the Supplementary Information section), which results in a higher energy accumulation. In sum, on the basis of these results, it can be concluded that the energy and power values of the supercapacitors developed from the synthesized materials are still low compared to what has been reported for other carbon materials [54]. It is clear, though, that the simultaneous presence of P- and S-containing groups in the rGO samples prepared by the simple route described here is beneficial for the promising electrochemical properties of these materials (such as good electrochemical stability in wide voltage windows, high capacitance per unit of surface area and low degradation after successive charge–discharge cycles), which stimulates the pursuit of further investigations aiming to improve the energy- and power-related properties for possible applications of these materials as supercapacitor electrodes.

4. Conclusions

By modifying the H_2SO_4/H_3PO_4 ratio used on the GO synthesis by the Marcano–Tour method [22], it was possible to synthesize graphene oxide samples with good degree of oxidation and varied phosphorus and sulfur contents. The thermal treatment of the P/S-GO samples at 150 °C led to the production of P/S-rGO samples containing 0.3–5 wt.% of phosphorus and 2–7 wt.% of sulfur depending on the synthesis conditions. These species were found to be present mostly as phosphate and sulfate groups, as evidenced by solid-state NMR and FTIR results. The P/S-rGO samples obtained at 150 °C had their electrochemical characteristics evaluated, and sample GOT91_150 (containing both phosphorus and sulfur) exhibited the best capacitance behavior, reaching a maximum C_s value of 155 F g^{-1} at 1 A g^{-1}, with a good rate capability at higher current density and good capacitance retention after 1000 charge–discharge cycles. Altogether, these results suggest that the simultaneous presence of sulfur and phosphorus plays a fundamental role in enhancing the capacitive properties of reduced graphene oxide samples. Even though there is still room for improvement of these features, especially regarding the energy- and power-related properties of the supercapacitors prepared from these materials, the presented methods represent a facile and low-cost alternative for the production of P/S-enriched rGO papers with promising properties (such as good electrochemical stability in wide voltage windows, high capacitance per unit of surface area and low degradation after successive charge–discharge cycles) for use in electrochemical energy-storage devices.

Supplementary Materials: The following supporting information can be downloaded at: https://www.mdpi.com/article/10.3390/c9020060/s1, Figure S1. X-ray diffractogram of the GOH sample after thermal treatment at 80 °C for 24 h (protocol C), compared to the XRD pattern of the graphite precursor (BG). Figure S2. X-ray diffractograms of the GOH sample after thermal treatments at various temperatures and times, according to the protocols described in Table S1. Figure S3. TG curve of the GOH sample. The dotted lines indicate the temperatures used in the thermal treatment protocols described in Table S1. The TG experiments were conducted in a Shimadzu TGA-50H instrument, using alumina pan at a heating rate of 5 °C/min, under O2 flow (50 mL/min). Figure S4. 13C NMR spectra of the GOH sample and of the rGO samples D24 and E24, which were prepared by thermally treating GOH aliquots at 100 and 150 °C, respectively, for 24 h. Figure S5. FITR spectra of the GOH sample and of the rGO samples D24 and E24, which were prepared by thermally treating GOH aliquots at 100 and 150 °C, respectively, for 24 h. Figure S6. SEM images of GO samples and rGO papers prepared at 150 °C. Figure S7. Galvanostatic charge-discharge curves of the rGO samples obtained at a constant current density of (a) 1 A g^{-1} and (b) 8 A g^{-1}. Table S1. Description of the thermal reduction protocols used to obtain rGO from the GOH sample. Table S2. S contents of the GO sample synthesized by the modified Hummers method and the corresponding reduced materials. Table S3. Specific surface area of the rGO samples, determined from N2 adsorption isotherms at 77 K. Refs. [57–73] are cited in the supplementary material.

Author Contributions: Conceptualization, M.A.V. and J.C.C.F.; methodology, M.A.V. and T.L.G.C.; formal analysis, M.A.V., T.L.G.C., G.R.G., D.F.C. and E.L.d.S., data curation, M.A.V., J.C.C.F, A.C. and M.A.S.J., writing—original draft preparation, M.A.V.; writing—review and editing, M.A.V., T.L.G.C., J.C.C.F. and E.L.d.S.; visualization, M.A.V., T.L.G.C., J.C.C.F., E.L.d.S. and A.C.; supervision, J.C.C.F.

and A.C.; project administration, J.C.C.F.; funding acquisition, J.C.C.F. and A.C. All authors have read and agreed to the published version of the manuscript.

Funding: This research was financed in part by the Coordenação de Aperfeiçoamento de Pessoal de Nível Superior—Brasil (CAPES)—Finance Code 001. The authors acknowledge also the support from the Brazilian agencies FAPES (grants 280/2021, 495/2021 and 418/2022) and CNPq (grant 310528/2022-4).

Data Availability Statement: Data sharing not applicable. No new data were created or analyzed in this study. Data sharing is not applicable to this article.

Acknowledgments: Laboratory for Research and Development of Methodologies for Crude Oil Analysis (LabPetro), Laboratory of Thermal Plasma (LPT), Laboratory of Air Quality, located at the Federal University of Espírito Santo (UFES), Brazil, for the use of their experimental facilities. The support from the Comisión Académica de Posgrado (CAP), Universidad de la República, Uruguay, is also gratefully acknowledged.

Conflicts of Interest: The authors declare no conflict of interest.

References

1. Novoselov, K.S.; Geim, A.K.; Morozov, S.V.; Jiang, D.; Zhang, Y.; Dubonos, S.V.; Grigorieva, I.V.; Firsov, A.A. Electric Field Effect in Atomically Thin Carbon Films. *Science* **2004**, *306*, 666–669. [CrossRef] [PubMed]
2. Brodie, B.C. On the Atomic Weight of Graphite. *Philos. Trans. R. Soc. Lond.* **1859**, *149*, 249–259.
3. Pei, S.; Cheng, H.M. The Reduction of Graphene Oxide. *Carbon* **2012**, *50*, 3210–3228. [CrossRef]
4. Huang, H.; Shi, H.; Das, P.; Qin, J.; Li, Y.; Wang, X.; Su, F.; Wen, P.; Li, S.; Lu, P.; et al. The Chemistry and Promising Applications of Graphene and Porous Graphene Materials. *Adv. Funct. Mater.* **2020**, *30*, 1909035. [CrossRef]
5. Dreyer, D.R.; Park, S.; Bielawski, C.W.; Ruoff, R.S. The chemistry of graphene oxide. *Chem. Soc. Rev.* **2009**, *39*, 228–240. [CrossRef] [PubMed]
6. Brisebois, P.P.; Siaj, M. Harvesting Graphene Oxide—Years 1859 to 2019: A Review of Its Structure, Synthesis, Properties and Exfoliation. *J. Mater. Chem. C* **2020**, *8*, 1517–1547. [CrossRef]
7. Park, S.; An, J.; Potts, J.R.; Velamakanni, A.; Murali, S.; Ruoff, R.S. Hydrazine-Reduction of Graphite-and Graphene Oxide. *Carbon* **2011**, *49*, 3019–3023. [CrossRef]
8. De Silva, K.K.H.; Huang, H.H.; Joshi, R.K.; Yoshimura, M. Chemical Reduction of Graphene Oxide Using Green Reductants. *Carbon* **2017**, *119*, 190–199. [CrossRef]
9. Zhou, Q.; Zhao, Z.; Chen, Y.; Hu, H.; Qiu, J. Low Temperature Plasma-Mediated Synthesis of Graphene Nanosheets for Supercapacitor Electrodes. *J. Mater. Chem.* **2012**, *22*, 6061–6066. [CrossRef]
10. Ghadim, E.E.; Rashidi, N.; Kimiagar, S.; Akhavan, O.; Manouchehri, F.; Ghaderi, E. Pulsed Laser Irradiation for Environment Friendly Reduction of Graphene Oxide Suspensions. *Appl. Surf. Sci.* **2014**, *301*, 183–188. [CrossRef]
11. Dumée, L.F.; Feng, C.; He, L.; Allioux, F.M.; Yi, Z.; Gao, W.; Banos, C.; Davies, J.B.; Kong, L. Tuning the Grade of Graphene: Gamma Ray Irradiation of Free-Standing Graphene Oxide Films in Gaseous Phase. *Appl. Surf. Sci.* **2014**, *322*, 126–135. [CrossRef]
12. Okhay, O.; Gonçalves, G.; Tkach, A.; Dias, C.; Ventura, J.; Ribeiro Da Silva, M.F.; Valente Gonçalves, L.M.; Titus, E. Thin Film versus Paper-like Reduced Graphene Oxide: Comparative Study of Structural, Electrical, and Thermoelectrical Properties. *J. Appl. Phys.* **2016**, *120*, 051706. [CrossRef]
13. Khan, F.; Khan, S.; Kamal, S.; Arshad, M. Recent Advances in Graphene Oxide and Reduced Graphene Oxide Based Nanocomposites for the Photodegradation of Dyes. *J. Mater. Chem. C* **2020**, *8*, 15940–15955. [CrossRef]
14. Zhu, Y.; Murali, S.; Cai, W.; Li, X.; Suk, J.W.; Potts, J.R.; Ruoff, R.S. Graphene and Graphene Oxide: Synthesis, Properties, and Applications. *Adv. Mater.* **2010**, *22*, 3906–3924. [CrossRef]
15. Valentini, C.; Montes-Garcia, V.; Livio, P.A.; Chudziak, T.; Raya, J.; Ciesielski, A.; Samorì, P. Tuning the electrical properties of graphene oxide through low-temperature thermal annealing. *Nanoscale* **2023**, *15*, 5743–5755. [CrossRef] [PubMed]
16. Kumar, N.A.; Baek, J. Doped Graphene Supercapacitors. *Nanotechnology* **2015**, *26*, 492001. [CrossRef]
17. Yu, X.; Feng, L.; Park, H.S. Highly Flexible Pseudocapacitors of Phosphorus-Incorporated Porous Reduced Graphene Oxide Films. *J. Power Sources* **2018**, *390*, 93–99. [CrossRef]
18. Qiao, X.; Liao, S.; You, C.; Chen, R. Phosphorus and Nitrogen Dual Doped and Simultaneously Reduced Graphene Oxide with High Surface Area as Efficient Metal-Free Electrocatalyst for Oxygen Reduction. *Catalysts* **2015**, *5*, 981–991. [CrossRef]
19. Ghafuri, H.; Talebi, M. Water-Soluble Phosphated Graphene: Preparation, Characterization, Catalytic Reactivity, and Adsorption Property. *Ind. Eng. Chem. Res.* **2016**, *55*, 2970–2982. [CrossRef]
20. Niu, F.; Tao, L.-M.; Deng, Y.-C.; Wang, Q.-H.; Song, W.-G. Phosphorus Doped Graphene Nanosheets for Room Temperature NH3 Sensing. *N. J. Chem.* **2014**, *38*, 2269. [CrossRef]
21. Sun, X.; Cheng, P.; Wang, H.; Xu, H.; Dang, L.; Liu, Z.; Lei, Z. Activation of Graphene Aerogel with Phosphoric Acid for Enhanced Electrocapacitive Performance. *Carbon* **2015**, *92*, 1–10. [CrossRef]

22. Marcano, D.C.; Kosynkin, D.V.; Berlin, J.M.; Sinitskii, A.; Sun, Z.; Slesarev, A.; Alemany, L.B.; Lu, W.; Tour, J.M. Improved Synthesis of Graphene Oxide. *ACS Nano* **2010**, *4*, 4806–4814. [CrossRef] [PubMed]
23. Hummers, W.S.; Offeman, R.E. Preparation of Graphitic Oxide. *J. Am. Chem. Soc.* **1958**, *80*, 1339. [CrossRef]
24. Chua, C.K.; Sofer, Z.; Pumera, M. Graphite Oxides: Effects of Permanganate and Chlorate Oxidants on the Oxygen Composition. *Chem.-A Eur. J.* **2012**, *18*, 13453–13459. [CrossRef] [PubMed]
25. Costa, T.L.G.; Vieira, M.A.; Gonçalves, G.R.; Cipriano, D.F.; Lacerda Jr., V.; Gonçalves, A.S.; Scopel, W.L.; de Siervo, A.; Freitas, J.C.C. Combined computational and experimental study about the incorporation of phosphorus into the structure of graphene oxide. *Phys. Chem. Chem. Phys.* **2023**, *25*, 6927–6943. [CrossRef]
26. Cory, D.G.; Ritchey, W.M. Suppression of Signals from the Probe in Bloch Decay Spectra. *J. Magn. Reson.* **1988**, *32*, 128–132. [CrossRef]
27. Sanderson, K. Carbon Makes Super-Tough Paper. *Nature* **2007**. [CrossRef]
28. Eigler, S.; Dotzer, C.; Hof, F.; Bauer, W.; Hirsch, A. Sulfur Species in Graphene Oxide. *Chem.-A Eur. J.* **2013**, *19*, 9490–9496. [CrossRef]
29. Farjadian, F.; Abbaspour, S.; Sadatlu, M.A.A.; Mirkiani, S.; Ghasemi, A.; Hoseini-Ghahfarokhi, M.; Mozaffari, N.; Karimi, M.; Hamblin, M.R. Recent Developments in Graphene and Graphene Oxide: Properties, Synthesis, and Modifications: A Review. *ChemistrySelect* **2020**, *5*, 10200–10219. [CrossRef]
30. Moreno-Fernández, G.; Gómes-Urbano, J.L.; Enterría, M.; Cid, R.; López del Amo, J.M.; Mysyk, R.; Carriazo, D. Understanding enhanced charge storage of phosphorus-functionalized graphene in aqueous acidic electrolytes. *Electrochim. Acta* **2020**, *361*, 136985. [CrossRef]
31. Dreyer, D.R.; Todd, A.D.; Bielawski, C.W. Harnessing the Chemistry of Graphene Oxide. *Chem. Soc. Rev.* **2014**, *43*, 5288–5301. [CrossRef]
32. Aliyev, E.; Filiz, V.; Khan, M.M.; Lee, Y.J.; Abetz, C.; Abetz, V. Structural Characterization of Graphene Oxide: Surface Functional Groups and Fractionated Oxidative Debris. *Nanomaterials* **2019**, *9*, 1180. [CrossRef]
33. Skákalová, V.; Kotrusz, P.; Jergel, M.; Susi, T.; Mittelberger, A.; Vretenár, V.; Šiffalovič, P.; Kotakoski, J.; Meyer, J.C.; Hulman, M. Chemical Oxidation of Graphite: Evolution of the Structure and Properties. *J. Phys. Chem. C* **2018**, *122*, 929–935. [CrossRef]
34. Gao, W.; Alemany, L.B.; Ci, L.; Ajayan, P.M. New Insights into the Structure and Reduction of Graphite Oxide. *Nat. Chem.* **2009**, *1*, 403–408. [CrossRef] [PubMed]
35. He, H.; Riedl, T.; Lerf, A.; Klinowski, J. Solid-State NMR Studies of the Structure of Graphite Oxide. *J. Phys. Chem.* **1996**, *100*, 19954–19958. [CrossRef]
36. Rawal, A.; Man, S.H.C.; Agarwal, V.; Yao, Y.; Thickett, S.C.; Zetterlund, P.B. Structural Complexity of Graphene Oxide: The Kirigami Model. *ACS Appl. Mater. Interfaces* **2021**, *13*, 18255–18263. [CrossRef]
37. Freitas, J.C.C.; Emmerich, F.G.; Cernicchiaro, G.R.C.; Sampaio, L.C.; Bonagamba, T.J. Magnetic Susceptibility Effects on ^{13}C MAS NMR Spectra of Carbon Materials and Graphite. *Solid State Nucl. Magn. Reson.* **2001**, *20*, 61–73. [CrossRef] [PubMed]
38. Vieira, M.A.; Gonçalves, G.R.; Cipriano, D.F.; Schettino, M.A.; Silva Filho, E.A.; Cunha, A.G.; Emmerich, F.G.; Freitas, J.C.C. Synthesis of Graphite Oxide from Milled Graphite Studied by Solid-State ^{13}C Nuclear Magnetic Resonance. *Carbon* **2016**, *98*, 496–503. [CrossRef]
39. Puziy, A.M.; Poddubnaya, O.I.; Socha, R.P.; Gurgul, J.; Wisniewski, M. XPS and NMR Studies of Phosphoric Acid Activated Carbons. *Carbon* **2008**, *46*, 2113–2123. [CrossRef]
40. Wang, Y.; Zuo, S.; Yang, J.; Yoon, S.H. Evolution of Phosphorus-Containing Groups on Activated Carbons during Heat Treatment. *Langmuir* **2017**, *33*, 3112–3122. [CrossRef] [PubMed]
41. Lopes, T.R.; Cipriano, D.F.; Gonçalves, G.R.; Honorato, H.A.; Schettino, M.A.; Cunha, A.G.; Emmerich, F.G.; Freitas, J.C.C. Multinuclear Magnetic Resonance Study on the Occurrence of Phosphorus in Activated Carbons Prepared by Chemical Activation of Lignocellulosic Residues from the Babassu Production. *J. Environ. Chem. Eng.* **2017**, *5*, 6016–6029. [CrossRef]
42. Zhou, Y.; Ma, R.; Candelaria, S.L.; Wang, J.; Liu, Q.; Uchaker, E.; Li, P.; Chen, Y.; Cao, G. Phosphorus/Sulfur Co-Doped Porous Carbon with Enhanced Specific Capacitance for Supercapacitor and Improved Catalytic Activity for Oxygen Reduction Reaction. *J. Power Sources* **2016**, *314*, 39–48. [CrossRef]
43. Yu, X.; Kang, Y.; Park, H.S. Sulfur and Phosphorus Co-Doping of Hierarchically Porous Graphene Aerogels for Enhancing Supercapacitor Performance. *Carbon* **2016**, *101*, 49–56. [CrossRef]
44. Ying, K.; Tian, R.; Zhou, J.; Li, H.; Dugnani, R.; Lu, Y.; Duan, H.; Guo, Y.; Liu, H. A Three Dimensional Sulfur/Reduced Graphene Oxide with Embedded Carbon Nanotubes Composite as a Binder-Free, Free-Standing Cathode for Lithium-Sulfur Batteries. *RSC Adv.* **2017**, *7*, 43483–43490. [CrossRef]
45. Thomas, H.R.; Marsden, A.J.; Walker, M.; Wilson, N.R.; Rourke, J.P. Sulfur-Functionalized Graphene Oxide by Epoxide Ring-Opening. *Angew. Chem. Int. Ed.* **2014**, *53*, 7613–7618. [CrossRef]
46. Bi, Z.; Huo, L.; Kong, Q.; Li, F.; Chen, J.; Ahmad, A.; Wei, X.; Xie, L.; Chen, C. Structural Evolution of Phosphorus Species on Graphene with a Stabilized Electrochemical Interface. *Appl. Mater. Interfaces* **2019**, *11*, 11421–11430. [CrossRef]
47. Al, M.; Oh, P.O.; Frost, R.L.; Scholz, R.; López, A.; Xi, Y. A Vibrational Spectroscopic Study of the Phosphate Mineral Whiteite CaMn($^{++}$)Mg$_2$Al$_2$(PO$_4$)$_4$(OH)$_2$·8(H$_2$O). *Spectrochim. Acta Part A Mol. Biomol. Spectrosc.* **2014**, *124*, 243–248.
48. Gurusamy, L.; Anandan, S.; Liu, N.; Wu, J.J. Synthesis of a Novel Hybrid Anode Nanoarchitecture of Bi$_2$O$_3$/Porous-RGO Nanosheets for High-Performance Asymmetric Supercapacitor. *J. Electroanal. Chem.* **2020**, *856*, 113489. [CrossRef]

49. Zhao, X.; Li, W.; Kong, F.; Chen, H.; Wang, Z.; Liu, S.; Jin, C. Carbon Spheres Derived from Biomass Residue via Ultrasonic Spray Pyrolysis for Supercapacitors. *Mater. Chem. Phys.* **2018**, *219*, 461–467. [CrossRef]
50. Pandolfo, A.G.; Hollenkamp, A.F. Carbon Properties and Their Role in Supercapacitors. *J. Power Sources* **2006**, *157*, 11–27. [CrossRef]
51. Cuña, A.; Ortega Vega, M.R.; da Silva, E.L.; Tancredi, N.; Radtke, C.; Malfatti, C.F. Nitric Acid Functionalization of Carbon Monoliths for Supercapacitors: Effect on the Electrochemical Properties. *Int. J. Hydrogen Energy* **2016**, *41*, 12127–12135. [CrossRef]
52. Liu, Y.; Chang, Z.; Yao, L.; Yan, S.; Lin, J.; Chen, J.; Lian, J.; Lin, H. Nitrogen/Sulfur Dual-Doped Sponge-like Porous Carbon Materials Derived from Pomelo Peel Synthesized at Comparatively Low Temperatures for Superior-Performance Supercapacitors. *J. Electroanal. Chem.* **2019**, *847*, 113111. [CrossRef]
53. Qiang, Z.; Dan, D.; Wang, M.; Xiong, C.; Ying, X.; Jie, Z. Sulfur Modification of Carbon Materials as Well as the Redox Additive of Na_2S for Largely Improving Capacitive Performance of Supercapacitors. *J. Electroanal. Chem.* **2020**, *856*, 113678.
54. Béguin, F.; Frackowiak, E. *Supercapacitors: Materials, Systems, and Applications*; Wiley-VCH: Hoboken, NJ, USA, 2013.
55. Mishra, A.K.; Ramaprabhu, S. Functionalized Graphene-Based Nanocomposites for Supercapacitor Application. *J. Phys. Chem. C* **2011**, *115*, 14006–14013. [CrossRef]
56. Cuña, A.; Tancredi, N.; Bussi, J.; Deiana, A.C.; Sardella, M.F.; Barranco, V.; Rojo, J.M.E. Grandis as a Biocarbons Precursor for Supercapacitor Electrode Application. *Waste Biomass Valorization* **2014**, *5*, 305–313. [CrossRef]
57. Vieira, M.A.; Frasson, C.M.R.; Costa, T.L.G.; Cipriano, D.F.; Schettino, M.A.; Cunha, A.G.; Freitas, J.C.C. Solid state 13C NMR study on the synthesis of graphite oxide from different graphitic precursors. *Quim. Nova* **2017**, *40*, 1164–1171.
58. Jeong, H.K.; Jin, M.H.; So, K.P.; Lim, S.C.; Lee, Y.H. Tailoring the characteristics of graphite oxides by different oxidation times. *J. Phys. D Appl. Phys.* **2009**, *42*, 065418. [CrossRef]
59. Wang, H.; Hu, Y.H. Effect of oxygen content on structures of graphite oxides. *Ind. Eng. Chem. Res.* **2011**, *50*, 6132–6137. [CrossRef]
60. Mu, S.J.; Su, Y.C.; Xiao, L.H.; Liu, S.D.; Hu, T.; Tang, H.B. X-ray Difraction Pattern of Graphite Oxide. *Chin. Phys. Lett.* **2013**, *30*, 096101. [CrossRef]
61. Rattana, T.; Chaiyakun, S.; Witit-Anun, N.; Nuntawong, N.; Chindaudom, P.; Oaew, S.; Kedkeaw, C.; Limsuwan, P. Preparation and characterization of graphene oxide nanosheets. *Procedia Eng.* **2012**, *32*, 759–764. [CrossRef]
62. Huh, S.H. *Thermal Reduction of Graphene Oxide*; Mikhailov, S., Ed.; InTech: London, UK, 2011; Volume 19, pp. 73–90.
63. Stankovich, S.; Dikin, D.A.; Piner, R.D.; Kohlhaas, K.A.; Kleinhammes, A.; Jia, Y.; Wu, Y.; Nguyen, S.T.; Ruoff, R.S. Synthesis of graphene-based nanosheets via chemical reduction of exfoliated graphite oxide. *Carbon* **2007**, *45*, 1558–1565. [CrossRef]
64. Krishnamoorthy, K.; Veerapandian, M.; Yun, K.; Kim, S.J. The chemical and structural analysis of graphene oxide with different degrees of oxidation. *Carbon* **2013**, *53*, 38–49. [CrossRef]
65. Tegou, E.; Pseiropoulos, G.; Filippidou, M.K.; Chatzandroulis, S. Low-temperature thermal reduction of graphene oxide films in ambient atmosphere: Infra-red spectroscopic studies and gas sensing applications. *Microelectron. Eng.* **2016**, *159*, 146–150. [CrossRef]
66. Hontoria-Lucas, C.; López-Peinado, A.J.; López-González, J.D.; Rojas-Cervantes, M.L.; Martín-Aranda, R.M. Study of oxygen-containing groups in a series of graphite oxides: Physical and chemical characterization. *Carbon* **1995**, *33*, 1585–1592. [CrossRef]
67. Cai, W.; Piner, R.D.; Stadermann, F.J.; Park, S.; Shaibat, M.A.; Ishii, Y.; Yang, D.; Velamakanni, A.; An, S.J.; Stoller, M.; et al. Synthesis and Solid-State NMR Structural Characterization of 13C-Labeled Graphite Oxide. *Science* **2008**, *321*, 1815–1817. [CrossRef] [PubMed]
68. Lerf, A.; He, H.; Forster, M.; Klinowski, J. Structure of Graphite Oxide Revisited. *J. Phys. Chem. B* **1998**, *102*, 4477–4482. [CrossRef]
69. Dimiev, A.M.; Eigler, S. *Graphene Oxide: Fundamentals and Applications*, 1st ed.; John Wiley & Sons, Ltd.: Chichester, UK, 2016.
70. Sorokina, N.E.; Shornikova, O.N.; Avdeev, V.V. Stability limits of graphite intercalation compounds in the systems graphite-$HNO_3(H_2SO_4)$-H_2O-$KMnO_4$. *Inorg. Mater.* **2007**, *43*, 822–826. [CrossRef]
71. Dimiev, A.M.; Polson, T.A. Contesting the two-component structural model of graphene oxide and reexamining the chemistry of graphene oxide in basic media. *Carbon* **2015**, *93*, 544–554. [CrossRef]
72. Brunauer, S.; Emmett, P.H.; Teller, E. Adsorption of Gases in Multimolecular Layers. *J. Am. Chem. Soc.* **1938**, *60*, 309–319. [CrossRef]
73. Ouyang, Z.; Lei, Y.; Chen, Y.; Zhang, Z.; Jiang, Z.; Hu, J. Preparation and Specific Capacitance Properties of Sulfur, Nitrogen Co-Doped Graphene Quantum Dots. *Nanoscale Res. Lett.* **2019**, *14*, 1–9. [CrossRef] [PubMed]

Disclaimer/Publisher's Note: The statements, opinions and data contained in all publications are solely those of the individual author(s) and contributor(s) and not of MDPI and/or the editor(s). MDPI and/or the editor(s) disclaim responsibility for any injury to people or property resulting from any ideas, methods, instructions or products referred to in the content.

Review

Carbon Fibers: From PAN to Asphaltene Precursors; A State-of-Art Review

Hossein Bisheh [1,2,*] and Yasmine Abdin [2]

[1] Composites Research Network, Department of Mechanical Engineering, The University of British Columbia, Vancouver, BC V6T 1Z4, Canada
[2] Composites Research Network, Department of Materials Engineering, The University of British Columbia, Vancouver, BC V6T 1Z4, Canada
* Correspondence: hossein.bisheh@ubc.ca or h.bisheh83@gmail.com

Abstract: Due to their outstanding material properties, carbon fibers are widely used in various industrial applications as functional or structural materials. This paper reviews the material properties and use of carbon fiber in various applications and industries and compares it with other existing fillers and reinforcing fibers. The review also examines the processing of carbon fibers and the main challenges in their fabrication. At present, two main precursors are primarily utilized to produce carbon fibers, i.e., polyacrylonitrile (PAN) and petroleum pitch. Each of these precursors makes carbon fibers with different properties. However, due to the costly and energy-intensive processes of carbon fiber production based on the existing precursors, there is an increasingly growing need to introduce cheaper precursors to compete with other fibers on the market. A special focus will be given to the most recent development of manufacturing more sustainable and cost-effective carbon fibers derived from petroleum asphaltenes. This review paper demonstrates that low-cost asphaltene-based carbon fibers can be a substitute for costly PAN/pitch-based carbon fibers at least for functional applications. The value proposition, performance/cost advantages, potential market, and market size as well as processing challenges and methods for overcoming these will be discussed.

Keywords: asphaltenes; carbon fiber; PAN; pitch; precursor

Citation: Bisheh, H.; Abdin, Y. Carbon Fibers: From PAN to Asphaltene Precursors; A State-of-Art Review. *C* **2023**, *9*, 19. https://doi.org/10.3390/c9010019

Academic Editors: Olena Okhay and Gil Gonçalves

Received: 22 December 2022
Revised: 23 January 2023
Accepted: 28 January 2023
Published: 4 February 2023

Copyright: © 2023 by the authors. Licensee MDPI, Basel, Switzerland. This article is an open access article distributed under the terms and conditions of the Creative Commons Attribution (CC BY) license (https://creativecommons.org/licenses/by/4.0/).

1. Introduction

Composite materials are made up of two main constituents or phases, i.e., reinforcement and matrix phases. Generally, the reinforcement phase, which is stronger and stiff, imparts the strength and stiffness properties of the composite materials, while the matrix phase, which is weaker and soft, binds the reinforcing fibers, transfers loads between fibers, and makes the net shape of the composite. The reinforcing component of a composite can be in the form of short or continuous fibers, particles, or whiskers [1]. Based on the application and importance of composite materials, different materials of reinforcement are available on the market depending on the desired performance and cost of the resultant composite materials. Hence, developing the reinforcing fiber materials design and processing to reduce the cost of manufacturing with acceptable material properties can attract the interest of industrial owners to deliver more reliable products to the market.

The aim of fabricating composite materials is to produce lightweight structures with higher mechanical properties and performance in comparison to conventional materials such as metals. A composite material generally presents superior properties to its ingredients utilized individually. Structural composites aim to optimize the performance of the structure during the service life. Various materials can be employed for the matrix phase such as polymers, ceramics, metals, etc. However, polymeric materials, due to their easy processing, low cost of production, and high productivity are the favorite material for the matrix phase of composites. Ceramics and metal matrices are usually utilized in very high-temperature environments such as engines. Typical polymers used as the matrix material are epoxy, phenolic, and polyester resins. The most commonly used reinforcing materials in composites are carbon, graphite, Kevlar, and glass, which can be dispersed in the matrix in various forms. Carbon fiber-reinforced polymeric composites deliver a composite material with higher performance and lower weight than other fiber-reinforced composites with a higher cost of production. Recently, polymeric composites reinforced with nanofibers have attracted the attention of many researchers, but the cost of nanomaterials is high, and this is the main obstacle to producing nanocomposites on commercial scales [2]. However, additive manufacturing technology can be a cost-effective approach to fabricating fiber-reinforced composites [3].

Fiber-reinforced polymer composites are widely used in many industrial applications from high-performance structural applications such as ships, spacecraft, aircraft, buildings, bridges, off-shore platforms, etc. to low-performance structural applications such as boats, automotive parts, sports goods, etc. [4–7]. The demands for fiber-reinforced polymer composites are increasing due to their outstanding properties compared with conventional materials, and they are capturing other markets such as biomedical devices, energy storage devices, microelectronic devices, etc. as examples of functional applications. Carbon fibers are commonly used as the reinforcing material in the fabrication of advanced composite materials for both structural and functional applications due to their lower weight, high stiffness, high strength, and high fatigue resistance. Other commercially available reinforcing fibers such as Kevlar, boron, and glass are typically utilized in various structural and functional applications in which high strength and performance are not required.

Although carbon fibers have more advantages than other fibers, their high cost of production is a barrier for commercial production purposes. Hence, developing the process of carbon fibers production through deriving from inexpensive natural resources can decrease their production cost to be comparable with other low-cost fibers such as glass at least for functional and structural applications when lower strength is needed. Accordingly, the main objective of this paper is to review and discuss the process of carbon fiber production from common precursors and then introduce a cost-effective process for carbon fiber production from low-cost petroleum asphaltene with the corresponding challenges.

This paper mainly reviews commercially available fibers and their applications and outlook on their markets. In addition, the review briefly discusses the production processes of existing fibers, especially carbon fibers. In addition, both the structural and functional applications of carbon fibers are reviewed and discussed, and then, the carbon fiber production from petroleum asphaltene is introduced and investigated. Lastly, we discuss a market assessment of asphaltene-derived carbon fibers and their advantages over other commercially available fibers as well as challenges in commercializing the asphaltene-based carbon fibers.

2. Existing Reinforcing Fibers

There is a large variety of reinforcing fibers for composites. The favorite properties of reinforcing fibers are their high stiffness, high strength, and relatively low density, where these characteristics can be chosen based on the application of a composite material as well as its fabrication cost. Each type of reinforcing fiber has its advantages and disadvantages, as presented in Table 1 [1]. Most fibers display a linear behavior to failure. The ultimate strain of fibers affects greatly the strength of the composite laminate. High specific stiffness (modulus to density ratio) and high specific strength (strength to density ratio) lead to high-performance composites. These two properties strongly depend on the fibers [1]. In the following, some commercially available fibers in the market are introduced.

Table 1. Advantages and disadvantages of reinforcing fibers (adapted with permission from Ref. [1]).

Fiber	Advantages	Disadvantages
E-glass, S-glass	High strength Low cost	Low stiffness Short fatigue life High-temperature sensitivity
Aramid (Kevlar)	High tensile strength Low density	Low compressive strength High moisture absorption
Boron	High stiffness High compressive strength	High cost
Carbon (AS4, T300, IM7)	High strength High stiffness	Moderately high cost
Graphite (GY-70, Pitch)	Very high stiffness	Low strength High cost
Ceramic (Silicon, Carbide, Alumina)	High stiffness High use temperature	Low strength High cost

2.1. Carbon Fibers

Carbon fibers are widely used in the fabrication of advanced composites with different forms and ranges of stiffness and strength. The mechanical properties of carbon fibers are strongly dependent on how they are treated and manufactured by the organic precursor and processing conditions used [1]. The diameter of carbon fibers is about 5 to 10 μm (0.00020–0.00039 in), and they are composed mostly of carbon atoms (92 wt%). The main advantages of carbon fibers are their high tensile strength, high stiffness, low weight-to-strength ratio, high-temperature tolerance, low thermal expansion, and high chemical resistance, which have made carbon fibers the most widely used and very popular reinforcing fiber in various industries such as aerospace, civil, and motorsports. However, they are costly and expensive in comparison to similar fibers, such as basalt fibers, glass fiber, or plastic fibers [8]. In terms of the overall application, the association of composite companies and research institutes, Carbon Composites e.V. (CCeV), reported that defense and aerospace were the largest consumers of carbon fiber followed by sports/leisure sectors and wind turbines in the year 2013, as displayed in Figure 1 [9]. Due to the extraordinary properties of carbon fibers, they can be ideal reinforcing and matrix phases for composite materials requiring high specific strength (strength/weight ratio). As a carbon fiber, it may be dispersed in polymer matrices to deliver carbon/polymer composites and/or embedded in a carbonaceous matrix to construct carbon/carbon composites. Carbon fibers, either in form of unwoven or woven into fabric sheets, have been widely used in many applications such as aerospace, marine, and automotive industries [1].

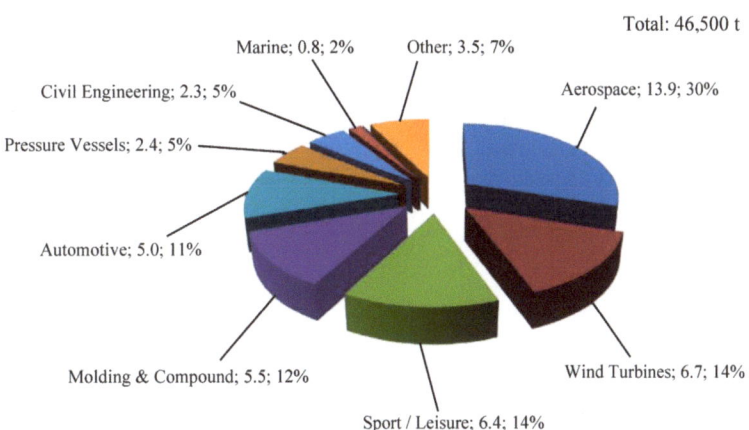

Figure 1. Carbon fiber global demand in the year 2013 (adapted with permission from Ref. [10]).

2.2. Glass Fibers

Glass fibers are commonly used in the fabrication of composites with low to medium performance because of their high tensile strength and low cost. However, they are not recommended to be utilized in composites with high performance due to their relatively low fatigue endurance, low stiffness, and rapid property degradation. Glass fibers are fabricated by the extrusion of a molten mixture of silica (SiO_2) and other oxides through small holes of a platinum bushing. Glass fiber diameters are in the range of 10–20 μm (0.4×10^{-3}–0.8×10^{-3} in). Glass fibers are amorphous and considered isotropic [1].

2.3. Kevlar (or Aramid) Fibers

Kevlar (or Aramid) fibers are organic fibers produced by dissolving the polymer (aromatic polyamide) in sulfuric acid and extruding it through small holes in a rotating device. Kevlar fiber diameter for composite application is typically 12 μm (0.5×10^{-3} in). Kevlar fibers deliver higher stiffness than glass fibers with low density (about half that of glass), excellent toughness, high tensile strength, and impact resistance, but they deliver very low transverse tensile strength and longitudinal compressive strength. They are very anisotropic mechanically and thermally due to their high molecular orientation [1].

2.4. Ceramic Fibers

Boron and other ceramic fibers, such as alumina (Al_2O_3) and silicon carbide (SiC), have high use temperature, high stiffness, and reasonably high strength. Ceramic fibers are not commonly blended or dispersed in polymeric matrices but are used with ceramic or metal matrices for high-temperature applications. Boron fiber-reinforced composites have limited usage for local stiffening and repair patching due to their high stiffness [1].

3. Carbon Fiber Processing from Precursors

To produce carbon fibers, a precursor is needed, and the choice of a precursor can be influenced by a variety of factors such as availability, cost, renewability, inorganic content, ease activation, and carbon yield. In the following, carbon fiber production from precursors is demonstrated.

3.1. PAN-Based Carbon Fibers

Polyacrylonitrile (PAN) is a synthetic and semicrystalline organic polymer resin. PAN is the most commonly used precursor for carbon fiber production and theoretically yields 68% carbon and delivers carbon fibers with a high elastic modulus (344 GPa) and high tensile strength (2070 MPa). PAN-based carbon fibers are stretched initially from 500% to 1300% and then thermostabilized in an oxygen atmosphere between 200 and 300 °C under tension. Afterward, fiber is carbonized (heat treatment under an inert atmosphere between 1000 and 1700 °C); then, the graphitization is conducted (heat treatment between 2500 and 3000 °C), and finally, through surface treatment and epoxy sizing, carbon fibers will be ready for use (Figure 2). Currently, PAN-based carbon fibers, due to their higher strength and moderate elastic modulus, occupied 90% of the carbon fiber market, and the remaining market is supported by carbon fibers derived from other precursors [11–13]. However, the main disadvantage of PAN-based carbon fibers is their high cost of processing.

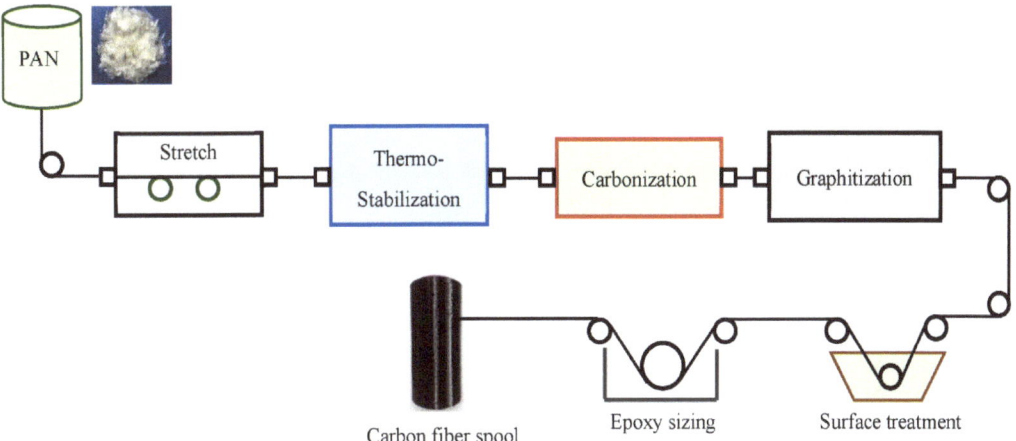

Figure 2. Carbon fiber production using PAN precursor.

3.2. Pitch-Based Carbon Fibers

Petroleum pitch, as a viscoelastic polymer, is another common precursor used for carbon fiber production. Both isotropic and mesophase pitches are utilized for the production of carbon fibers. The theoretical carbon yield of pitch precursors is 80%. Pitch-based carbon fibers are also produced with the same processes used for PAN-based fibers production without an expensive stretching process during heat treatment to have aligned crystallites (Figure 3) [14]. Pitch-based carbon fibers lead to lower tensile strength and higher elastic modulus (about 1050 GPa) than PAN-based carbon fibers [15–18]. Moreover, pitch-based carbon fibers present better thermal and electrical properties than carbon fibers produced by PAN precursors [11]. However, internal voids, surface defects, and other contaminations in the pitch structure lead to a decrease in the mechanical properties of produced carbon fibers [19].

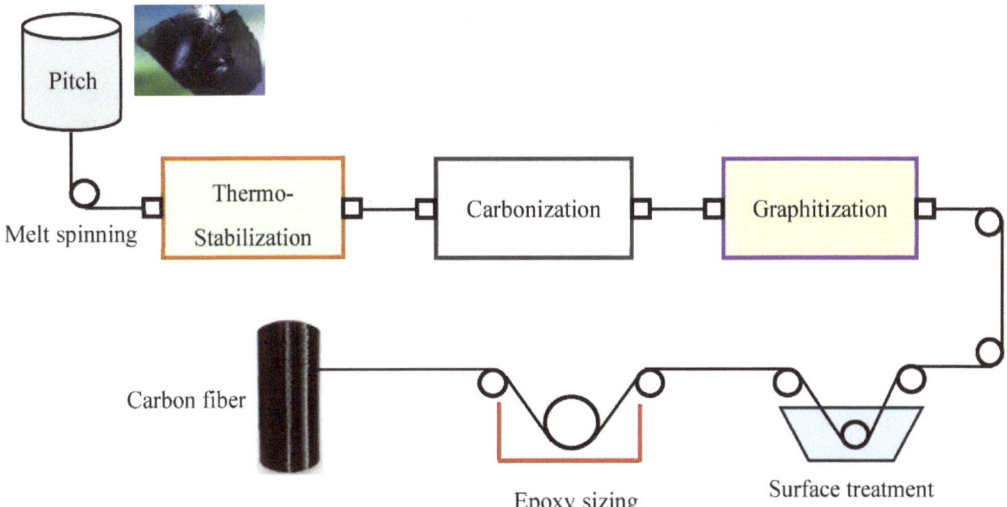

Figure 3. Carbon fiber production using petroleum pitch precursor.

3.3. Lignin-Based Carbon Fibers

Lignin is an organic polymer forming key structural materials in the tissues of most plants. Lignins are polymers made by cross-linking phenolic precursors [20]. Lignin contains a high carbon percentage (60–65%), which leads to high carbon yield after fiber processing, hence making it an alternative to PAN precursor for carbon fiber production. Lignin-based carbon fibers are produced by the melt spinning under an inert atmosphere. The lignin fiber is then oxidatively thermostabilized and carbonized and finally graphitized with the surface treatment (Figure 4).

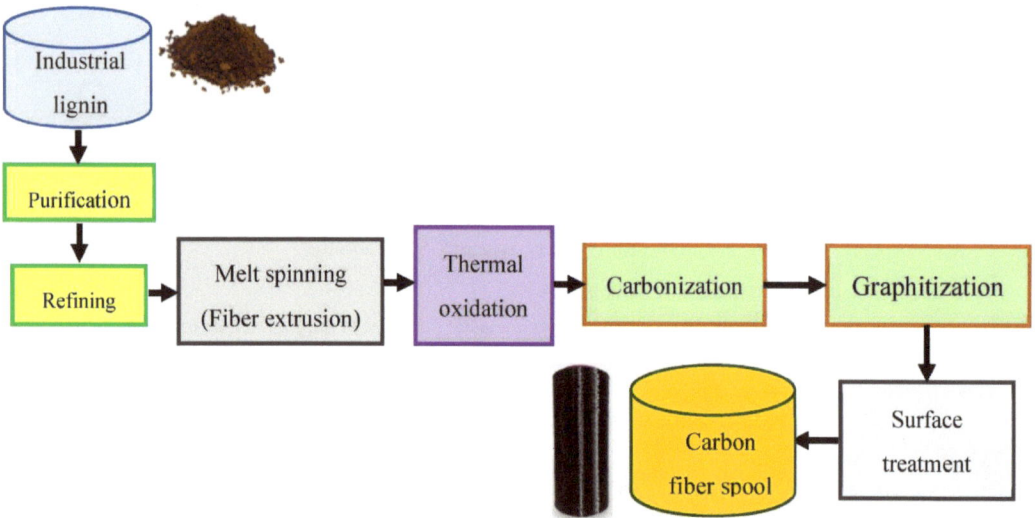

Figure 4. Process of fiber production from industrial lignin.

The mechanical properties of lignin-based carbon fibers are not high enough to meet some criteria for structural applications such as the requirement set by the US automotive industry [21]. It is found that the lignin-based carbon fibers presented a tensile strength of 388–1060 MPa and elastic modulus of 40 GPa [22,23]; however, they exhibited a weak blending with the polymer matrices [24,25].

In addition, the long stabilization time (over 100 h) [26] is problematic as well as the challenges in the melt spinning of kraft lignin without plasticizing additives [22,27]. The process of obtaining high-strength carbon fiber from the lignin is complex and needs careful control of melt spinning conditions, ramping profiles, and treatment temperatures. The lignin should have a low enough melt flow temperature to be melt spun without polymerizing during extrusion, but a high enough glass transition temperature for fiber stabilization is required to proceed at an acceptable rate. Although lignin has some limitations and does not lead to high-strength carbon fibers obtained from other precursors (PAN and pitch), it is renewable, very inexpensive, and is already oxidized, leading to being oxidatively thermostabilized at higher rates than either PAN or pitch [21].

3.4. Cellulose-Based Carbon Fibers

Cellulose, the most abundant organic polymer on earth, is abundantly found in the primary cell wall of green plants, the oomycetes, and many forms of algae. Overall, 90% of cotton fiber content, 40–50% of wood content, and approximately 57% of dried hemp content is cellulose [28]. Cellulose-based carbon fibers can be extracted from cotton, wood, hemp, flax, sisal, rayon, and linen. However, among them, rayon has been utilized commercially and extensively studied. The molecular orientation of cellulose, in contrast to lignin, significantly influences the mechanical properties of carbon fibers [29].

Similar to the process of PAN- and pitch-based carbon fibers, thermal oxidation/stabilization, carbonization, and an optional graphitization as well as the surface treatment are utilized to convert cellulose (from plants) to carbon fibers [30]. Heating the fiber at $T > 400\ °C$ leads to cellulose pyrolyzing and then by heating to $T > 1000\ °C$, the carbonization is completed. Finally, the fiber is graphitized by heating at $T > 2000\ °C$ with 100% carbon for all practical applications. Cellulose-based carbon fibers have a low elastic modulus; for example, rayon-based carbon fibers have a low elastic modulus of 27.6 GPa. To obtain high-modulus carbon fibers from cellulose precursors such as rayon, carbon fibers should be stretched at the final heat treatment temperature, which is a costly process [14].

The production of cellulose-based carbon fibers is mainly inhibited due to the low carbon content of cellulose (44.4%) and delivering low-yield carbon fibers (10–30% after carbonization) because of releasing carbon-containing gases such as CO and CO_2 in the process [30]. A comparison of material properties of carbon fibers produced using different precursors is shown in Table 2 [31].

Table 2. Properties of carbon fibers from different precursors (Adapted with permission from Ref. [31]).

Carbon Fiber	Diameter (μm)	Density (g/cm^3)	Elastic Modulus (GPa)	Tensile Strength (MPa)	Elongation at Break (%)
PAN-based carbon fiber	5–10	1.7–1.8	200–500	3500–6300	0.8–2.2
Pitch-based carbon fiber	10–11	1.8–2.2	150–900	1300–3100	0.3–0.9
Lignin-based carbon fibers	—	—	40	388–1060	—
Rayon-based carbon fiber	5–10	1.4–1.5	40–100	500–1200	—
Lyocell-based carbon fiber	8	—	90–100	900–1100	1–1.1

4. Applications of Carbon Fibers

After the end of classic wars in the world and changes in the political situation, the usage of carbon fibers in military industries has decreased because of a major cut in defense. Hence, commercial applications of carbon fibers have grown extensively in many industries. Due to rapid development and advance in the composite field, some applications of carbon fibers may now have been discontinued and replaced with new applications in new technologies. Figure 5 portrays, briefly, various applications of carbon fibers. In the following, functional and structural applications of carbon fibers are explained in detail with some practical examples.

4.1. Functional Applications of Carbon Fibers

Carbon fibers are widely used as a functional material with applications in various industries as explained in the following.

4.1.1. Molecular Sieves

Molecular sieves are materials with uniform size pores or very small holes. These pore diameters are similar in size to small molecules, and thus, large molecules cannot enter or be adsorbed, while smaller molecules can [32]. Figure 6 shows a carbon molecular sieve. Molecular sieves produced from carbon fiber composites are able to absorb CO_2 emitted from gas turbines and coal-fired power plants. To produce this product capable of absorbing CO_2, a pitch-based chopped fiber-reinforced phenolic resin composite is activated in steam, O_2, or CO_2 at 850 °C. The molecular sieve has a pore volume with mesopores of 2–50 nm and a large surface area. It also has macropores (50–100 mm) allowing sufficient fluid flow with low-pressure drop. The molecular sieve can also be used for the removal of CO_2 from fuel cells or natural gases [33].

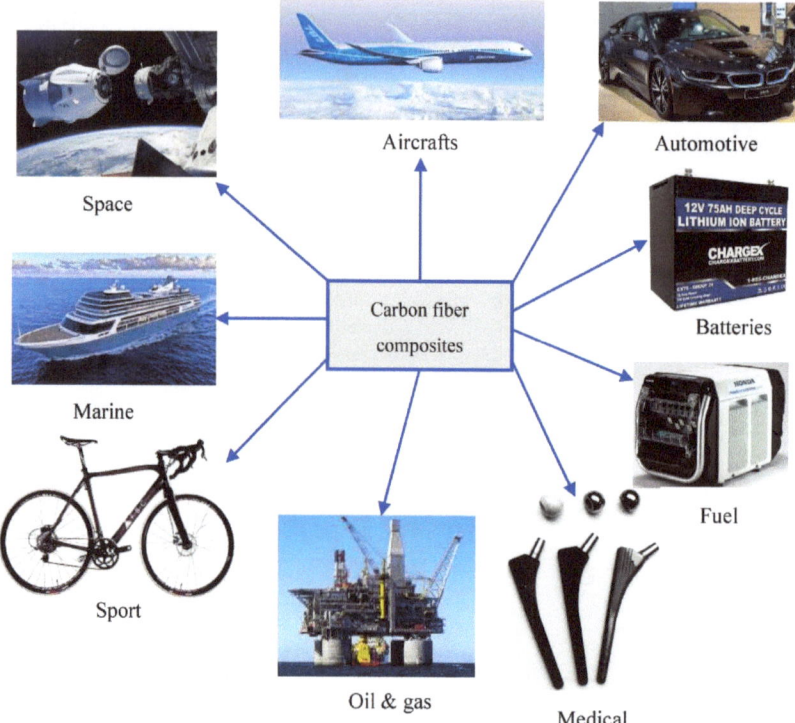

Figure 5. Various applications of carbon fibers.

4.1.2. Catalysts

Carbon fibers have the potential to be used as catalyst support by making a porous carbon fiber carbon composite with a density of 40.2 g/cm^3, a significant volume of mesopores (2–50 nm), and macropores (50–100 mm) allowing excellent fluid flow with minimal pressure drop. The procedure to reach this product include: slurring Fortafil P200 PAN-based carbon fiber in water with a phenolic resin, vacuum molding, drying at 50 °C, curing for 3 h at 130 °C and carbonization in a flow of N_2 at 650 °C [33–35].

4.1.3. Electrical Conduction

The early application of PAN-based carbon fiber, when it was developed in the 1960–1970 era, was in wall panels to prevent heat loss and keep the room warm. Nowadays, with the advances in technology, the PAN-based woven carbon fiber can be used as a large area temperature sensor, a portable heating unit, a flexible heating element, an electrical switching function, a warning and control device, and a temperature management system [36].

Figure 6. Carbon molecular sieve (reprinted with permission from Ref. [37]).

4.1.4. Electrodes

Carbon fibers are used in the fabrication of electrodes. For example, carbon fiber microelectrodes are utilized to extracellularly record neuronal action potentials [38] and to detect electrochemical signals in vivo and in vitro [39]. Furthermore, they have been used for the detection of catecholamines such as norepinephrine or dopamine and other oxidizable biological species such as nitric oxide [40]. Carbon fiber microelectrodes can be used in sensing tissue oxygen levels at a micrometer scale. In another application, the immobilization of DNA molecules or carbon nanotubes onto carbon fiber microelectrodes leads to making microsensors for various analytes [41].

The carbon fiber microelectrodes are graphite monofilaments with a 7 µm diameter. To construct carbon fiber microelectrodes, carbon filaments are placed in a mechanically supportive and electrically insulating borosilicate glass tube or plastic sheathing and an uninsulated carbon tip protruded from the sheathing by 10 µm to a few 100 µm (Figure 7). The carbon tip creates an electroactive surface for picking up spikes from the near vicinity neurons and/or surface for electron transfer in micro-biosensors applications and electrochemical measurements [41].

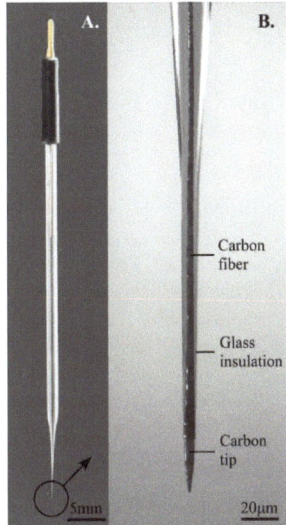

Figure 7. (**A**) View of a carbon fiber microelectrode, and (**B**) microstructure of the tip (reprinted with permission from Ref. [41]).

4.1.5. Energy Storage Devices

Electrochemical energy storage devices, such as fuel cells, batteries, and electrochemical capacitors, act as portable or stationary stores of electric power for later use and thereby are crucial for expanding the contribution of sustainable and renewable energy resources. Carbon fibers, due to their exceptional properties, can be used in the construction of electrodes for energy storage devices.

Rechargeable lithium-ion batteries (LIBs), due to their lightweight, high energy density, long lifespan, and environmentally friendly nature, have been widely used in portable electronics, communication devices, transportation, hybrid electric vehicles, and grid-scale applications. However, with the rapid development of electric vehicles and consumer electronics as well as the increasing demand for clean energy, more advanced LIBs with longer life, higher capacity and performance, enhanced charging speed, and improved safety are urgently required. Amorphous carbon fiber, with a proper heat treatment, has a high discharging capacity for the anode material of LIBs. Pitch-based carbon fiber has been utilized for anodes of rechargeable LIBs [42,43]. Figure 8 displays a typical Li-ion cell using a Li_2O cathode and a carbon compound anode which is separated by a microporous membrane, utilizing a non-aqueous electrolyte such as a Li salt dispersed in a mixture of alkyl carbonates [36]. LIBs generate DC power by using chemical reactions. When batteries are charged and discharged, lithium ions move back and forth between the electrodes (anode and cathode). Generally, the cathode material is made of cobalt-, nickel- or manganese-based transition metal oxides, and the anode material is made of graphite. Both the anode and cathode are fabricated using a stacked structure, and the lithium ions are placed between layers. Within charging, the lithium ions move from the cathode to the anode, while within discharging, the lithium ions move from the anode to the cathode (Figure 8) [44].

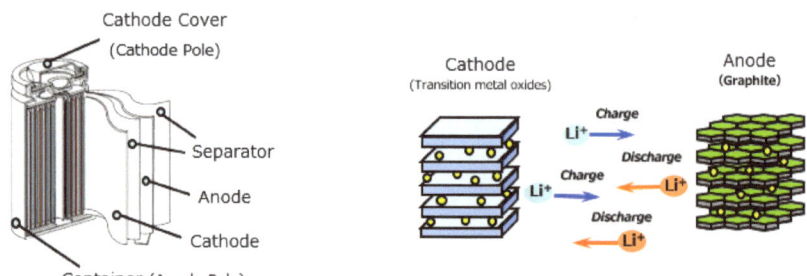

Figure 8. A rechargeable LIB with its components and function during charging and recharging (reprinted with permission from Ref. [44]).

The polymer electrolyte membrane fuel cell (PEMFC) can be a good candidate as a power source for future passenger vehicles due to its high-power density at a relatively low operating temperature of about 80 °C. Figure 9 depicts a layered PEMFC made of various components including end plates, bipolar plates, the gas diffusion layer (GDL), and the membrane electrode assembly (MEA). The bipolar plate is the main component of the PEMFC stack, and its development has a significant effect on the performance of the PEMFC. Hence, a carbon fiber composite can be utilized to develop the bipolar plate of the PEMFC due to the high thermal and electrical conductivities of carbon/epoxy composite as well as its high specific stiffness and strength [45].

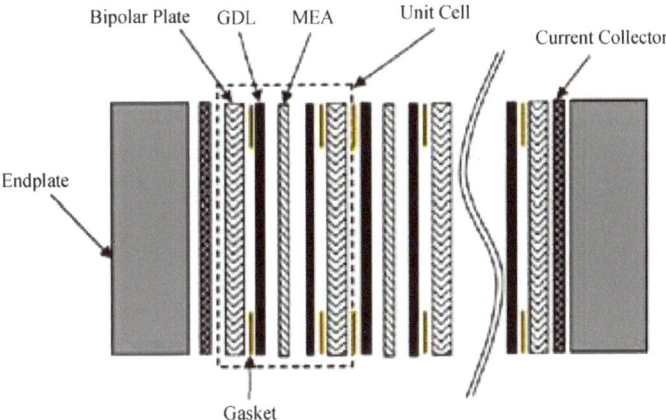

Figure 9. Schematic view of the PEMFC (adapted with permission from Ref. [45]).

Fiber-shaped supercapacitors due to their higher performance are promising energy storage devices for future portable electronic devices. A fiber-shaped asymmetric supercapacitor (ASC) device is composed of metal oxides and directly grown on a flexible and conductive carbon fiber substrate which makes a large work function difference. Specifically, carbon fiber/MoO_3 (CF/MoO_3) and carbon fiber/MnO_2 (CF/MnO_2) are produced using a simple electrodeposition approach. The solid fiber-shaped ASC device is then assembled with CF/MoO_3 as the negative electrode and CF/MnO_2 as the positive electrode. The high work function difference between the high conductivity of the carbon fiber substrate and the metal oxides leads to the ASC device with notable performance. Figure 10 illustrates the overall procedure to assemble the ASC device based on CF/MoO_3 as the negative electrode and CF/MnO_2 as the positive electrode [46].

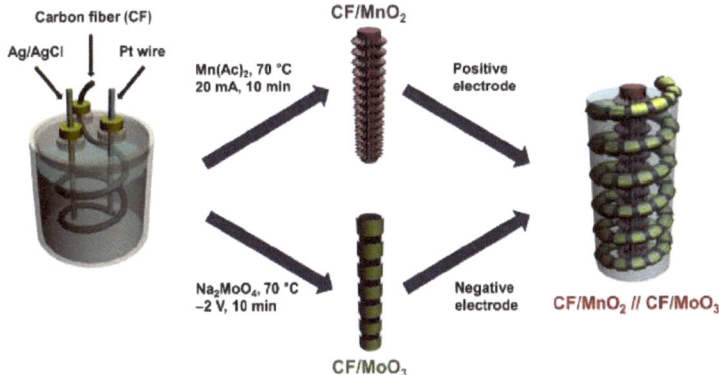

Figure 10. A schematic assembly of a fiber-shaped asymmetric supercapacitor (ASC) device based on CF/MnO_2 as the positive electrode and CF/MoO_3 as the negative electrode, respectively (reprinted with permission from Ref. [46]).

4.1.6. Insulation

Carbon fibers are fire resistant and present high thermal insulation with low electrical conductivity and low smoke emission as well as weight saving. Current applications of carbon fibers as an insulator are in aircraft fire blockers, aircraft fuselage thermal insulation, personal insulation, fire protective clothing, and fire-retardant insulation boards for special lightweight applications. Carbon fibers provide a measure of sound insulation in an aircraft.

They can also be used in packing materials and gaskets, since they have higher thermal and oxidative stability [36]. They can also be used as electromagnetic interface shields in cement matrices for building purposes [47].

4.2. Structural Applications of Carbon Fibers

Using high-strength carbon fiber-reinforced composites for structural applications is economical by reducing the weight of final structures. Hence, carbon fibers in thermoset matrices can be used in many structural applications as described in the following.

4.2.1. Aerospace

Carbon fibers are widely used in the fabrication of aircraft components. Airbus Industries was the first civil aircraft manufacturer in the world that used carbon fiber-reinforced prepreg (CFRP) for the fabrication of parts of the primary structures of the Airbus A300. For example, in the Airbus 350 XWB, 53% of the used materials are CFRP including the wings, center wing box and keel beam, skin panels, tail cone, frames, doors, stringers and doublers (Figure 11) [48]. In 1990, CFRP was also adopted by Boeing as the primary airframe structure material. Overall, 50% of the total weight of the Boeing 787, including the frame and wings, was made of CFRP. Figure 12 displays a comparison between the Boeing 787 and Boeing 767 with aluminum as the main material (77% of the weight) [49].

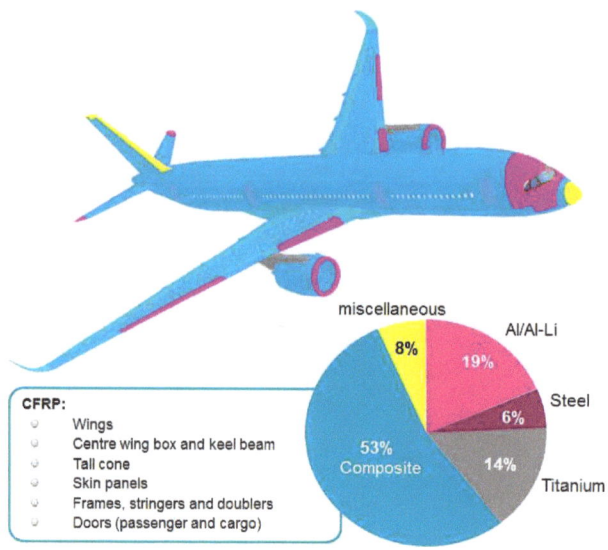

Figure 11. Materials used in the Airbus 350 XWB (reprinted with permission from Ref. [48]).

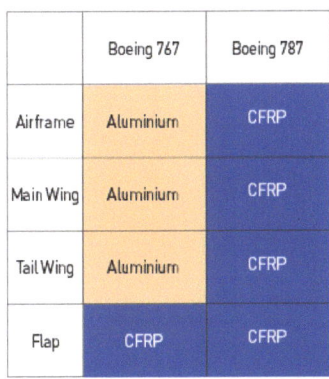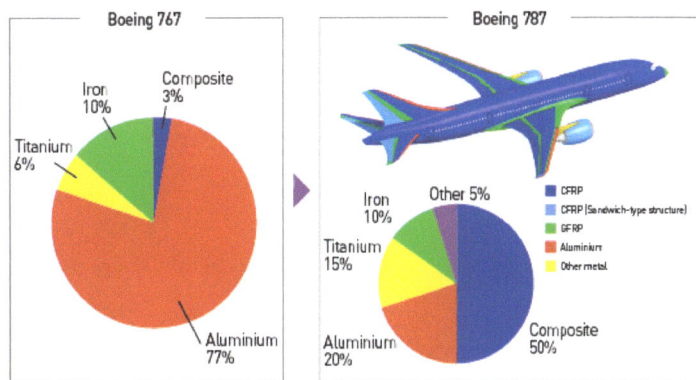

Figure 12. A comparison between the Boeing 767 and the Boeing 787 in using CFRP (reprinted with permission from Ref. [49]).

In past decades, composites have been used in space applications, and due to their outstanding material properties, their use is growing. Composite materials are used in components of human spaceflight vehicles, payloads, satellites, and launch vehicles used to throw these into space. Pressure vessels for fuel and gas storage and solid rocket motors are made of composite materials. Carbon fiber laminates are extensively utilized on satellites and payload support structures. Special high-strength carbon composites are used for the hottest components in rocket nozzles such as exit cones and throat. Carbon–carbon panels are utilized on the wing leading edge and the nose of space shuttles to protect them from high temperatures exceeding 2300 °F experienced during re-entry. Carbon fiber-reinforced phenolic is used to make ablative composites to absorb heat by changing states. The ablative heat shield was utilized in Apollo and Orion capsules, which will return humans to the moon and beyond [50]. Typical structural components of a space vehicle fabricated from CFRP are displayed in Figure 13 [51]. Furthermore, carbon fiber-reinforced composites are commonly utilized in aero engines, propeller blades, and Unmanned Aerial Vehicles (UAVs).

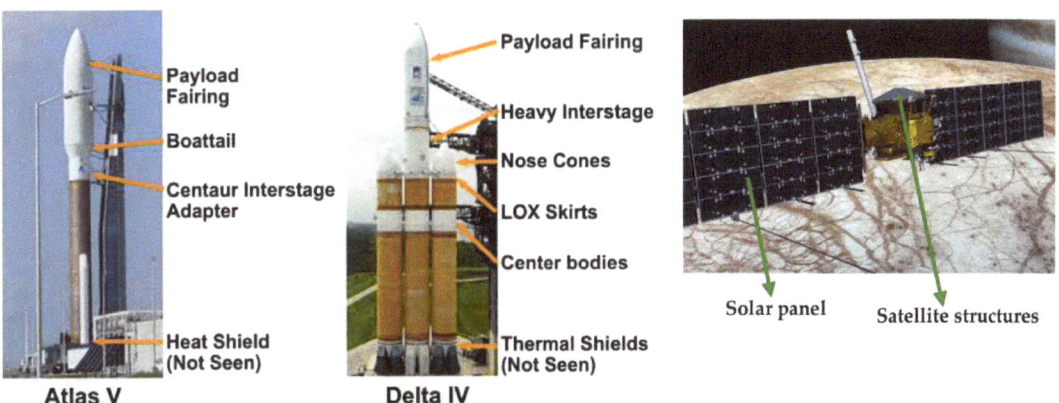

Figure 13. Schematic space vehicle (adapted with permission from Refs. [51,52]).

4.2.2. Marine

Sport boats are made of CFRP with a honeycomb core based on the dry prepreg approaches. A carbon fiber-reinforced composite racing catamaran (Team Philips) (Figure 14) made by Goss Challenges is one of the largest carbon composite structures fabricated in Europe (36.5 m long and 21 m wide, with an unstayed mast of 39 m high) [53].

Figure 14. A Team Philips catamaran (reprinted with permission from Ref. [54]).

In another example, the hull and beams of the 38 m long catamaran PlayStation were built from CFRP/AL honeycomb, and its mast was made of carbon fiber, which was 45 m above the water. This catamaran led to a new world record for boating from Miami to New York. The Consolidated Yacht company fabricated a tall mast (53.33 m) with two halves for the yacht from the carbon/epoxy prepreg. Generally, tall masts are built in several sections because of limitations of the curing oven, but a 59 m long mast made of carbon fiber prepreg was fabricated in one piece for the yacht Hyperion (Figure 15) [36].

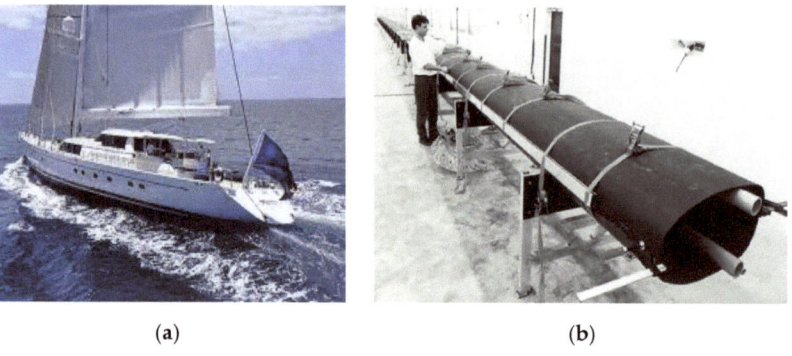

(a)   (b)

Figure 15. (a) Hyperion 2 yacht and (b) Hyperion 2 yacht mast with 59 m length fabricated from CFRP (adapted with permission from Refs. [36,55]).

4.2.3. Automotive

Recently, carbon fiber-reinforced composites have been used to fabricate automotive parts. It was reported that a bonnet fabricated from carbon fiber-reinforced composites can decrease its weight from 18 to 7.25 kg. In another example, carbon fiber was used in fabricating the hood of GM's Corvette to reduce its weight to 9.3 kg, saving 4.8 kg from the standard fiberglass SMC [56]. Furthermore, carbon fiber-reinforced composites can be used in the fabrication of the chassis, the body, and the interior of cars, where many automaker companies have used it in their products [36]. Carbon–carbon is also utilized in the fabrication of brakes and clutches. Suspension systems of cars also benefit from carbon fiber composites: for instance, BMW uses carbon fiber in the BMW Z22 model to reinforce the roof, tailgate, flooring, and side frames, which leads to 20 parts replacing 80 components with 50% less weight than a steel body [36]. CFRP can also be utilized to fabricate pushrods with 70% less weight than metal pushrods and decreasing noise and increasing engine efficiency [57].

Carbon fibers have been used in the fabrication of drive shafts for many years with more advantages than metallic drive shafts including less weight, improved mechanical properties, excellent torsional strength, good corrosion resistance, improved damping characteristics, high fatigue resistance and torsional compliance reducing shock loads on gears and universal joints. It was reported that the weight savings of pure aluminum, Al/carbon/epoxy composite, E-glass/epoxy composite, Kevlar/epoxy composite, and carbon/epoxy composite drive shafts were obtained, respectively, 46.157%, 53.865%, 36.87%, 64.615%, and 69.236% of the weight of the conventional steel drive shaft, where the carbon fiber-reinforced composite provides high strength and lighter components meeting the design requirements [58]. Initially, the cost of carbon fiber-reinforced composite drive shafts was a significant concern and drawback for industrial sectors to produce on commercial scales, and with the development of fibers and reduction in the fabrication cost, thousands of composite shafts are in service now in the industry today. Heavy goods vehicles and buses are also beneficial in using carbon fibers in their components. For example, CFRP leaf springs are 80% lighter than steel springs with the same spring rate and load-carrying capacity. A hybrid of carbon and glass bumper is used in buses and heavy trucks with higher corrosion resistance, superior vibration resistance, and much less weight than metallic ones [36]. Figure 16 describes briefly automotive parts which can be replaced with composite materials.

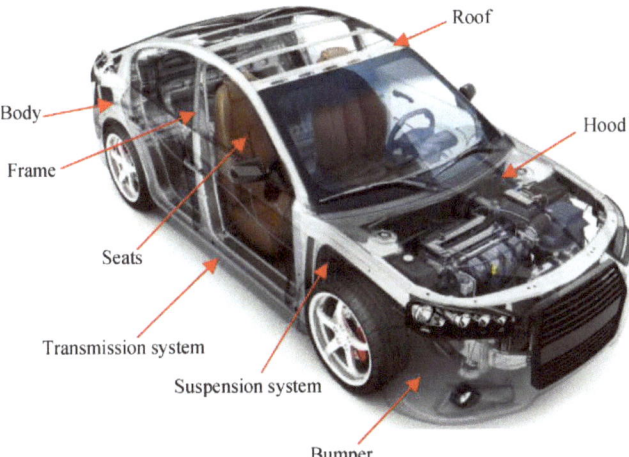

Figure 16. Automotive parts can be replaced with composites (adapted with permission from Ref. [59]).

4.2.4. Oil and Gas Extraction and Transmission Pipelines

The oil industry is now interested to extract oil from deep water at a depth of more than 1 km in areas such as the Gulf of Mexico, the Gulf of Guinea, the Caspian Sea, and Brazil. In the Brazil location, Petrobras/RB Falcon could reach a depth of 2777 m, and obviously, the risk factor increases with the depth increase, where the oil companies will need smart systems composed of composite structures. In offshore oil drilling installations, carbon fiber-reinforced composites can be used in drilling risers, as shown in Figure 17 [1].

At deep water, CFRP risers, rig, and tendons components lead to lower costs and offer considerable savings over conventional materials. The standard platform installation is not suitable, and its modification is extremely costly. In water deeper than 1600 m, a tension leg platform, based on carbon fiber composite cables as tethers, leads to lower costs. Spencer composites through over-winding carbon fiber and fiberglass onto Ti tubing fabricates drilling rise about 15 m long and 560 mm diameter, weighing 20% less, costing 40% less, and with an extended fatigue life [36].

Figure 17. Composite drilling riser for offshore oil drilling: 15 m long, 59 cm inside diameter, 315 bar pressure; manufactured for Norske Conoco A/S and other oil companies (reprinted with permission from Ref. [1]).

There are high lengths of oil and gas pipelines in the world, and most of the highest ones are in North American countries. Including these pipelines, it is estimated that 60% of the world's oil and gas transmission pipelines have been used for more than 40 years and are at risk of defects. External corrosion, internal corrosion, erosion, abrasion, dents and cracks are typical defects that may occur in oil and gas transmission pipelines and potentially lead to disasters and catastrophic incidents. If we do not monitor and repair these pipelines, these defects could make expensive and potentially deadly outcomes to the operators and owners of the pipelines as well as disasters to civilians. Therefore, to prevent further disasters and damages to other intact sections of the pipelines, an urgent repair or replacement of defects at the damaged locations is required, depending on the severity of the defects. Advanced composite wraps have been widely used within oil and gas transmission pipelines over the past two decades for the permanent repair and reinforcement of sections of the pipe wall, which have been weakened due to the defects such as corrosion, cracks, etc. (Figure 18a) [60]. Unidirectional carbon fiber-reinforced composite can be used to fabricate composite wraps or sleeves with high strength to withstand high pressure applied by ongoing oil or gas (Figure 18b). The composite wrap exceeds the yield strength of the original pipe and delivers a more economical repair solution than other approaches [61].

(a) Defect   (b) Carbon composite repair

Figure 18. (a) Typical defects on transmission pipelines [60], and (b) carbon fiber-reinforced composite wraps to repair defected sections of pipelines [61] (adapted with permission from Refs. [60,61]).

4.2.5. Biomedical Devices and Sport Goods

Carbon fibers have been used in biomedical applications such as prosthetic devices, artificial limb parts, and implants (Figure 19). Furthermore, CFRP has been used in leisure and sports products such as golf clubs, skis, fishing poles, tennis rackets, and bicycles. An example of a carbon fiber-reinforced composite bicycle frame is displayed in Figure 20. For instance, the Applied Composite Technology company fabricates artificial feet from carbon fiber epoxy prepreg with high efficiency for athletes where it was reported that a sports event participant with an artificial foot ran 100 m in 11.3 s in the 1996 Atlanta Paralympic Games. In another example, Ossur, an Icelandic company, used braided carbon fiber to fabricate a custom-made socket for a limb amputee with a high rate of conformity. Carbon fibers have been widely used in dental restorations, implants and prostheses as well as other medical devices [36].

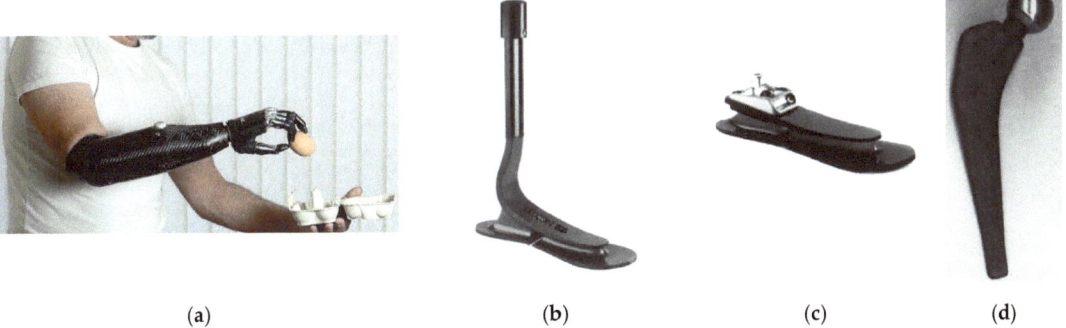

(a)　　　　　(b)　　　　　(c)　　　　　(d)

Figure 19. (a) Carbon composite mechanical hand [62], (b) carbon/epoxy composite leg prosthesis, (c) carbon/epoxy foot prosthesis, and (d) carbon/polysulfone hip prosthesis [1] (adapted with permission from Refs. [1,62]).

Figure 20. Carbon/epoxy composite bicycle frame weighing much less than the corresponding steel frame (reprinted with permission from Ref. [63]).

5. Challenges in Carbon Fibers Production

The main obstacle in the production of carbon fibers is the high cost of precursors which includes more than 50% of the total cost of carbon fiber production [21]. The production of commercial-grade carbon fibers from pitch and PAN precursors is expensive due to the high cost of raw materials and complex processing steps with USD 15–20 per kg for pitch-based carbon fibers and USD 18–35 per kg for PAN-based carbon fibers [64]. Meanwhile, Kevlar and glass fibers cost, respectively, USD 23 and USD 2 per kg. It was also reported that PAN-based carbon fibers cost USD 35 per kg for use in the automotive industry [65]. If the cost of carbon fiber fabrication decreases to USD 11 per kg, carbon fiber-based automotive parts can be cost-competitive with steel-based automotive components with an average cost of USD 5 per kg. By using carbon fiber-reinforced composites in the automotive industry in the fabrication of automobile bodies, engine parts, transmission shafts, interior components, suspension systems, brakes, etc., the automobile manufacturing cost can be driven down by 80% due to a decrease in tooling and simpler manufacturing and assembling procedures as well as reducing the fuel consumption and cost by having a lighter weight automobile than a steel-based one. Using lightweight carbon fibers in batteries of electric vehicles could also be cost-effective by reducing the electricity consumption by cutting the weight of batteries where the lightest produced batteries by Tesla weigh over 450 kg [66].

To use carbon fibers in industrial applications, two essential criteria need to be considered: the cost (price per kilogram) and performance (mechanical/thermal/electrical properties) in comparison to conventionally used materials such as steel, aluminum, etc. The US Department of Energy [67] published the accepted minimum properties of carbon fibers which can be used as a reference in the production of low-cost carbon fibers. Over the past two decades, many research studies have been dedicated to producing low-cost carbon fibers with sufficient mechanical properties from inexpensive resources such as biomass (cellulose and lignin), coal, and petroleum by-products [22,68–70]. One of the natural materials that could be used as a precursor in carbon fiber production is petroleum asphaltene; its potential and applications, as well as carbon fiber processing from asphaltene, will be discussed in the next sections.

6. Asphaltene-Based Carbon Fibers

Asphaltene is a molecular substance found in crude oil, along with resins, aromatic hydrocarbons, and saturates. Asphaltene consists of carbon, hydrogen, nitrogen, oxygen, sulfur, vanadium, and nickel. Heavy oils, oil sands, and bitumen have higher proportions of asphaltene than light oils. Asphaltene can be an ideal candidate to be used as a precursor to producing carbon fibers due to their low cost (USD 0.05 per kg) with abundant resources and high carbon (C) to hydrogen (H) ratio (1:1.2), depending on the asphaltene source [71]. Hence, low-cost and high-performance carbon fibers can be derived from asphaltene precursors for wide usages from functional to structural applications. As described before, carbon fiber production generally consists of several main procedures, i.e., spinning, oxidation/stabilization, carbonization, and graphitization. Stabilization is the slowest and most energy-consuming process affecting both mechanical properties and the cost of carbon fiber production [72–74].

Due to its high carbon content, heteroatom, aromaticity, double bond equivalent (DBE), and polar functional groups, asphaltene potentially can be used for the synthesis and development of functional carbonaceous materials and structures [75,76]. Asphaltene-based carbon fibers could be fabricated through optimization of the process of polymer blend and nano-reinforcement for applications in functional composites (such as energy storage devices, gas adsorbents, water treatment, etc.) and structural composites (such as automotive components, aerospace structures, marine structure, biomedical devices, sports goods, etc.). Producing carbon fibers from inexpensive feedstock such as petroleum asphaltene could reduce the precursor cost by about 90% and cut down the carbon fiber production cost from about USD 18–35 per kg (PAN-based carbon fibers) to less than USD 9 per kg [77].

Pitch (including asphaltene)-based precursors have a lower cost than PAN-based precursors with a higher elastic modulus and extremely lower tensile strength and failure strain than PAN-based carbon fibers. To have a low-cost carbon fiber with acceptable mechanical properties required for various applications, the development of a new carbon fiber precursor through a blend of asphaltene and PAN precursors can enhance fiber spinning, which results in a combination of lower cost, better ductility, and improved modulus and strength without the brittleness observed in the pitch-based carbon fibers. This hybrid (asphaltene/PAN) precursor can be further reinforced with nanofillers, such as carbon nanotubes (CNTs), graphene, and nanocrystal cellulose (NCC) to improve the mechanical properties.

7. Asphaltene-Based Carbon Fiber Processing Steps

The raw asphaltene is provided from natural resources without further purification. Then, the elemental composition is determined, and samples are finely ground before the measurement and weight into crucibles. Green fibers are produced through a melt spinning process where the raw asphaltene is melted at a temperature of 197 °C. It is necessary to prevent fiber from melting during carbonization at high temperatures. Then, an oxidative stabilization process is conducted on asphaltene fibers. It was found that by oxidative stabilization without acid pretreatment, the fibers will not retain their fibrous shape, while with acid pretreatment (OF_{HNO_3}), a visible fibrous shape of asphaltene fibers is obtained (Figure 21) [78].

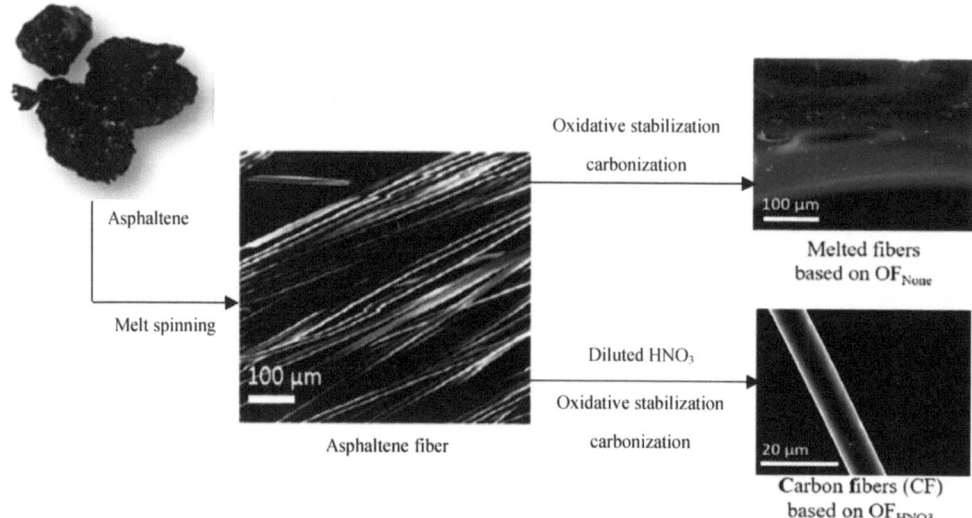

Figure 21. Asphaltene-based carbon fiber manufacturing process (reprinted with permission from Ref. [78]).

8. Main Challenges of Asphaltene-Based Carbon Fibers

Similar to any newly developed fibers, producing asphaltene-based carbon fibers will have some challenges, which should be addressed and solved before commercialization. The main challenges that we may face in producing asphaltene-based carbon fibers are:

- Increasing mechanical properties and enhancing the physical properties for novel applications.
- Meeting productivity requirements for novel applications.
- Purification of asphaltene precursors and the defects mitigation of fiber during the extrusion process.
- The impurities and sulfur content can highly impact the spinnability and mechanical properties of the resulting fibers.

To solve and remove these problems, it is suggested to consider the following solutions:

- Using fractionation process of asphaltene by using solvent (such tetrahydrofuran) to separate impurities/insoluble.
- Applying electrospinning to fabricate nanofibers to reduce the defects.
- Exploring an economical approach for reducing/eliminating these impurities and using greener solvents that can be readily recycled and do not cause any health concerns.

9. Required Performance and Properties of Asphaltene-Based Carbon Fibers

In structural composites, the high specific stiffness (stiffness per unit weight) and high specific strength (strength per unit weight) of the reinforcing fibers lead to a high-performance and lightweight structure. Therefore, it is necessary to develop and optimize asphaltene-based carbon fibers to meet the requirements of structural applications. Moreover, to use asphaltene-based carbon fibers in functional applications such as batteries, supercapacitors, and fuel cells, they should have enough electrical conductivity required for the electrodes of energy storage devices.

In addition to the above technical requirements of carbon fibers, some techno-economical challenges should be overcome. For instance, by reducing the cost of the precursor materials (down to USD 10 per kg) as well as improving their mechanical properties, we can attract the attention of the automotive industry. Asphaltene-based carbon fibers by reducing the density by 30% (1.8 g/cm^3 vs. 2.54 g/cm^3) and elastic (tensile) modulus of 70 GPa can compete favorably in the glass fiber reinforcement market. To compete with other fibrous composites, the cost of carbon fiber production should be reduced where we can achieve this goal through the newly developed asphaltene-based carbon fibers with the lower cost of the precursor and processing. It was also found that an electrical conductivity of 5 S/cm can be sufficient for the electrodes of energy storage devices. Accordingly, if we could, firstly, produce asphaltene-based carbon fibers with composite tensile strength: 50–500 MPa, composite elastic modulus: 50–100 GPa, and composite electrical conductivity: >5 S/cm, we can compete with other low-performance fiber-reinforced composites as well as lowering the processing and fabrication costs.

10. Target Markets of Asphaltene-Based Carbon Fibers
10.1. Functional Composites
10.1.1. Energy Storage Devices

Graphite (as carbon fiber) commercially is used as an anode material of LIBs delivering a limited capacity (372 mAhg^{-1}) [79]. Conventional graphite provides limited storage capacity and ion channels because of its stacked sheets. To remove the limitations of graphite, Zn, Mn, Co or Fe alloyed lithium, and metal oxides (e.g., Fe_2O_3, Co_3O_4, Mn_2O_3 and $ZnMn_2O_4$) with relatively higher capacities can be employed to replace the graphite anode [80]. Although metal oxides as LIB anodes lead to high capacity, their industrial application is hindered by the rapid capacity fading and electrode disintegration during Li$^+$ insertion and extraction due to their inherent poor electrical conductivity.

To solve these problems, significant research studies have been dedicated to stabilizing metal oxides and reducing their pulverization during cycling by accommodating the volume change. It is found that metal oxide/porous carbon composites are a reliable electrode material with superior electrochemical properties and mechanical stability due to their high electrical conductivity, buffering effect, and low activity of the carbon support [81]. In previous research works, porous carbon as a support has an important role to relax stresses and prevent the pulverization and aggregation of metal oxide nanoparticles. $ZnMn_2O_4$, due to the low oxidation potentials of manganese (1.5 V) and zinc (1.2 V) and its high capacity (784 mAh g^{-1}), has attracted much attention to be used as the electrode material to increase the output voltage of LIBs [81]. Furthermore, the low cost and environmental friendliness of $ZnMn_2O_4$, in comparison to Co or Fe-based oxide, are attractive for both governmental and industrial sectors. The industrial application of $ZnMn_2O_4$/porous carbon as electrodes of LIBs can develop a low-cost and environmentally friendly mechanism for the delivery of $ZnMn_2O_4$/porous carbon for LIBs. Asphaltene-based carbon fiber is a good candidate to construct a $ZnMn_2O_4$/porous carbon framework through the template synthesis of a 3D porous carbon framework and incorporation with $ZnMn_2O_4$ (Figure 22) [82]. Asphaltene-based carbon fibers, due to their lower cost, lead to the fabrication of inexpensive LIBs for clean energy production.

It was shown that high energy and power values can be acquired from built electrochemical double-layer supercapacitor (EDLS) cells extracted from asphaltene precursors [83]. N, S-codoped activated carbon extracted from the asphaltene has been prepared and characterized to be used as an electrode material for an electric double-layer capacitor. The derived activated carbon contained both nitrogen (1.17 wt%) and sulfur (0.32 wt%) with a high surface area of 2558 m^2/g and a mesopore volume of 0.98 cm^3/g compared to the raw asphaltene. The asphaltene-based activated carbon delivered a high specific capacitance of 128 F/g at 0.5 A/g in 1 M tetraethylammonium tetrafluoroborate electrolyte with acceptable fatigue stability after 5000 cycles. Therefore, asphaltene-based carbon is able to be used as an electrode material to fabricate high-performance and efficient supercapacitors [84]. Asphaltene was used as a precursor to synthesize porous carbon fibers to be used as an electrode material for high-performance supercapacitors with superior capacitance related to the synergistic effects of the high specific area and abundant micropores of porous carbon fibers [85].

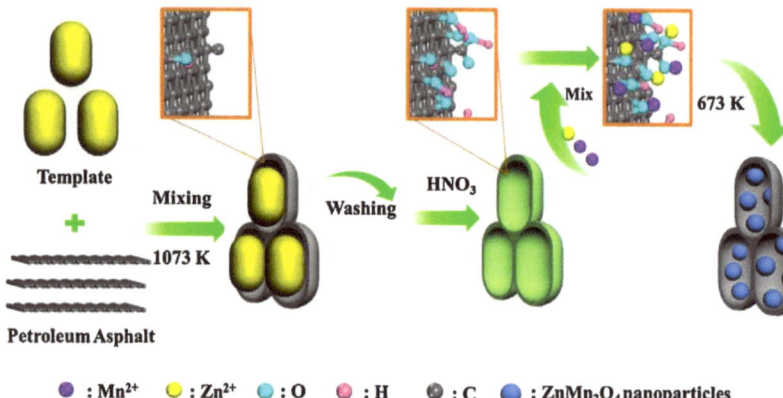

Figure 22. Schematic synthesis of 3D porous carbon framework made of asphaltene and incorporation with ZnMn$_2$O$_4$ (reprinted with permission from Ref. [82]).

10.1.2. Oil and Gas Absorption

The natural graphene of the raw asphaltene can be utilized to prepare a graphene–polyurethane sponge (GPU) to separate oil from water through a facile and inexpensive route of dip-coated sponge carbonization. In this process, low-value petroleum asphaltene and polyurethane sponges were used, respectively, as the dip-coating reagent and template (Figure 23). The GPU presents an excellent oil absorption performance, which is higher than other oil absorbents, as well as good recyclability. It also can be used in the pollution control of split oil [86].

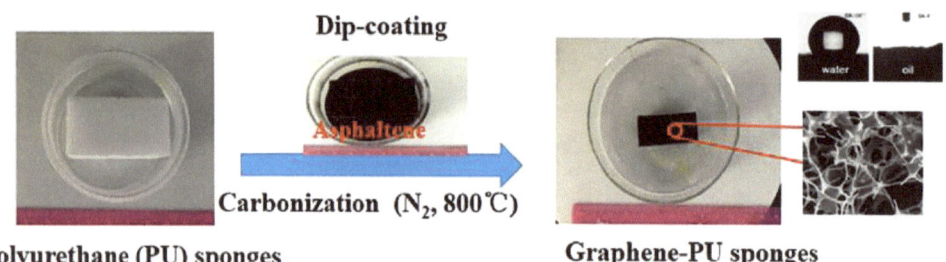

Figure 23. High-quality asphaltene-based graphene–polyurethane sponges (GPU) as absorbent (reprinted with permission from Ref. [86]).

Asphaltene, as a low-cost and abundant crude oil by-product, has also the potential for the production of high-quality carbon nanomaterials for the application of gas separation. It was found that asphaltene can be used as a precursor for the fabrication of microporous activated carbon absorbents to capture CO_2 [87]. A novel nitrated asphaltene-derived absorbent (Asf-Nitro) was produced using facile isolation and modification procedures. Asf-Nitro absorbent presented superior dispersive interactions (197.50 ± 1.12 mJm^{-2} at 423 K) in comparison to unmodified asphaltenes [88]. Nitrogen-doped asphaltene-based porous carbon nanosheets have an excellent ability to absorb CO_2 because of their developed pore structure and surface nitrogen-containing groups [89]. Hence, abundant and economically low-value asphaltene obtained from the petroleum industry could be a valuable source for the production of a variety of low-cost and highly effective gas absorbents and separators in various industrial applications.

10.1.3. Insulation

Asphaltene-based carbon fibers can be used for the thermal insulation of oil pipelines. A formation of asphaltene-based carbon fiber reinforced resin and paraffin layer can be utilized for oil pipelines in permafrost, and it can provide (1) anti-corrosion insulation and (2) thermal insulation because of the low coefficient of thermal conductivity of the asphaltene-based carbon fibers [90]. Therefore, asphaltene-based carbon fibers may be considered as a replacement for PAN-based carbon fibers for both structural and functional applications.

10.2. Structural Composites

In addition to the widespread use of carbon fibers in aerospace structures, their use in the automotive industry has increased dramatically due to increasing demand for fuel-efficient and lightweight vehicles meeting the new emission standard set by the European Union (95 g CO_2/km) and the United States (114 g CO_2/km) for 2020 [91,92]. The lower grades of carbon fibers than those used in the aerospace industry can be utilized in the automotive industry. Carbon fiber-reinforced composites can decrease the weight of automotive components by up to 60% and fuel consumption by up to 36–84%. However, the high cost of carbon fibers (about USD 18–35 per kg) is a barrier to the fabrication of automotive parts [93]. If we could decrease this cost to about USD 11–15 per kg, using carbon fiber-reinforced composites will be economical for automotive industries. Carbon fiber-reinforced composites in automobiles can be used as primary and secondary structures such as drive shafts, suspension parts, bumpers, dashboards, other interior components, etc. Hence, asphaltene-based carbon fibers as low-cost carbon fibers can be used as lightweight reinforcement for automotive composites.

Carbon fiber-reinforced composites can be used in building and construction as components of reinforced concrete or heat-resistive insulation. Fiber-reinforced polymer composites are an effective means for strengthening shear-deficient reinforced concrete flexural members. Asphaltene-based carbon fibers, due to their lower cost as well as relative strength and stiffness, can be a good replacement for carbon fibers used in building applications.

Moreover, carbon fiber-reinforced composites are widely used in the fabrication of medical prostheses and implants. For example, prosthetic rehabilitation costs USD 5000 to USD 50,000 and requires replacement every 3–5 years due to wear and tear [94]. Hence, low-cost asphaltene-based carbon-fiber reinforced composites are promising carbon fibers for medical applications to reduce the costs of prosthetic rehabilitation and implantable devices.

In addition, a coupling of structural and functional applications of asphaltene-based carbon fibers could be used in the construction of smart laminated composite structures with both core laminated composite structures and integrated piezoelectric patches made of asphaltene-derived carbon fibers for the applications of energy-harvesting and structural health monitoring [95–105].

Therefore, at the initial development of asphaltene-based carbon fibers, due to their low cost of processing, they can be a good candidate to be utilized in low-performance structural applications as the replacement of other existing fibers.

11. Innovation's Value Proposition

To have high-quality and low-cost carbon fibers, it is suggested to blend PAN with asphaltene for electrospun carbon nanofibers (CNFs) and blend thermal plastic polyurethane (TPU) with asphaltene for melt-spun carbon fibers. For electrospun nanofiber, it can be applied as continuous non-woven fabrics in composite form. PAN/asphaltene nanofiber can be directly electrospun without the pre-treatment or a slight treatment of asphaltenes, and the CNFs diameter is around 500 nm. PAN/asphaltene could have a higher surface area with good flexibility, and it can be applied as binder-less and free-standing carbon electrodes for different energy storage devices such as supercapacitors or batteries. The electrochemical performance could be further enhanced through the activation process. Electromagnetic interface (EMI) shielding is also a promising application for the PAN/asphaltene electrospun nanofiber. The application of PAN/asphaltene CNFs on energy storage devices and EMI shielding could create new paths of applications and opportunities for asphaltene-based carbon fibers.

The asphaltene/polymer blending electrospun and melt-spun carbon fibers can provide various carbon fiber forms from continuous non-woven fabrics to milled fibers with the fiber diameter ranging from 500 nm to the micron level. The carbon fiber properties can also be tailored and modified through tuning the asphaltene/polymer blending ratios or modification depending on the requirement of the end-users. The combination of asphaltene–polymer carbon nanofiber and melt-spun/melt-blown carbon fibers can increase the versatility and options of the asphaltene-based carbon fiber products to adapt for different applications. Accordingly, low-cost fabrication with various fiber diameters makes asphaltene-based carbon fibers an ideal candidate to create carbon fiber-reinforced composites for different applications by convincing clients to pay for this newly developed carbon fiber.

12. Market Potential Analysis

Since a newly developed carbon fiber derived from petroleum asphaltene is introduced, its market potential should be investigated before commercialization. In the market potential, performance (P) and cost (C) are two main parameters that should be considered carefully. Any new material commonly presents enhanced performance with higher cost or is cheaper with lower performance. To have an accurate analysis, a market potential diagram is needed. Figure 24 shows a trade-off plot with the market potential analysis comparing the asphaltene-derived carbon fiber with existing reinforcing fibers on the market based on their cost and performance. By producing asphaltene-derived carbon fibers with composite tensile strength: 50–500 MPa, composite elastic modulus: 50–100 GPa, and composite electrical conductivity: >5 S/cm, we can compete with other fiber-reinforced composites as well as lowering the fabrication processes and costs.

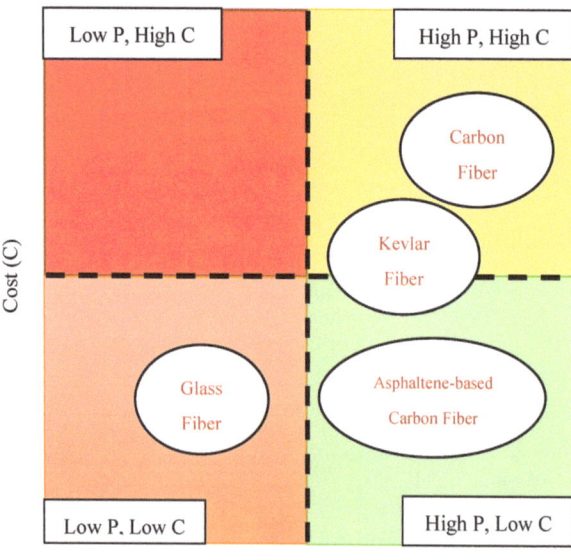

Figure 24. The cost/performance assessment for reinforcing fibers using a market potential diagram.

13. Target Market Size

In Table 3, a comparison between three widely used reinforcing fibers is displayed in view of their applications, cost, performance, annual growth rate, and market size.

Table 3. Target application and market assessment of typical reinforcing fibers.

Fiber	Application	Cost per Kilogram	Cost / Performance	Compound Annual Growth Rate (CAGR)	Size of the Target Market
Carbon	Aerospace, Civil Engineering, and Motorsports	USD 35	High/High	8.6%	USD 3.7 billion in 2020 USD 8.9 billion by 2031
Aramid (Kevlar)	Telecommunication, Aerospace, and Mechanical Rubbers	USD 23	Medium/Medium	7.5%	USD 3.28 billion in 2018 USD 5.78 billion by 2024
Glass	Automotive, Construction materials, Insulators, Oil and Gas, Boat hulls	USD 2	Low/Low	6.8%	USD 65.9 billion in 2019 USD 91.4 billion by 2024

Carbon fibers have several advantages in comparison to other fibers such as high tensile strength, high stiffness, high chemical resistance, low weight, low thermal expansion, and high-temperature tolerance. These outstanding properties have led to carbon fibers having very popular applications in aerospace, civil engineering, and motorsports. However, they are relatively costly in comparison to similar fibers such as plastic fibers and glass fibers. The global carbon fiber market size is expected to grow from USD 3.7 billion in 2020 to USD 8.9 billion by 2031 with a Compound Annual Growth Rate (CAGR) of 8.6%. Increasing demand from aerospace and wind energy industries is expected to make the growth of the market from 2021 to 2031 [106].

Aramid (kevlar) fibers as synthetic polymers have unique properties such as low density and high strength bearing capacity and can endure corrosive environments and very high temperatures. Aramid fibers are on average five times stronger than steel, without melting point, functorially efficient between 400 and 600 °C, and with dielectric properties. Due to these unique material properties, aramid fibers can meet the requirements in industries such as aerospace, telecommunication, and mechanical rubber goods, among others. These fibers are commonly utilized for protection and security applications due to their low density, thermal resistance, and high load-bearing capacity. The aramid fiber market size is estimated at USD 3.28 billion in 2018 and is expected to reach USD 5.78 billion by 2024 at a CAGR of 7.5%. By volume, the aramid fiber market was estimated to be 96.8 kilotons in 2018 and is projected to reach 149.5 kilotons in 2024. Due to regulations related to the reduction in carbon emissions and on the other hand, an increased need for lightweight and flexible materials for the automotive and aerospace industries, aramid fibers can be a great replacement and alternative to carbon fibers [106].

In 2019, the global market size of glass fibers was USD 65.9 billion, and it is expected to reach USD 91.4 billion by 2024 with a CAGR of 6.8% from 2019 to 2024. Glass fibers are widely used in automotive, construction and building materials, the oil and gas industry, boat hulls, etc. The glass and especially synthetic fibers market is increasing due to the rise in the demand for high-performance and lightweight materials globally [106].

Although carbon fibers have higher performance than Kevlar and glass fibers, their market is lower. However, the interest to use carbon fibers is increasing with a higher CAGR (8.6%) than Kevlar and glass fibers. Therefore, by producing low-cost asphaltene-derived carbon fibers with relatively high stiffness and strength and lower density, it will be able to at least catch the market of other cheaper and low-performance reinforcing fibers (such as glass fibers) as well as deliver high-performance and lighter weight fiber-reinforced composites.

14. Conclusions

This paper reviewed and discussed the production of carbon fibers from different precursors, from PAN to petroleum asphaltene (a new precursor), in comparison to other existing fibers on the market. Existing developed reinforcing fibers such as carbon fiber, Kevlar fiber, and glass fiber present different levels of performance and cost, and they are chosen based on the needs of clients. Although performance is the most important parameter for all clients, the cost always becomes a limitation for them in purchasing high-performance reinforcing fibers such as carbon fibers. If we reduce the cost of high-performance fibers by mitigating some unnecessary properties, at least for functional applications, we can deliver lightweight components to the clients which have a high impact on the performance of the final products such as lighter batteries in electric vehicles. Using inexpensive precursors from natural resources such as petroleum asphaltene could be a solution to reduce the production cost of carbon fibers.

Petroleum asphaltenes are abundantly available, and they are much cheaper than PAN and pitch and can be a great candidate as a precursor to produce low-cost carbon fibers. However, we need to optimize the process of carbon fiber production from the asphaltene to have the same quality of PAN- or pitch-based carbon fibers for both functional and structural applications.

Producing low-cost asphaltene-derived carbon fibers, at the first step of development, can target low-strength fibrous composites used in functional components, i.e., molecular sieves, catalysts, electrical conductors, electrodes, fuel cells, batteries, supercapacitors, insulators, absorbents, gaskets, etc. which occupied extensive markets. Then, as a long-term plan, the mechanical and thermal properties of asphaltene-based carbon fibers can be enhanced to be utilized in thermoset matrices for structural applications such as aerospace, marine, oil and gas, automotive parts, railway, pressure vessels, medical implants and prostheses, sports and leisure goods, etc. by satisfying the design requirements which leads to arousing the interests of industrial sectors to invest in this newly developed carbon fiber with relatively high performance and lower cost of production than typical carbon fibers.

Author Contributions: Conceptualization, data curation, investigation, data analysis, visualization, writing, and original draft preparation, H.B.; supervision, funding acquisition, writing review and editing, Y.A. All authors have read and agreed to the published version of the manuscript.

Funding: This research was funded by Alberta Innovates with grant number 202103138.

Data Availability Statement: Not applicable.

Conflicts of Interest: The authors declare no conflict of interest.

References

1. Daniel, I.M.; Ishai, O. *Engineering Mechanics of Composite Materials*, 2nd ed.; Oxford University Press: New York, NY, USA, 2006; ISBN 978-0-19-515097-1.
2. Das, T.K.; Ghosh, P.; Das, N.C. Preparation, development, outcomes, and application versatility of carbon fiber-based polymer composites: A review. *Adv. Compos. Hybrid Mater.* **2019**, *2*, 214–233. [CrossRef]
3. Khosravani, M.R.; Frohn-Sörensen, P.; Reuter, J.; Engel, B.; Reinicke, T. Fracture studies of 3D-printed continuous glass fiber reinforced composites. *Theor. Appl. Fract. Mech.* **2022**, *119*, 103317. [CrossRef]
4. Chowdhury, P.; Sehitoglu, H.; Rateick, R. Damage tolerance of carbon-carbon composites in aerospace application. *Carbon* **2018**, *126*, 382–393. [CrossRef]
5. Elanchezhian, C.; Vijaya Ramnath, B.; Ramakrishnan, G.; Sripada Raghavendra, K.N.; Muralidharan, M.; Kishore, V. Review on metal matrix composites for marine applications. *Mater. Today Proc.* **2018**, *5*, 1211–1218. [CrossRef]
6. Baschnagel, F.; Härdi, R.; Triantafyllidis, Z.; Meier, U.; Terrasi, G.P. Fatigue and Durability of Laminated Carbon Fibre Reinforced Polymer Straps for Bridge Suspenders. *Polymers* **2018**, *10*, 169. [CrossRef] [PubMed]
7. US Patent for Shafts with Reinforcing Layer for Sporting Goods and Methods of Manufacture Patent (Patent # 10,907,942 Issued 2 February 2021)—Justia Patents Search. Available online: https://patents.justia.com/patent/10907942 (accessed on 10 February 2022).
8. Wikipedia. Carbon Fibers. 2022. Available online: https://en.wikipedia.org/wiki/Carbon_fibers (accessed on 20 December 2022).

9. Othman, R.; Ismail, N.I.; Ab, M.; Pahmi, H.; Hisyam, M.; Hisyam Basri, M.; Sharudin, H.; Hemdi, A. Application of carbon fiber reinforced plastics in automotive industry: A review. *J. Mech. Manuf.* **2019**, *1*, 144–154.
10. Holmes, M. Global carbon fibre market remains on upward trend. *Reinf. Plast.* **2014**, *58*, 38–45. [CrossRef]
11. MInus, M.; Kumar, S. The processing, properties, and structure of carbon fibers. *JOM* **2005**, *57*, 52–58. [CrossRef]
12. Dubary, N.; Taconet, G.; Bouvet, C.; Vieille, B. Influence of temperature on the impact behavior and damage tolerance of hybrid woven-ply thermoplastic laminates for aeronautical applications. *Compos. Struct.* **2017**, *168*, 663–674. [CrossRef]
13. Khayyam, H.; Jazar, R.N.; Nunna, S.; Golkarnarenji, G.; Badii, K.; Fakhrhoseini, S.M.; Kumar, S.; Naebe, M. PAN precursor fabrication, applications and thermal stabilization process in carbon fiber production: Experimental and mathematical modelling. *Prog. Mater. Sci.* **2020**, *107*, 100575. [CrossRef]
14. Buckley, J.D.; Edie, D.D. *Carbon-Carbon Materials and Composites*; Noyes Publication: Park Ridge, NJ, USA, 1993.
15. Naito, K.; Yang, J.-M.; Tanaka, Y.; Kagawa, Y. Tensile properties of carbon nanotubes grown on ultrahigh strength polyacrylonitrile-based and ultrahigh modulus pitch-based carbon fibers. *Appl. Phys. Lett.* **2008**, *92*, 231912. [CrossRef]
16. Naito, K.; Yang, J.-M.; Tanaka, Y.; Kagawa, Y. The effect of gauge length on tensile strength and Weibull modulus of polyacrylonitrile (PAN)- and pitch-based carbon fibers. *J. Mater. Sci.* **2012**, *47*, 632–642. [CrossRef]
17. Gao, Q.; Jing, M.; Zhao, S.; Wang, Y.; Qin, J.; Yu, M.; Wang, C. Effect of spinning speed on microstructures and mechanical properties of polyacrylonitrile fibers and carbon fibers. *Ceram. Int.* **2020**, *46*, 23059–23066. [CrossRef]
18. Shirasu, K.; Nagai, C.; Naito, K. Mechanical anisotropy of PAN-based and pitch-based carbon fibers. *Mech. Eng. J.* **2020**, *7*, 19-00599. [CrossRef]
19. Bacon, R.; Modlen, G.F.; McEnaney, B.; Blakeley, T.H.; Manfre, G.; Frank, F.C.; Watt, W.; Harris, B.; Ham, A.C. Carbon fibres from mesophase pitch. *Philos. Trans. R. Soc. Lond. Ser. Math. Phys. Sci.* **1980**, *294*, 437–442. [CrossRef]
20. Wikipedia. Lignin. 2021. Available online: https://en.wikipedia.org/wiki/Lignin (accessed on 20 December 2022).
21. Baker, D.A.; Rials, T.G. Recent advances in low-cost carbon fiber manufacture from lignin. *J. Appl. Polym. Sci.* **2013**, *130*, 713–728. [CrossRef]
22. Kubo, S.; Kadla, J.F. Lignin-based Carbon Fibers: Effect of Synthetic Polymer Blending on Fiber Properties. *J. Polym. Environ.* **2005**, *13*, 97–105. [CrossRef]
23. Braun, J.L.; Holtman, K.M.; Kadla, J.F. Lignin-based carbon fibers: Oxidative thermostabilization of kraft lignin. *Carbon* **2005**, *43*, 385–394. [CrossRef]
24. Thunga, M.; Chen, K.; Grewell, D.; Kessler, M.R. Bio-renewable precursor fibers from lignin/polylactide blends for conversion to carbon fibers. *Carbon* **2014**, *68*, 159–166. [CrossRef]
25. Culebras, M.; Beaucamp, A.; Wang, Y.; Clauss, M.M.; Frank, E.; Collins, M.N. Biobased Structurally Compatible Polymer Blends Based on Lignin and Thermoplastic Elastomer Polyurethane as Carbon Fiber Precursors. *ACS Sustain. Chem. Eng.* **2018**, *6*, 8816–8825. [CrossRef]
26. Mainka, H.; Hilfert, L.; Busse, S.; Edelmann, F.; Haak, E.; Herrmann, A.S. Characterization of the major reactions during conversion of lignin to carbon fiber. *J. Mater. Res. Technol.* **2015**, *4*, 377–391. [CrossRef]
27. Norberg, I.; Nordström, Y.; Drougge, R.; Gellerstedt, G.; Sjöholm, E. A new method for stabilizing softwood kraft lignin fibers for carbon fiber production. *J. Appl. Polym. Sci.* **2013**, *128*, 3824–3830. [CrossRef]
28. Wikipedia. Cellulose. 2022. Available online: https://en.wikipedia.org/wiki/Cellulose (accessed on 20 December 2022).
29. Bengtsson, A.; Bengtsson, J.; Sedin, M.; Sjöholm, E. Carbon Fibers from Lignin-Cellulose Precursors: Effect of Stabilization Conditions. *ACS Sustain. Chem. Eng.* **2019**, *7*, 8440–8448. [CrossRef]
30. Huang, X. Fabrication and Properties of Carbon Fibers. *Materials* **2009**, *2*, 2369–2403. [CrossRef]
31. Dumanlı, A.G.; Windle, A.H. Carbon fibres from cellulosic precursors: A review. *J. Mater. Sci.* **2012**, *47*, 4236–4250. [CrossRef]
32. Wikipedia. Molecular Sieve. 2021. Available online: https://en.wikipedia.org/wiki/Molecularsieve (accessed on 20 December 2022).
33. Burchell, T.D. Carbon fiber composite molecular sieves. In Proceedings of the 8th Annual Fossil Energy Materials Conference, Washington, DC, USA, 10–12 May 1994; Oak Ridge National Laboratory: Oak Ridge, TN, USA, 1994; pp. 63–70.
34. Klett, J.W. Carbon fiber carbon composites for catalyst supports. In Proceedings of the 22nd Conference on Carbon, San Diego, CA, USA, 16–21 July 1995; Then.
35. Kimber, G.M.; Fei, Y.Q. *Physical Properties of Carbon Fiber Composites for Catalytic Applications*; American Chemical Society: Washington, DC, USA, 1996; p. 211.
36. Morgan, P. *Carbon Fibers and Their Compsoites*; CRC Press: New York, NY, USA, 2005.
37. Carbon Molecular Sieve | Manufacturer & Supplier. Sorbead India n.d. Available online: https://www.sorbeadindia.com/product/carbon-molecular-sieve/ (accessed on 11 February 2022).
38. Armstrong-James, M.; Millar, J. Carbon fibre microelectrodes. *J. Neurosci. Methods* **1979**, *1*, 279–287. [CrossRef]
39. Ponchon, J.L.; Cespuglio, R.; Gonon, F.; Jouvet, M.; Pujol, J.F. Normal pulse polarography with carbon fiber electrodes for in vitro and in vivo determination of catecholamines. *Anal. Chem.* **1979**, *51*, 1483–1486. [CrossRef] [PubMed]
40. Malinski, T.; Taha, Z. Nitric oxide release from a single cell measured in situ by a porphyrinic-based microsensor. *Nature* **1992**, *358*, 676–678. [CrossRef] [PubMed]
41. Budai, D. Electrochemical responses of carbon fiber microelectrodes to dopamine in vitro and in vivo. *Acta Biol. Szeged.* **2010**, *54*, 155–160.

42. Takami, N.; Satoh, A.; Hara, M.; Ohsaki, T. Rechargeable Lithium-Ion Cells Using Graphitized Mesophase-Pitch-Based Carbon Fiber Anodes. *J. Electrochem. Soc.* **1995**, *142*, 2564. [CrossRef]
43. Suzuki, K.; Iijima, T.; Wakihara, M. Electrode characteristics of pitch-based carbon fiber as an anode in lithium rechargeable battery. *Electrochim. Acta* **1999**, *44*, 2185–2191. [CrossRef]
44. Lithium-ion Batteries—Industrial Devices & Solutions—Panasonic. Available online: https://industrial.panasonic.com/ww/products/pt/lithium-ion (accessed on 13 February 2022).
45. Hwang, I.U.; Yu, H.N.; Kim, S.S.; Lee, D.G.; Suh, J.D.; Lee, S.H.; Ahn, B.K.; Kim, S.H.; Lim, T.W. Bipolar plate made of carbon fiber epoxy composite for polymer electrolyte membrane fuel cells. *J. Power Sources* **2008**, *184*, 90–94. [CrossRef]
46. Noh, J.; Yoon, C.-M.; Kim, Y.K.; Jang, J. High performance asymmetric supercapacitor twisted from carbon fiber/MnO_2 and carbon fiber/MoO_3. *Carbon* **2017**, *116*, 470–478. [CrossRef]
47. Min Jeon, S.; Hongyue, J.; Jae Song, Y.; Lee, S. Enhanced Electromagnetic Absorption of Cement Composites by Controlling the Effective Cross-sectional Area of MXene Flakes with Diffuse Reflection Based on Carbon Fibers. *Constr. Build. Mater.* **2022**, *348*, 128711. [CrossRef]
48. Bachmann, J.; Hidalgo, C.; Bricout, S. Environmental analysis of innovative sustainable composites with potential use in aviation sector—A life cycle assessment review. *Sci. China Technol. Sci.* **2017**, *60*, 1301–1317. [CrossRef]
49. Changing the World with New Materials. A Half-Century History of CFRP. | MITSUBISHI MATERIALS CORPORATION. Available online: http://www.mitsubishicarbide.com/it/magazine/article/vol05/tec_vol05 (accessed on 13 February 2022).
50. Team, S.G.; SAMPE Global. Composites Applications for Space. 2018. Available online: https://www.sampe.org/composites-applications-for-space/ (accessed on 13 February 2022).
51. Rockets—Are Composites Used by Space Programs? Available online: https://space.stackexchange.com/questions/3077/are-composites-used-by-space-programs (accessed on 13 February 2022).
52. Howell, S.M.; Pappalardo, R.T. NASA's Europa Clipper—A mission to a potentially habitable ocean world. *Nat. Commun.* **2020**, *11*, 1311. [CrossRef]
53. Jacob, A. Racing catamaran relies on carbon fibre. *Reinf. Plast.* **2000**, *44*, 36–42.
54. Team Philips Catamaran. MIC [800 × 525]. R/Boatporn. 2016. Available online: https://www.pinterest.ca/pin/810225789212669008 (accessed on 20 December 2022).
55. HYPERION Yacht for Sale is a 155'7" Royal Huisman Cutter. Available online: https://www.worthavenueyachts.com/yachts-for-sale/hyperion/ (accessed on 13 February 2022).
56. Adam, H. Carbon fibre in automotive applications. *Mater. Des.* **1997**, *18*, 349–355. [CrossRef]
57. Gilchrist, M.; Curley, L. Manufacturing and ultimate mechanical performance of carbon fibre-reinforced epoxy composite suspension push-rods for a Formula 1 racing car. *Fatigue Fract. Eng. Mater. Struct.* **1999**, *22*, 25–32. [CrossRef]
58. Bisheh, H.; Bisheh, M.; Wu, N. An Investigation into Laminated Composite Automotive Drive Shafts. *Int. J. Mech. Prod. Eng.* **2020**, *8*, 89–95.
59. 24,439 Transparent Car Stock Photos, Pictures & Royalty-Free Images—iStock. Available online: https://www.istockphoto.com/photos/transparent-car (accessed on 20 January 2023).
60. Patrick, A.J. Composites—Case studies of pipeline repair applications. *Pigging Prod. Serv. Assoc.* **2004**, *12*. Available online: https://ppsa-online.com (accessed on 20 January 2023).
61. Pipeline Repair—Wrap It and Forget It. Available online: https://www.napipelines.com/composite-wraps-permanent-pipeline-repair/ (accessed on 14 February 2022).
62. Pinterest. Available online: https://www.pinterest.ca/pin/562738915909089504/ (accessed on 14 February 2022).
63. Carbon Frame, Carbon Bike Frame, Bicycle Frame—FIBERTEK COMPOSITE (XIAMEN) Co., Ltd. Available online: https://www.ecplaza.net/products/carbon-framecarbon-bike-framebicycle-frame_2005247 (accessed on 14 February 2022).
64. Edie, D.D.; Dunham, M.G. Melt spinning pitch-based carbon fibers. *Carbon* **1989**, *27*, 647–655. [CrossRef]
65. RMI. Available online: https://rmi.org/ (accessed on 14 February 2022).
66. Tesla Battery Weight Overview—All Models. EnrgIo. 2020. Available online: https://enrg.io/tesla-battery-weight-overview-all-models/ (accessed on 15 February 2022).
67. Department of Energy. Available online: https://www.energy.gov/ (accessed on 15 February 2022).
68. Yang, C.Q.; Simms, J.R. Infrared spectroscopy studies of the petroleum pitch carbon fiber—I. The raw materials, the stabilization, and carbonization processes. *Carbon* **1993**, *31*, 451–459. [CrossRef]
69. Alcañiz-Monge, J.; Cazorla-Amorós, D.; Linares-Solano, A.; Oya, A.; Sakamoto, A.; Hosm, K. Preparation of general purpose carbon fibers from coal tar pitches with low softening point. *Carbon* **1997**, *35*, 1079–1087. [CrossRef]
70. Wu, Q.; Pan, D. A New Cellulose Based Carbon Fiber from a Lyocell Precursor. *Text. Res. J.* **2002**, *72*, 405–410. [CrossRef]
71. Zhang, L.; Yang, G.; Wang, J.-Q.; Li, Y.; Li, L.; Yang, C. Study on the polarity, solubility, and stacking characteristics of asphaltenes. *Fuel* **2014**, *128*, 366–372. [CrossRef]
72. Arbab, S.; Zeinolebadi, A. A procedure for precise determination of thermal stabilization reactions in carbon fiber precursors. *Polym. Degrad. Stab.* **2013**, *98*, 2537–2545. [CrossRef]
73. Xue, Y.; Liu, J.; Liang, J. Correlative study of critical reactions in polyacrylonitrile based carbon fiber precursors during thermal-oxidative stabilization. *Polym. Degrad. Stab.* **2013**, *98*, 219–229. [CrossRef]

74. Liu, J.; Chen, X.; Liang, D.; Xie, Q. Development of pitch-based carbon fibers: A review. *Energy Sources Part Recovery Util. Environ. Eff.* **2020**, 1–21. [CrossRef]
75. Kamkar, M.; Natale, G. A review on novel applications of asphaltenes: A valuable waste. *Fuel* **2021**, *285*, 119272. [CrossRef]
76. Mullins, O.C. The Asphaltenes. *Annu. Rev. Anal. Chem.* **2011**, *4*, 393–418. [CrossRef] [PubMed]
77. Saad, S.; Zeraati, A.S.; Roy, S.; Shahriar Rahman Saadi, M.A.; Radović, J.R.; Rajeev, A.; Miller, K.A.; Bhattacharyya, S.; Larter, S.R.; Natale, G.; et al. Transformation of petroleum asphaltenes to carbon fibers. *Carbon* **2022**, *190*, 92–103. [CrossRef]
78. Leistenschneider, D.; Zuo, P.; Kim, Y.; Abedi, Z.; Ivey, D.G.; de Klerk, A.; Zhang, X.; Chen, W. A mechanism study of acid-assisted oxidative stabilization of asphaltene-derived carbon fibers. *Carbon Trends* **2021**, *5*, 100090. [CrossRef]
79. Li, X.-H.; He, Y.-B.; Miao, C.; Qin, X.; Lv, W.; Du, H.; Li, B.; Yang, Q.-H.; Kang, F. Carbon coated porous tin peroxide/carbon composite electrode for lithium-ion batteries with excellent electrochemical properties. *Carbon* **2015**, *81*, 739–747. [CrossRef]
80. Poizot, P.; Laruelle, S.; Grugeon, S.; Dupont, L.; Tarascon, J.-M. Nano-sized transition-metal oxides as negative-electrode materials for lithium-ion batteries. *Nature* **2000**, *407*, 496–499. [CrossRef]
81. Courtel, F.M.; Abu-Lebdeh, Y.; Davidson, I.J. $ZnMn_2O_4$ nanoparticles synthesized by a hydrothermal method as an anode material for Li-ion batteries. *Electrochim. Acta* **2012**, *71*, 123–127. [CrossRef]
82. Li, P.; Liu, J.; Liu, Y.; Wang, Y.; Li, Z.; Wu, W.; Wang, Y.; Yin, L.; Xie, H.; Wu, M.; et al. Three-dimensional $ZnMn_2O_4$/porous carbon framework from petroleum asphalt for high performance lithium-ion battery. *Electrochim. Acta* **2015**, *180*, 164–172. [CrossRef]
83. Abedi, Z.; Leistenschneider, D.; Chen, W.; Ivey, D.G. Superior Performance of Electrochemical Double Layer Supercapacitor Made with Asphaltene Derived Activated Carbon Fibers. *Energy Technol.* **2020**, *8*, 2000588. [CrossRef]
84. Lee, K.S.; Park, M.; Choi, S.; Kim, J.-D. Preparation and characterization of N, S-codoped activated carbon-derived asphaltene used as electrode material for an electric double layer capacitor. *Colloids Surf. Physicochem. Eng. Asp.* **2017**, *529*, 107–112. [CrossRef]
85. Ni, G.; Qin, F.; Guo, Z.; Wang, J.; Shen, W. Nitrogen-doped asphaltene-based porous carbon fibers as supercapacitor electrode material with high specific capacitance. *Electrochim. Acta* **2020**, *330*, 135270. [CrossRef]
86. Zhao, P.; Wang, L.; Ren, R.; Han, L.; Bi, F.; Zhang, Z.; Han, K.; Weifeng, G. Facile fabrication of asphaltene-derived graphene-polyurethane sponges for efficient and selective oil-water separation. *J. Dispers. Sci. Technol.* **2018**, *39*, 977–981. [CrossRef]
87. Kueh, B.; Kapsi, M.; Veziri, C.M.; Athanasekou, C.; Pilatos, G.; Reddy, K.S.K.; Raj, A.; Karanikolos, G.N. Asphaltene-Derived Activated Carbon and Carbon Nanotube Membranes for CO_2 Separation. *Energy Fuels* **2018**, *32*, 11718–11730. [CrossRef]
88. Plata-Gryl, M.; Momotko, M.; Makowiec, S.; Boczkaj, G. Highly effective asphaltene-derived adsorbents for gas phase removal of volatile organic compounds. *Sep. Purif. Technol.* **2019**, *224*, 315–321. [CrossRef]
89. Qin, F.; Guo, Z.; Wang, J.; Qu, S.; Zuo, P.; Shen, W. Nitrogen-doped asphaltene-based porous carbon nanosheet for carbon dioxide capture. *Appl. Surf. Sci.* **2019**, *491*, 607–615. [CrossRef]
90. Ivanova, I.K.; Semenov, M.E.; Koryakina, V.V. Research into the possibility of using asphaltene-resin-paraffin deposits for thermal insulation of oil pipelines in permafrost. *AIP Conf. Proc.* **2018**, *2053*, 040034. [CrossRef]
91. Mock, P. EU CO_2 emission standards for passenger cars and light-commercial vehicles. *Commer. Veh.* **2014**, 1–9. Available online: https://theicct.org (accessed on 20 December 2022).
92. Wikipedia. United States Vehicle Emission Standards. 2022. Available online: https://en.wikipedia.org/wiki/United_States_vehicle_emission_standards (accessed on 20 December 2022).
93. How to Turn Pitch into Carbon Fiber for Automotive Applications. Available online: https://www.azom.com/article.aspx?ArticleID=19200 (accessed on 15 February 2022).
94. Shahar, F.S.; Sultan, M.T.H.; Md Shah, A.U.; Safri, S.N.A. Natural Fibre for Prosthetic and Orthotic Applications—A Review. In *Structural Health Monitoring System for Synthetic, Hybrid and Natural Fiber Composites*; Jawaid, M., Hamdan, A., Hameed Sultan, M.T., Eds.; Composites Science and Technology; Springer: Singapore, 2021; pp. 51–70. ISBN 9789811588402.
95. Bisheh, H.; Wu, N. Wave propagation characteristics in a piezoelectric coupled laminated composite cylindrical shell by considering transverse shear effects and rotary inertia. *Compos. Struct.* **2018**, *191*, 123–144. [CrossRef]
96. Bisheh, H.K.; Wu, N. Analysis of wave propagation characteristics in piezoelectric cylindrical composite shells reinforced with carbon nanotubes. *Int. J. Mech. Sci.* **2018**, *145*, 200–220. [CrossRef]
97. Bisheh, H.K.; Wu, N. Wave propagation in piezoelectric cylindrical composite shells reinforced with angled and randomly oriented carbon nanotubes. *Compos. Part B Eng.* **2019**, *160*, 10–30. [CrossRef]
98. Bisheh, H.; Wu, N. Wave propagation in smart laminated composite cylindrical shells reinforced with carbon nanotubes in hygrothermal environments. *Compos. Part B Eng.* **2019**, *162*, 219–241. [CrossRef]
99. Bisheh, H.; Wu, N. On dispersion relations in smart laminated fiber-reinforced composite membranes considering different piezoelectric coupling effects. *J. Low Freq. Noise Vib. Act. Control* **2019**, *38*, 487–509. [CrossRef]
100. Bisheh, H.; Wu, N.; Hui, D. Polarization effects on wave propagation characteristics of piezoelectric coupled laminated fiber-reinforced composite cylindrical shells. *Int. J. Mech. Sci.* **2019**, *161*, 105028. [CrossRef]
101. Bisheh, H.; Wu, N.; Rabczuk, T. Free vibration analysis of smart laminated carbon nanotube-reinforced composite cylindrical shells with various boundary conditions in hygrothermal environments. *Thin-Walled Struct.* **2020**, *149*, 106500. [CrossRef]
102. Bisheh, H.; Rabczuk, T.; Wu, N. Effects of nanotube agglomeration on wave dynamics of carbon nanotube-reinforced piezocomposite cylindrical shells. *Compos. Part B Eng.* **2020**, *187*, 107739. [CrossRef]

103. Bisheh, H.; Civalek, Ö. Vibration of smart laminated carbon nanotube-reinforced composite cylindrical panels on elastic foundations in hygrothermal environments. *Thin-Walled Struct.* **2020**, *155*, 106945. [CrossRef]
104. Bisheh, H.; Wu, N.; Rabczuk, T. A study on the effect of electric potential on vibration of smart nanocomposite cylindrical shells with closed circuit. *Thin-Walled Struct.* **2021**, *166*, 108040. [CrossRef]
105. Bisheh, H. Wave dispersion relations in laminated fiber-reinforced composite plates with surface-mounted piezoelectric materials. *Eng. Comput.* **2022**, 1–11. [CrossRef]
106. Market Research Reports, Marketing Research Company, Business Research by MarketsandMarkets. Available online: https://www.marketsandmarkets.com/ (accessed on 15 February 2022).

Disclaimer/Publisher's Note: The statements, opinions and data contained in all publications are solely those of the individual author(s) and contributor(s) and not of MDPI and/or the editor(s). MDPI and/or the editor(s) disclaim responsibility for any injury to people or property resulting from any ideas, methods, instructions or products referred to in the content.

Article

Hands-On Quantum Sensing with NV^- Centers in Diamonds

J. L. Sánchez Toural [1], V. Marzoa [1], R. Bernardo-Gavito [1], J. L. Pau [2] and D. Granados [1,*]

[1] IMDEA Nanociencia, Faraday, 9, 28049 Madrid, Spain
[2] Facultad de Ciencias, Universidad Autónoma de Madrid, 28049 Madrid, Spain
* Correspondence: daniel.granados@imdea.org

Abstract: The physical properties of diamond crystals, such as color or electrical conductivity, can be controlled via impurities. In particular, when doped with nitrogen, optically active nitrogen-vacancy centers (NV), can be induced. The center is an outstanding quantum spin system that enables, under ambient conditions, optical initialization, readout, and coherent microwave control with applications in sensing and quantum information. Under optical and radio frequency excitation, the Zeeman splitting of the degenerate states allows the quantitative measurement of external magnetic fields with high sensitivity. This study provides a pedagogical introduction to the properties of the NV centers as well as a step-by-step process to develop and test a simple magnetic quantum sensor based on color centers with significant potential for the development of highly compact multisensor systems.

Keywords: quantum sensing; diamond; magnetometry at room temperature; color centers; NV centers; microwaves; nanotechnology

Citation: Sánchez Toural, J.L.; Marzoa, V.; Bernardo-Gavito, R.; Paul, J.L.; Granados, D. Hands-On Quantum Sensing with NV^- Centers in Diamonds. C **2023**, 9, 16. https://doi.org/10.3390/c9010016

Academic Editors: Olena Okhay and Gil Gonçalves

Received: 14 November 2022
Revised: 16 December 2022
Accepted: 18 December 2022
Published: 29 January 2023

Copyright: © 2023 by the authors. Licensee MDPI, Basel, Switzerland. This article is an open access article distributed under the terms and conditions of the Creative Commons Attribution (CC BY) license (https://creativecommons.org/licenses/by/4.0/).

1. Introduction

The word diamond comes from the Greek "adamantem" which means "invincible". The diamond is an electrical insulator with strong covalent bonds that make it a material with extraordinary hardness, broadband optical transparency, and extremely high thermal conductivity. In addition, it can withstand large electric fields and, when doped, behaves like a semiconductor.

Diamonds are associated with the idea of perfection. However they are rarely perfect, and lattice irregularities or impurities are very common. By using artificial growth techniques, the nature and density of impurities can be controlled. This alters their physical properties, such as color or electrical conductivity. Optically active defects are called color centers.

In the NV color center, a nitrogen atom substitutes a carbon atom and a vacancy, in one of four adjacent positions, replaces another carbon atom. In this configuration, one electron is unpaired and remains trapped inside the vacancy. The center is charged negatively when it captures an additional electron, usually from a nitrogen atom donor in the lattice. The spin state of the two-electron quantum system can be controlled by using microwave pulses and optically addressed by measuring the photoluminescence [1].

All this, together with the long coherence time of the quantum state, and possibility of working at room temperature, makes them an ideal physical platform for the development of a magnetic sensor with unprecedented performance.

The spin orientation of the two electrons trapped inside the center is aligned with the axis of symmetry (the line joining the vacancy and the nitrogen).

Under a laser light illumination pulse in the range of 465 nm to 565 nm, the center fluoresces, emitting a photon in the red spectrum. When excited with microwaves, the fluorescence changes in such a way that it is possible to determine the external magnetic field [2].

Current technologies providing a high magnetic sensitivity, such as optical pumped magnetometry (OPM) [3], superconducting quantum interference devices (SQUID) [4], microelectromechanical systems (MEMS) [5], and magnetic resonance force microscopy (MRFM) [6], are highly successful technologies that have made possible the measurement of the magnetic field generated by neuronal activity with great precision. These systems have been previously described and compared [7].

A SQUID is based on superconducting loops containing Josephson junctions [8]. It is a very sensitive magnetometer used to measure extremely small magnetic fields, sensitive enough to measure fields as low as 5×10^{-14} T with a noise equivalent field of approximately 3 fT·Hz$^{-1/2}$. For the sake of comparison, it is important to notice that a common, small neodymium magnet produces a magnetic field of about 10^{-2} T, and neural activity in animals produces magnetic fields between 10^{-6} T and 10^{-9} T. Spin exchange relaxation-free (SERF) magnetometers measure magnetic fields by using lasers to detect the interaction of the magnetic field with alkali metal atoms in a vapor. They are potentially more sensitive and do not require cryogenic refrigeration but are orders of magnitude larger in size (1 cm^3).

SQUID requires cryogenics, OPM reduces the sensitivity when the device reduces its dimension [9] due to the atomic collisions that alter the spin, and they require operation in a near-zero magnetic field; consequently, the ambient 50 µT Earth magnetic field must be properly screened. These technologies show limitations for miniaturization, because they either require special conditions and bulky instrumentation or their temporal resolution decreases when trying to reduce their dimensions.

In contrast, the quantum state of a color center in a diamond can be read out optically because the fluorescence is spin dependent, allowing its use in high-precision magnetometry at room temperature [10,11] and under an ambient magnetic field, such as the Earth's magnetic field. Its properties have demonstrated temporal resolution [12] and ultrahigh sensitivity [13] to measure fields as low as 10^{-12} T [14] while allowing device miniaturization down to the millimeter, micrometer, or even the nanoscale.

The crystal structure of diamond consists of tetrahedral covalent bonds between an atom and its four nearest neighbors, linked in a face-centered cubic Bravais lattice. This strongly bonded, tightly packed, dense, and rigid structure gives rise to its outstanding properties.

An *NV* center [15] is a point defect in a diamond with an axial, trigonal C$_{3v}$ [16] symmetry [17] (three vertical reflection planes, and two 120° rotations about the Z axis), caused by a nitrogen impurity. The center has three possible configurations, a positive charge NV^+, a neutral NV^0 (Figure 1a), and a negative charge NV^- (Figure 1d). These states have been well studied: NV^+ is nonfluorescent, NV^0 [18,19] has only one electron unpaired, is paramagnetic, and its luminescence intensity is lower than in the case of NV^-. The NV^- occurs when the center captures an additional electron, normally from a nitrogen donor in the lattice, and they form a pair with an integer quantum spin number of 0 or ± 1, and it is the one used for magnetometry [20].

The electronic structure of the center (Figure 2) consists of a triplet ground state, a triplet excited state and two singlet states [21]. In the singlet state the spins are anti-aligned (up-down or down-up with $m_s = 0$). In the triplet state, the spins can be aligned (up-up with $m_s = +1$ or down-down with $m_s = -1$) or anti-aligned (with $m_s = 0$). The state with the spins aligned is degenerated and requires more energy due to the electron-electron magnetic interaction.

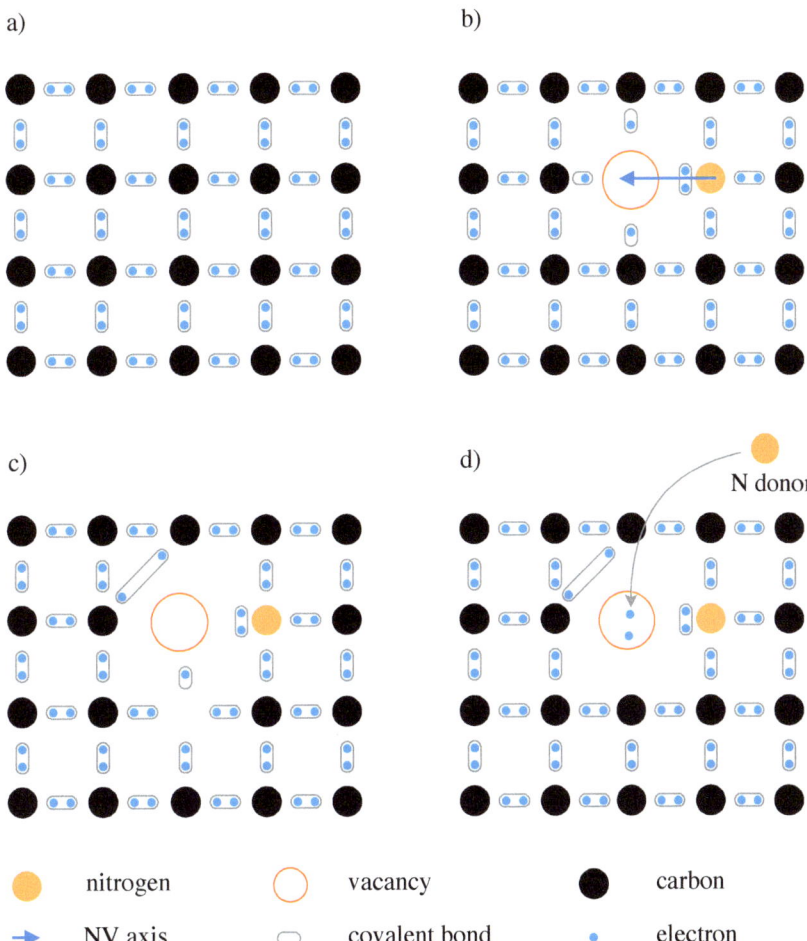

Figure 1. Simplified, flat representation of the diamond lattice. (**a**) A missing carbon atom, or vacancy, and the nearest neighbor, a substitutional nitrogen atom, create the NV center where the axis aligns the nitrogen atom with the vacancy, in one of four possible orientations. (**b**) The nitrogen atom has five electrons in the valence band, three of them form three covalent bonds with the three neighbor carbon atoms, leaving a lone pair. From the carbon atoms surrounding the vacancy, three electrons from dangling bonds are also part of the center, one remains unpaired, and the other two form a covalent bond (**c**). A total of five electrons leads to the neutral NV^0 state which does not exhibit the magnetic activity as the negative charge state NV^- does. When the center captures a sixth electron [22], normally from another nitrogen atom donor in the lattice (**d**), the center is charged negatively and shows the expected behavior useful in magnetometry.

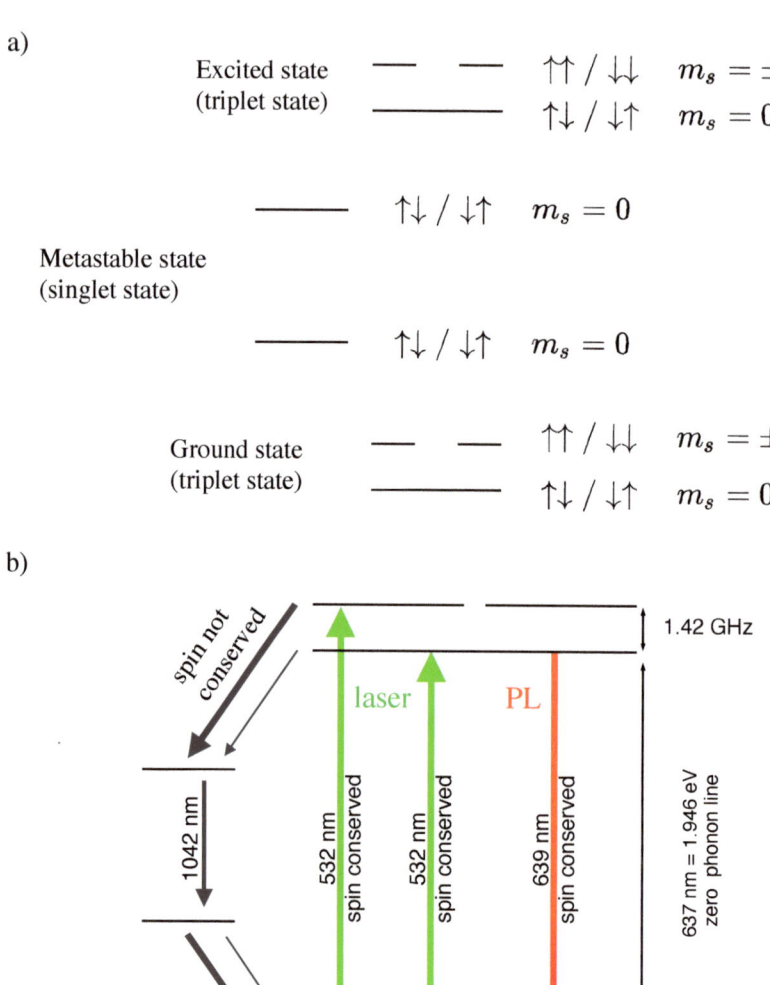

Figure 2. Energy levels scheme of a NV^- center. (**a**) Ground and excited states are split by spin interaction into a triplet. The electron is a fermion, its spin quantum number has a magnitude of $s = 1/2$ and two possible magnetic component $m_s = \pm 1/2$. The two electrons in the NV^- center exist in a triplet state: one state where electrons have opposite spins ($m_s = 0$), and two degenerated states with electrons spin pointing to the same direction ($m_s = \pm 1$) and higher energy. (**b**) The excitation (green arrows) conserves the spin and also the decay from $m_s = 0$ excited to $m_s = 0$ ground, however, when the center state is $m_s = \pm 1$ excited, one of the electrons flips and the state changes to $m_s = 0$, and the decay follows (preferentially) a non radiative path through the metastable singlet states, with the emission of a 639 nm (red arrow in the picture) photon corresponding to the zero phonon line (ZPL) and the decay from $m_s = \pm 1$ with the emission of a 1042 nm (infrared, black arrows in the picture), decreasing the observed PL intensity. The energy gap between $m_s = 0$ and $m_s = \pm 1$, in the ground and excited states, with zero field splitting, corresponds to 2.87 GHz and 1.42 GHz respectively.

Magnetometry under ambient conditions with color centers in a diamond is based on the optical measurement of the eigenvalues of the two-spin Hamiltonian of the electrons trapped in the center under the application of an external magnetic field. The measurement can be performed optically because the $m_s = \pm 1$ degenerated states preferentially follow a nonradiative path through the metastable singlet states, resulting in a reduction in the observed photoluminescence [23] compared to that corresponding to the $m_s = 0$ state. Spin-state initialization [24] is achieved by applying properly tuned laser light, which causes optical pumping to the $m_s = 0$ state. The microwave radiation at the resonance frequency populates the $m_s = \pm 1$ states that, under an external magnetic field, will split, giving rise to a decrease in the PL observed at the resonance frequencies.

If the center is shined on with laser light (we used a 532-nm or 2.33-eV, diode-pumped solid-state laser), the electrons are excited from the ground to the excited state, where these transitions are predominantly spin-conserving. The electrons immediately decay to the ground state and emit a photon of a lower frequency (red color); this photoluminescence (PL) is more intense when the center is in the $m_s = 0$ state because the transition is closed (from $m_s = 0$ ground state to the $m_s = 0$ excited state and then back again to $m_s = 0$ ground state). In the case of $m_s = \pm 1$, the excitement conserves the spin but not the decay because one of the electrons flips and the state changes to $m_s = 0$, decaying through the metastable singlet states and emitting a photon in an infrared frequency [25], and therefore reducing the observed PL.

In a continuous optical excitation shining with a laser (465 nm to 565 nm in range) [26], the electron population, initially in the $m_s = \pm 1$ state, is pumped to the $m_s = 0$ state; this is the method by which to initialize the system, with all the electrons set on the $m_s = 0$ state.

In this situation, it is not possible to observe the effect of an external magnetic field on the PL because the degenerated states $m_s = \pm 1$ are not populated. Applying a microwave radiation at the 2.87 GHz resonance frequency, the $m_s = \pm 1$ ground states are populated again, allowing the excitement to the $m_s = \pm 1$ excited states and the subsequent decay through the nonradiative route showing a decrease in the PL.

Under this continuous microwave radiation, applying an external magnetic field with strength B_0, the magnetic moment of the electron aligns itself as either antiparallel ($m_s = -1$) to the magnetic field component corresponding to the NV axis, with a specific energy level, or parallel ($m_s = +1$) with a different energy, due to the Zeeman shift of the spin sublevels. Therefore, now, the PL decreases at two different resonance frequencies. This separation of the sublevels is linear and proportional to the applied magnetic field component corresponding to the NV axis and produces a decrease in the observed PL at those frequencies. Increasing the magnetic field and due to the hyperfine structure and different orientations of different NV centers, other levels are also split and different dips on the PL are observed.

2. Materials and Methods

Color centers in diamonds can be characterized experimentally by using [27,28] optically detected magnetic resonance (ODMR) or electron spin resonance (ESR) techniques together with the photoluminescence signal emitted during the relaxation of an excited state to its ground state. In this study, ODMR is used for the characterization of the color centers in diamonds.

Throughout this research and for all the experiments, two synthetic-type Ib diamond samples have been used, manufactured using the high-pressure, high-temperature (HPHT) synthesis process by the supplier element6 [29]. The luminescence has been studied by using such samples for a set of NV^- centers in which each NV^- axis is randomly oriented in the four possible directions.

It is important, for nanoscale applications, that the centers are located as close to the surface as possible. Type Ib diamonds are artificially fabricated [30] to contain up to 500 ppm nitrogen, absorb green light, and have a dark yellow or brown color. The samples used in our investigation have a central NV concentration of approximately 1 ppm.

With the sensor as the final target, the first step is the characterization of the two diamonds, developed for quantum sensing purposes. Both samples—the polycrystalline and the single crystal—have been characterized under ambient conditions by using a 488 nm wavelength argon ion laser in a low vibration optical setup (Figure 3), achieving almost identical results in PL intensity and peaks observed.

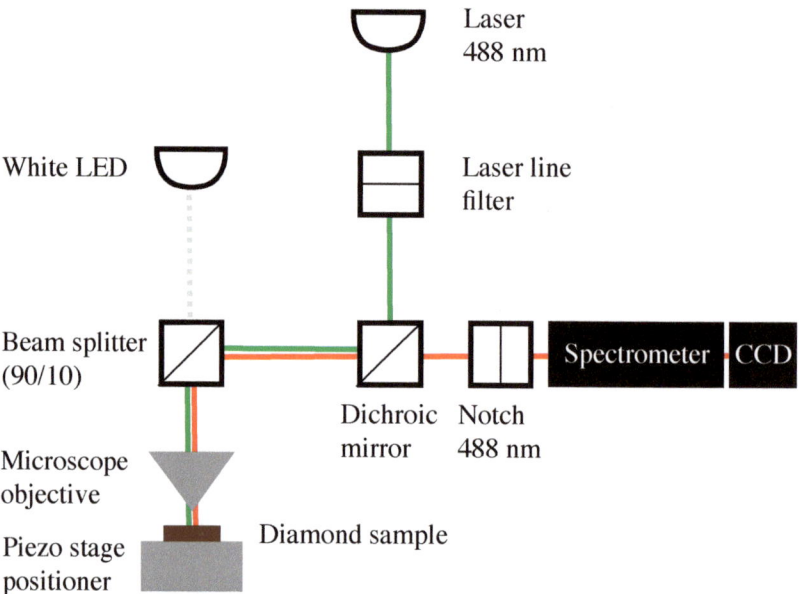

Figure 3. Experimental setup used for the characterization of the samples. It includes a 488 nm laser source, a piezo-positioner, a microscope objective lens, a dichroic mirror (94:06), a 488 nm notch filter, a beam splitter (90:10) a CCD camera, a spectrometer and other optical elements as shown in the diagram. The laser power source is 40 µW which is partially absorbed by the dichroic mirror and the splitter, reducing the power incident on the sample to 11.25 µW. The white LED is used for lighting and navigating the sample.

3. Results and Discussion
3.1. Optical Characterization

At low power, the PL intensity is linear to the excitation power, and at high power it may reach saturation. Due to the vibrational and rotational effects on atomic levels, emission does not occur at a single wavelength but in a range from 600 nm to 800 nm. The photoluminescence spectra (Figure 4) shows the four characteristic frequency peaks: the NV^- zero phonon line (ZPL) at 637 nm, the NV^0 ZPL at 576 nm and the wide phonon sideband with two visible peaks at 662 nm and 684 nm [27]. A small Raman peak corresponding to the vibration of the sp3 (tetrahedral) diamond lattice was also identified.

The mentioned zero-phonon line and the phonon sidebands constitute the spectra of the single-color centers in diamond-absorbing and emitting light. For the ensemble of color centers, each NV center contributes with a zero-phonon line and a phonon sideband to the total absorption and emission spectra, which is considered not homogeneously broadened because each NV center is surrounded by a different environment in the lattice, which modifies in a different way the energy required for an electronic transition, shifting and overlapping each zero phonon and vibrionic phonon sideband positions [31].

The zero-phonon line is located at a frequency determined by the energy gap between the NV center ground and excited state and also determined by the local environment.

The phonon sideband is shifted to a higher frequency in absorption and to a lower frequency in fluorescence (see Appendix B. Characterization details).

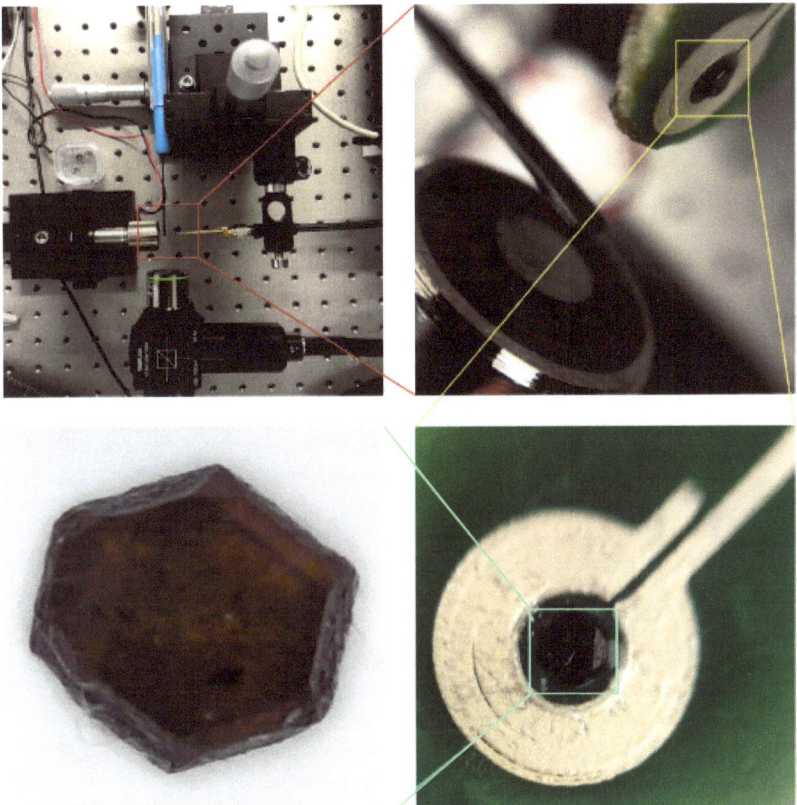

Figure 4. Four views of the experimental setup at different levels of detail. Upper left: experimental setup including the teslameter probe (blue), electromagnet (left), objective (down), XYZ positioner (up) and resonator with diamond (center); Upper right: details of the electromagnet, the teslameter probe and the resonator with the diamond; Lower right: the resonator silver printed on the PCB with a diamond in the middle; Lower left: a zoom image of the single crystal diamond sample.

3.2. Optically Detected Magnetic Resonance (ODMR)

Because the magnetic field is detected optically through the splitting of the degenerated states due to the Zeeman effect, the next step in the process requires the $m_s = \pm 1$ center state populated. For that purpose the setup is extended by adding a microwave source and a resonator.

As shown before, by continuously applying laser light (532 nm diode), the electrons in the centers are pumped to the $m_s = 0$ ground state (Figure 2). By applying a resonant microwave field with an energy of 2.87 GHz, corresponding to the gap between the quantum levels $m_s = 0$ and $m_s = \pm 1$, the electrons in the center are excited to the $m_s = \pm 1$ state.

The comparative measurement of the 1042 nm transition, by applying or not applying the resonant microwave field under continuous laser excitation, allows the detection of the PL attenuation, due to the greater absorption in the infrared frequency that occurs in the transition between the singlet states as well as the long coherence time of those singlet states (NV centers inhomogeneous spin relaxation time was assessed to be in the order of 300 ns, [32,33]). When the microwave field is retired, the electrons return to the $m_s = 0$ state and therefore that transition does not take place, resulting in an increase in the photoluminescence.

To irradiate the diamond with microwaves pulses, the research team manufactured a resonator with the topology and characteristics described in Figure 5, based on a previous design by Man Zhao et al. [34] and the study of Eisuke Abe et al. [35] by silver printing on a PCB by using a Voltera V-One PCB printer [36].

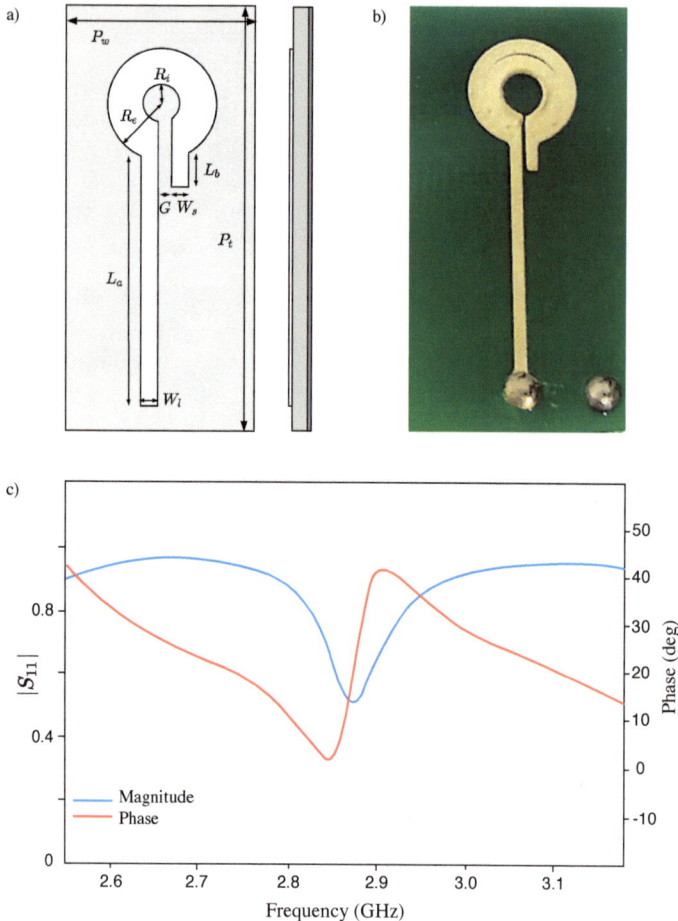

Figure 5. Schema of a top view and a cross section of the resonator (**a**). The flat ring resonator is made up of a radiating patch, a dielectric substrate, and a grounding plate (**b**). The radiating ring structure was printed with silver on a substrate 1.3 mm PCB flat laminated composite substrate, made from a non-conductive material, and the grounding plate, a 0.08 mm thick copper foild. The geometric dimensions (mm) of the resonator are: $L_a = 16$, $L_b = 2.1$, $W_l = 0.5$ $W_s = 0.5$, $G = 0.2$, $P_w = 10.2$ mm and $P_l = 30$ mm. The inner and outer radius of the ring are: $R_i = 1.3$ mm and $R_e = 3.7$ mm respectively. S_{11} is the resonator reflection coefficient (**c**), it is a ratio and therefore, a non-dimensional parameter but usually the magnitude is specified in dB ($20 log |S_{11}|$), for instance 0 dB indicates a magnitude $|S_{11}| = 1$ which means that all the radiation applied is reflected and it is the case for this specific resonator at all frequencies different to the 2.87 Gz. The phase in red and the magnitude in blue show the response as designed.

The diamond sits on top of the resonator. The resonator generates a spatially uniform and concentrated field over an area of approximately 1 mm² with a resonance frequency of 2.87 GHz and a bandwidth of 100 MHz, permitting it to work with different resonances generated by the hyperfine structure.

3.3. Photoluminescence Intensity in Proportion to the Applied Microwave Frequency

The new experimental setup (Figure 6a) includes a microwave source, a spectrometer, and a computer to process the signal (all of them detailed in Appendix A. Component detail), allowed us to visualize the effect of the microwaves in the PL

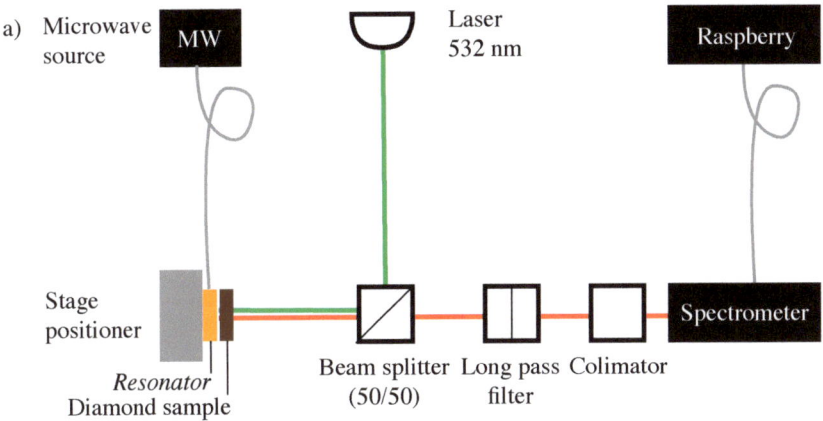

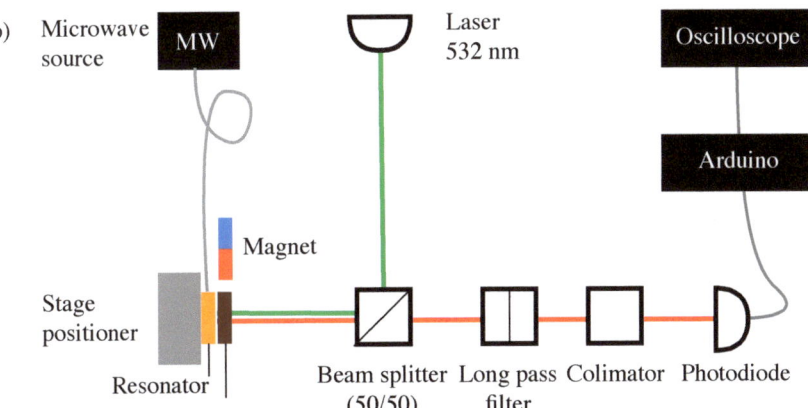

Figure 6. (**a**) Experimental measurement setup used to study the effect of the microwaves on the observed PL. The setup is simplified, the confocal microscope and other elements are replaced by the spectrometer and a Raspberry computer running a Python code to process and visualize the results. The microwave source and the resonator remain the same. (**b**) Experimental and simplified setup adding an electromagnet and using a photodiode (AMS TSL257-LF Amplified Si PD) and a Portenta H7 Arduino board instead of the spectrometer and the Raspberry. The microwave source, the resonator, laser, and optical elements remain the same.

For this study we programmed the microwave (MW) source to sweep [37] in a range of frequencies from 2.7 GHz to 3.0 GHz (10 MHz step; 5.0 dBm constant power; 26 s sweep duration) and a code to integrate the total luminescence detected by the spectrometer per time unit (100 ms integration time).

As predicted by the theory, we observed a dip right at the resonance frequency of 2.87 GHz.

Under the laser beam (532 nm), the two electrons in the color center, are excited from the $m_s = 0$ ground state to the $m_s = 0$ excited state and then recombined again to the $m_s = 0$ ground state. The microwaves have no effect except for the resonance frequency. At this frequency, the electrons absorb the energy from the MW pulse and populate the $m_s = \pm 1$ ground state; from that, the laser beam excites them to the $m_s = \pm 1$ excited state to finally decay through the spin nonconservative and nonradiative path across the metastable states, resulting in a reduced fluorescence signal (Figure 7a).

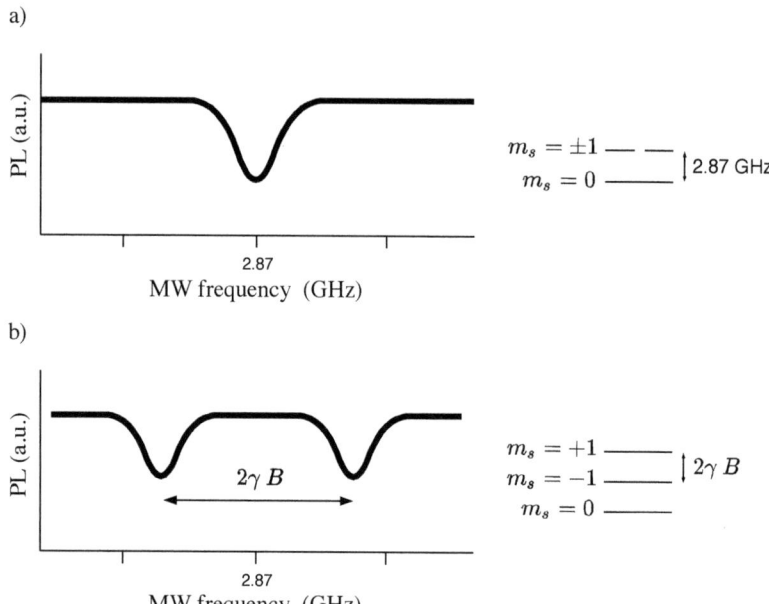

Figure 7. Observed dip for a bulk sample in the direction of the magnetic field in the crystallographic direction (other directions are also present but not represented in this case) of the photoluminescence at the resonance frequency (**a**) and split of the degenerated state because of the Zeeman effect (**b**).

At this point, the research team was able to move forward and explore the effect of the magnetic field in the photoluminescence.

3.4. PL under an External Magnetic Field

The spin-triplet ground state has a zero field splitting at 2.87 GHz between $m_s = 0$ and $m_s = \pm 1$ sublevels, splitting and shift of the levels appear due to electron spin–spin interaction [38]. Intersystem crossing (ISC), plays a key role in the excited state decay dynamics. It is the mechanism by which the center changes its spin state from excited state $m_s = \pm 1$ to the metastable $m_s = 0$.

Due to the higher ISC rate for the $m_s = \pm 1$ state under the laser light, the center is polarized to the $m_s = 0$ state, being the lifetime for the $m_s = 0$ excited state of 23 ns, and almost double the one corresponding to the $m_s = \pm 1$ excited state of 12.5 ns. The metastable state emits an infrared photon with a wavelength of 1046 nm and has a long lifetime of 300 ns, which is used to optically read the state of the center [39].

Considering the NV axis aligned with the Z-axis and an external magnetic field coupled to the center through the Zeeman effect, the triplet ground state has an effective spin Hamiltonian given by

$$H_{NV} = DS_Z^2 + \gamma_e B \cdot S,$$

where D = 2.87 GHz is the zero field splitting, S represents the electron spin projection operator, γ_e is the electron gyromagnetic ratio with a value of 28.0249 GHz/T, and B is the vector of the magnetic field.

The magnetic field vector component aligned with the NV^- axis couples with the center, and then the gap between the energies of the degenerated states, initially the same energy at zero field, now separates linearly with the magnetic field by $2\gamma_e B$ and therefore, the effect can be used to measure the external magnetic field strength because the splitting of the energy levels is directly proportional to the strength of the applied magnetic field component aligned with the NV axis, as shown in Figure 8.

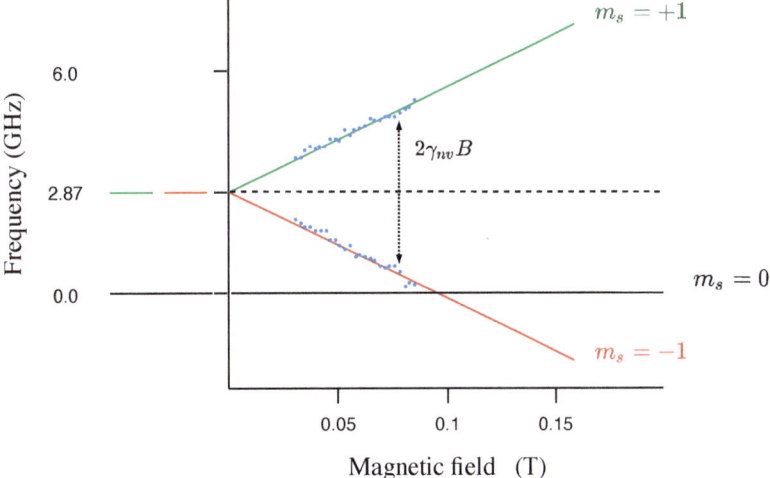

Figure 8. Zeeman splitting of the degenerated state under the effect of an external magnetic field B and experimental results (blue dots), observed for a bulk sample in the direction of the magnetic field in the crystallographic direction (other directions are also represented in Figures 9 and 10). The energy splitting of the states $m_s = +1$ and $m_s = -1$ (green and red lines) is $2\gamma_e B$, linear and proportional to B with 2γ the gyromagnetic ratio of the NV center. The zero splitting happens at the resonance frequency of 2.87 GHz. When the magnetic field reaches 0.1 T for the ground state and 0.05 T for the excited state, and then, due to the ground or excited state level anticrossing, the $m_s = 0$ and $m_s = -1$ states become equal in energy.

To continue the experiment, the experimental setup was extended by including an electromagnet placed close to the diamond as seen in Figure 4. The electromagnet was connected to a precision power supply that allowed us to vary the voltage and thus the external magnetic field that acted on the material. Then the process consisted of taking measurements of the observed PL for the different voltage values between 0 and 12 volts (blue dots in Figure 8).

Once having demonstrated the control on the measurement of the magnetic field, it was possible to proceed with the steps toward improving the sensitivity and reduce the size of the device. For that purpose, a new setup, more compact and simplified, was designed to replace the spectrometer by a photodiode connected to an Arduino board to read the signal and visualize it on an oscilloscope (Figure 6b).

With this new setup, it was possible to represent in more detail the PL landscape in proportion to the microwave frequency and the magnetic field, producing the series represented in the Figure 9.

To more accurately measure the relationship between the magnetic field and the observed luminescence, the research team used a teslameter (FH-55 Teslameter [40]). The FH-55 tip probe was placed at the point where later the diamond would be placed; then, the value of the magnetic field, to the nearest hundredth of a millitesla, was recorded for each position of the magnet (moved at steps of 0.5 mm).

Once these measurements were collected, the study continued following the usual procedure, that is, first, shining with a 532-nm laser light to polarize the centers, leaving at least 30 min to ensure laser stabilization (otherwise, the observed PL could be displaced, although it would not affect dips separation as a result of the splitting). Then, once the system stabilizes, a frequency sweep was performed from 2.7 to 3.0 GHz for each of the magnetic field values, corresponding to a specific position of the magnet with respect of the diamond.

Figure 9 presents the results of the photoluminescence curves plotting intensity as a function of wavelength. Figure 9a shows the magnetic resonance spectra of the ensemble of NV^- centers corresponding to the 24 scans for different electromagnet voltage values between 0 and 12 volts, increasing 0.5 V at each step. In the figure, the darker the curve, the smaller the magnetic field. It can be seen how, with the increase of the magnetic field, the number of dips (corresponding to the transitions between the states $m_s = 0$ and $m_s = \pm 1$) increases due to how the different possible orientations of the NV axis within the crystal structure perceive different projections of the magnetic field.

Figure 9b also presents the PL as a function of the frequency for magnetic field values, increasing at steps of 0.005 mT from 0.23 mT to 0.01 mT; in this case, the lighter the color of the curve, the lower the magnetic field.

In both cases, the pattern is clear because it shows the expected symmetry as a consequence of the Zeeman splitting, and the separation of the two global minima in the PL shows the linear dependence of the applied magnetic field and the distance between dips.

Finally, Figure 10a,b depicts the same data, but in a 3D representation combining PL intensity, magnetic field, and frequency.

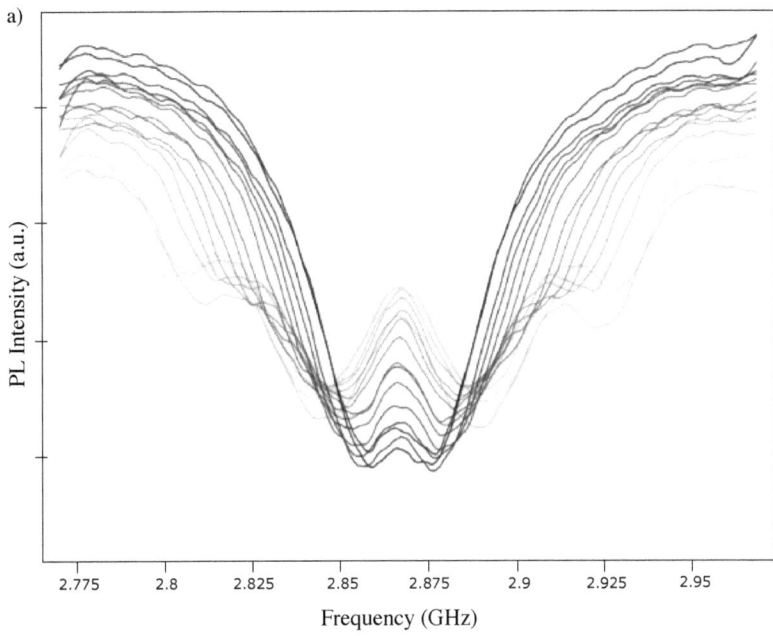

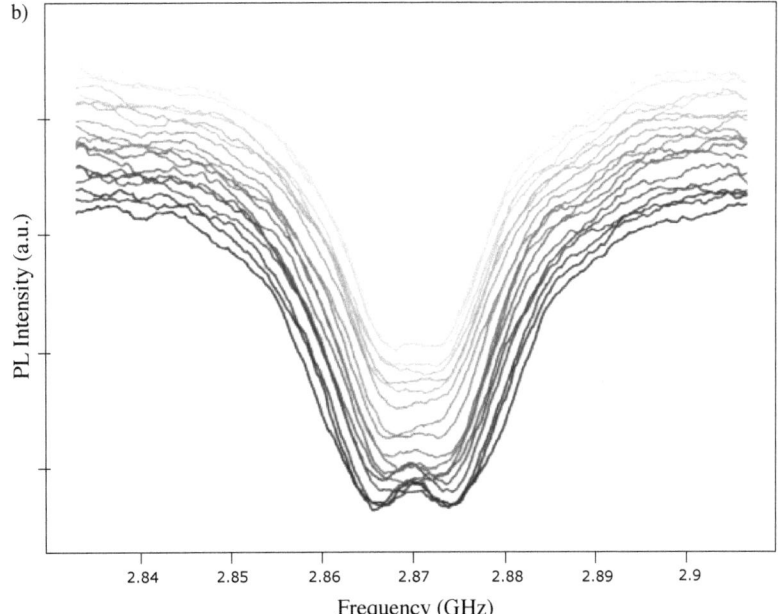

Figure 9. Experimental measurement of photoluminescence as a function of the magnetic field for frequencies in the range of 2.7 to 3 GHz where it is observed how the gap between the resonance holes grows when the magnetic field increases. (**a**) is obtained by using the described electromagnet, changing the voltage and (**b**) is the result of the characterization moving the permanent magnet away at each step.

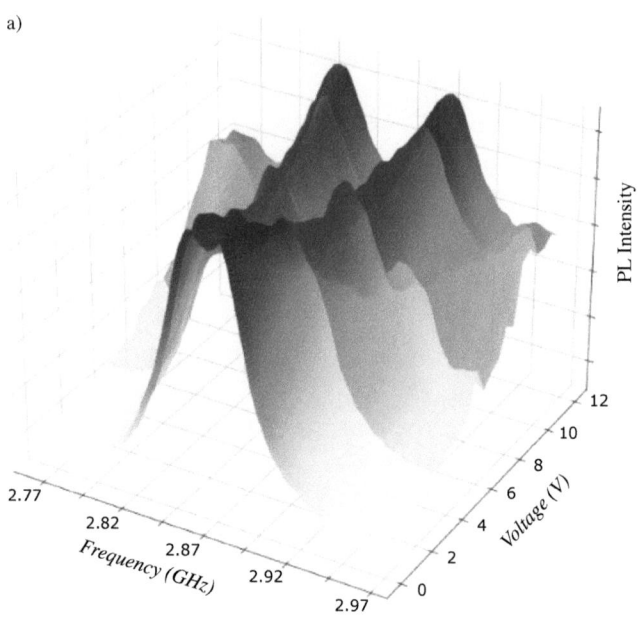

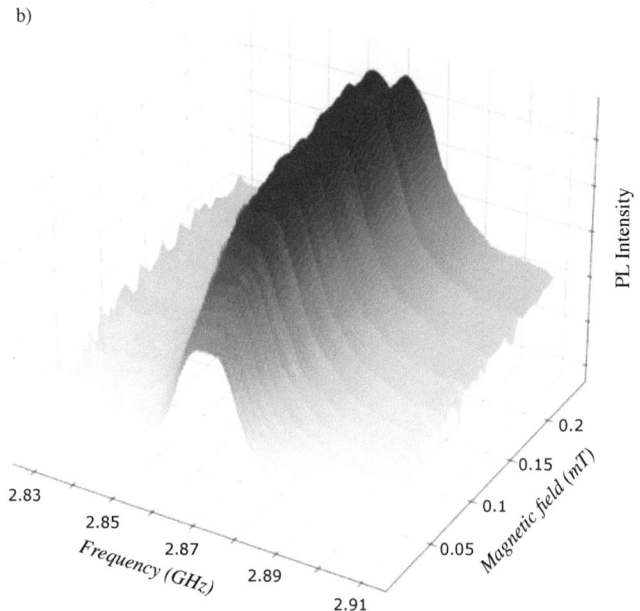

Figure 10. Experimental measurement of photoluminescence as a function of the magnetic field for frequencies in the range of 2.7 to 3 GHz where it is observed how the gap between the resonance holes grows when the magnetic field increases. A 3D view, combining the three dimensions, photoluminescence, frequency, and magnetic field. (**a**) is obtained by using the described electromagnet, changing the voltage and (**b**) is the result of the characterization moving the permanent magnet away at each step.

4. Conclusions

In a series of experiments, two samples of commercially available synthetic Ib diamonds from element6 vendor, a polycrystalline diamond and a single crystal, have been used. These samples were grown by the high-pressure, high-temperature (HPHT) synthesis processes and characterized in a low-vibration optical setup showing the characteristic spectra of an ensemble of the NV centers, both NV^0 and NV^-, as well as the sidebands. The photoluminescence was represented as a function of wavelength, as well as its evolution in time under different conditions.

By using the previous material, the research team has prepared an experimental setup in order to initialize the system, pumping the electrons that form the color centers to the $m_s = 0$ state, by applying laser light, observing the expected behavior.

The development of a resonator, specifically designed to emit at the resonance frequency, has made it possible to apply a microwave pulse to the diamond under continuous laser illumination. These conditions have repopulated the degenerated states $m_s = \pm 1$ making it possible to observe the decrease in the observed PL. This happens because the electrons follow the nonradiative decay path through the metastable state. The frequency sweep has verified the resonance frequency or energy gap between the states $m_s = 0$ and $m_s = \pm 1$. With this experimental setup, it has been possible not only to initialize the state but also to identify the resonance frequency and verify the evolution of the PL when applying microwave pulses, laying the groundwork for the next experiment.

In a last experiment, a magnetic field has been applied in a controlled way with the aim of measuring the displacement of the resonance frequencies due to the Zeeman effect and which constitutes the basis of magnetometry with color centers. The steps to develop this new setup, as well as the results, are also part of this work, leaving the way open to the miniaturization phase.

Author Contributions: J.L.S.T. conceived the idea, designed the experiments, acquired and analysed experimental data, and wrote the manuscript. V.M. participated in the acquisition of the PL and Raman measurements. R.B.-G. contributed to the design of the experiments, the use of instruments and helped with the attainment of experimental data. J.L.P. contributed to the idea and supervised the experimental findings. D.G. obtained financial support, managed the research team and contributed to the analysis of the experimental data. All the authors contributed to the editing of the final manuscript. All authors have read and agreed to the published version of the manuscript.

Funding: This research was funded by MICIN-AEI: Grants DETECTA ESP2017-86582-C4-3-R and EQC2018-005134-P Comunidad de Madrid: Grant TEC2SPACE-CM P2018/NMT-4291, ONR-G: G#N62909-19-1-2053 (DEFROST), MADE-MICINN: PID2019-105552RB-C44. Garantía Juvenil n°201701520868, R.B.-G. would like to thank Comunidad de Madrid for the funding through the grant 2019-T2/IND-13367.

Acknowledgments: We thank M.A. Ramos, G. García, N. Gordillo and A. Redondo from CMAM (Centre for Micro Analysis of Materials) for fruitful discussions on diamond characteristics and its possible modification by using ion irradiation in order to improve sensitivity. Matthew Markham from Element6 for providing the diamond samples. IMDEA Nanoscience nanofabrication facility, for assistance in many aspects, analysis of materials, and fabrication of some of the components.

Conflicts of Interest: The authors declare no conflict of interest.

Appendix A. Component Detail

Figure A1 shows the list of components organized in four main blocks: optical components, including diodes, filters, lenses, splitters, and objectives; electronic components and instruments, including computers, oscilloscopes, spectrometers, CCD cameras, teslameter, microwaves sources, electromagnets and power supplies; structural components, including positioners, posts, brackets, and other elements; and finally the diamond samples used for the experiments.

The component name, the supplier, the part number when possible, and the approximate market price in USD.

Component	Supplier	Part Number	Price (USD)
Optical components			
Compact Laser Diode Module	Thorlabs	CPS532	$164.86
Argon Laser	Modu-Laser	Stellar Pro 488/50	$9815.00
Laser Line Filter	Thorlabs	FL488-10	$144.28
Long-Pass Dichroic Mirror	Semrock	LPD02-488RU	$630.00
Long-Pass Edge Filter	Semrock	LP02-488RU	$710.00
Plano-Convex Lens	Thorlabs	LA1027 N-BK7	$22.34
Objective 20x/0.35	Olympus	SLMPlan	$1050.00
Beam Splitter (50/50)	Thorlabs	BSW26R	$295.21
Beam Splitter (90/10)	Thorlabs	-	$178.00
Electronic components			
Dual SMU System	Keithley	2614B	-
Portenta H7	Arduino	ABX00042	$99.00
Raspberry Pi	Raspberry Pi	3 Model B+	$35.00
Oscilloscope	Tektronics	TBS1202B	$1509.53
Spectrometer	Andor	Shamrock 500i	-
CCD Camera	Thorlabs	DCC1645C-HQ	-
Amplified Si Phtodiode	Texas Instruments	TSL257-LF	$2.09
Teslameter FH55	MAGNET-PHYSIK	2000550EBA01	-
Microwave Source	Rohde & Schwarz	SMC100 A	-
Spectrometer	Ocean Optics	AVS USB2000	-
Electromagnet	RS PRO	7393264	$14.15
Structural components			
XYZ Positioner	Thorlabs	LNR50M	3x$946.00
Base Plate	Thorlabs	LNR50P1	$74.54
Right-Angle Bracket	Thorlabs	LNR50P2	$114.64
Optical Post x5	Thorlabs	TR20/M-P5	$21.30
Samples			
Polycrystaline diamond	Element Six		-
Single crystal diamond	Element Six		-

Figure A1. List of components.

Appendix B. Characterization Details

Figure A2 shows the photoluminescence spectra for an NV spin ensemble embedded in a single-crystal diamond (red) and a polycrystalline diamond (blue).

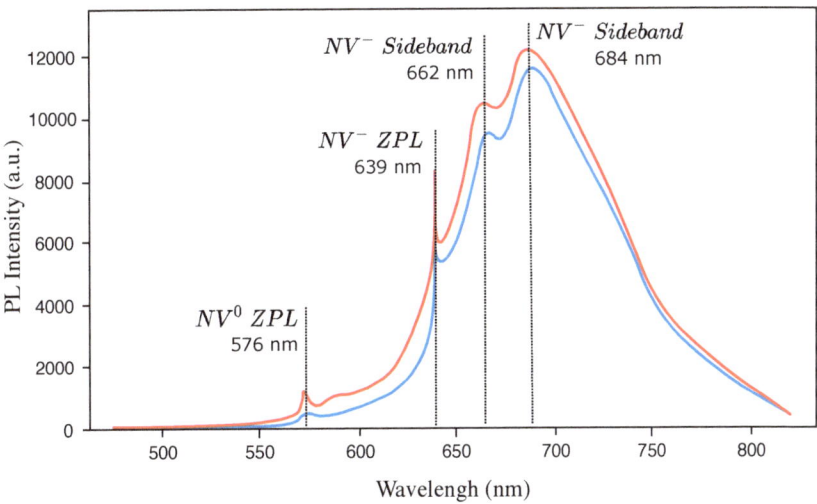

Figure A2. Photoluminescence spectra smoothed out for an NV spin ensemble.

These curves show the typical luminescence representing the intensity as a function of the wavelength, highlighting the zero-phonon lines (ZPL) and the phonon sidebands.

NV^- and NV^0 centers can be optically distinguished by their different zero phonon lines at 639 nm and 575 nm, respectively. The NV^- zero-phonon line is determined by the intrinsic difference in energy levels between the spin triplet ground and excited states.

The NV^- phonon sideband is shifted to a higher frequency in absorption and to a lower frequency in fluorescence.

The two samples were characterized at room temperature under different conditions of laser light intensities and at different depths of focus with an integration time of 0.25 s, getting almost identical results. Specifically, Figure A3a shows the luminescence curve plotting intensity as a function of wavelength for the diamond single crystal.

In the setup, the objective can be translated upward or downward, changing the focus. The lines represent the spectra measured directly with focus on the surface (green), and the rest of the lines in blue, orange, red, and gray. Those measured focusing at the different depths—that is, 4 µm, 8 µm, 12 µm and 16 µm—are very similar. The greater the depth, the greater the observed intensity of the luminescence due to the volume of the stimulated material is greater and therefore, so is the number of activated centers. As a consequence, the number of emitted photons increases.

Similarly Figure A3b shows the luminescence curve plotting intensity as a function of wavelength for the diamond single crystal but, in this case, the lines represent the spectra measured for different laser intensities: 20 µW (green line), 40 µW (blue), 60 µW (orange), 75 µW (red) and 80 µW (grey). In general, the greater the laser power, the greater the observed intensity of the luminescence because of the linearity between laser intensity and PL.

Figure A3c,d show results for the diamond polycrystalline sample, very similar to the ones from the single-crystal diamond.

The four Figure A3a–c show an oscillation after 650 nm due to the optical interference in the spectrometer sensor which is back illuminated.

Finally, Figure A3e,f show the zoom in the frequency range from 480 nm to 550 nm for the single-crystal and polycrystalline samples, respectively. In both cases, a small Raman peak corresponding to the vibration of the $sp3$ diamond lattice [41] was observed at 523 nm or 1371 cm^{-1}.

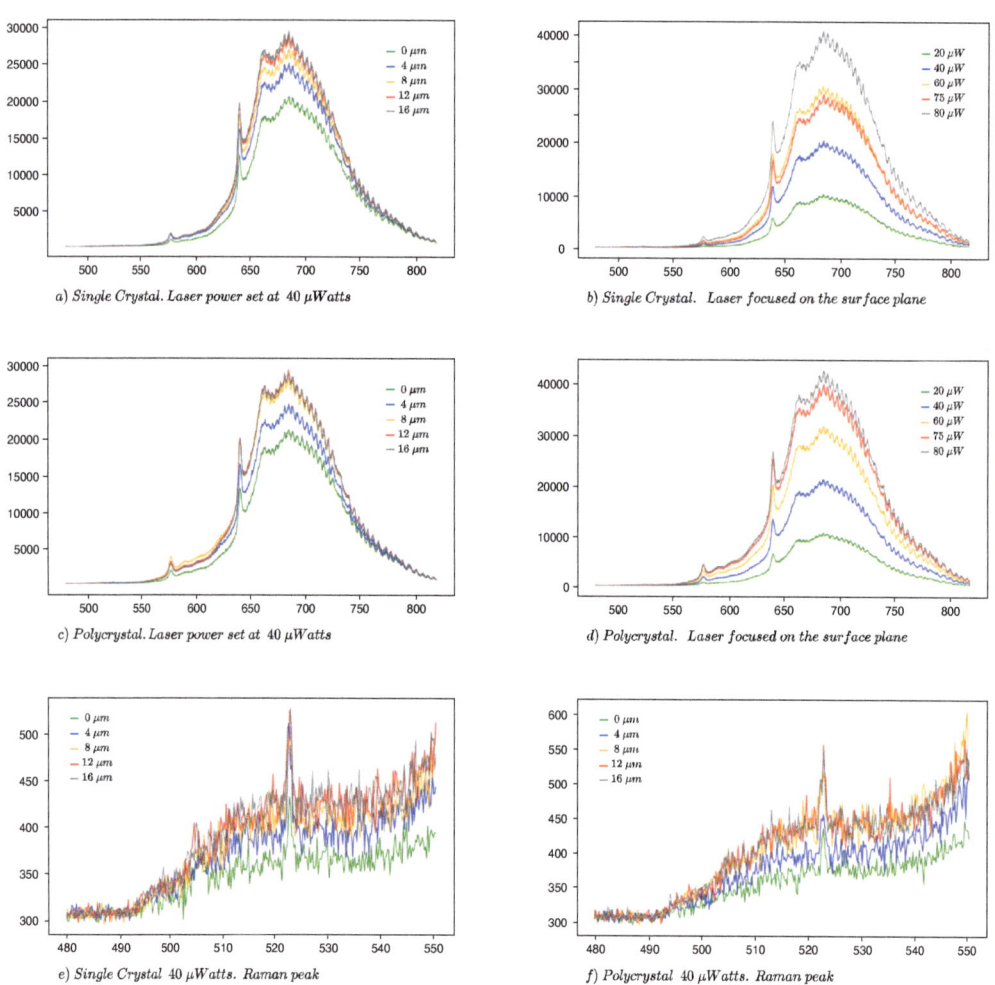

Figure A3. Photoluminescence spectra for the single crystal (**a**,**b**) and detail of the observed Raman peak (**e**). Photoluminescence spectra for the polycrystal (**c**,**d**) and detail of the observed Raman peak (**f**).

References

1. Doherty, M.W.; Manson, N.B.; Delaney, P.; Jelezko, F.; Wrachtrup, J.; Hollenberg, L.C. The nitrogen-vacancy colour centre in diamond. *Phys. Rep.* **2013**, *528*, 1–45. [CrossRef]
2. Taylor, J.M.; Cappellaro, P.; Childress, L.; Jiang, L.; Budker, D.; Hemmer, P.R. High-sensitivity diamond magnetometer with nanoscale resolution. *Nat. Phys.* **2008**, *4*, 810–816. [CrossRef]
3. Dang, H.B.; Maloof, A.C.; Romalis, M.V. Ultrahigh sensitivity magnetic field and magnetization measurements with an atomic magnetometer. *Appl. Phys. Lett.* **2010**, *97*, 151110. [CrossRef]
4. Drung, D.; Assmann, C.; Beyer, J.; Kirste, A.; Peters, M.; Ruede, F.; Schurig, T. Highly Sensitive and Easy-to-Use SQUID Sensors. *IEEE Trans. Appl. Supercond.* **2007**, *17*, 699–704. [CrossRef]

5. Todaro, M.; Sileo, L.; Vittorio, M. Magnetic Field Sensors Based on Microelectromechanical Systems (MEMS) Technology. *Magn. Sens.* **2012**. [CrossRef]
6. Degen, C.; Poggio, M.; Mamin, H.; Rettner, C.; Rugar, D. Nanoscale magnetic resonance imaging. *Proc. Natl. Acad. Sci. USA* **2009**, *106*, 1313–1317. [CrossRef]
7. Amir Borna, T.R.C. *Non-Invasive Functional-Brain-Imaging with an OPM-based Magnetoencephalography System*; Technical report; Sandia National Laboratories: Albuquerque, NM, USA, 2020.
8. Khaneja, N. SQUID Magnetometers, Josephson Junctions, Confinement and BCS Theory of Superconductivity. In *Magnetometers*; Curilef, S., Ed.; IntechOpen: Rijeka, Croatia, 2019; Chapter 5. [CrossRef]
9. Tierney, T.; Holmes, N.; Mellor, S. Optically pumped magnetometers: From quantum origins to multi-channel magnetoencephalography. *NeuroImage* **2019**, *199*, 598–608. [CrossRef]
10. Degen, C.; Reinhard, F.; Cappellaro, P. Quantum sensing. *Rev. Mod. Phys.* **2017**, *89*, 035002. [CrossRef]
11. Maze, J.; Stanwix, P.; Hodges, J.; Hong, S.; Taylor, J.; Cappellaro. Nanoscale magnetic sensing with an individual electronic spin in diamond. *Nature* **2008**, *455*, 644–647. [CrossRef]
12. Hall, L.; Beart, G.; Thomas, E.; Simpson, D.; Mcguinness. High spatial and temporal resolution wide-field imaging of neuron activity using quantum NV-diamond. *Sci. Rep.* **2012**, *2*, 401. [CrossRef]
13. Wolf, T.; Neumann, P.; Nakamura, K.; Sumiya, H.; Ohshima, T.; Isoya, J.; Wrachtrup, J. Subpicotesla Diamond Magnetometry. *Phys. Rev. X* **2015**, *5*, 041001. [CrossRef]
14. Mitchell, M.W.; Alvarez, S.P. Quantum limits to the energy resolution of magnetic field sensors. *Rev. Mod. Phys.* **2020**, *92*, 21001. [CrossRef]
15. Marcus, W.; Doherty, N.B.M. *The Nitrogen-Vacancy Colour Centre in Diamond*; Technical report; School of Physics, University of Melbourne: Melbourne, VIC, Australia, 2013.
16. Wikipedia Contributors. Molecular symmetry Wikipedia, The Free Encyclopedia. 2022. Available online: https://en.wikipedia.org/w/index.php?title=Element_Six&oldid=1118487397 (accessed on 23 May 2022).
17. Humphreys, J.; Liu, Q.; Humphreys, J.; Erne, R. *A Course in Group Theory*; Oxford Graduate Texts in Mathematics, Oxford University Press: Oxford, UK, 1996.
18. Ferrari, A.M.; D'Amore, M.; El-Kelany, K.E.; Gentile, F.S.; Dovesi, R. The NV0 defects in diamond: A quantum mechanical characterization through its vibrational and Electron Paramagnetic Resonance spectroscopies. *J. Phys. Chem. Solids* **2022**, *160*, 110304. [CrossRef]
19. Gali, A. Theory of the neutral nitrogen-vacancy center in diamond and its application to the realization of a qubit. *Phys. Rev. B* **2009**, *79*, 235210. [CrossRef]
20. Kasper Jensen, P.K. *Magnetometry with Nitrogen-Vacancy Centers in Diamond*; Technical report; Niels Bohr Institute, University of Copenhagen: Copenhagen, Denmark, 2017.
21. van Oort, E.; Glasbeek, M. Cross-relaxation dynamics of optically excited *NV* centers in diamond. *Phys. Rev. B* **1989**, *40*, 6509–6517. [CrossRef]
22. Alkahtani, M.; Hemmer, P. Charge Stability of Nitrogen-Vacancy Color Center in Organic Nanodiamonds. *Opt. Mater. Express* **2020**, *10*. [CrossRef]
23. Gaebel, T.; Domhan, M.; Wittmann, C. Photochromism in single nitrogen-vacancy defect in diamond. *Appl. Phys.* **2006**, *455*, 243–246. [CrossRef]
24. Choi, S.; Jain, M.; Louie, S. Mechanism for optical initialization of spin in *NV* center in diamond. *Phys. Rev. B* **2012**, *86*, 041202. [CrossRef]
25. Rogers, L.J.; Armstrong1, S.; Sellars, M.L.; Manson, N.B. *New Infrared Emission of the NV Centre in Diamond: Zeeman and Uniaxial Stress Studies*; Technical report; Laser Physics Center, Australian National University: Canberra, Australia, 2008.
26. Robledo, L.; Bernien, H.; van Weperen, I.; Hanson, R. Control and Coherence of the Optical Transition of Single Nitrogen Vacancy Centers in Diamond. *Phys. Rev. Lett.* **2010**, *105*, 177403. [CrossRef]
27. Subedi, S.D.; Fedorov, V.V.; Peppers, J.; Martyshkin, D.V.; Mirov, S.B.; Shao, L.; Loncar, M. Laser spectroscopic characterization of negatively charged nitrogen-vacancy (NV^-) centers in diamond. *Opt. Mater. Express* **2019**, *9*, 2076–2087. [CrossRef]
28. Neumann, P.; Kolesov, R.; Jacques, V.; Beck, J. Excited-state spectroscopy of single *NV* defects in diamond using optically detected magnetic resonance. *New J. Phys.* **2009**, *11*, 013017. [CrossRef]
29. Element Six. Synthetic Diamond and Tungsten Carbide Experts. 2022.
30. Vavilov, V.S. The properties of natural and synthetic diamond. *Physics-Uspekhi* **1993**, *36*, 1083–1084. [CrossRef]
31. Jelezko, F.; Wrachtrup, J. *Single Defect Centres in Diamond: A Review*; Technical report; DPhysikalisches Institut, Universität Stuttgart: Stuttgart, Germany, 2006.
32. Childress, L.; Dutt, M.G.; Taylor, J.M.; Zibrov, A.S.; Jelezko, F. Coherent Dynamics of Coupled Electron and Nuclear Spin Qubits in Diamond. *Science* **2006**, *314*, 281–285. [CrossRef]
33. Trofimov, S.D.; Tarelkin, S.A.; Bolshedvorskii, S.V. Spatially controlled fabrication of single *NV* centers in IIa HPHT diamond. *Opt. Mater. Express* **2020**, *10*, 198–207. [CrossRef]
34. Man Zhao, Q.L. *Antenna for Microwave Manipulation of NV Colour Centres*; Technical report; The Institution of Engineering and Technology: London, UK, 2020.

35. Eisuke Abe, K.S. *Tutorial: Magnetic Resonance with Nitrogen-Vacancy Centers in Diamond Microwave Engineering, Materials Science, and Magnetometry*; Technical report; Spintronics Research Center, Keio University: Tokyo, Japan, 2018.
36. Voltera v-one PCB Printer. 2022. Available online: https://www.voltera.io/store/v-one (accessed on 24 July 2022).
37. Maria Simanovskaia, K.J. *Sidebands in Optically Detected Magnetic Resonance Signals of Nitrogen Vacancy Centers in Diamond*; Technical report; Department of Physics, University of California: Berkeley, CA, USA, 2013.
38. Harrison, J.; Sellars, M.; Manson, N. Optical spin polarisation of the N-V centre in diamond. *J. Lumin.* **2004**, *107*, 245–248. [CrossRef]
39. Rogers, L.; Jahnke, K.; Metsch, M.; Sipahigil, A.; Binder, J.; Teraji. All-Optical Initialization, Readout, and Coherent Preparation of Single Silicon-Vacancy Spins in Diamond. *Phys. Rev. Lett.* **2014**, *113*, 263602. [CrossRef]
40. Magnetic Field Strength Meter Gauss-/Teslameter FH 55. 2022. Available online: https://www.magnet-physik.de/upload/31925018-FH-55-e-3157.pdf (accessed on 24 July 2022).
41. Gilkes, K.; Prawer, S.; Nugent, K.; Robertson, J.; Sands, H.; Lifshitz, Y.; Shi, X. Direct quantitative detection of the sp3 bonding in diamond-like carbon films using ultraviolet and visible Raman spectroscopy. *J. Appl. Phys.* **2000**, *87*, 7283–7289. [CrossRef]

Disclaimer/Publisher's Note: The statements, opinions and data contained in all publications are solely those of the individual author(s) and contributor(s) and not of MDPI and/or the editor(s). MDPI and/or the editor(s) disclaim responsibility for any injury to people or property resulting from any ideas, methods, instructions or products referred to in the content.

Article

Carbon Dots versus Nano-Carbon/Organic Hybrids—Divergence between Optical Properties and Photoinduced Antimicrobial Activities

Audrey F. Adcock [1], Ping Wang [2], Elton Y. Cao [3], Lin Ge [2], Yongan Tang [4], Isaiah S. Ferguson [1], Fares S. Abu Sweilem [1], Lauren Petta [2], William Cannon [2], Liju Yang [1,*], Christopher E. Bunker [3,*] and Ya-Ping Sun [2,*]

1. Department of Pharmaceutical Sciences, Biomanufacturing Research Institute and Technology Enterprise, North Carolina Central University, Durham, NC 27707, USA
2. Department of Chemistry, Clemson University, Clemson, SC 29634, USA
3. Air Force Research Laboratory, Propulsion Directorate, Wright-Patterson Air Force Base, Dayton, OH 45433, USA
4. Department of Mathematics and Physics, North Carolina Central University, Durham, NC 27707, USA
* Correspondence: lyang@nccu.edu (L.Y.); christopher.bunker@us.af.mil (C.E.B.); syaping@clemson.edu (Y.-P.S.)

Abstract: Carbon dots (CDots) are generally defined as small-carbon nanoparticles with surface organic functionalization and their classical synthesis is literally the functionalization of preexisting carbon nanoparticles. Other than these "classically defined CDots", however, the majority of the dot samples reported in the literature were prepared by thermal carbonization of organic precursors in mostly "one-pot" processing. In this work, thermal processing of the selected precursors intended for carbonization was performed with conditions of 200 °C for 3 h, 330 °C for 6 h, and heating by microwave irradiation, yielding samples denoted as CS200, CS330, and CS_{MT}, respectively. These samples are structurally different from the classical CDots and should be considered as "nano-carbon/organic hybrids". Their optical spectroscopic properties were found comparable to those of the classical CDots, but very different in the related photoinduced antibacterial activities. Mechanistic origins of the divergence were explored, with the results suggesting major factors associated with the structural and morphological characteristics of the hybrids.

Keywords: carbon dots; nano-carbon/organic hybrids; thermal carbonization; optical spectroscopy; photoinduced antibacterial

1. Introduction

Small-carbon nanoparticles, mostly amorphous without defined crystal structures, have suddenly attracted significant attention as a newly "discovered" class of carbon nanomaterials, even with an edge over the pretty fullerene molecules in the competition for the title of carbon at zero dimension in the family of nanoscale carbon allotropes [1]. The fortune of these otherwise ugly nanoparticles is credited to their spectacularly enhanced optical properties upon the effective nanoparticle surface passivation by organic functionalization, namely the formation of carbon "quantum" dots or, more appropriately, carbon dots, due to the lack of any evidence for the classical quantum confinement in small-carbon nanoparticles [1–3]. Carbon dots of such a structural configuration (Figure 1) are commonly referred to as classically defined carbon dots, often denoted as CDots, from the preparation considered as the "top-down" method. In CDots, the photoexcited state properties and redox characteristics are attributed to the effective passivation and stabilization of the abundant surface defect sites on small-carbon nanoparticles via organic functionalization [1,4]. However, despite the rather simple structure of the classically defined CDots and their straightforward synthesis, the rapidly expanding research community relevant to these nanomaterials has apparently decided that the simple surface functionalization of preexisting small-carbon nanoparticles is not facile enough, opting instead for the "bottom-up" carbonization approach, with the thermal processing of organic precursors to produce dot samples in "one-pot" [1,5–19]. The implicit assumption for the approach must be such that organic precursors could be thermally carbonized into nanostructures analogous to that of the classically defined CDots (Figure 1) [19], though there has been a general lack of any serious efforts on the nanoscale structural elucidation of the samples produced by the thermal carbonization, let alone any concrete experimental evidence for the validation of the implicit assumption.

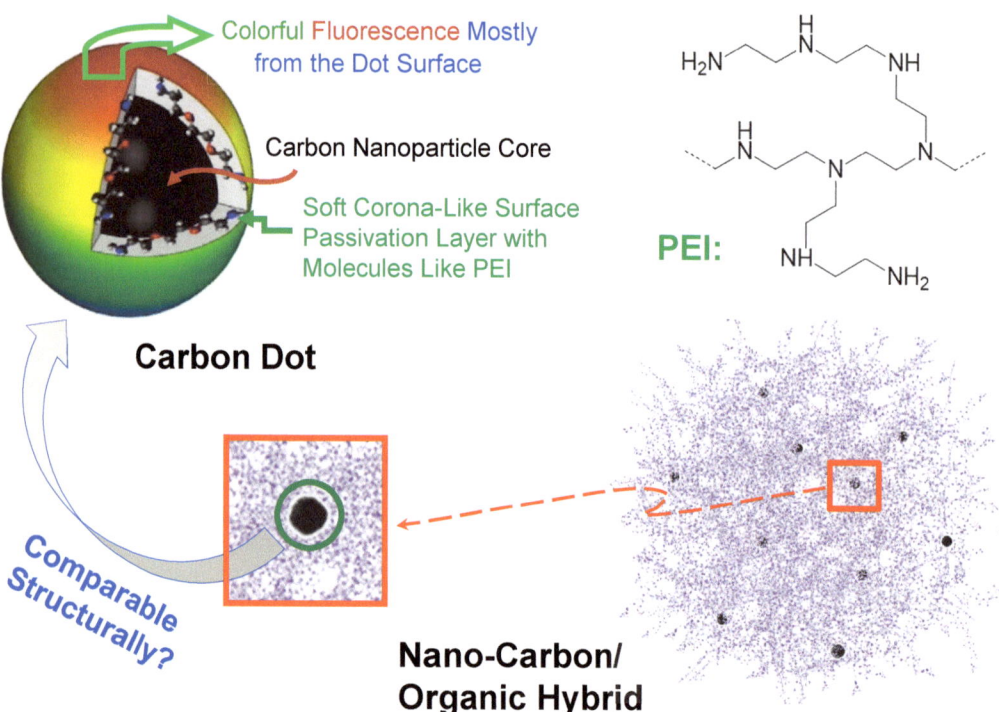

Figure 1. Carbon dots (PEI-CDots as a specific example, classically defined and prepared by the functionalization of preexisting small-carbon nanoparticles with PEI molecules) versus the nanocarbon/organic hybrids from the thermal processing of selected organic precursors.

Among popular organic precursors for the carbonization approach are mixtures of citric acid (CA) with oligomeric polyethylenimine (PEI) [20–26]. The "classical" thermal processing conditions for the CA-PEI precursor mixtures, similar to those for most other organic precursors, are thermal or hydrothermal treatments at temperatures up to 200 °C for a few hours.

The attraction of such an extremely user-friendly synthetic approach is obvious, though the chemical and nanoscale structures of the samples, thus prepared, are most likely very different from those of CDots (Figure 1) [27,28]. There is growing experimental evidence suggesting that the commonly employed processing conditions are generally insufficient for the intended carbonization, yielding samples of little nanoscale carbon (non-molecular carbon) and abundant organic species, often vulnerable to the sample contamination or even spectroscopic dominance by the molecular dyes/chromophores produced in the thermally induced chemical reactions [29–34]. In the much-less-popular synthesis, with more vigorous processing conditions, such as higher temperatures coupled with longer treatment times, more nanoscale carbon contents in the resulting samples could be achieved [27], though the sample structures are likely crosslinked mixtures of the nanoscale carbon with a significant amount of organic species, thus, more appropriately characterized as "nano-carbon/organic hybrids" (Figure 1) [19,28]. It may be further extrapolated that in such sample structures, there are local areas of nano-carbon domains passivated by the surrounding organic species, thus, a structural configuration analogous to that of classically defined CDots (Figure 1), which may account for their observed similar optical spectroscopic features [19].

In this work, the samples from thermal processing of the CA-PEI precursor mixture under different conditions, including hydrothermal at 200 °C and 330 °C and heating by microwave irradiation, were compared among themselves and with PEI-CDots (classically defined CDots of PEI-functionalized preexisting small-carbon nanoparticles) for their optical spectroscopic properties and photoinduced antimicrobial activities. The results show major divergence, such that there are significant similarities among the samples in optical spectroscopic properties, but their visible-light activated antibacterial activities are very different. Possible structural and mechanistic origins of the divergence are explored and implications of the different structures of the carbonization produced samples from those of the classically defined CDots are discussed.

2. Experimental Section

2.1. Materials

Citric acid was purchased from Alfa Aesar (Tewksbury, MA, USA), oligomeric polyethylenimine (PEI, branched, average molecular weight ~600) from Polysciences, Inc. (Warrington, PA, USA) and N,N-diethylaniline and 2,4-dinitrotoluene from Sigma-Aldrich. Dialysis membrane tubing (molecular weight cut-off ~500 or ~1000) was supplied by Spectrum Laboratories. Water was deionized and purified by being passed through a Labconco WaterPros water purification system (Labconco, Kansas City, MO, USA).

2.2. Measurement

UV/vis absorption spectra were recorded on Shimadzu UV2501 and UV-3600 spectrophotometers (Shimadzu Scientific Instruments, Inc., Columbia, MD, USA). Fluorescence spectra were measured on a Jobin-Yvon emission spectrometer equipped with a 450 W xenon source, Gemini-180 excitation and Triax-550 emission monochromators, and a photon-counting detector (Hamamatsu R928P PMT at 950 V) (Hamammatsu Photonics, Bridgewater Township, NJ, USA). Atomic force microscopy (AFM) images were acquired using a Nanosurf CoreAFM instrument (Liestal, Switzerland). Sample specimens for AFM were prepared on clean Si wafer pieces, with blank images obtained under the same conditions using the same Si wafers.

2.3. CS200 Sample

Citric acid (1 g) and PEI (0.5 g) were mixed in water (10 mL) and the resulting mixture was loaded into a stainless-steel tube reactor. The reactor was sealed for heating in a tube furnace at 200 °C for 3 h. Then, the reactor was cooled back to ambient and the reaction mixture in the reactor was collected. The mixture was placed in a membrane tubing (molecular weight cut-off ~1000) for dialysis against fresh water, followed by concentration. A light-brown aqueous solution was obtained as the sample from the carbonization synthesis at 200 °C, denoted as CS200.

2.4. CS330 Sample

Citric acid (1 g) was dissolved in water (2 mL) with brief sonication and the resulting solution was mixed well with PEI (0.5 g). Then, the mixture was loaded into a stainless-steel tube reactor. The reactor was sealed for heating in a tube furnace at 330 °C for 6 h. Upon the cooling of the reactor back to ambient temperature, the reaction mixture in the reactor was collected by washing with water. The aqueous mixture was placed in a membrane tubing (molecular weight cut-off ~1000) for dialysis against fresh water. Upon concentration and then centrifugation, the colored supernatant was collected as the sample from the carbonization synthesis at 330 °C, denoted as CS330.

2.5. CS_{MT} Sample

CA (1 g) in water (2 mL) was mixed well with PEI (3 g), followed by evaporation to remove water for a solid-like mixture. Separately, a bath of silicon carbide (150 g) in a silica crucible casting dish (about 8 cm in diameter and 2.5 cm in height) was prepared and pre-heated in a conventional microwave oven at 500 W for 3 min. The solid-like mixture of CA-PEI in a scintillation vial was immersed in the pre-heated silicon carbide bath for treatments with microwave irradiation, first at 200 W for ~4 min until no more bubbles in the sample and with the sample color turning dark orange and then 1000 W for 3 min, during which there were 5 brief pauses each of a few seconds. Post-treatment, the reaction mixture was cooled to ambient temperature and dispersed in deionized water with sonication. The resulting aqueous dispersion was centrifuged at 5000× g to keep the supernatant, followed by dialysis in a membrane tubing (molecular weight cut-off ~1000) against fresh water to obtain a colored aqueous solution of the sample denoted as CS_{MT}.

2.6. PEI-CDots

Details on the preparation and characterization of PEI-CDots were reported previously [35,36]. Copies of the synthesis protocol and results with references are provided as Supplementary Materials.

2.7. Antibacterial Evaluations

Listeria monocytogenes 10403S cells were used as the target bacteria for evaluations on photoinduced antibacterial activities of the PEI/CA samples and PEI-CDots. *L. monocytogenes* cells were grown in 10 mL brain heart infusion (BHI) by inoculating with a single colony streaked on BHI plate at 37 °C overnight with constant agitation at 225 rpm in an Excella E24 incubator shaker (New Brunswick Scientific, Edison, NJ, USA). The cells were centrifuged in an Eppendorf 5424 microcentrifuge at 4000× g for 5 min and then washed twice with phosphate-buffer saline (PBS). The cell pellet was re-suspended in PBS and diluted to the desired cell concentration for further experiments.

Treatments of the cells with the PEI/CA samples and PEI-CDots were performed in 96-well plates. Aliquots of 100 µL of bacteria cell suspensions were placed into the wells and 100 µL of aqueous solutions of the PEI/CA samples and PEI-CDots in various concentrations were added to reach the desired final concentrations for treating the cells. The bacterial cell concentration in each well was in the order of 10^8 CFU/mL. The control samples were bacterial suspension with PBS. The samples for each treatment were duplicated in each experiment. The plates were placed on orbital shaker (BT Lab Systems) at 350 rpm, with exposure to visible light from a commercial 8 W daylight LED lamp (CREE, omnidirectional 815 lumens) placed at a distance of ~10 cm away from the plate surface (the light intensity of ~4.8 mW/cm^2 experienced by the samples) for 1 h. For all treatment experiments, at least three independent experiments were performed.

After the treatments, the viable cell numbers in the treated and the control samples were determined using the traditional surface plating method. Briefly, the samples were 1:10 serially diluted with PBS and aliquots of 100 µL of appropriate dilutions were surface plated on BHI agar plates. The plates were incubated at 37 °C overnight and the colonies were counted and calculated in colony-forming units per mL (CFU/mL) for the viable cell numbers in the samples. The reductions in logarithmic viable cell number in the treated samples comparing to the control samples were used as the measures for the photoinduced antibacterial effects of the PEI/CA samples and PEI-CDots at a given concentration.

3. Results and Discussion

As reported previously [27], the precursor mixture of citric acid (CA) and oligomeric polyethylenimine (PEI) was used for the thermal processing intended for carbonization, thus, the popular "bottom-up" approach for dot samples [1,5–19]. The processing included the heating with the presence of water in a sealed reactor at 200 °C for 3 h and 330 °C for 6 h to obtain the PEI/CA samples denoted as CS200 and CS330 (with CS referring to "carbonization synthesis"), respectively. The same precursor mixture was also processed by more efficient heating via microwave irradiation, again intended for carbonization, yielding the PEI/CA carbonization product denoted as CS_{MT} (with MT referring to "microwave thermal"). For comparison, the classically defined CDot sample PEI-CDots was prepared by using preexisting small-carbon nanoparticles for surface functionalization with PEI molecules, as previously reported [35,36]. All these samples were cleaned post-synthesis to remove residual precursors and other impurities.

On the comparison of optical absorption spectra of the different samples in aqueous solutions (Figure 2), CS200 is only weakly absorptive in visible light (400 nm and longer wavelengths), obviously much weaker than CS330 at comparable sample solution concentrations, consistent with the conclusion that the thermal processing conditions for CS200 are much too mild for the intended substantial carbonization of the precursor mixture [27,28]. However, largely similar are their fluorescence spectral profiles for 400 nm excitation (Figure 2), where the optical absorptions for the excitation of both samples must be due to the nano-carbon domains produced as a result of the carbonization. In fact, the spectral similarity between the fluorescence emissions of these samples and that of PEI-CDots (Figure 2) may also support the more generalized notion that the nano-carbon domains in carbonization-produced samples responsible for the observed optical spectroscopic properties are probably somewhat comparable in structure and morphology to the preexisting small-carbon nanoparticles in classically defined CDots [19], such as PEI-CDots here.

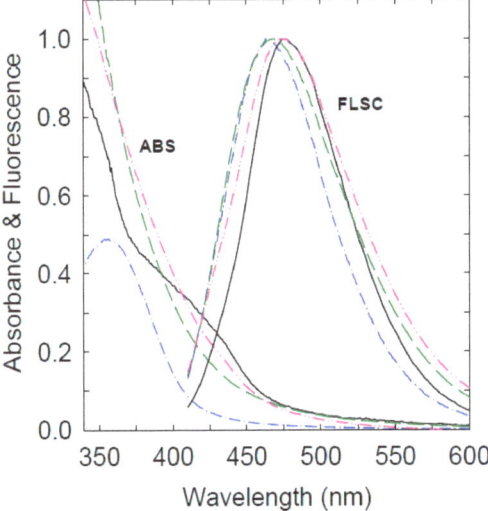

Figure 2. Optical absorption (ABS) and fluorescence (FLSC, 400 nm excitation) spectra of CS200 (dash-dot line), CS330 (solid line), CS_{MT} (dash-dot-dot line), and PEI-CDots (dash line), all in aqueous solutions.

The absorption and fluorescence emission spectra of the CS_{MT} sample in aqueous solution are similar to those of CS330 (Figure 2), for which one may rationalize, such that the microwave irradiation is known for highly efficient heating to result in carbonization of the CA-PEI precursor mixture, comparable to that by thermal processing at high temperatures and/or for longer time periods. If only the optical spectroscopic properties are targeted, one may even argue that the microwave processing should be the method of choice in the carbonization synthesis of dot samples. Interestingly and surprisingly, however, there are major differences between CS330 and CS_{MT} in their other photoexcited state properties and processes or in the consequences of such properties and processes, specifically with respect to the photoinduced antimicrobial function [37,38].

It is known that PEI-CDots with exposure to visible light are effective and efficient in the inactivation of various model bacteria and bacterial pathogens [39,40]. In this work, the three PEI/CA samples (CS200, CS330, and CS_{MT}) were evaluated for visible-light-activated antibacterial activities against *L. monocytogenes*, in reference to those of PEI-CDots. In the evaluation, *L. monocytogenes* cells in PBS suspensions (~10^8 CFU/mL) were treated with CS200, CS330, CS_{MT}, and PEI-CDots in different concentrations and all with the visible-light exposure for 1 h, except for the dark controls. After the treatments, the viable cell numbers in the treated samples and the control samples were determined. The dose–response curve for *L. monocytogenes* in response to each of the tested dot samples is plotted in Figure 3 in terms of the observed logarithmic viable cell numbers against the concentrations of the individual PEI/CA samples and PEI-CDots used in the treatments.

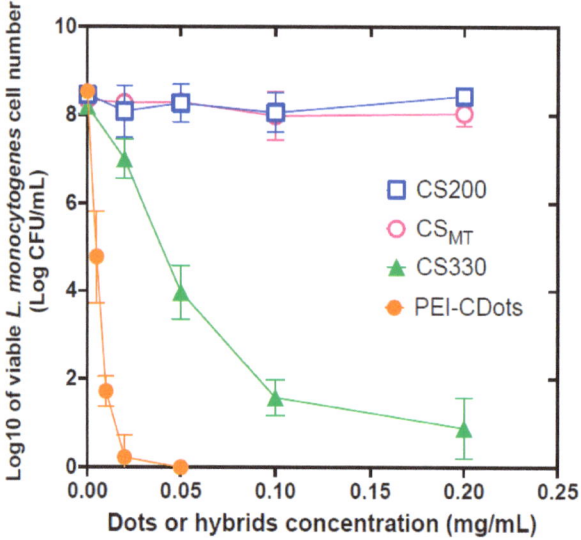

Figure 3. The antibacterial activities of CS200, CS330, CS_{MT}, and PEI-CDots with visible-light exposure (all for 1 h) against *L. monocytogenes* cells, expressed by the reduction in logarithmic viable cell number with increasing concentrations of the hybrids and PEI-CDots.

The CS200 sample clearly lacks the light-activated antibacterial function (Figure 3), which one might argue should be expected for the low content of nano-carbon domains in the sample due to insufficient carbonization, associated with the much-too-mild thermal processing conditions. The more appropriately carbonization-produced CS330 sample with the same visible-light exposure could inactivate *L. monocytogenes* significantly, though still much less effective than PEI-CDots (Figure 3). For the CS_{MT} sample, however, even with its microwave-assisted thermal processing conditions designed for sufficient carbonization and its similar optical spectroscopic properties to those of CS330, its lack of any meaningful antibacterial action, the same as that of the CS200 sample (Figure 3), is really unexpected and striking, suggesting that there are other major factors beyond the content of nano-carbon domains, such as the structural details and/or morphological characteristics of the sample, determining the photoinduced antimicrobial properties.

On the origin of the striking difference between CS330 and CS_{MT} samples in their photoinduced antibacterial activities, photoexcited state properties of the two samples were compared, including the excitation-wavelength-dependent fluorescence emissions and the fluorescence quenching by known electron donor and acceptor molecules, *N,N*-diethylaniline (DEA) and 2,4-dinitrotoluene (DNT), respectively. The electron transfer quenching of fluorescence emissions provides insight into the photoinduced redox processes in the PEI/CA samples, thus, more directly relevant to the photoinduced antibacterial function. As compared in Figure 4, the characteristic excitation wavelength dependence of fluorescence emissions, as generally found in classically defined CDots [41], is apparently similar between CS330 and CS_{MT} samples. Further, also similar between the two samples are the results from redox quenching of fluorescence emissions with electron donor DEA (Figure 5) and acceptor DNT as quenchers (Figure 6) [34,41,42], suggesting no major differences in the photoexcited state redox characteristics that might account for the strikingly different antibacterial outcomes (Figure 3).

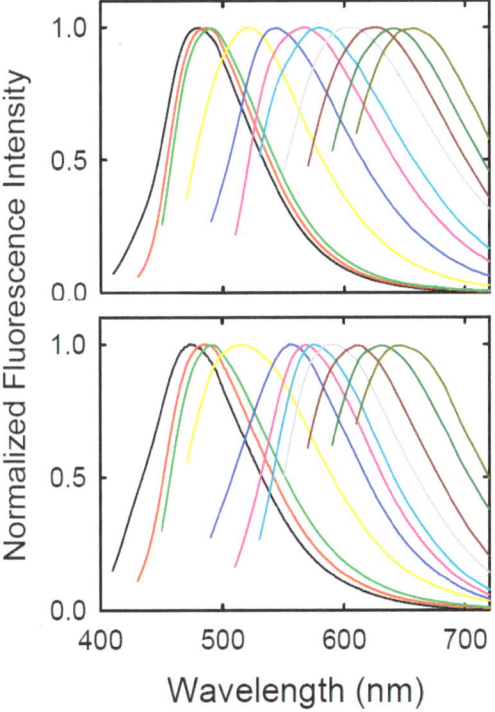

Figure 4. The excitation wavelength dependence of fluorescence emissions for CS330 (upper) and CS_{MT} (lower) in aqueous solutions, with the excitations of 400 nm and (from left to right) in 20 nm increments (thus 600 nm excitation for the last spectrum at right).

The structural and morphological characteristics of the three PEI/CA samples were probed by atomic force microscopy (AFM) imaging. According to the AFM images shown in Figure 7, the CS200 sample is apparently composed of smaller nano-carbon domains in lower population, consistent with the consequence of insufficient carbonization due to the much-too-mild thermal processing conditions for the sample. The CS330 sample contains more defined nano-carbon domains, with features somewhat similar to those found in the AFM image of PEI-CDots derived from preexisting small-carbon nanoparticles (Supplementary Materials). The CS_{MT} sample seems morphologically different from the other two samples (Figure 7). However, the sample characteristics revealed in AFM images may not be sufficient to explain the observed different photoinduced antibacterial activities, though not contradictory to the observed behaviors of the samples either. Further, an unfortunate consequence of the results is that they may serve to disqualify AFM from being a viable tool for more quantitative probing and understanding of the structural and morphological details in these PEI/CA samples. One may naturally suggest high-resolution transmission electron microscopy (HR-TEM) for the task, but the reality is that HR-TEM is intrinsically unsuitable for the probing of these soft composite-like materials on their structures and morphologies at the length scale (or resolution) of sub-nanometer to a few nanometers.

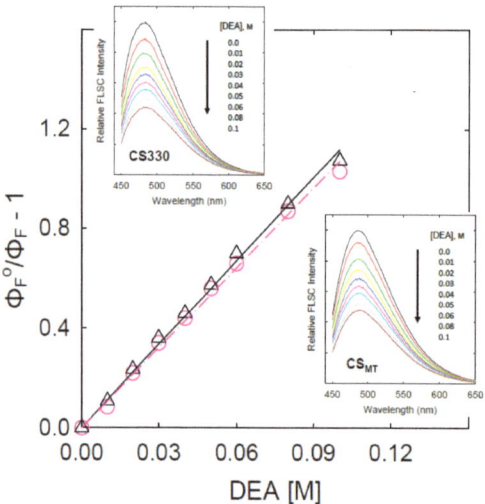

Figure 5. Stern–Volmer plots for the quenching of fluorescence emission intensity by DEA for CS330 (triangle, solid line) and CS_{MT} (circle, dash-dot line) in ethanol solutions, with the slopes (Stern–Volmer constants) of 11.2 M-1 and 10.7 M-1, respectively. Insets: The corresponding fluorescence emission spectra with the quenching at different DEA concentrations.

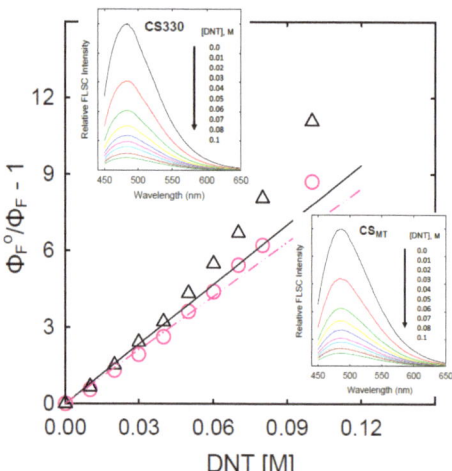

Figure 6. Stern–Volmer plots for the quenching of fluorescence emission intensity by DNT for CS330 (triangle, solid line) and CS_{MT} (circle, dash-dot-dot line) in ethanol solutions, with the slopes (Stern–Volmer constants for the linear portion with low DNT concentrations) of 79 M-1 and 70 M-1, respectively. Insets: The corresponding fluorescence emission spectra with the quenching at different DNT concentrations.

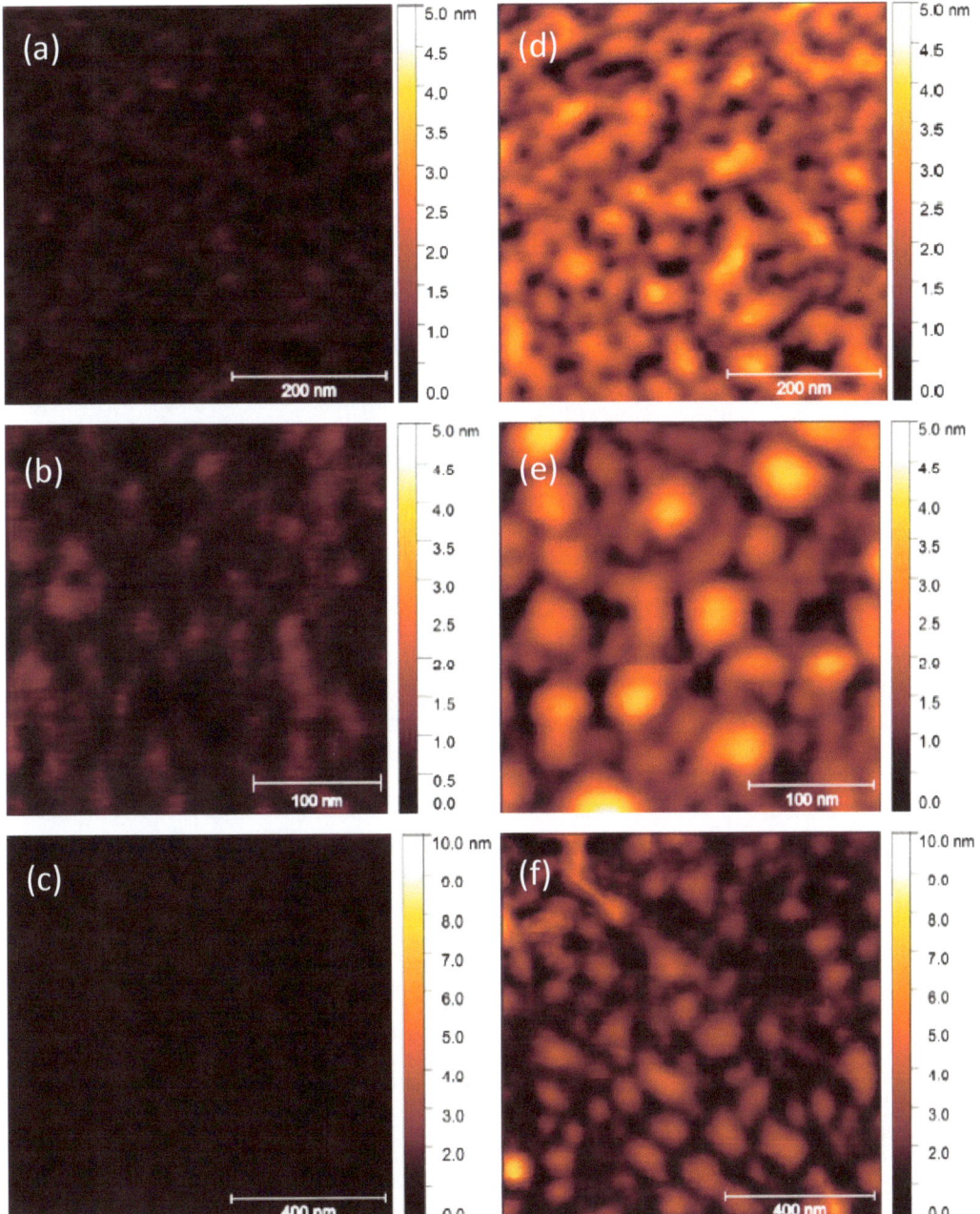

Figure 7. AFM images of the CS200 (**d**), CS330 (**e**), and CSMT (**f**) samples on Si wafers, with the corresponding images of clean Si wafers on the left (**a**–**c**).

Mechanistically, the optical absorptions of all these samples over the visible spectrum must be due to electronic transitions in the small-carbon nanoparticles or, analogously, the nano-carbon domains. There is experimental evidence for the insensitive nature of the absorptions to the surface organic functionalization of the small-carbon nanoparticles in CDots [1,36]. The same is likely applicable to the absorptions of the carbonization-produced samples, in which the nano-carbon domains are immersed in various organic materials (Figure 1). In the classically defined CDots derived from preexisting small-carbon nanoparticles, upon photoexcitation, there must be rapid charge transfers and separation for the generation of separated electrons and holes that are trapped at abundant surface defect sites of the small-carbon nanoparticles (Figure 8) [1]. The difference between "naked" small-carbon nanoparticles and their surface functionalized counterpart CDots must be such that in the latter, the effective stabilization of the surface defect sites is made possible by the organic functionalization, a phenomenon almost identical to that well established for conventional semiconductor quantum dots (QDs, CdSe nanoparticles, for example) in their early development [43,44]. The radiative recombinations of the separated redox pairs in CDots are responsible for the observed bright and colorful fluorescence emissions (Figure 8), with experimentally determined quantum yields orders of magnitude higher than those of solvent-dispersed small-carbon nanoparticles [1,45]. Such a photoexcited state mechanistic framework may also be considered for the observed spectroscopic properties of the carbonization-produced samples.

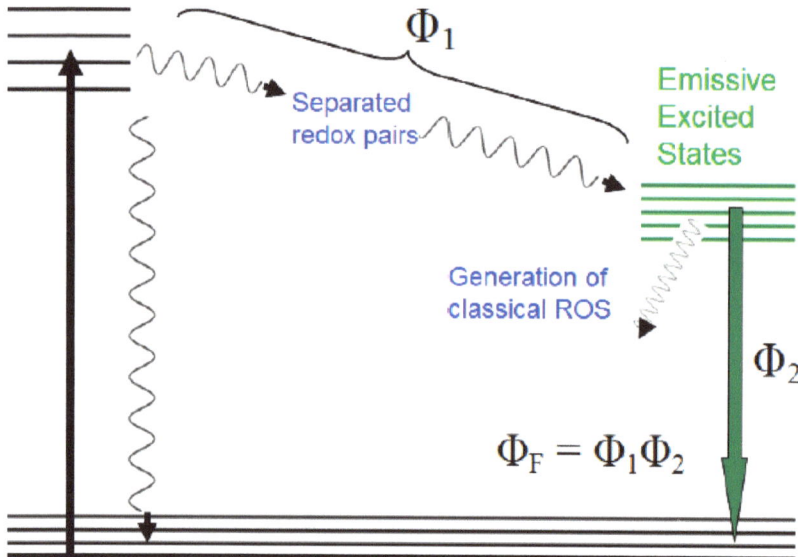

Figure 8. A state energy diagram on the photoexcited states and processes of CDots, highlighting the two sets of highly reactive species: the separated redox pairs from the initial charge transfers and separation and the generation of classical ROS as a part of the nonradiative deactivation of the emissive excited states. Φ_F denotes observed fluorescence quantum yields (reprinted with permission from [40]. Copyright 2020 Elsevier).

The three PEI/CA samples were from the same precursor mixture and thermal carbonization, but with different processing conditions, thus, the logical expectation that they should all be mixtures of nano-carbon (namely non-molecular carbon produced in the thermal carbonization) and organic species (precursor molecules crosslinked in the thermal processing with or without involving the nano-carbon) in different compositions and structural arrangements. As suggested previously [19], samples of such mixtures are more appropriately considered and denoted as "nano-carbon/organic hybrids" (Figure 1). Locally, in the structure of these hybrids, the configuration of a nano-carbon domain immersed in the matrix of organic species might be effectively analogous or equivalent to a small-carbon nanoparticle with surface passivation by the organic functionalization in classically defined CDots (Figure 1) [19]. Such an equivalency might account for the observed similar fluorescence emission properties of the hybrids and CDots. Nevertheless, the composition of the CS200 sample is too low in nano-carbon content due to the much-too-mild thermal processing conditions in the synthesis, likely only marginally capable of the properties dominated by the passivated nano-carbon domains found in the other hybrids. Despite the equivalency argument that would place the hybrids and classically defined CDots under the same mechanistic umbrella, the photoinduced antibacterial activities of the hybrids (Figure 3) obviously diverge from the optical spectroscopic properties, not only in terms of the comparison with PEI-CDots but also for the very different antibacterial outcomes between the CS330 and CD_{MT} samples.

In the same mechanistic framework discussed above for the photoexcited state properties and processes of CDots (Figure 8), the related photoinduced antimicrobial function can be attributed to the combined actions by two sets of reactive species, the separated redox pairs formed upon photoexcitation and the classical reactive oxygen species (ROS) produced in the emissive excited states (Figure 8) [40]. The former is analogous to the "light-activated redox species" (LARS) proposed for conventional semiconductor QDs in the mechanistic account for their photoinduced antimicrobial properties [46] and the latter is equivalent to the ROS generated by classical molecular dye photosensitizers [47,48]. In this regard, one might argue that these same reactive species could also be found in the photoexcited "nano-carbon/organic hybrids" (Figure 1), in light of the above discussed equivalency of their local structural configuration to that of classically defined CDots, so that the same or similar photoinduced antibacterial function should be expected. Nevertheless, a counter argument is that the formation of the reactive species may not be equivalent to the realization of their reactive function, which would depend significantly on the interactions of the reactive species with the targeted bacteria. The fact of the matter is that the reactive species are, by nature, short lived, especially so for the separated redox pairs, thus, requiring close-range actions against the bacterial cells [40]. The structural and morphological characteristics of the hybrids, with the nano-carbon domains as the centers for the photoinduced generation of reactive species immersed in the matrix of abundant organic species, must be unfavorable to the required close-range (or near-neighbor) antibacterial action in general (versus PEI-CDots, for example, Figure 3) or practically impossible for the CS_{MT} and CS200 samples (Figure 3) in particular.

In summary, the dot samples prepared by the thermal processing of selected organic precursors, if the processing conditions are adequate for the intended carbonization, may be considered and denoted as "nano-carbon/organic hybrids" for their sample structures, characterized by nano-carbon domains in crosslinked mixtures with abundant organic species. In the hybrids, there may be nanoscale structural features, specifically individual nano-carbon domains immersed in organic matrices for the effect similar to passivation, which are analogous or equivalent to the organic functionalization of pre-existing small-carbon nanoparticles in the classically defined CDots. Such structural features of the hybrids make their optical spectroscopic properties comparable to those of the classically defined CDots. For the photoinduced antimicrobial function associated with the same photoexcited state properties and processes, the sample structural and morphological characteristics of the hybrids may hinder the required close-range or near-neighbor action (due to the short-lived nature of the reactive species) against the targeted bacteria, resulting in poor outcomes or no activities at all.

Supplementary Materials: The following supporting information can be downloaded at: https://www.mdpi.com/article/10.3390/c8040054/s1. Details on the preparation and characterization of PEI-CDots. Figure S1: Representative AFM image for PEI-CDots. Figure S2: Representative TEM imaging results for PEI-CDots. Figure S3: Absorption spectrum of PEI-CDots in aqueous solution (solid line) is compared with that of the aqueous dispersed small carbon nanoparticles (dash-dot-dot line). Inset: Fluorescence spectra of PEI-CDots at different excitation wavelengths (from left to right with excitation at 400 nm and in 20 nm increments).

Author Contributions: Data acquisition and analyses, A.F.A., P.W., E.Y.C., L.G., Y.T., I.S.F., F.S.A.S., L.P. and W.C.; Funding acquisition, L.Y., Y.T. and Y.-P.S.; Investigation, A.F.A., P.W., E.Y.C. and L.G.; Project administration, L.Y., C.E.B. and Y.-P.S.; Supervision, L.Y., C.E.B. and Y.-P.S.; Validation, A.F.A., P.W. and E.Y.C.; Writing—original draft, L.Y., C.E.B. and Y.-P.S.; Writing—review and editing, L.Y., C.E.B. and Y.-P.S. All authors have read and agreed to the published version of the manuscript.

Funding: National Science Foundation, USDA, and Air Force Research Laboratory.

Data Availability Statement: All data is contained within the article or Supplementary Material.

Acknowledgments: Financial support from NSF (2102021 and 2102056, and 1855905), USDA (2019-67018-29689), and Air Force Research Laboratory is gratefully acknowledged. L.P. and W.C. were participants in the Palmetto Academy, a summer undergraduate research program of the South Carolina Space Grant Consortium.

Conflicts of Interest: The authors declare no conflict of interest.

References

1. Sun, Y.-P. *Carbon Dots—Exploring Carbon at Zero-Dimension*; Springer International Publishing: Berlin/Heidelberg, Germany, 2020.
2. Sun, Y.-P.; Zhou, B.; Lin, Y.; Wang, W.; Fernando, K.A.S.; Pathak, P.; Meziani, M.J.; Harruff, B.A.; Wang, X.; Wang, H.; et al. Quantum-Sized Carbon Dots for Bright and Colorful Photoluminescence. *J. Am. Chem. Soc.* **2006**, *128*, 7756–7757. [CrossRef] [PubMed]
3. Sun, Y.-P. Fluorescent Carbon Nanoparticles. U.S. Patent 7,829,772, 9 November 2010.
4. Cao, L.; Meziani, M.J.; Sahu, S.; Sun, Y.-P. Photoluminescence Properties of Graphene *versus* Other Carbon Nanomaterials. *Acc. Chem. Res.* **2013**, *46*, 171–180. [CrossRef] [PubMed]
5. Luo, P.G.; Sahu, S.; Yang, S.-T.; Sonkar, S.K.; Wang, J.; Wang, H.; LeCroy, G.E.; Cao, L.; Sun, Y.-P. Carbon "Quantum" Dots for Optical Bioimaging. *J. Mater. Chem. B* **2013**, *1*, 2116–2127. [CrossRef] [PubMed]
6. Ding, C.; Zhu, A.; Tian, Y. Functional Surface Engineering of C-Dots for Fluorescent Biosensing and in Vivo Bioimaging. *Acc. Chem. Res.* **2014**, *47*, 20–30. [CrossRef]
7. Luo, P.G.; Yang, F.; Yang, S.-T.; Sonkar, S.K.; Yang, L.; Broglie, J.J.; Liu, Y.; Sun, Y.-P. Carbon-Based Quantum Dots for Fluorescence Imaging of Cells and Tissues. *RSC Adv.* **2014**, *4*, 10791–10807. [CrossRef]
8. Lim, S.Y.; Shen, W.; Gao, Z. Carbon Quantum Dots and Their Applications. *Chem. Soc. Rev.* **2015**, *44*, 362–381. [CrossRef]
9. Fernando, K.A.S.; Sahu, S.; Liu, Y.; Lewis, W.K.; Guliants, E.A.; Jafariyan, A.; Wang, P.; Bunker, C.E.; Sun, Y.-P. Carbon Quantum Dots and Applications in Photocatalytic Energy Conversion. *ACS Appl. Mater. Interfaces* **2015**, *7*, 8363–8376. [CrossRef]
10. LeCroy, G.E.; Yang, S.-T.; Yang, F.; Liu, Y.; Fernando, K.A.S.; Bunker, C.E.; Hu, Y.; Luo, P.G.; Sun, Y.-P. Functionalized Carbon Nanoparticles: Syntheses and Applications in Optical Bioimaging and Energy Conversion. *Coord. Chem. Rev.* **2016**, *320*, 66–81. [CrossRef]
11. Peng, Z.; Han, X.; Li, S.; Al-Youbi, A.O.; Bashammakh, A.S.; El-Shahawi, M.S.; Leblanc, R.M. Carbon Dots: Biomacromolecule Interaction, Bioimaging and Nanomedicine. *Coord. Chem. Rev.* **2017**, *343*, 256–277. [CrossRef]
12. Hutton, G.A.M.; Martindale, B.C.M.; Reisner, E. Carbon Dots as Photosensitisers for Solar-Driven Catalysis. *Chem. Soc. Rev.* **2017**, *46*, 6111–6123. [CrossRef]
13. Xu, D.; Lin, Q.; Chang, H.-T. Recent Advances and Sensing Applications of Carbon Dots. *Small Methods* **2020**, *4*, 1900387. [CrossRef]
14. Das, R.; Bandyopadhyay, R.; Pramanik, P. Carbon Quantum Dots from Natural Resource: A Review. *Mater. Today Chem.* **2018**, *8*, 96–109. [CrossRef]
15. Du, J.; Xu, N.; Fan, J.; Sun, W.; Peng, X. Carbon Dots for In Vivo Bioimaging and Theranostics. *Small* **2019**, *15*, 1805087. [CrossRef]
16. Li, Y.; Xu, X.; Wu, Y.; Zhuang, J.; Zhang, X.; Zhang, H.; Lei, B.; Hu, C.; Liu, Y. A Review on the Effects of Carbon Dots in Plant Systems. *Mater. Chem. Front.* **2020**, *4*, 437–448. [CrossRef]
17. Indriyati; Primadona, I.; Permatasari, F.A.A.; Irham, M.A.; Nasir, D.E.M.; Iskandar, F. Recent Advances and Rational Design Strategies of Carbon Dots towards Highly Efficient Solar Evaporation. *Nanoscale* **2021**, *13*, 7523–7532. [CrossRef]
18. Đorđević, L.; Arcudi, F.; Cacioppo, M.; Prato, M.A. Multifunctional Chemical Toolbox to Engineer Carbon Dots for Biomedical and Energy Applications. *Nat. Nanotech.* **2022**, *17*, 112–130. [CrossRef] [PubMed]
19. Yuan, D.; Wang, P.; Yang, L.; Quimby, J.L.; Sun, Y.-P. Carbon "Quantum" Dots for Bioapplications. *Exp. Bio. Med.* **2022**, *247*, 300–309. [CrossRef]
20. Dong, Y.; Wang, R.; Li, G.; Chen, C.; Chi, Y.; Chen, G. Polyamine-Functionalized Carbon Quantum Dots as Fluorescent Probes for Selective and Sensitive Detection of Copper Ions. *Anal. Chem.* **2012**, *84*, 6220–6224. [CrossRef]
21. Dong, Y.; Wang, R.; Li, H.; Shao, J.; Chi, Y.; Lin, X.; Chen, G. Polyamine-Functionalized Carbon Quantum Dots for Chemical Sensing. *Carbon* **2012**, *50*, 2810–2815. [CrossRef]
22. Wang, R.; Li, G.; Dong, Y.; Chi, Y.; Chen, G. Carbon Quantum Dot-Functionalized Aerogels for NO_2 Gas Sensing. *Anal. Chem.* **2013**, *85*, 8065–8069. [CrossRef]

23. Dong, Y.; Wang, R.; Tian, W.; Chi, Y.; Chen, G. "Turn-on" Fluorescent Detection of Cyanide Based on Polyamine-Functionalized Carbon Quantum Dots. *RSC Adv.* **2014**, *4*, 3685–3689. [CrossRef]
24. Liu, J.; Liu, X.; Luo, H.; Gao, Y. One-Step Preparation of Nitrogen-Doped and Surface-Passivated Carbon Quantum Dots with High Quantum Yield and Excellent Optical Properties. *RSC Adv.* **2014**, *4*, 7648. [CrossRef]
25. Wang, C.; Xu, Z.; Zhang, C. Polyethyleneimine-Functionalized Fluorescent Carbon Dots: Water Stability, PH Sensing, and Cellular Imaging. *ChemNanoMat* **2015**, *1*, 122–127. [CrossRef]
26. Pierrat, P.; Wang, R.; Kereselidze, D.; Lux, M.; Didier, P.; Kichler, A.; Pons, F.; Lebeau, L. Efficient in Vitro and in Vivo Pulmonary Delivery of Nucleic Acid by Carbon Dot-Based Nanocarriers. *Biomaterials* **2015**, *51*, 290–302. [CrossRef] [PubMed]
27. Hou, X.; Hu, Y.; Wang, P.; Yang, L.; Al Awak, M.M.; Tang, Y.; Twara, F.K.; Qian, H.; Sun, Y.-P. Modified Facile Synthesis for Quantitatively Fluorescent Carbon Dots. *Carbon* **2017**, *122*, 389–394. [CrossRef]
28. Wang, P.; Meziani, M.J.; Fu, Y.; Bunker, C.E.; Hou, X.; Yang, L.; Msellek, H.; Zaharias, M.; Darby, J.P.; Sun, Y.-P. Carbon Dots *versus* Nano-Carbon/Organic Hybrids—Dramatically Different Behaviors in Fluorescence Sensing of Metal Cations with Structural and Mechanistic Implications. *Nanoscale Adv.* **2021**, *3*, 2316–2324. [CrossRef]
29. Khan, S.; Sharma, A.; Ghoshal, S.; Jain, S.; Hazra, M.K.; Nandi, C.K. Small Molecular Organic Nanocrystals Resemble Carbon Nanodots in Terms of Their Properties. *Chem. Sci.* **2018**, *9*, 175–180. [CrossRef]
30. Hinterberger, V.; Damm, C.; Haines, P.; Guldi, D.M.; Peukert, W. Purification and Structural Elucidation of Carbon Dots by Column Chromatography. *Nanoscale* **2019**, *11*, 8464–8474. [CrossRef]
31. Liang, W.; Ge, L.; Hou, X.; Ren, X.; Yang, L.; Bunker, C.E.; Overton, C.M.; Wang, P.; Sun, Y.-P. Evaluation of Commercial "Carbon Quantum Dots" Sample on Origins of Red Absorption and Emission Features. *C J. Carbon Res.* **2019**, *5*, 70. [CrossRef]
32. Liang, W.; Wang, P.; Meziani, M.J.; Ge, L.; Yang, L.; Patel, A.K.; Morgan, S.O.; Sun, Y.-P. On the Myth of "Red/Near-IR Carbon Quantum Dots" from Thermal Processing of Specific Colorless Organic Precursors. *Nanoscale Adv.* **2021**, *3*, 4186–4195. [CrossRef]
33. Liang, W.; Wang, P.; Yang, L.; Overton, C.M.; Hewitt, B.; Sun, Y.-P. Chemical Reactions in Thermal Carbonization Processing of Citric Acid—Urea Mixtures. *Gen. Chem.* **2021**, *7*, 210011–210017.
34. Bartolomei, B.; Bogo, A.; Amato, F.; Ragazzon, G.; Prato, M. Nuclear Magnetic Resonance Reveals Molecular Species in Carbon Nanodot Samples Disclosing Flaws. *Angew. Chem. Int. Ed.* **2022**, *61*, e202200038. [CrossRef] [PubMed]
35. Hu, Y.; Al Awak, M.M.; Yang, F.; Yan, S.; Xiong, Q.; Wang, P.; Tang, Y.; Yang, L.; LeCroy, G.E.; Bunker, C.E.; et al. Photoexcited State Properties of Carbon Dots from Thermally Induced Functionalization of Carbon Nanoparticles. *J. Mater. Chem. C* **2016**, *4*, 10554–10561. [CrossRef]
36. Ge, L.; Pan, N.; Jin, J.; Wang, P.; LeCroy, G.E.; Liang, W.; Yang, L.; Teisl, L.R.; Tang, Y.; Sun, Y.-P. Systematic Comparison of Carbon Dots from Different Preparations—Consistent Optical Properties and Photoinduced Redox Characteristics in Visible Spectrum, and Structural and Mechanistic Implications. *J. Phys. Chem. C* **2018**, *122*, 21667–21676. [CrossRef]
37. Meziani, M.J.; Dong, X.; Zhu, L.; Jones, L.P.; LeCroy, G.E.; Yang, F.; Wang, S.; Wang, P.; Zhao, Y.; Yang, L.; et al. Visible-Light-Activated Bactericidal Functions of Carbon "Quantum" Dots. *ACS Appl. Mater. Interfaces* **2016**, *8*, 10761–10766. [CrossRef] [PubMed]
38. Dong, X.; Liang, W.; Meziani, M.J.; Sun, Y.-P.; Yang, L. Carbon Dots as Potent Antimicrobial Agents. *Theranostics* **2020**, *10*, 671–686. [CrossRef] [PubMed]
39. Abu Rabe, D.I.; Mohammed, O.O.; Dong, X.; Patel, A.K.; Overton, C.M.; Tang, Y.; Kathariou, S.; Sun, Y.-P.; Yang, L. Carbon Dots for Highly Effective Photodynamic Inactivation of Multidrug-Resistant Bacteria. *Mater. Adv.* **2020**, *1*, 321–325. [CrossRef]
40. Dong, X.; Ge, L.; Abu Rabe, D.I.; Mohammed, O.O.; Wang, P.; Tang, Y.; Kathariou, S.; Yang, L.; Sun, Y.-P. Photoexcited State Properties and Antibacterial Activities of Carbon Dots Relevant to Mechanistic Features and Implications. *Carbon* **2020**, *170*, 137–145. [CrossRef]
41. LeCroy, G.E.; Messina, F.; Sciortino, A.; Bunker, C.E.; Wang, P.; Fernando, K.A.S.; Sun, Y.-P. Characteristic Excitation Wavelength Dependence of Fluorescence Emissions in Carbon "Quantum" Dots. *J. Phys. Chem. C* **2017**, *121*, 28180–28186. [CrossRef]
42. LeCroy, G.E.; Fernando, K.A.S.; Bunker, C.E.; Wang, P.; Tomlinson, N.; Sun, Y.-P. Steady-State and Time-Resolved Fluorescence Studies on Interactions of Carbon "Quantum" Dots with Nitrotoluenes. *Inorg. Chim. Acta* **2017**, *468*, 300–307. [CrossRef]
43. Kortan, A.R.; Hull, R.; Opila, R.L.; Bawendi, M.G.; Steigerwald, M.L.; Carroll, P.J.; Brus, L.E. Nucleation and Growth of CdSe on ZnS Quantum Crystallite Seeds, and Vice Versa, in Inverse Micelle Media. *J. Am. Chem. Soc.* **1990**, *112*, 1327–1332. [CrossRef]
44. Murray, C.B.; Norris, D.J.; Bawendi, M.G. Synthesis and Characterization of Nearly Monodisperse CdE (E = sulfur, selenium, tellurium) Semiconductor Nanocrystallites. *J. Am. Chem. Soc.* **1993**, *115*, 8706–8715. [CrossRef]
45. Cao, L.; Anilkumar, P.; Wang, X.; Liu, J.-H.; Sahu, S.; Meziani, M.J.; Myers, E.; Sun, Y.-P. Reverse Stern-Volmer Behavior for Luminescence Quenching in Carbon Nanoparticles. *Can. J. Chem.* **2011**, *89*, 104–109. [CrossRef]
46. Courtney, C.M.; Goodman, S.M.; McDaniel, J.A.; Madinger, N.E.; Chatterjee, A.; Nagpal, P. Photoexcited Quantum Dots for Killing Multidrug-Resistant Bacteria. *Nat. Mater.* **2016**, *15*, 529–534. [CrossRef]
47. Hara, K.; Holland, S.; Woo, J. Effects of Exogenous Reactive Oxygen Species Scavengers on the Survival of Escherichia coli B23 during Exposure to UV-A Radiation. *J. Exp. Microbiol. Immunol.* **2004**, *12*, 62–66.
48. Ishiyama, K.; Nakamura, K.; Ikai, H.; Kanno, T.; Kohno, M.; Sasaki, K.; Niwano, Y. Bactericidal Action of Photogenerated Singlet Oxygen from Photosensitizers Used in Plaque Disclosing Agents. *PLoS ONE* **2012**, *7*, e37871. [CrossRef] [PubMed]

Article

Easy and Low-Cost Method for Synthesis of Carbon–Silica Composite from Vinasse and Study of Ibuprofen Removal

Yuvarat Ngernyen [1,*], Thitipong Siriketh [2], Kritsada Manyuen [2], Panta Thawngen [2], Wipha Rodtoem [2], Kritiyaporn Wannuea [2], Jesper T. N. Knijnenburg [3] and Supattra Budsaereechai [4,*]

Citation: Ngernyen, Y.; Siriketh, T.; Manyuen, K.; Thawngen, P.; Rodtoem, W.; Wannuea, K.; Knijnenburg, J.T.N.; Budsaereechai, S. Easy and Low-Cost Method for Synthesis of Carbon–Silica Composite from Vinasse and Study of Ibuprofen Removal. C 2022, 8, 51. https://doi.org/10.3390/c8040051

Academic Editors: Olena Okhay and Gil Goncalves

Received: 11 September 2022
Accepted: 3 October 2022
Published: 7 October 2022

Publisher's Note: MDPI stays neutral with regard to jurisdictional claims in published maps and institutional affiliations.

Copyright: © 2022 by the authors. Licensee MDPI, Basel, Switzerland. This article is an open access article distributed under the terms and conditions of the Creative Commons Attribution (CC BY) license (https://creativecommons.org/licenses/by/4.0/).

[1] Biomass & Bioenergy Research Laboratory, Department of Chemical Engineering, Faculty of Engineering, Khon Kaen University, Khon Kaen 40002, Thailand
[2] Lahan Sai Ratchadaphisek School, Lahansai District, Buri Ram 31170, Thailand
[3] Biodiversity and Environmental Management Division, International College, Khon Kaen University, Khon Kaen 40002, Thailand
[4] Department of Mechanical Engineering, Faculty of Engineering and Industrial Technology, Chaiyaphum Rajabhat University, Chaiyaphum 36000, Thailand
* Correspondence: nyuvarat@kku.ac.th (Y.N.); pla.supattra.cpru@gmail.com (S.B.)

Abstract: Vinasse was successfully utilized to synthesize carbon–silica composite with a low-cost silica source available in Thailand (sodium silicate, Na_2SiO_3) and most commonly used source, tetraethyl orthosilicate (TEOS). The composites were prepared by a simple one-step sol–gel process by varying the vinasse (as carbon source) to silica source (Na_2SiO_3 or TEOS) weight ratio. The resulting composites were characterized by N_2 adsorption, moisture and ash contents, pH, pH_{pzc}, bulk density, Fourier transform infrared spectroscopy (FTIR), thermogravimetric analysis (TGA) and scanning electron microscopy-energy dispersive X-ray analysis (SEM-EDX). The composites had highest surface area of 313 and 456 m^2/g, with average mesopore diameters of 5.00 and 2.62 nm when using Na_2SiO_3 and TEOS as the silica sources, respectively. The adsorption of a non-steroidal anti-inflammatory drug, ibuprofen, was investigated. The contact time to reach equilibrium was 60 min for both composites. The adsorption kinetics were fitted by a pseudo-second-order model with the correlation coefficient $R^2 > 0.997$. The adsorption isotherms were well described by the Langmuir model ($R^2 > 0.992$), which indicates monolayer adsorption. The maximal adsorption capacities of the Na_2SiO_3- and TEOS-based composites were as high as 406 and 418 mg/g at pH 2, respectively. The research results indicate that vinasse and a low-cost silica source (Na_2SiO_3) show great potential to synthesize adsorbents through a simple method with high efficiency.

Keywords: carbon–silica composite; vinasse; mesoporous; adsorption; ibuprofen

1. Introduction

Nonsteroidal anti-inflammatory drugs (NSAIDs) represent one of the most widely used pharmaceutical products available without prescription. Ibuprofen (IBP) (Figure 1) is one of the most commonly and widely used NSAIDs in the treatment of fever, muscle pain and inflammation [1]. IBP is released into the environment through hospital and medical effluents, pharmaceutical wastewater and veterinary use [1]. NSAIDs become a great threat to aquatic ecosystems when left untreated before being discharged into the environment. Due to the inhibition of NSAIDs on microbial growth, it is difficult to treat them using anaerobic and aerobic sewage treatment [2]. Among many treatment techniques, adsorption is a simple and practical method. Many types of solid adsorbents are available and mesoporous materials are particularly attractive in the pharmaceutical field due to their large pore diameter, high thermal stability and high surface area as well as large pore volume [3,4].

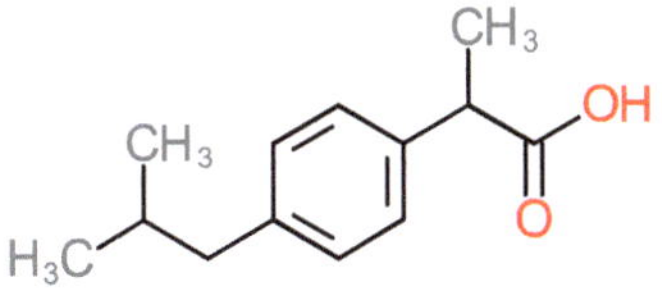

Figure 1. Chemical structure of ibuprofen.

Mesoporous carbon–silica composites are materials that can combine the beneficial properties of carbon and silica, resulting in high surface area, high porosity, chemical stability and high conductivity [5]. They can be used in many applications, such as adsorbents, insulation materials, lithium-ion battery anode materials and biosensors [6]. Carbon–silica composites can be derived from many carbon and silica sources, for example, resorcinol–formaldehyde [5], sucrose [7], furfuryl alcohol [7], activated carbon [8], tetraethyl orthosilicate (TEOS) [5,8] and mesoporous silica MCM-41 [7]. However, these chemicals have a high cost in Thailand. This motivated us to use waste from industry and a low-cost chemical to synthesize carbon–silica composites.

Wu et al. [6] reviewed that there are three methods to synthesize carbon–silica composites: hybrid pyrolysis carbonization, hydrothermal carbonization and sol–gel method. The first method involves the pyrolysis of organic precursors followed by carbonization to obtain carbonaceous residues with high carbon content. During the preparation process, the carbon precursor is homogeneously mixed with the silica material that retains a relatively rigid structure during the carbonization process due to its high melting point. As the pyrolytic carbonization is completed, the two components (carbon and silica) will be closely combined to form a carbon–silica composite. Typical pyrolysis temperatures used in the literature are 800–850 °C. Secondly, the hydrothermal carbonization process takes place in aqueous solution in a closed vessel at a moderate temperature of 180–250 °C and high pressure of 2–10 MPa. The carbon and silica precursors are fully mixed in an aqueous solution, transferred into a high-pressure reactor for hydrothermal reaction and, finally, a carbon–silica composite is obtained. Finally, the sol–gel method provides a suitable route to combine the organic and inorganic compounds into a homogeneous hybrid in a chemically linking or physically mixing state. This process can be generally described in five steps: hydrolysis, polycondensation, aging, drying and thermal decomposition. The catalysts used in the literature for hydrolysis reaction are formic acid, hydrochloric acid and nitric acid, while ethylenediamine, ammonia water and ammonium hydroxide are used as catalysts for the polycondensation reaction [6].

In this study, vinasse, a by-product from ethanol production, was used as carbon source and locally produced sodium silicate (Na_2SiO_3) as well as TEOS were used as silica sources. Carbon–silica composites were synthesized via sol–gel method without using high-temperature processing and catalyst for polycondensation, instead utilizing a hydrolysis procedure with sulfuric acid. It is important to note that the cost of Na_2SiO_3 was USD 0.68 per liter while TEOS costs as much as USD 132 per liter, which is a difference in cost by a factor of almost 200. Moreover, to the best of our knowledge, vinasse has not been previously used as a carbon source to synthesize carbon–silica composite, apart from our previous work that used a heating process at 85 °C [9]. Various techniques were used to study the morphology and surface properties of the composites and the adsorption behavior for IBP was evaluated.

2. Materials and Methods

2.1. Materials

Vinasse was collected from Mitr Phol Bio Fuel Plant (Phukhieo), Chaiyaphum province, Thailand. The major components of sugarcane vinasse are typically sugars (15.7% C), acids (13.1% C), alcohols (0.8% C) and amino acids [10]. TEOS ($C_8H_{20}O_4Si$, 99.0%) was purchased from Sigma-Aldrich (St. Louis, MO, USA). Sulfuric acid (H_2SO_4, 98%) and hydrochloric acid (HCl, 37%) were purchased from ANaPURE (New Zealand) and NaOH from LOBA

Chemie (Maharashtra, India). Sodium chloride (NaCl) was acquired from RCI Labscan (Bangkok, Thailand). Ibuprofen ($C_{13}H_{18}O_2$, 98%) was purchased from Sigma-Aldrich. All chemicals and reagents described above were of analytical grade and were used without any further purification. Sodium silicate (Na_2SiO_3) was locally produced by C. Thai Chemicals (Samutsakhon, Thailand).

2.2. Carbon–silica Composite Synthesis

The simple method to synthesize the carbon–silica composites (CSCs) was as follows: first, vinasse was mixed with 50% H_2SO_4 (volume ratio 1:2) to make the carbon source. In contrast to previous studies where concentrated H_2SO_4 was used [11,12], we used a lower H_2SO_4 concentration to reduce cost. Second, a fixed amount of 5 g silica source (either Na_2SiO_3 or TEOS) was mixed into the carbon source. The mixture was stirred at room temperature for 6 h. Finally, the sample was washed several times with distilled water until the pH of the washed water was equal to the pH of the starting distilled water, followed by drying at 130 °C for 3 h in an oven. The amount of vinasse was varied from 5 to 100 g to study the effect of vinasse:silica source between 1:1 to 20:1 by weight. The yield of the CSCs was calculated from the weight of the obtained composite divided by the sum of vinasse weight and silica source weight.

2.3. Materials Characterization

Surface area and pore volume analysis was conducted on a Micromeritics ASAP 2460 volumetric gas adsorption apparatus using N_2 as gas adsorbate at 77 K. Before gas adsorption measurements, the composites were outgassed under vacuum at 150 °C until the pressure was <100 mTorr. Surface area (S_{BET}) and micropore volume (V_{mic}) were determined by applying Brunauer–Emmet–Teller (BET) and Dubinin–Radushkevich (DR) methods, respectively. The total pore volume (V_T) and average pore size (D_P) were determined at $P/P°^{(U+b0)}$ = 0.99 and Barrett–Joyner–Halenda (BJH) method of the desorption branch. Finally, the mesopore volume (V_{meso}) was found from the difference between total pore volume and micropore volume.

To determine the moisture content, 1 g of composite was heated in an oven at 150 °C for 3 h. The moisture content was calculated from the weight loss during heating divided by the initial sample weight. The ash content of the composites was determined by heating the sample in a muffle furnace at 800 °C for 2 h. After heating, the sample was placed in a desiccator for cooling and was subsequently weighed to determine the ash content. The bulk density was determined by filling a 10 mL measuring cylinder with each composite and dividing the sample weight by the volume of the measuring cylinder. The cylinder was carefully tapped during filling to ensure that no voids were created.

The pH of the composites was determined by adding 1 g of sample into 100 mL hot distilled water followed by boiling for 5 min. Then, 100 mL of distilled water was added to the mixture and was left to cool down to room temperature. The pH of the solution was measured using a digital pH meter (OHAUS, ST3100-F). The point of zero charge of the composites (pH_{pzc}) was determined by the pH drift method. This method consisted of preparing six 0.01 M NaCl solutions with initial pH values between 2 and 12 by adjusting with 0.1 M NaOH or HCl. Then, 0.15 g of composite was added to 50 mL of each solution and the mixture was shaken by using orbital shaker (GALLENKAMP) for 48 h and filtered. The final pH of each filtrate was measured using a digital pH meter and plotted against the initial pH. The pH_{pzc} point was determined at the point where pH_{final} was equal to $pH_{initial}$.

The surface functional groups of the synthesized composites were analyzed by Fourier Transform Infrared Spectroscopy (FTIR, Bruker, ALPHA II, Billercia, MA, USA) using ATR accessory with diamond crystal. The thermal stability was studied using a thermogravimetric analyzer (TGA, Shimadzu, DTG-60H, Kyoto, Japan). The TGA analysis was carried out in both nitrogen and air atmosphere at a flow rate of 40 mL/min and the samples were heated from room temperature to 950 °C at a heating rate of 10 °C/min. The surface morphology of the composites was investigated by Scanning Electron Microscope-Energy

Dispersive X-ray Spectroscopy (SEM-EDX, FEI, Helios NanoLab G3 CX, Hillsboro, OR, USA). The samples were sputter coated with gold prior to analysis.

2.4. Adsorption of Ibuprofen

The adsorption experiments were performed using batch mode at room temperature. A stock solution of IBP (100 mg/L) containing 10% v of ethanol (95%) in order to increase the IBP solubility was diluted with distilled water to various concentrations. The absorbance of the IBP solution was measured using a UV-Vis spectrophotometer (Analytik Jena AG, Jena, Germany) at 222 nm.

The contact time was varied from 5 to 300 min with adsorbent dosage of 0.01 g/50 mL of IBP solution and initial IBP concentration of 30 mg/L. The influence of pH was studied from 2 to 12 with IBP concentration 30 mg/L at equilibrium time obtained from the adsorption time experiment. Here, 0.1 M HCl or NaOH was used to adjust the solution pH. To determine the equilibrium isotherm, adsorption tests were carried out by varying the initial IBP concentration from 5 to 100 mg/L with equilibrium time and optimum pH obtained from above experiments. The amount of IBP adsorbed (q, mg/g) was calculated from the following equation:

$$q = \frac{(C_o - C_t)V}{m} \quad (1)$$

where C_o is the initial concentration of IBP solution (mg/L), C_t is the concentration of IBP at any time t, V is the volume of IBP solution (L) and m is the mass of composites used in the experiment (g).

3. Results

3.1. Characterization of Carbon–silica Composite

3.1.1. Specific Surface Area of Composites under Different Ratio

The effect of vinasse:silica source ratio on the specific surface area of the CSCs was investigated and the results are shown in Figure 2. It is important to note that a solid material could not be synthesized when using only a silica source, while using only vinasse resulted in a solid with very low surface area of 0.35 m^2/g. When adding increasing amounts of silica source (i.e., for higher vinasse:silica source ratios), an increase in surface area of CSCs was observed for both silica sources (Na_2SiO_3 and TEOS). While the vinasse did not contain any structure directing agent, the H_2SO_4 treatment of vinasse in the gel mixture acted as a structure directing agent and its further interaction with silica species facilitated the formation of a porous structure. After reaching an optimum, the surface area decreased with a further increase in the source ratio. The highest surface areas were obtained at vinasse:silica source ratios of 5:1 and 10:1 when using Na_2SiO_3 and TEOS, respectively. At these optimal conditions, the composites prepared from Na_2SiO_3 and TEOS were assigned as CSC_Na and CSC_TEOS, respectively. The other porous properties of these composites are discussed in more detail below.

Figure 3 illustrates the N_2 adsorption–desorption isotherms of composites prepared at optimum conditions (CSC_Na and CSC_TEOS). The isotherms belonged to Type IV according to IUPAC classification. The hysteresis loop for $P/P^{\circ\,(U+b0)} > 0.4$ presents evidence of capillary condensation, which indicated the formation of mesopores in both composites. Table 1 shows the porous properties of the optimum CSCs, which showed the largest surface area and total pore volume of 313 m^2/g and 0.39 cm^3/g and 456 m^2/g and 0.30 cm^3/g for CSC_Na and CSC_TEOS, respectively. It can be inferred that the low-cost silica source (Na_2SiO_3) can produce a composite with comparable surface area and pore volume to the high-cost silica source (TEOS). Using Na_2SiO_3 resulted in an adsorbent with mainly mesopore volume (69%), while TEOS yielded an adsorbent with mainly micropores (67%). The average pore sizes of both composites were between 2 and 5 nm, which indicated mesoporous materials.

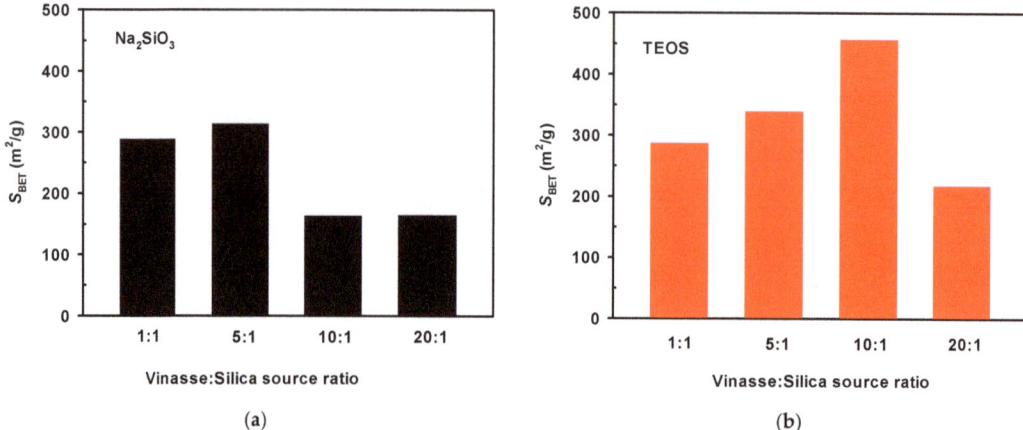

Figure 2. Effect of vinasse to silica source ratio on specific surface area of carbon–silica composites synthesis from: (**a**) Na$_2$SiO$_3$; (**b**) TEOS.

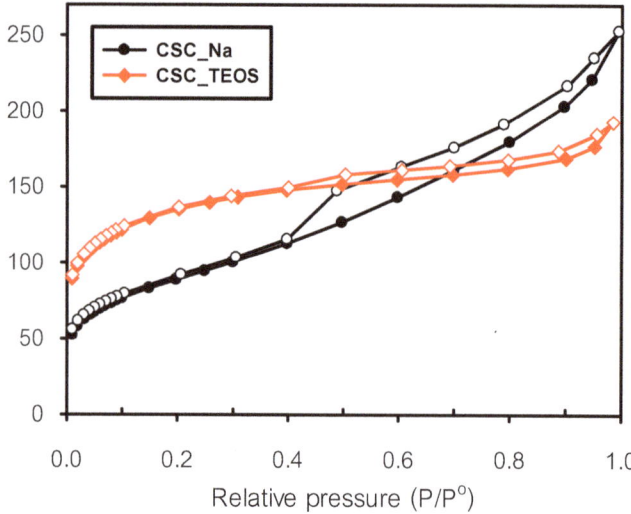

Figure 3. N$_2$ adsorption (closed symbol)/desorption (open symbol) isotherms of CSCs synthesized under optimum conditions.

Table 1. Textural properties of composites synthesized under optimum conditions.

Composite	S_{BET} (m^2/g)	V_{mic} (cm^3/g)	V_{meso} (cm^3/g)	V_T (cm^3/g)	D_P (nm)
CSC_Na	313	0.12 (31%)	0.27 (69%)	0.39	5.00
CSC_TEOS	456	0.20 (67%)	0.10 (33%)	0.30	2.62

3.1.2. Physicochemical Properties of Composites under Optimum Conditions

The CSC yield for both silica sources is presented in Table 2. Both yields were quite low because vinasse contains around 93 wt.% water [13] that was evaporated during drying at 130 °C in the synthesis process. The moisture content is the amount of water bound to samples under normal condition, while the ash content refers to residual minerals in samples. The moisture contents of CSC_Na and CSC_TEOS were 12.97 and 16.67 wt.%, respectively. Generally, a high moisture content decreases the adsorption capacity of an adsorbent. As shown in Table 2, the ash contents of both composites were quite high, likely due to the present of silica.

Table 2. Physicochemical properties of composites synthesized under optimum conditions.

Composite	Yield (wt.%)	Moisture (wt.%)	Ash (wt.%)	pH	pH_{pzc}	Bulk Density (g/cm^3)
CSC_Na	6.33	12.97	44.39	1.85	2.20	0.66
TEOS	4.00	16.67	30.08	2.82	2.70	0.86

The CSCs had a pH of 1.85 (CSC_Na) and 2.82 (CSC_TEOS) and pH_{pzc} values of 2.20 (CSC_Na) and 2.70 (CSC_TEOS), which confirmed the acidic character of the surface. The bulk densities of composites were 0.66 and 0.86 g/cm^3 for, respectively, CSC_Na and CSC_TEOS. These bulk densities are within the values reported for other adsorbents, for example, 0.67 g/cm^3 for activated carbon prepared from mango seed [14] or 0.74 g/cm^3 for 5A molecular sieve [15].

The FTIR spectra of the samples show the interaction between carbon and silica in both composites (Figure 4). The broad band at 3329 cm^{-1} was due to the presence of –OH stretching vibrations [11]. The band at 1705 cm^{-1} was attributed to the stretching vibrations of C=O of –COOH groups [11]. The low-intensity bands at 1622 and 1632 cm^{-1} were assigned to water physically bound to silica [16]. The peak at 1000–1200 cm^{-1} revealed the presence of characteristic bonds of Si–O–Si [16] and the peak observed at 955 cm^{-1} corresponded to Si–OH vibrations [16]. Finally, the peak at 797 cm^{-1} is related to O–Si–O vibrations [11].

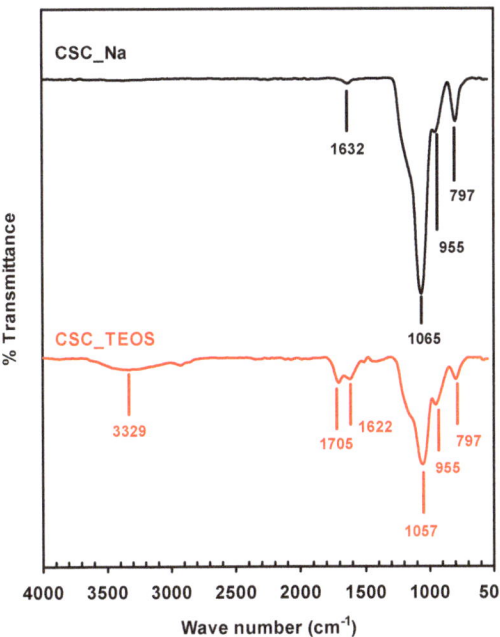

Figure 4. FTIR spectra of synthesized carbon–silica composites.

The results of the thermogravimetric analysis are shown in Figure 5. These plots show the percent weight loss of composites synthesized at optimum conditions as a function of temperature under N$_2$ and air atmospheres. As seen in Figure 5a,b, approximately 10% and 20% weight loss occurred at approximately 100 °C, which was attributed to the molecular water adsorbed on the composite surface. The final weight of both composites under N$_2$ atmosphere was higher than in air atmosphere. In the case of N$_2$, pyrolysis takes place, whereas an oxidation phenomenon occurs under air. The mass loss of the composites between 250 and 650 °C under air atmosphere was due to the combustion of carbon [17]. The TGA curves indicated a 19% mass loss for CSC_Na and 58% for CSC_TEOS. The higher mass loss for CSC_TEOS was ascribed to its higher carbon content, which is consistent with the higher amount of vinasse used (vinasse:silica source ratio 10:1) compared to CSC_Na sample (vinasse:silica source ratio 5:1) at optimum conditions. From the residual weight, it is reasonable to assume that the composites contained 34 wt.% and 23 wt.% silica for CSC_Na and CSC_TEOS, respectively. In both atmospheres, the residual weight of CSC_Na was higher than CSC_TEOS, which illustrated a better thermal stability.

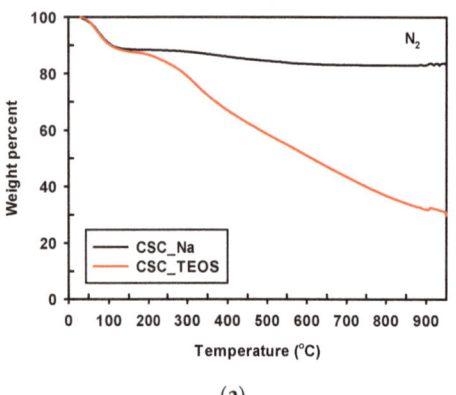
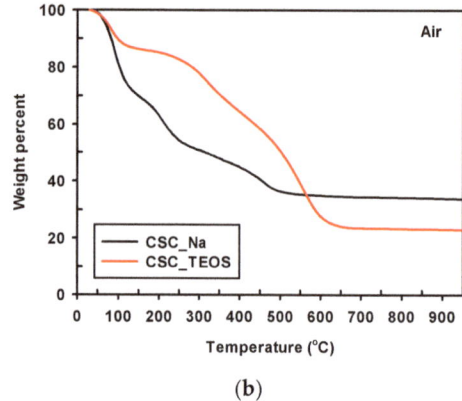

(a)　　　　　　　　　　　　　　　　　　(b)

Figure 5. Thermogravimetric analysis of carbon–silica composites under: (**a**) N$_2$; (**b**) air atmosphere.

3.1.3. Surface Morphology

The morphological structures of the composites were analyzed by SEM (Figure 6). It can be clearly seen that both composites had cavities, in agreement with the high surface area and porosity. EDX analysis indicated the presence of C and Si in the synthesized composites. The atomic percentage of C and Si calculated from the quantification of the peaks gave values of 22.4 and 36.0 wt.% for CSC_Na and 54.1 and 17.7 wt.% for CSC_TEOS, respectively. This higher Si content for CSC_Na is consistent with is higher stability, as seen by TGA analysis and its lower carbon content.

3.2. Adsorption Study

3.2.1. Effect of Adsorption Time

The effect of contact time on the IBP adsorption capacities for CSC_Na and CSC_TEOS is shown in Figure 7. The adsorption of IBP onto the CSCs increased with increasing contact times and the process reached equilibrium within about 60 min. Further experiments were conducted at this equilibrium contact time. Figure 7 also shows that a rapid increase in the IBP adsorption capacities occurred during the first 30 min. The fast adsorption at the initial stage might be due to the availability of the active sites on the composite. After that, the adsorption increased slowly because the active sites became saturated.

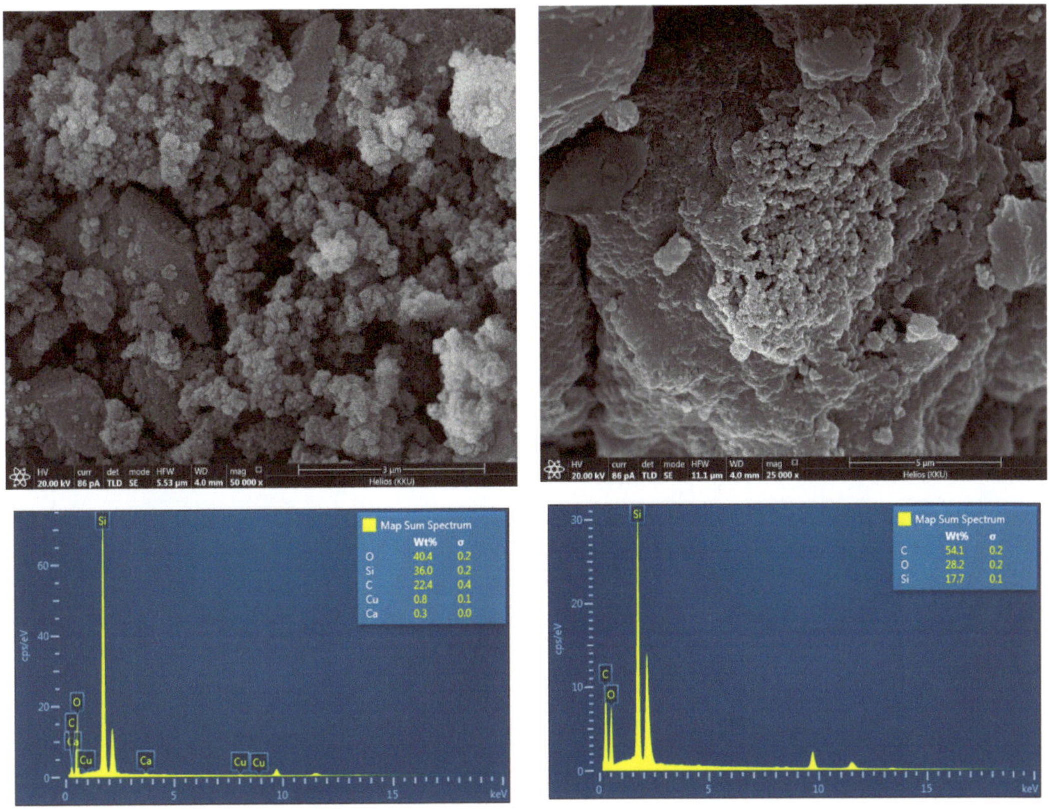

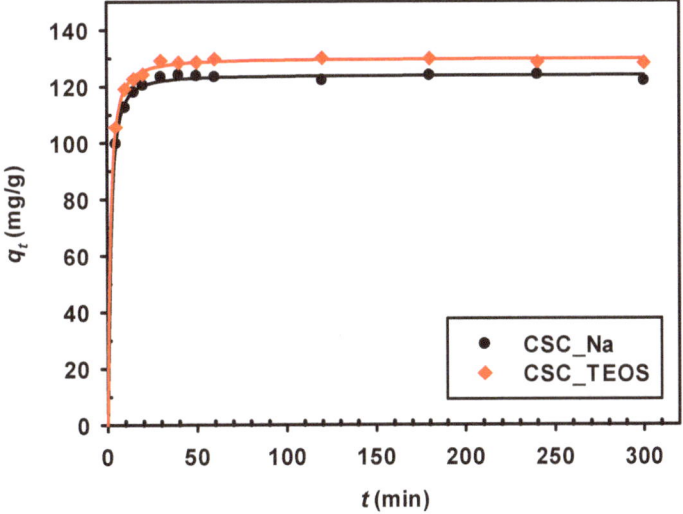

Figure 6. SEM micrographs and EDX analysis of carbon–silica composites synthesized from: (**a**) Na$_2$SiO$_3$; (**b**) TEOS.

Figure 7. Effect of adsorption time on ibuprofen adsorption (C_0 = 30 mg/L).

3.2.2. Effect of Initial pH

The effect of pH on IBP adsorption on both composites is presented in Figure 8. Both composites showed similar results regarding the pH effect, namely that a decrease in pH favored IBP adsorption and the highest adsorption was found at pH 2. The surface of an adsorbent is neutral at pH = pH_{pzc}, negatively charged for pH > pH_{pzc} and positively charged at pH < pH_{pzc}. The charge of the adsorbate will be based on its pKa value. Ibuprofen is an acidic drug with pKa = 4.91 [18] and the IBP is distributed in water as either molecular (IBP) and anionic (IBP$^-$) forms [19]. This drug is essentially neutral at pH < pKa and acquires a negative charge when pH > pKa. For pH < pH_{pzc} (pH = 2), both CSCs are positively charged while IBP has a neutral charge; thus, adsorption most likely involved hydrogen bonding and/or van der Waals interactions [18]. For pH > pH_{pzc}, the surfaces of the CSCs are negatively charged and a higher proportion of negatively charged IBP at pH > pKa leads to an electrostatic repulsion between the adsorbate anions and the CSC surface [18]. Consequently, the IBP adsorption decreased. Baccar et al. [18] and Fröhlich et al. [20] also found that the adsorption of IBP was favored at low pH values. Here, pH 2 was selected for subsequent adsorption experiments.

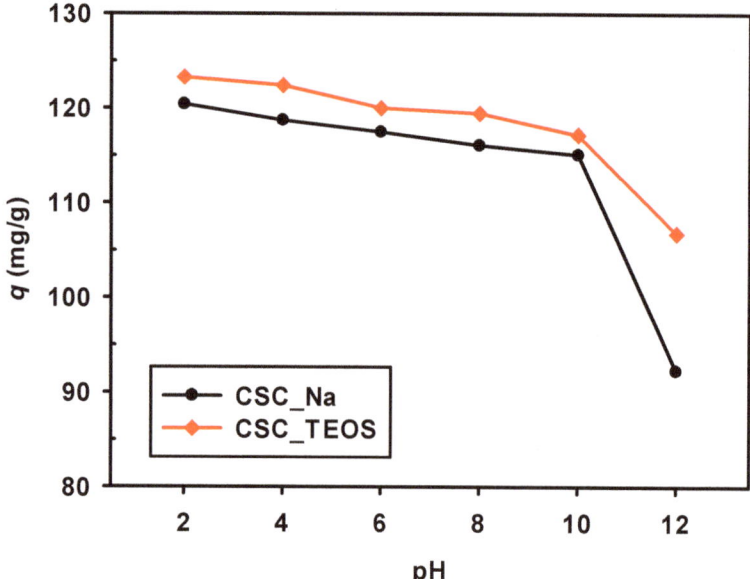

Figure 8. Effect of pH value on IBP adsorption (C_0 = 30 mg/L; t = 60 min).

3.2.3. Effect of Initial Solution Concentration

Figure 9 shows the IBP adsorption on the optimum CSCs with initial IBP concentrations of 5–100 mg/L. The adsorption capacity increased gradually with increasing IBP concentration. When the initial concentration was higher than 80 mg/L, the increase in adsorption capacity leveled off. The concentration provides an important driving force to overcome all mass transfer resistance of IBP between the aqueous and solid phase and a higher initial IBP concentration enhances the adsorption process. Due to the limitation of active sites on the adsorbent surface, the IBP molecules occupy the surface sites at critical concentration and no further capacity enhancement is possible. The CSC_Na composite prepared from Na_2SiO_3 had the highest equilibrium adsorption capacity of about 370 mg/g that was slightly lower than CSC_TEOS (400 mg/g), which was possibly due to the higher specific surface area of the latter.

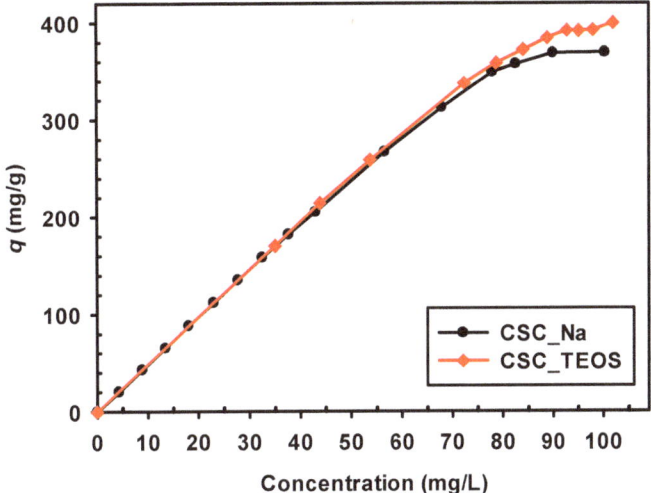

Figure 9. Effect of initial concentration on IBP adsorption (C_0 = 5–100 mg/L; pH = 2; t = 60 min).

3.2.4. Kinetic Models

The nonlinear models of pseudo-first-order and pseudo-second-order kinetics were investigated in this study. The use of nonlinear equations avoids the need for any further translation to a linear form and corresponding loss in accuracy [21]. Both models are generally expressed as [22]:

$$q_t = q_e(1 - e^{-k_1 t}) \quad (2)$$

$$q_t = \frac{k_2 q_e^2 t}{1 + k_2 q_e t} \quad (3)$$

where q_e and q_t are the amount of IBP adsorbed on CSCs in mg/g at equilibrium and at any time t (min), respectively, and k_1 and k_2 are the rate constant of the pseudo-first-order and (1/min) pseudo-second-order (g/(mg·min)) models, respectively.

Figure 10 demonstrates the results of both kinetic models and the rate constant and q_e values are given in Table 3. The suitability of the models depends on the similarity of the model and experimental q_e value and the regression coefficient (R^2). High R^2 values were obtained for both models, but the model and experimental q_e values were closer for the pseudo-second-order model. This indicated that both physical and chemical adsorption coexisted during the adsorption process and electrostatic interactions and hydrogen bonding were the main adsorption mechanisms [2].

3.2.5. Adsorption Isotherms

The equilibrium adsorption study provides information about the distribution of adsorbate molecules between the liquid and the solid phase. Several mathematical models were used to describe experimental data of adsorption isotherms, of which the most commonly used are the Langmuir and Freundlich isotherms. The Langmuir model assumes that adsorption is a monolayer process and a limited number of adsorption sites exists on the adsorbent surface. Once an adsorbate molecule is deposited onto an active site, no further adsorption occurs at that site because it can hold, at most, one adsorbate molecule. This model considers homogeneous active sites on the adsorbent surface and there are no interactions between two adsorbed molecules during the adsorption process. The Freundlich isotherm describes that a heterogeneous adsorbate surface is formed with multilayer adsorption with different adsorption energies [23]. The general nonlinear forms of the Langmuir and Freundlich equations are given below, respectively:

$$q_e = \frac{q_{max} K_L C_e}{1 + K_L C_e} \quad (4)$$

$$q_e = K_F C_e^{1/n} \quad (5)$$

where q_{max} is the maximum monolayer adsorption capacity (mg/g), C_e is the concentration of adsorbate in the solution at equilibrium (mg/L), K_L is the Langmuir constant (L/mg), K_F is the Freundlich constant ((mg/g)(L/mg)$^{1/n}$) and $1/n$ is the dimensionless Freundlich adsorption intensity parameter.

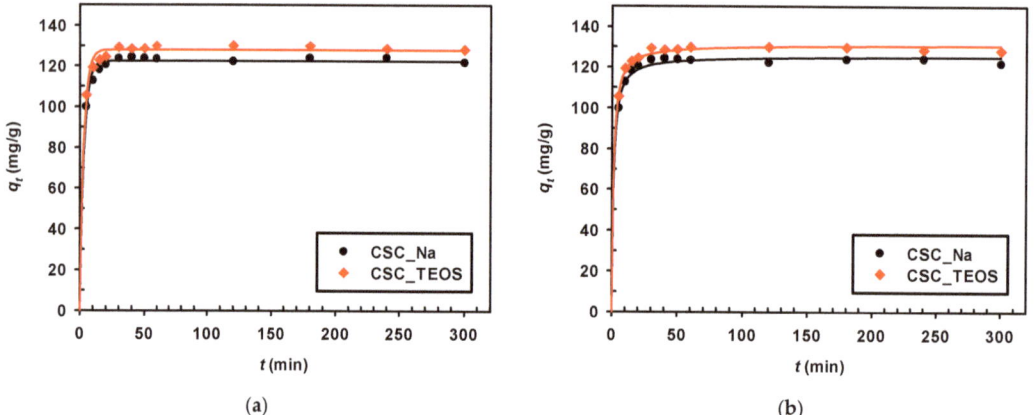

Figure 10. Nonlinear kinetic models of IBP adsorption using CSCs: (**a**) pseudo-first-order; (**b**) pseudo-second-order.

Table 3. Kinetic model parameters for the adsorption of IBP on CSCs.

Adsorbent	$q_{e,exp}$ (mg/g)	Pseudo-First-Order			Pseudo-Second-Order		
		q_e (mg/g)	k_1 (1/min)	R^2	q_e (mg/g)	k_1 (g/(mg·min))	R^2
CSC_Na	124	122.37	0.318	0.996	125.52	0.007	0.997
CSC_TEOS	130	127.86	0.332	0.995	131.14	0.007	0.998

Fitting of the experimental data to the isotherm models was conducted using nonlinear regression and the results are shown in Figure 11. The parameters and the correlation coefficients (R^2) are collected in Table 4. The highest R^2 values for both composites were obtained for the Langmuir model, suggesting monolayer adsorption and a relatively homogeneous adsorbent surface. The values of constant K_L were 0.601 and 0.880 L/mg for CSC_Na and CSC_TEOS, respectively. A higher K_L value indicates a strong adsorbate–adsorbent interaction, while a smaller K_L value indicates a weak interaction between adsorbate molecule and adsorbent surface. For the Freundlich isotherm model, the higher value for K_F of CSC_TEOS indicated a higher affinity for IBP. This is consistent with the smaller value of $1/n$ (larger value of n) that implies a stronger interaction between the adsorbent and the adsorbate [24]. Moreover, the values of $1/n$ are 0.333 and 0.211 for CSC_Na and CSC_TEOS, respectively, implying that the adsorption of IBP is favorable. The calculated q_{max} values from the Langmuir model were comparable for both CSCs. This indicated that the low-cost silica source (Na$_2$SiO$_3$) can be used to synthesize a carbon–silica composite with high potential with similar properties and performance as the high-cost silica source (TEOS).

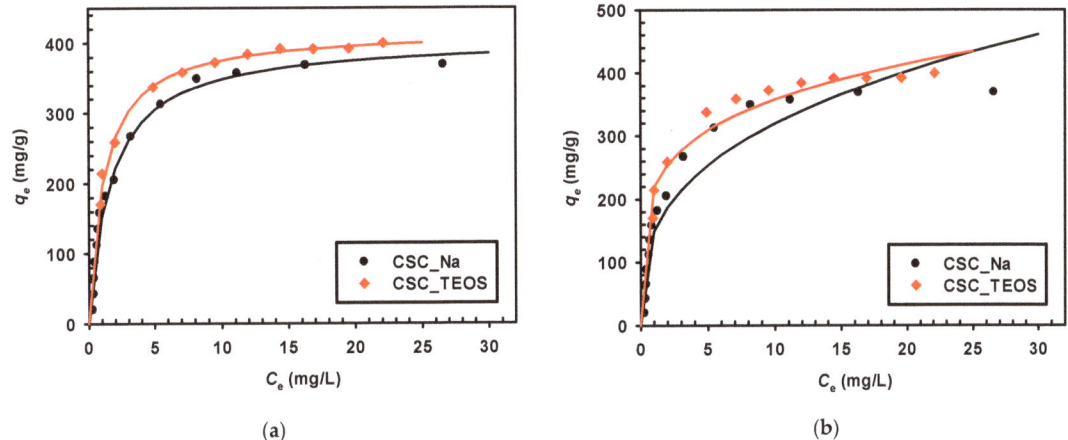

Figure 11. Adsorption isotherms for IBP adsorption: (**a**) Langmuir model; (**b**) Freundlich model.

Table 4. Langmuir and Freundlich adsorption isotherm constants for IBP adsorbed by CSCs.

Adsorbent	Langmuir			Freundlich		
	q_{max} (mg/g)	K_L (L/mg)	R^2	K_F ((mg/g)(L/mg)$^{1/n}$)	$1/n$	R^2
CSC_Na	406	0.601	0.992	148.560	0.333	0.869
CSC_TEOS	418	0.880	0.998	220.100	0.211	0.914

The adsorption performance of CSCs for IBP was compared with various other adsorbents reported in the literature. Table 5 shows that CSCs have comparable or higher adsorption capacities to other sorbents, indicating a good and promising adsorbent for IBP.

Table 5. Adsorption capacities of IBP for various adsorbents.

Adsorbent	Maximum Adsorption Capacity (mg/g)	Reference
CSC_Na	406	Present study
CSC_TEOS	418	Present study
Sonicated activated carbon	134	[20]
Alternanthera philoxeroides-based biochar	172	[25]
ATP@ZIF-8-DETA-20 composite	218	[2]
Surface modified activated carbon cloth	492	[26]
Cu-doped Mil-101(Fe)	497	[27]

4. Conclusions

In summary, a new, easy route to synthesize mesoporous carbon–silica composites (CSCs) is presented. Vinasse was used as a carbon source and Na_2SiO_3 was used as low-cost silica source. TEOS, a commonly used silica source, was used for comparison. The resulting Na_2SiO_3-based composite (CSC_Na) had a moderate BET surface area of 313 m^2/g, a moderate pore volume of 0.39 cm^3/g and an average pore diameter of 5.00 nm. The CSC_Na was thermally stable and the adsorption capacity of ibuprofen was strongly influenced by the porous structure and surface chemistry. The excellent capacity of this composite to act as an adsorbent of ibuprofen (406 mg/g) was confirmed. The similar physicochemical properties and adsorption performance of CSC_Na, when compared to CSC_TEOS, make the low-cost Na_2SiO_3 an attractive alternative silica source over more expensive TEOS. It can be envisaged that the synthesis of carbon–silica composite from biomass-based waste products paves the way towards low-cost production of biobased products. It is also expected that this simple synthetic method will encourage the extensive applications of carbon–silica composites for gas separation, electrode materials and other applications.

Author Contributions: Conceptualization, Y.N.; methodology, Y.N.; formal analysis, Y.N, J.T.N.K. and S.B.; investigation, T.S., K.M., P.T., W.R. and K.W.; writing—original draft preparation, Y.N.; writing—review and editing, J.T.N.K.; visualization, Y.N.; project administration, Y.N. and S.B.; funding acquisition, Y.N. All authors have read and agreed to the published version of the manuscript.

Funding: This work was funded by the Office of National Higher Education Science Research and Innovation Policy Council (NXPO) via PMU Flagship of grant number C10F630230 and Faculty of Engineering, Khon Kaen University.

Institutional Review Board Statement: Not applicable.

Informed Consent Statement: Not applicable.

Data Availability Statement: The data presented in this study are available on request from the corresponding author.

Acknowledgments: The authors acknowledge Nontipa Supanchaiyamat and Andrew J. Hunt, Department of Chemistry, Faculty of Science, Khon Kaen University for SEM-EDX characterization. The authors would like to thank Tippawan Ponnikorn, Chaiwat Rattanet and Ronnachai Songkran, Department of Chemical Engineering, Faculty of Engineering, Khon Kaen University, for research assistance.

Conflicts of Interest: The authors declare no conflict of interest.

References

1. Oba, S.N.; Ighalo, J.O.; Aniagor, C.O.; Igwegbe, C.A. Removal of Ibuprofen from Aqueous Media by Adsorption: A comprehensive review. *Sci. Total Environ.* **2021**, *780*, 146608. [CrossRef] [PubMed]
2. Fröhlich, A.C.; dos Reis, G.S.; Pavan, F.A.; Lima, É.C.; Foletto, E.L.; Dotto, G.L. Improvement of Activated Carbon Characteristics by Sonication and Its Application for Pharmaceutical Contaminant Adsorption. *Environ. Sci. Pollut. Res.* **2018**, *25*, 24713–24725. [CrossRef] [PubMed]
3. Ulfa, M.; Prasetyoko, D. Drug Loading-Release Behaviour of Mesoporous Materials SBA-15 and CMK-3 using ibuprofen molecules as drug model. *J. Phys. Conf. Ser.* **2019**, *1153*, 012065. [CrossRef]
4. Shen, S.-C.; Ng, W.K.; Chia, L.S.O.; Dong, Y.-C.; Tan, R.B.H. Applications of Mesoporous Materials as Excipients for Innovative Drug Delivery and Formulation. *Curr. Pharm. Des.* **2013**, *19*, 6270–6289. [CrossRef] [PubMed]
5. Xu, H.; Zhang, H.; Huang, W.; Wang, Y. Porous Carbon/silica Composite Monoliths Derived from Resorcinol-formaldehyde/TEOS. *J. Non-Cryst. Solids* **2010**, *356*, 971–976. [CrossRef]
6. Wu, T.; Ke, Q.; Lu, M.; Pan, P.; Zhou, Y.; Gu, Z.; Cui, G. Recent Advances in Carbon-Silica Composites: Preparation, Properties, and Applications. *Catalysts* **2022**, *12*, 573. [CrossRef]
7. Furtado, A.M.B.; Wang, Y.; LeVan, M.D. Carbon Silica Composites for Sulfur dioxide and Ammonia Adsorption. *Micropor. Mesopor. Mater.* **2013**, *165*, 48–54. [CrossRef]
8. Fu, L.; Zhu, J.; Huang, W.; Fang, J.; Sun, X.; Wang, X.; Liao, K. Preparation of Nano-Porous Carbon-Silica Composites and Its Adsorption Capacity to Volatile Organic Compounds. *Processes* **2020**, *8*, 372. [CrossRef]
9. Fernandes, B.S.; Vieira, J.P.F.; Contesini, F.J.; Mantelatto, P.E.; Zaiat, M.; da Cruz Pradella, J.G. High Value Added Lipids Produced by Microorganisms: A Potential Use of Sugarcane Vinasse. *Critical Reviews in Biotechnology* **2017**, *37*, 1048–1061. [CrossRef]
10. Ponnikorn, T.; Knijnenburg, J.T.N.; Macquarrie, D.J.; Ngernyen, Y. Novel Mesoporous Carbon-Silica Composites from Vinasse for the Removal of Dyes from Aqueous Silk Dyeing Wastes. *Eng. Appl. Sci. Res.* **2022**, *49*, 707–719.
11. Nandan, D.; Sreenivasulu, P.; Konathala, L.N.S.; Kumar, M.; Viswanadham, N. Acid Functionalized Carbon-Silica Composite and Its Application for Solketal Production. *Micropor. Mesopor. Mater.* **2013**, *179*, 182–190. [CrossRef]
12. Zhang, S.; Gao, Y.; Dan, H.; Xu, X.; Yue, Q.; Yan, J.; Wang, W.; Gao, B. Effect of Washing Conditions on Adsorptive Properties of Mesoporous Silica Carbon Composites by In-situ Carbothermal Treatment. *Sci. Total Environ.* **2020**, *716*, 136770. [CrossRef] [PubMed]
13. Noa-Bolaño, A.; Pérez-Ones, O.; Zumalacárregui-de Cárdenas, L.; Pérez-de los Ríos, J.L. Simulation of Concentration and Incineration as an Alternative for Vinasses' treatment. *Rev. Mex. Ing. Química* **2020**, *19*, 1265–1275. [CrossRef]
14. Abdus-Salam, N.; Buhari, M. Adsorption of Alizarin and Fluorescein Dyes on Adsorbent prepared from Mango Seed. *Pac. J. Sci. Technol.* **2014**, *15*, 232–244.
15. Asgari, M.; Anisi, H.; Mohammadi, H.; Sadighi, S. Designing a Commercial Scale Pressure Swing Adsorber for Hydrogen Purification. *Pet. Coal* **2014**, *56*, 552–561.

16. Shweta, K.; Jha, H. Rice Husk Extracted Lignin−TEOS biocomposites: Effects of Acetylation and Silane Surface Treatments for Application in Nickel Removal. *Biotechnol. Rep.* **2015**, *7*, 95–106. [CrossRef]
17. Yang, Z.; Xia, Y.; Mokaya, R. Periodic Mesoporous Organosilica Mesophases are Versatile Precursors for the Direct Preparation of Mesoporous Silica/Carbon Composites, Carbon and Silicon Carbide Materials. *J. Mater. Chem.* **2006**, *16*, 3417–3425. [CrossRef]
18. Baccar, R.; Sarrà, M.; Bouzid, J.; Feki, M.; Blánquez, P. Removal of Pharmaceutical Compounds by Activated Carbon Prepared from Agricultural By-product. *Chem. Eng. J.* **2012**, *211-212*, 310–317. [CrossRef]
19. Ai, T.; Jiang, X.; Zhong, Z.; Li, D.; Dai, S. Methanol-modified Ultra-fine Magnetic Orange Peel Powder Biochar as an Effective Adsorbent for Removal of Ibuprofen and Sulfamethoxazole from Water. *Adsorpt. Sci. Technol.* **2020**, *38*, 304–321. [CrossRef]
20. Altowayti, W.A.H.; Othman, N.; Al-Gheethi, A.; bini Mohd Dzahir, N.H.; Asharuddin, S.M.; Alshalif, A.F.; Nasser, I.M.; Tajarudin, H.A.; AL-Towayti, F.A.H. Adsorption of Zn^{2+} from synthetic Wastewater Using Dried Watermelon Rind (D-WMR): An Overview of Nonlinear and Linear Regression and Error Analysis. *Molecules* **2021**, *26*, 6176. [CrossRef]
21. Guo, L.; Li, G.; Liu, J.; Meng, Y.; Xing, G. Nonlinear Analysis of the Kinetics and Equilibrium for Adsorptive Removal of Cd(II) by starch Phosphate. *J. Dispers. Sci. Technol.* **2012**, *33*, 403–409. [CrossRef]
22. Ni, W.; Xiao, X.; Li, Y.; Li, L.; Xue, J.; Gao, Y.; Ling, F. DETA Impregnated Attapulgite Hybrid ZIF-8 Composite as an Adsorbent for the Adsorption of Aspirin and Ibuprofen in Aqueous Solution. *New J. Chem.* **2021**, *45*, 5637–5644. [CrossRef]
23. Salim, N.A.A.; Puteh, M.H.; Khamidun, M.H.; Fulazzaky, M.A.; Abdullah, N.H.; Yusoff, A.R.M.; Zaini, M.A.A.; Ahmad, N.; Lazim, Z.M.; Nuid, M. Interpretation of Isotherm Models for Adsorption of Ammonium onto Granular Activated Carbon. *Biointerface Res. Appl. Chem.* **2021**, *11*, 9227–9241. [CrossRef]
24. Anah, L.; Astrini, N. Isotherm Adsorption Studies of Ni(II) ion Removal from Aqueous Solutions by Modified Carboxymethyl Cellulose Hydrogel. *IOP Conf. Ser. Earth Environ. Sci.* **2018**, *160*, 012017. [CrossRef]
25. Du, Y.-D.; Zhang, X.-Q.; Shu, L.; Feng, Y.; Lv, C.; Liu, H.-Q.; Xu, F.; Wang, Q.; Zhao, C.C.; Kong, Q. Safety Evaluation and Ibuprofen Removal via an *Alternanthera philoxeroides*-based biochar. *Environ. Sci. Pollut. Res.* **2021**, *28*, 40568–40586. [CrossRef]
26. Guedidi, H.; Reinert, L.; Soneda, Y.; Bellakhal, N.; Duclaux, L. Adsorption of Ibuprofen from Aqueous Solution on Chemically Surface-Modified Activated Carbon Cloths. *Arab. J. Chem.* **2017**, *10*, S3584–S3594. [CrossRef]
27. Xiong, P.; Zhang, H.; Li, G.; Liao, C.; Jiang, G. Adsorption Removal of Ibuprofen and Naproxen from Aqueous Solution with Cu-doped Mil-101(Fe). *Sci. Total Environ.* **2021**, *797*, 149179. [CrossRef]

Article

Synthesis of Graphene Quantum Dots by a Simple Hydrothermal Route Using Graphite Recycled from Spent Li-Ion Batteries

Lyane M. Darabian, Tainara L. G. Costa, Daniel F. Cipriano, Carlos W. Cremasco, Miguel A. Schettino, Jr. and Jair C. C. Freitas *

Laboratory of Carbon and Ceramic Materials, Department of Physics, Federal University of Espírito Santo, Vitória 29075-910, ES, Brazil
* Correspondence: jairccfreitas@yahoo.com.br

Abstract: Graphene quantum dots (GQDs) are nanosized systems that combine beneficial properties typical of graphenic materials (such as chemical stability, biocompatibility and ease of preparation from low-cost precursors) with remarkable photoluminescent features. GQDs are well-known for their low cytotoxicity and for being promising candidates in applications, such as bioimaging, optoelectronics, electrochemical energy storage, sensing and catalysis, among others. This work describes a simple and low-cost synthesis of GQDs, starting from an alcoholic aqueous suspension of graphene oxide (GO) and using a hydrothermal route. GO was prepared using graphite recycled from spent Li-ion batteries, via a modified Hummers method. The GO suspension was submitted to hydrothermal treatments at different temperatures using a homemade hydrothermal reactor that allows the control of the heating program and the assessment of the internal pressure generated in the reaction. The synthesized GQDs exhibited bright blue/green luminescence under UV light; showing the success of the chosen route and opening the way for future applications of these materials in the field of optoelectronic devices.

Keywords: graphene quantum dots; hydrothermal synthesis; graphene oxide; graphite; recycling

Citation: Darabian, L.M.; Costa, T.L.G.; Cipriano, D.F.; Cremasco, C.W.; Schettino, M.A., Jr.; Freitas, J.C.C. Synthesis of Graphene Quantum Dots by a Simple Hydrothermal Route Using Graphite Recycled from Spent Li-Ion Batteries. C **2022**, *8*, 48. https://doi.org/10.3390/c8040048

Academic Editors: Olena Okhay and Gil Goncalves

Received: 23 August 2022
Accepted: 20 September 2022
Published: 22 September 2022

Publisher's Note: MDPI stays neutral with regard to jurisdictional claims in published maps and institutional affiliations.

Copyright: © 2022 by the authors. Licensee MDPI, Basel, Switzerland. This article is an open access article distributed under the terms and conditions of the Creative Commons Attribution (CC BY) license (https://creativecommons.org/licenses/by/4.0/).

1. Introduction

The research on carbon dots (CDs) has emerged in recent years as one of the most promising branches in the field of carbon nanomaterials. With outstanding optical properties, these materials have found a wide range of applications as biosensors and fluorescent probes, among many others [1,2]. Graphene quantum dots (GQDs) are considered an important sub-group of the CDs family, exhibiting in general morphological features typical of graphenic materials; such as the existence of basic structural units consisting of nanosized disk-shaped aggregates of stacked graphene-like layers [3,4]. The GQDs combine many beneficial properties typical of graphenic materials (such as chemical stability, biocompatibility and ease of preparation from low-cost precursors) with remarkable photoluminescent features [3,4]. GQDs are well-known for their low cytotoxicity [3] and for being promising candidates in applications such as bioimaging, optoelectronics, electrochemical energy storage, sensing and catalysis, among others [4]. The use of GQDs as fluorescent probes for cell imaging is particularly promising, as demonstrated in several recent reports [5–8].

GQDs can be obtained following different routes, broadly divided in two classes: top-down and bottom-up approaches. In the bottom-up methods, solution-based chemical routes allow the synthesis of GQDs starting from several types of organic precursors; such as malic acid, urea and other organic precursors [4,7,8]. In these cases, the size and colloidal stability of the GQD dispersions can be controlled; however, the yield is generally low and purification treatments are required in order to obtain the final product [5]. The top-down methods involve the size reduction of a carbon-rich precursor, leading to the achievement of small graphenic aggregates (nanodots) [2–4,9]. Several routes have been proposed to achieve this end, such as oxidative cutting [10], hydrothermal/solvothermal methods [5,6,11,12] and electrochemical processes [13], among others. Numerous precursors have also been used in these syntheses, such as graphite [9,12], graphene oxide (GO) [11,13–15], carbon fibers [16] and carbon nanotubes [12,15]. Nevertheless, the production of GQDs at large scale, with uniform size, high yield and low cost, remains a challenge.

This work involves the use of electronic waste (i.e., graphite recycled from the anode of spent Li-ion batteries) as the starting point to obtain technologically important materials such as the GQDs, following a simple and low-cost method. GO was prepared from the recycled graphite via a modified Hummers method; alcoholic aqueous GO suspensions (with no addition of acids or strong oxidizing agents) were submitted to hydrothermal treatments at different temperatures using a homemade hydrothermal reactor (constructed from recycled pieces of old instruments) that allows the control of the heating program and the assessment of the internal pressure generated in the reaction. The synthesized GQDs exhibited bright blue/green luminescence under UV light, showing the success of the chosen route and opening the way for future applications of these materials in the field of optoelectronic devices.

2. Materials and Methods

2.1. Samples Preparation

GO was prepared by a modified Hummers method [17,18], using recycled graphite derived from spent lithium-ion batteries (LIBs) [19]. The spent LIBs were first dismantled into their different parts (including cathode, anode, organic separator, plastic and metallic shell). The graphite in the anode was separated from the copper foil by dissolution in concentrated nitric acid solution; followed by filtration, washing with distilled water and drying at ca. 100 °C for 24 h. For the graphite oxidation reaction, about 1.0 g of recycled graphite was added to 70 mL of H_2SO_4; followed by the slow addition of 3.0 g of $KMnO_4$ and 0.5 g of $NaNO_3$. The mixture was stirred for 2 h in an ice bath. After that, 400 mL of water and 10 mL of H_2O_2 (35%) were added to the reaction medium. The obtained product was washed with distilled water until the filtrate reached a pH next to 5–6 and then dried at 50 °C for 24 h. After drying, the obtained GO thick film was manually cut into small flakes; then, a suspension was prepared using 100 mg of GO added to 50 mL of 99.8% ethanol and 50 mL of distilled water. Afterwards, this suspension was sonicated during 40 min to obtain a homogeneous dispersion.

The synthesis of the GQDs followed a hydrothermal route similar to that described elsewhere [14,15]. It is worth emphasizing, though, that in this work no acids or strong oxidizing agents (e.g., H_2O_2) were added to the suspension submitted to the hydrothermal treatments. The alcoholic aqueous GO suspension was transferred to a Teflon-coated stainless steel homemade hydrothermal reactor and thermally treated at programmed temperatures of 125 and 175 °C during 2 h. The hydrothermal reactor used in this work (illustrated in Figure 1) is a homemade apparatus constructed from recycled pieces of old instruments. The main part is a stainless-steel vessel (which was originally part of a bomb calorimeter) containing a Teflon cup fitted to its internal volume. The two openings on the top part are used to insert the temperature sensor (K-type thermocouple) and the pressure transducer (comprising a flexible tube, such as in Bourdon manometers). The pressure is measured with the help of a Hall probe and a permanent magnet; the voltage change detected by the Hall probe when displaced with respect to the magnet by the deformation of the flexible tube is externally calibrated to allow the determination of the internal pressure in the vessel. The system is heated using an electrical band heater, containing electrical resistances with mica insulation and a stainless-steel shield. The computer control of the heating program and the pressure monitoring are implemented using an Arduino prototyping platform [20], as a microcontroller, coupled to a LabVIEW graphical interface [21]. The heating rate, final temperature and residence time are adjusted by the user via a LabVIEW proportional-integral-derivative (PID) controller that sets the electrical power to be delivered; activating a TRIAC as a power switch via the Arduino board. The communication between the Arduino board and the LabVIEW interface is provided by a homemade program written in C (which is freely available upon request). With this system, it is possible not only to set the operational temperature and the residence time, but also to record in real time the pressure and the temperature of the reaction medium inside the reactor. The temperature and pressure data as a function of time are collected by the Arduino board and sent to the LabVIEW interface for graphical representation in the computer screen; in addition, they are also saved in a log file for posterior data processing.

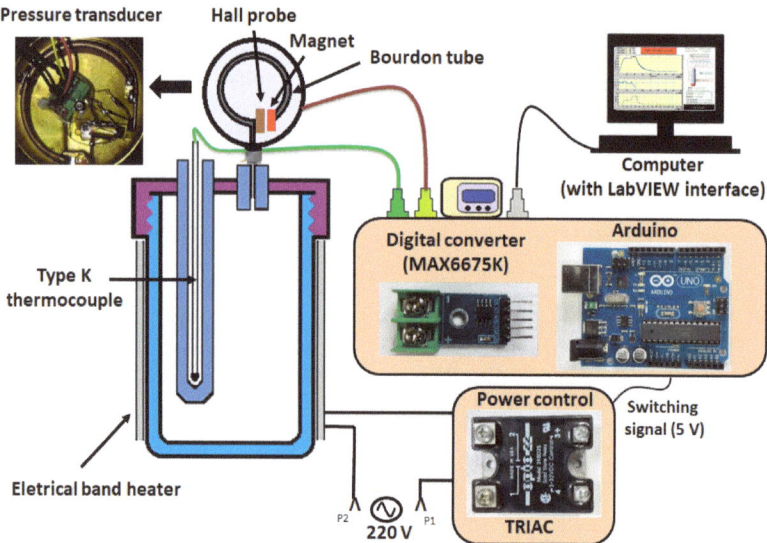

Figure 1. Scheme illustrating the operation of the homemade hydrothermal reactor.

After the conclusion of the reaction, the samples were left to cool down to room temperature naturally for 12 h. Afterwards, the obtained suspensions were centrifuged at 4000 rpm for 3 h; then, the supernatant was carefully collected and left to dry in a stove at 50 °C for 24 h. The dry material was then redispersed in 50 mL of distilled water using an ultrasonic bath for 20 min. The dispersion was double filtered to remove large particles/aggregates still present in the material and finally, the filtered dispersion was characterized by the methods described below. These samples were labeled by their temperature of synthesis; i.e., GQD_125 and GQD_175 are the labels of the samples synthesized at 125 and 175 °C, respectively. For comparison, a reference sample was also prepared by simply adding 2 mg of GO to 100 mL of distilled water (i.e., with no hydrothermal treatment); this suspension was sonicated for 40 min and filtered afterwards to remove the undispersed material. This product was labeled as GO_Ref.

2.2. Characterization

The produced GO sample was characterized by thermogravimetry (TG), X-ray diffraction (XRD) and solid-state nuclear magnetic resonance (NMR) spectroscopy. The TG curve was recorded using a TGA-50H Shimadzu thermobalance, with O_2 flow (50 mL/min) and a heating rate of 2 °C/min; the sample was mixed to alumina powder in order to avoid the loss of material following the strong combustion reaction. The XRD pattern was recorded in a XRD-6000 Shimadzu diffractometer, using Cu-Kα radiation (λ = 1.5418 Å); and with the diffraction angle 2θ ranging from 5 to 50° in 0.04° steps. The solid-state ^{13}C NMR spectrum was recorded at room temperature in a Varian/Agilent VNMR 400 MHz spectrometer

(NMR frequency of 100.52 MHz, magnetic field of 9.4 T), with the powdered sample packed into 4 mm diameter zirconia rotors for magic angle spinning (MAS) experiment at the spinning rate of 14 kHz; the spectrum was recorded using direct excitation of the ^{13}C nuclei ($\pi/2$ pulse of 4.3 µs), with a recycle delay of 15 s and accumulation of ca. 4000 transients. The chemical shifts were referenced to tetramethylsilane (TMS), using hexamethylbenzene as a secondary reference (signal at 17.3 ppm).

The produced GQDs and the GO_Ref sample were first analyzed by recording pictures of the aqueous suspensions in the dark under a 5 mW ultraviolet (UV) light with a wavelength of 365 nm produced by a commercial UV LED flashlight (Nitecore, model GEM 10 UV). The UV–visible (UV–Vis) optical absorption of the samples was analyzed using a Globlal Analyzer GTA-97 spectrophotometer. Fluorescence spectra were recorded in a Perkin Elmer LS 55 Fluorescence spectrometer. Finally, transmission electron microscopy (TEM) images were recorded using a JEOL microscope, model JEM-1400.

3. Results and Discussion
3.1. Chemical and Structural Characterization of GO

The TG curve obtained for the synthesized GO sample is shown in Figure 2. The weight losses typical of well-oxidized GO samples are clearly identified in this curve, including the water release from room temperature up to ~150 °C; the decomposition of oxygenated functions (such as epoxy, hydroxyl and carbonyl groups) between ca. 150 and 350 °C; and the oxidation of the remaining graphitic structure between ca. 400 and 700 °C [18,22]. The GO ash content (estimated from the final mass in the TG curve shown in Figure 2) is around 2.0 wt.%; which is attributed to the presence of metal impurities coming from the recycled graphite and/or introduced during the graphite oxidation reaction.

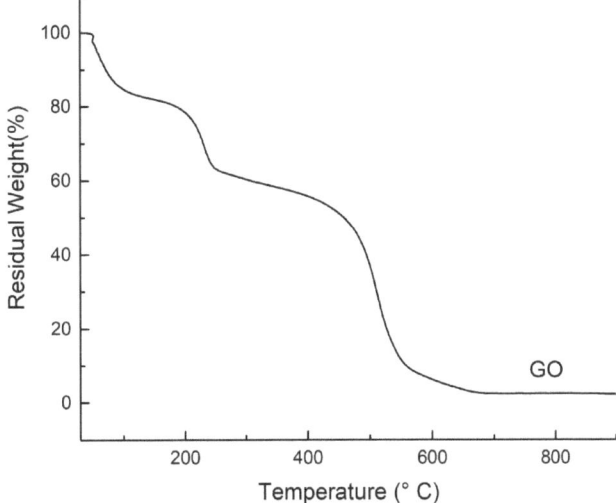

Figure 2. TG curve recorded under O$_2$ flow for the synthesized GO.

The well-oxidized nature of this GO sample is also revealed in the XRD pattern shown Figure 3, which is dominated by a peak at approximately 12°. The low value of this Bragg angle is associated with an increased interlayer spacing (around 7 Å) in comparison with graphite (3.35 Å); this is a consequence of the presence of oxygen-containing groups and intercalated water molecules in the GO structure [18,22,23].

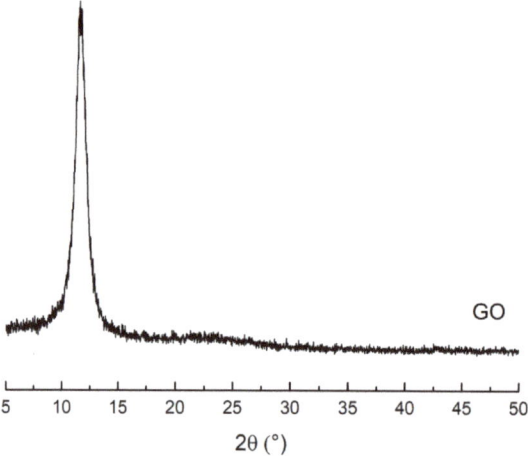

Figure 3. XRD pattern of the synthesized GO.

The nature of the oxygen-containing functional groups present in the GO structure was elucidated using solid-state ^{13}C NMR spectroscopy; the spectrum obtained for the synthesized GO sample is exhibited in Figure 4. The most characteristic ^{13}C NMR signals due to the oxygen-containing groups are observed at 70 ppm (C-OH groups) and 60 ppm (epoxy groups). In addition to these, the strong signal close to 130 ppm, along with the corresponding spinning sidebands at ca. 270 and −10 ppm, are assigned to sp^2-hybridized carbon atoms in hexagonal rings. A number of other weaker signals can also be observed, such as the ones associated with lactol (100 ppm), carbonyl (190 ppm) and carboxyl (167 ppm) groups [18,24,25].

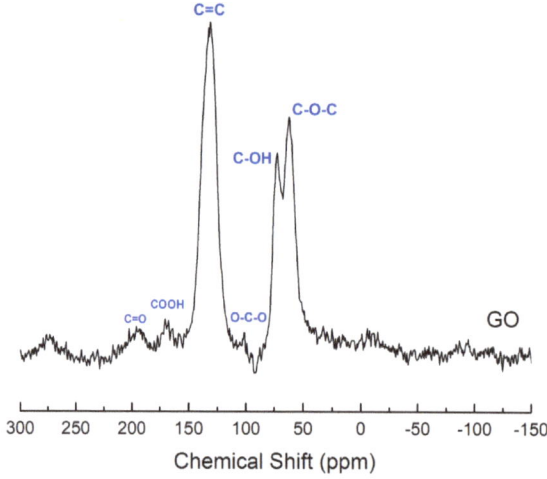

Figure 4. Solid-state ^{13}C NMR spectrum obtained for the GO sample.

Therefore, this set of results of chemical/structural characterization revealed that the synthesized GO sample exhibited indeed a well-oxidized structure, with plenty of oxygen-containing functional groups; thus, it is suitable to be used in the hydrothermal treatments to obtain the GQDs, as described below.

3.2. Hydrothermal Treatments

The apparatus used in this work revealed interesting findings on how the temperature and the pressure of the reaction medium inside the hydrothermal reactor evolved over time during the synthesis of the GQDs; this is shown in Figure 5. Regarding the temperature profile, it is possible to observe an overshoot of ~15–20 °C in the beginning of the temperature plateau (Figure 5a,c), even after careful adjustment of the temperature control parameters. A similar pattern was observed for the internal pressure (Figure 5b,d), which showed a significant increase in the beginning of the temperature plateau, before reaching a nearly stable value. The internal pressure fluctuations in this plateau closely match the temperature fluctuations, as expected. The maximum pressure reached values of ca. 3 and 23 bar; whereas the final values were around 0.5 and 11 bar, for the hydrothermal treatments performed at 125 and 175 °C, respectively.

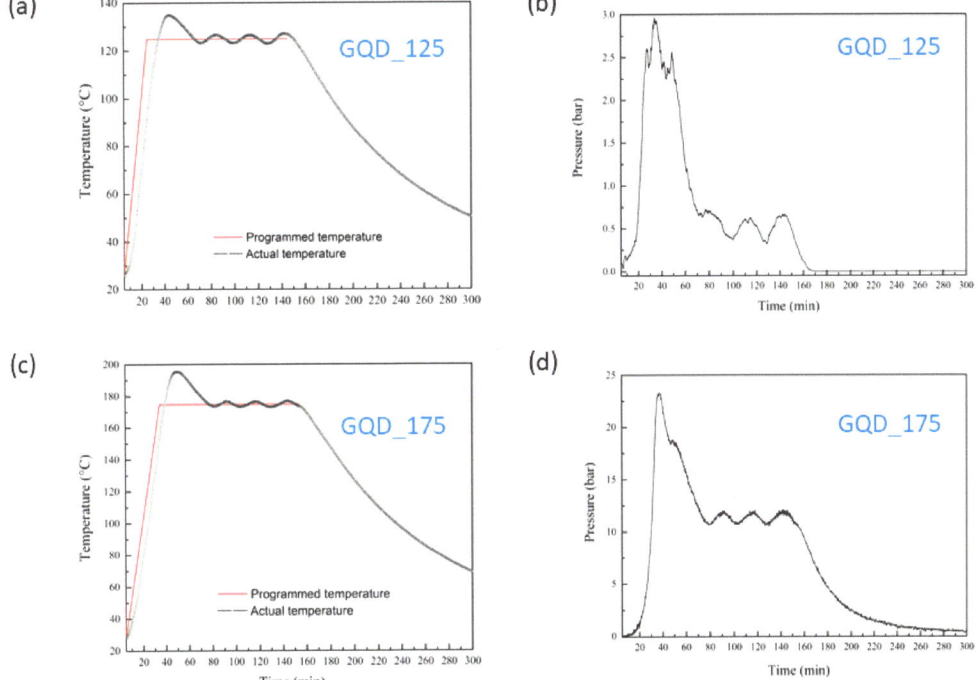

Figure 5. Evolution of temperature (parts (**a**,**c**)) and pressure (parts (**b**,**d**)) inside the hydrothermal reactor during the syntheses of the GQD_125 (parts (**a**,**b**)) and GQD_175 (parts (**c**,**d**)) samples.

3.3. Optical and Structural Properties of the GQDs

Figure 6 shows the pictures of the aqueous suspensions of the GQDs taken under 5mW, 365 nm UV light; these images exhibit clearly a bright bluish-green glow for both GQD samples (Figure 6b,d). For comparison, Figure 6f shows the corresponding picture obtained for the reference GO suspension (GO_Ref), where no glow is observed; pointing to the absence of quantum dots in this sample, as expected. On the other hand, all the samples (GO_Ref, GQD_125 and GQD_175) looked perfectly transparent under white light (Figure 6a,c,e). These results thus indicate that both produced GQD samples exhibit photoluminescent behavior when irradiated under UV light, which is a typical feature of aqueous GQD suspensions [3–5,9].

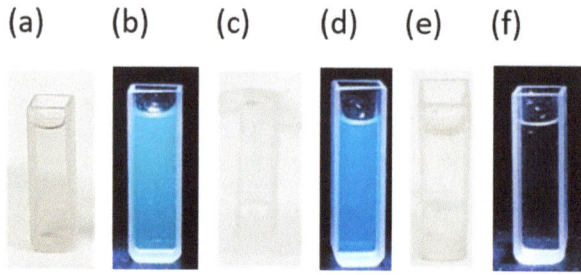

Figure 6. Images of aqueous suspensions of samples GQD_125 (parts (**a**,**b**)) and GQD_175 (parts (**c**,**d**)); and sample GO_Ref (parts (**e**,**f**)). The pictures shown in parts (**a**,**c**,**e**) were recorded under white light; whereas the ones in parts (**b**,**d**,**f**) were recorded under UV light.

The UV–Vis absorption spectra obtained for the produced samples are shown in Figure 7a; these spectra show a strong absorption band centered close to 225 nm, as usually observed in GO and undoped GQDs dispersions [9,14,15]. Even though the reference GO suspension (GO_Ref sample) also exhibits an absorption band in this region, it is important to highlight that it is not as intense as for the GQDs dispersions [14]. The fluorescence spectra recorded in the visible spectral region under 365 nm excitation for the produced samples are shown in Figure 7b; these spectra confirm the photoluminescent behavior qualitatively observed in the images shown in Figure 6. As expected, no fluorescence was detected for the GO_Ref sample, whereas the fluorescence signal was clearly observed with maximum intensity around 450 nm for both GQDs dispersions; this is in good agreement with previous reports in the literature [14,15].

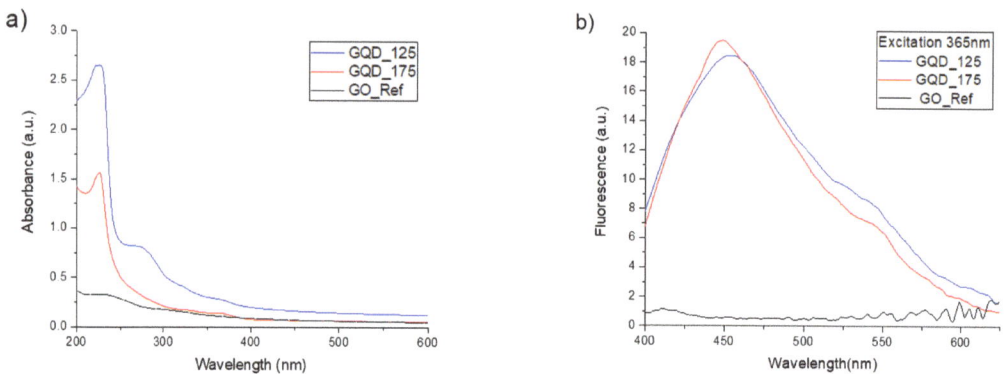

Figure 7. (**a**) UV–Vis absorption spectra and (**b**) fluorescence spectra (under 365 nm excitation) obtained for aqueous suspensions of the GQD_125, GQD_175 and GO_Ref samples.

Finally, the TEM images recorded for the produced samples (Figure 8) reveal that the GQD_125 and GQD_175 samples are composed of particles with average sizes of around 60 and 30 nm, respectively. The trend of smaller particles being produced at higher treatment temperatures is in agreement with previous reports [14,15]; as is the slight shift in the maximum of the fluorescence spectra obtained for these two samples (Figure 7), with the maximum fluorescence being observed at a slightly lower wavelength for the GQD_175 sample in comparison with the GQD_125 sample.

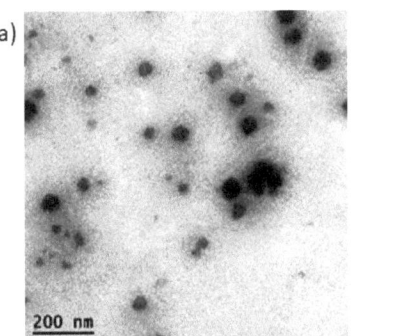 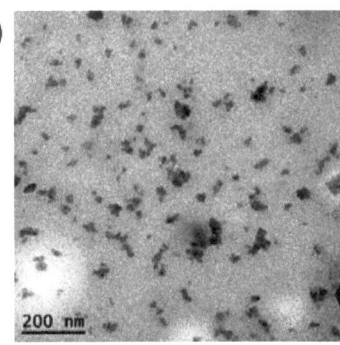

Figure 8. TEM images obtained for the GQD_125 (**a**) and GQD_175 (**b**) samples.

4. Conclusions

A simple and low-cost route for the synthesis of GQDs, starting from an alcoholic aqueous suspension of graphene oxide derived from recycled graphite and using a homemade hydrothermal reactor, was described in this work. The developed apparatus allowed the control of the heating program and the assessment of the internal pressure generated during the reaction at high temperatures. The synthesized GQDs were fully characterized, with the observation of bright blue/green luminescence under UV light, thus showing the success of the chosen route and opening the way for future applications of these materials in the field of optoelectronic devices.

Author Contributions: Conceptualization, J.C.C.F., M.A.S.J.; Data curation, L.M.D., T.L.G.C. and D.F.C.; Formal analysis, J.C.C.F.; Investigation, L.M.D., T.L.G.C., D.F.C., C.W.C. and M.A.S.J.; Methodology, L.M.D., T.L.G.C., C.W.C. and M.A.S.J.; Resources, J.C.C.F.; Writing—original draft, L.M.D. and M.A.S.J.; Writing—review & editing, J.C.C.F. All authors have read and agreed to the published version of the manuscript.

Funding: The authors acknowledge the financial support from the Brazilian agencies FAPES (grants 345/2019, 280/2021 and 495/2021), CAPES and CNPq. The authors are also grateful to the Laboratory of Cellular Ultrastructure Carlos Alberto Redins (LUCCAR) and the Laboratory for Research and Development of Methodologies for Crude Oil Analysis (LabPetro), both at the Federal University of Espírito Santo (UFES), for the use of their experimental facilities.

Data Availability Statement: The data are available upon request from the authors.

Conflicts of Interest: The authors declare no conflict of interest.

References

1. Liu, J.; Li, R.; Yang, B. Carbon dots: A new type of carbon-based nanomaterial with wide applications. *ACS Cent. Sci.* **2020**, *6*, 2179–2195. [CrossRef]
2. Li, M.; Chen, T.; Gooding, J.; Liu, J. Review of carbon and graphene quantum dots for sensing. *ACS Sens.* **2019**, *4*, 1732–1748. [CrossRef]
3. Tajik, S.; Dourandish, Z.; Zhang, K.; Beitollahi, H.; Van Le, Q.; Jang, H.W.; Shokouhimehr, M. Carbon and graphene quantum dots: A review on syntheses, characterization, biological and sensing applications for neurotransmitter determination. *RSC Adv.* **2020**, *10*, 15406. [CrossRef]
4. Jorns, M.; Pappas, D. A review of fluorescent carbon dots, their synthesis, physical and chemical characteristics, and applications. *Nanomaterials* **2021**, *11*, 1448. [CrossRef]
5. Wang, L.; Li, W.; Wu, B.; Li, Z.; Wang, S.; Liu, Y.; Pan, D.; Wu, M. Facile synthesis of fluorescent graphene quantum dots from coffee grounds for bioimaging and sensing. *Chem. Eng. J.* **2016**, *300*, 75–82. [CrossRef]
6. Li, W.; Jiang, N.; Wu, B.; Liu, Y.; Zhang, L.; He, J. Chlorine modulation fluorescent performance of seaweed-derived graphene quantum dots for long-wavelength excitation cell-imaging application. *Molecules* **2021**, *26*, 4994. [CrossRef]
7. Wang, L.; Li, W.; Yin, L.; Liu, Y.; Guo, H.; Lai, J.; Han, Y.; Li, G.; Li, M.; Zhang, J.; et al. Full-color fluorescent carbon quantum dots. *Sci. Adv.* **2020**, *6*, eabb6772. [CrossRef]
8. Li, W.; Guo, H.; Li, G.; Chi, Z.; Chen, H.; Wang, L.; Liu, Y.; Chen, K.; Le, M.; Han, Y.; et al. White luminescent single-crystalline chlorinated graphene quantum dots. *Nanoscale Horiz.* **2020**, *5*, 928–933. [CrossRef]

9. Shen, S.; Wang, J.; Wu, Z.; Du, Z.; Tang, Z.; Yang, J. Graphene quantum dots with high yield and high quality synthesized from low cost precursor of aphanitic graphite. *Nanomaterials* **2020**, *10*, 375. [CrossRef]
10. Yang, S.; Sun, J.; Li, X.; Zhou, W.; Wang, Z.; He, P.; Ding, G.; Xie, X.; Kang, Z.; Jiang, M. Large-scale fabrication of heavy doped carbon quantum dots with tunable-photoluminescence and sensitive fluorescence detection. *J. Mater. Chem. A* **2014**, *2*, 8660–8667. [CrossRef]
11. Pan, D.; Zhang, J.; Li, Z.; Wu, M. Hydrothermal route for cutting graphene sheets into blue-luminescent graphene quantum dots. *Adv. Mater.* **2010**, *22*, 734–738. [CrossRef]
12. Shi, Y.; Park, J.; Hyun, D.; Yang, J.; Lee, J.-H.; Kim, J.-H.; Lee, H. Acid-free and oxone oxidant-assisted solvothermal synthesis of graphene quantum dots using various natural carbon materials as resources. *Nanoscale* **2015**, *7*, 5633–5637. [CrossRef]
13. Ahirwar, S.; Mallick, S.; Bahadur, D. Electrochemical method to prepare graphene quantum dots and graphene oxide quantum dots. *ACS Omega* **2017**, *2*, 8343–8353. [CrossRef]
14. Tian, R.; Zhong, S.; Wu, J.; Jiang, W.; Wang, T. Facile hydrothermal method to prepare graphene quantum dots from graphene oxide with different photoluminescences. *RSC Adv.* **2016**, *6*, 40422–40426. [CrossRef]
15. Xie, J.-D.; Lai, G.-W.; Huq, M.M. Hydrothermal route to graphene quantum dots: Effects of precursor and temperature. *Diam. Relat. Mater.* **2017**, *79*, 112–118. [CrossRef]
16. Peng, J.; Gao, W.; Gupta, B.K.; Liu, Z.; Romero-Aburto, R.; Ge, L.; Song, L.; Alemany, L.B.; Zhan, X.; Gao, G.; et al. Graphene quantum dots derived from carbon fibers. *Nano Lett.* **2012**, *12*, 844–849. [CrossRef]
17. Hummers, W.S., Jr.; Offeman, R.E. Preparation of graphitic oxide. *J. Am. Chem. Soc.* **1958**, *80*, 1339. [CrossRef]
18. Vieira, M.A.; Gonçalves, G.R.; Cipriano, D.F.; Schettino, M.A., Jr.; Silva Filho, E.A.; Cunha, A.G.; Emmerich, F.G.; Freitas, J.C.C. Synthesis of graphite oxide from milled graphite studied by solid-state ^{13}C nuclear magnetic resonance. *Carbon* **2016**, *98*, 496–503. [CrossRef]
19. Ribeiro, J.S.; Freitas, M.B.; Freitas, J.C.C. Recycling of graphite and metals from spent Li-ion batteries aiming the production of graphene/CoO-based electrochemical sensors. *J. Environ. Chem. Eng.* **2021**, *9*, 104689. [CrossRef]
20. Kondaveeti, H.K.; Kumaravelu, N.K.; Vanambathina, S.D.; Mathe, S.E.; Vappangi, S. A systematic literature review on prototyping with Arduino: Applications, challenges, advantages, and limitations. *Comput. Sci. Rev.* **2021**, *40*, 100364. [CrossRef]
21. What Is LabVIEW? Available online: https://www.ni.com/pt-br/shop/labview.html (accessed on 14 August 2022).
22. Dreyer, D.R.; Park, S.; Bielawski, C.W.; Ruoff, R.S. The chemistry of graphene oxide. *Chem. Soc. Rev.* **2009**, *39*, 228–240. [CrossRef]
23. Mu, S.-J.; Su, Y.-C.; Xiao, L.-H.; Liu, S.-D.; Hu, T.; Tang, H.-B. X-ray diffraction pattern of graphite oxide. *Chin. Phys. Lett.* **2013**, *30*, 096101. [CrossRef]
24. Vieira, M.A.; Frasson, C.M.R.; Costa, T.L.G.; Cipriano, D.F.; Schettino, M.A., Jr.; Cunha, A.G.; Freitas, J.C.C. Solid state ^{13}C NMR study on the synthesis of graphite oxide from different graphitic precursors. *Quim. Nova* **2017**, *40*, 1164–1171. [CrossRef]
25. Rawal, A.; Man, S.H.C.; Agarwal, V.; Yao, Y.; Thickett, S.C.; Zetterlund, P.B. Structural complexity of graphene oxide: The Kirigami model. *ACS Appl. Mater. Interfaces* **2021**, *13*, 18255–18263. [CrossRef]

Article

Investigation of Electron Transfer Mechanistic Pathways of Ferrocene Derivatives in Droplet at Carbon Electrode

Sidra Ayaz [1], Afzal Shah [1,*] and Shamsa Munir [2]

1. Department of Chemistry, Quaid-Azam University, Islamabad 45320, Pakistan
2. School of Applied Sciences and Humanities, National University of Technology (NUTECH), Islamabad 44000, Pakistan
* Correspondence: afzals_qau@yahoo.com or afzalshah@qau.edu.pk

Abstract: The results of cyclic, differential pulse and square wave voltammetric studies of four ferrocene derivatives, i.e., 4-ferrocenyl-3-methyl aniline (FMA), 3-Chloro-4-ferrocenyl aniline (CFA), 4-ferrocenyl aniline (FA) and ferrocenyl benzoic acid (FBA) on carbon electrode, revealed that the redox behavior of these compounds is sensitive to pH, concentration, scan number and scan rate. One electron, diffusion controlled, with a quasi-reversible redox signal displaying ferrocene/ferrocenium couple was observed for each of the studied ferrocenyl derivatives. Quasi-reversibility of this signal is evidenced by ΔE_p, I_a/I_c current ratio and k_{sh} values. Another one electron and one proton irreversible oxidation signal was noticed in the voltammograms of these compounds except FBA. This signal corresponds to the electro-oxidation of the amine group and its irreversibility, as supported by ΔE_p, I_a/I_c current ratio and k_{sh} values, is due to the influence of the electron donating nature of the amine group. A number of electrochemical parameters such as D, k_{sh}, LOD and LOQ were evaluated for the targeted ferrocene derivatives. The obtained parameters are expected to provide insights into the redox mechanism for understanding their biochemical actions. The electrochemistry presented in this work is done using a unique environmentally benign and cost-effective droplet electrochemical approach.

Keywords: ferrocenes; electron transfer rate constant; electrochemistry in droplet; redox mechanism

Citation: Ayaz, S.; Shah, A.; Munir, S. Investigation of Electron Transfer Mechanistic Pathways of Ferrocene Derivatives in Droplet at Carbon Electrode. *C* **2022**, *8*, 45. https://doi.org/10.3390/c8030045

Academic Editors: Olena Okhay and Gil Goncalves

Received: 13 August 2022
Accepted: 7 September 2022
Published: 9 September 2022

Publisher's Note: MDPI stays neutral with regard to jurisdictional claims in published maps and institutional affiliations.

Copyright: © 2022 by the authors. Licensee MDPI, Basel, Switzerland. This article is an open access article distributed under the terms and conditions of the Creative Commons Attribution (CC BY) license (https://creativecommons.org/licenses/by/4.0/).

1. Introduction

Ferrocene is an organometallic compound of the general class *metallocene*, with the formula Fe(η^5-C$_5$H$_5$)$_2$, where iron is sandwiched between two cyclopentadienyl rings [1–3]. Ferrocene finds applications in the design and fabrication of sensors, biosensors, electrochemically active supramolecular switches, catalysis, drugs and fuel additives etc. It has been reported that ferrocenes could fit in the cavity of water-soluble β-cyclodextrin (βCD) [4–10]. The diverse applications of ferrocene derivatives are mainly attributed to their electrochemical properties, which govern their chemical and biological action. Therefore, the family of ferrocene has been rapidly developing in the last 60+ years, with synthesis of its derivatives having applications in homogeneous asymmetric catalysis, chemical sensors, biosensing, molecular electronics and electrocatalysis [11–17]. Ferrocene is also used in biological treatments, because it is chemically stable, neutral and able to cross the cell membrane.

Owing to its favorable and reversible redox peaks, a huge fraction of research is devoted to the electron transfer studies of ferrocene derivatives. Electrochemical sensors based on ferrocene as mediators have been extensively documented [18–25]. They are also considered as ideal redox mediators, due to the stability of each form of the redox couple and their insensitivity to physiological oxygen... It serves as a reference material due to its chemical and electrochemical reversibility in organic electrolytes and its invariant redox potential [26]. Abraham et al., have reported that ferrocene and its derivatives to be potentially beneficial redox reagents for the chemical overcharge protection of rechargeable lithium and lithium ion batteries [27] Other applications of ferrocene derivatives include their role as combustion regulators, radiation absorbers and components of various redox systems.

There is a wealth of research on ferrocene derivatives in medicinal chemistry, where they are documented as anticancerous drugs with antiproliferative activities [6]. Taking advantage of their favorable bioorganometallic chemistry, they are frequently employed in bioelectronics and in making glucose biosensors [7]. Moreover, ferrocene has been reported to improve the activity of various drugs such as ferrocene aspirin, the anticancer drug, *ferrocifen*, and an anti-malarial drug, *ferroquine* [28]. Similarly ferrocene-conjugated pepstatin, bioconjugate-3, is used for the detection of HIV-I protease [29].

Use of ferrocene and its derivatives in various fields of science have made them an essential class of compounds for further investigation [30]. The role of ferrocene and its derivatives for the welfare of mankind is obvious and is believed to relate to their favorable electron transfer behavior. The electrochemical studies of new ferrocene derivatives are crucial for the evaluation of their properties and potential biomedical applications. In this regard, we have assessed the electrochemical properties of some novel ferrocene derivatives using cyclic voltammetry (CV), differential pulse voltammetry (DPV) and square wave voltammetry (SWV) in different pH media to propose their redox mechanisms. The electrochemical measurements are performed using a unique environmentally benign and cost-effective droplet electrochemical approach. The chemical structures of the investigated ferrocenes can be seen in Scheme 1.

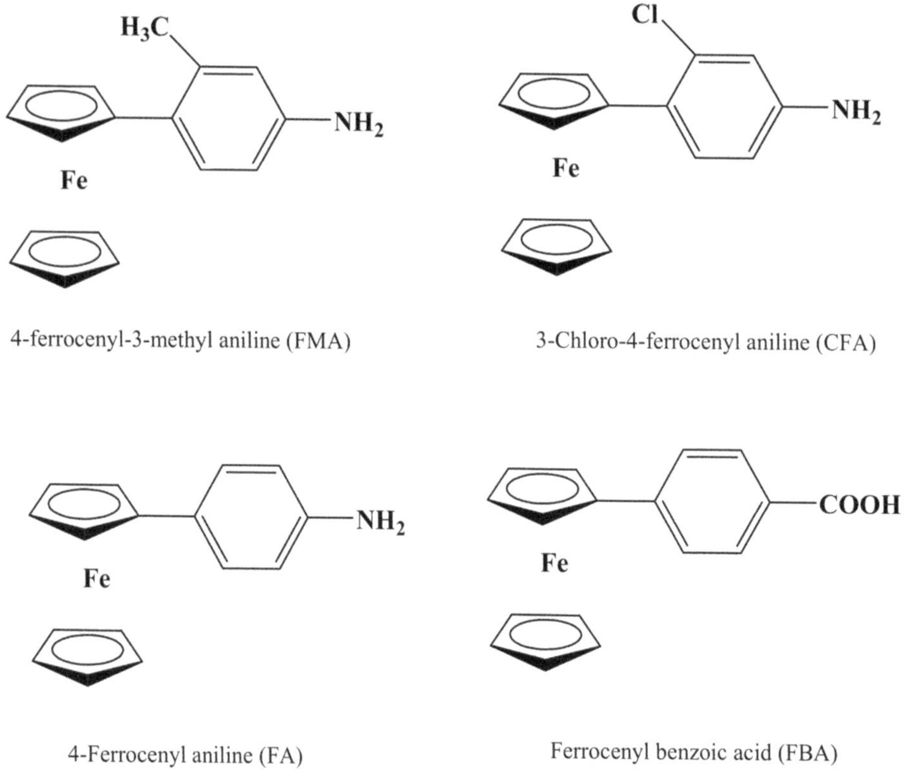

4-ferrocenyl-3-methyl aniline (FMA)

3-Chloro-4-ferrocenyl aniline (CFA)

4-Ferrocenyl aniline (FA)

Ferrocenyl benzoic acid (FBA)

Scheme 1. Structures and names of studied ferrocenyl derivatives.

2. Experimental Section

2.1. Materials and Reagents

The derivatives of ferrocene used in the current work were kindly gifted by Professor Dr. Amin Badshah [31–34]. 2 mM stock solutions of the compounds were prepared in an analytical grade ethanol solvent purchased from Sigma Aldrich. All supporting electrolytes in the pH range of 2–12 were prepared using a Britton–Robinson (BR) buffer solution prepared by using sodium hydroxide and a mixture of 0.04 mM boric acid, 0.04 mM acetic acid and 0.04 mM phosphoric acid. All the components of BR buffer were obtained from Sigma Aldrich. For pH measurements, an INOLAB pH meter with model no. pH 720 was used. Micro volume measurements were done by EP-10 and EP-100 plus motorized μL pipettes.

2.2. Voltammetric Parameters and Electrochemical Cells

Voltammetric experiments were performed using the three-electrode system. The instrument used was a three-electrode Digi-ivy potentiostat received from Austin, TX, USA. The potentiostat was controlled using the DY2100 series software version. The reference electrode Ag/AgCl and the counter electrode, a piece of Pt foil, were attached to each other and hung with the stand. A large area (0.07 cm^2) glassy carbon electrode was polished with 0.3-micron micro polish powder before each experiment and placed parallel to the two upper electrodes. A small droplet of different solutions containing a 2 mM concentration with supporting electrolyte was placed on the surface of the carbon electrode. All electrodes were then connected to the instrument to record the voltammograms. The distance between the upper electrodes and working electrode was controlled by the screw of stand and volume of droplet of solution was controlled using micropipette.

The operations of placing the droplet on the surface of the GC electrode and precise positioning of upper electrode's tip were observed with a magnifying glass, which allowed for magnification of the object. To make the droplet sit well on the GC electrode, the surface of the pipette tip was in contact with the carbon surface at start, and then was slowly pulled away while the droplet grew.

Conditions selected for differential pulse voltammetric experiment were pulse amplitude 50 mV, pulse width 50 ms, pulse period 200 ms and scan rate 100 mV s^{-1}. For SWV, the experimental conditions were as follows. The frequency of 5 Hz and potential increment 0.02 V, corresponding to an effective scan rate of 100 mV s^{-1}, unless stated otherwise.

2.3. Acquisition and Presentation of Voltammetric Data

All the voltammograms recorded were background subtracted and baseline corrected by using the moving average with a step window of 2 mV. This treatment helps in improving visualization and identification of peaks over the baseline without introducing any artefact, although the peak height in some cases is reduced (<10%) relative to that of the untreated curve. The peak current (I_p) values reported were taken from the original untreated voltammograms before the subtraction of the baseline.

3. Results and Discussion

3.1. Cyclic Voltammetry

Cyclic voltammetry (CV) was employed to determine the diffusion coefficient (D) and heterogeneous electron transfer rate constant (k_{sh}) of the compounds. The CVs of the 2 mM solution of all compounds were initially recorded between −0.1 and +1.5 V in 90% aqueous ethanol, buffered at pH 7 with a scan rate of 0.1 V/s. A redox couple labelled as 1a and 1c was observed for all derivatives. A 2nd oxidation peak, labeled as 2a, was also observed for FMA, CFA and FMA. The CVs were also recorded by limiting the potential range between −0.1 and +1.5 to clearly represent the 2a peaks of CFA and FMA, as displayed in Figure 1. The voltammetric response of all the selected ferrocenes featured well-defined and stable oxidation and reduction peaks (1a and 1c), which are assigned to the oxidation of the ferrocene nucleus in analogy with other ferrocene molecules [35,36].

An observation of these overlayed voltammograms reflect that the anodic signal of FA is obtained at least positive potential, displaying its favored oxidation. Most facile oxidation of FA is attributed to the electron donating effect of the attached aniline group. Whereas for CFA, oxidation peak is shifted to more positive potential displaying comparatively difficult oxidation than FA, mainly because of the electron withdrawing inductive effect of the attached chloro group. Similarly, the voltammetric signal of FBA, noticed at most positive potential, suggested the most difficult electron abstraction from its ferrocene moiety, due to the strong electron withdrawing effect of the benzoic acid group. This variation in the oxidation behavior of closely related ferrocene derivations demonstrated the sensitivity of ferrocene to the electronic effect of the attached substituents [37,38]. The CV of the pristine ferrocene molecule is reported to be 0.29 V, very close to FMA, with a peak separation of 0.074 V in aqueous solvent at glassy carbon electrode. However, the peak separation in case of FMA is larger, i.e., 0.12 V, compared to the pristine ferrocene molecule. The well-defined redox peaks 1a and 1c correspond to the oxidation of Fe as reported in the previous studies [16,39,40].

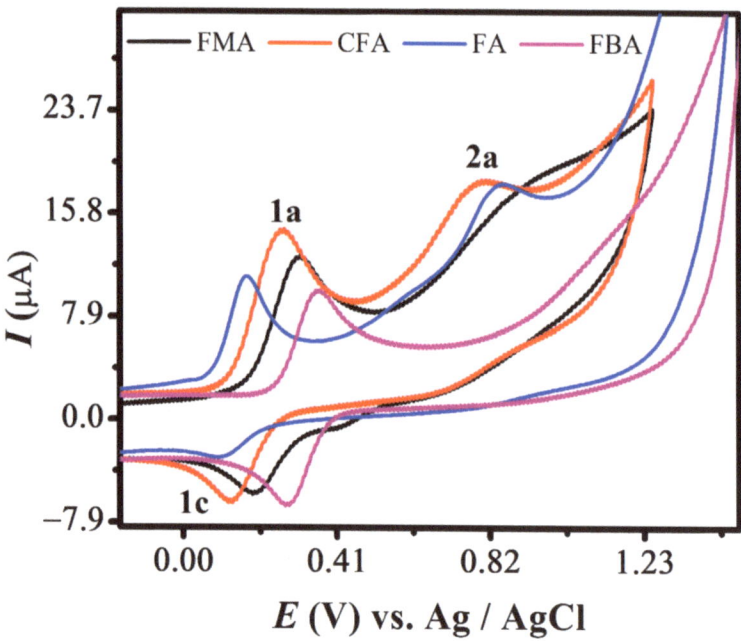

Figure 1. CVs of 2 mM ferrocene derivatives in pH 7 at 0.1 V/s.

ΔE_p values much higher than the expected Nernstian value of 59 mV per electron and I_{pa}/I_{pc} current ratio deviating from 1, predicted that the redox process is electrochemically quasi-reversible for all the derivatives of ferrocene [36]. A literature review reveals that ferrocene alone shows a well-defined, reversible redox signal [41]. This disparity is attributed to solution resistance and slow electron transfer [38]. Moreover, slightly displaced potential values observed for Fc/Fc$^+$ couple for these compounds in comparison to unsubstituted ferrocene reveals that the redox behavior of ferrocene is dependent on the electronic properties of the attached ligands [35]. A second anodic signal, 2a, was also noted for these derivatives, except FBA, representing the oxidation of the para-amine moiety present in these derivatives. Absence of its corresponding reduction signal in the reverse scan shows the stability of this moiety to reduction and indicates the irreversible nature of this 2nd oxidation step. $E_{pa}-E_{pa/2} > 60$ mV also supports the irreversibility of this electrode process. Summary of electrochemical data of these four ferrocene derivatives is listed in Table 1.

Table 1. Summary of electrochemical data obtained from CV.

Compound	Redox Signal	Voltammetric Parameter							Reversibility
		I_{pa} (μA)	E_p (V)	$E_p - E_{p1/2}$ (V)	ΔE_p (V)	I_{pa}/I_{pc}	$D/10^{-7}$ cm^2 s^{-1}	H$^+$s	
FMA	1a	13.2	0.31	-	0.12	4.12	9.03	-	QR
	1c	−3.2	0.23	-					
	2a	15	0.84	0.08	-	-		1	IR
CFA	1a	14.6	0.27	-	0.13	2.92	9.18	-	QR
	1c	−5	0.11	-					
	2a	16	0.85	0.07	-	-			IR
FA	1a	8.32	0.17	-	0.07	2.97	5.53	-	QR
	1c	−2.8	0.1	-					
	2a	18	0.85	0.07					IR
FBA	1a	9.90	0.36	0.05	0.07	1.5	5.64	-	QR
	1c	6.22	0.27	0.06					

Cyclic voltammograms of FBA (1a & 1c peak) obtained at different scan rates are shown in Figure 2. The CVs depict that peak current varies linearly with the increasing scan rate. The reversible signal for Fc/Fc$^+$ couple principally occurs at the same potential. [41] However, the slight variation in anodic and its corresponding reduction signal observed here is due to the presence of finite solution resistance and slightly slow electron transfer kinetics [37]. This ensures the quasi-reversible nature of redox peaks, presumably due to the bulky substituent attached to it. A similar trend with increasing scan rate was observed for redox signal 1a and 1c in all studied ferrocene derivatives (Figure S1A–C).

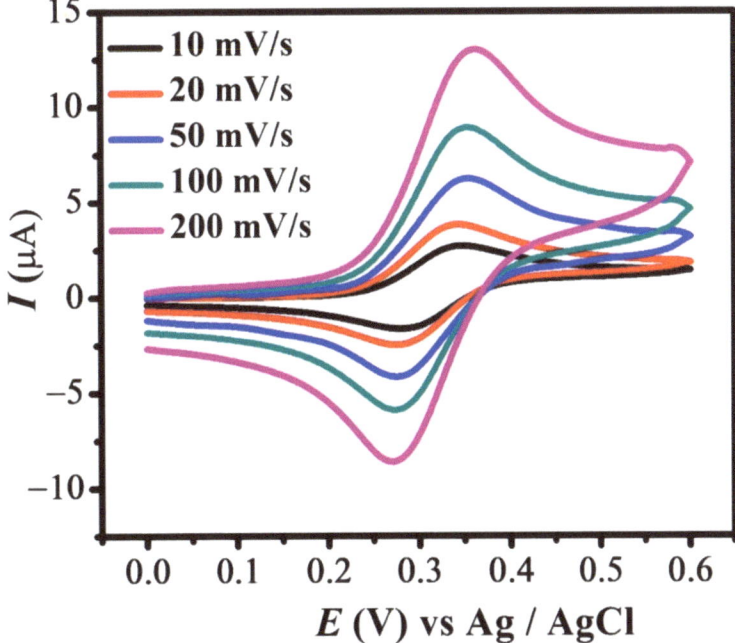

Figure 2. CVs of 2 mM FBA at different scan rates.

This study of the effect of scan rate was employed in determining an important parameter, i.e., diffusion coefficient (*D*). This is done by plotting log I_{pa} vs. log ν, as demonstrated in all compounds, by the procedure mentioned elsewhere. Figure 3 shows the plot of log I_{pa} vs. log ν for all compounds. The diffusion coefficient was calculated using the slope of straight lines by applying Randles–Sevcik equation [42]. For FBA, the slope value of 0.56 confirmed the diffusion limited nature of the first redox signal [43].

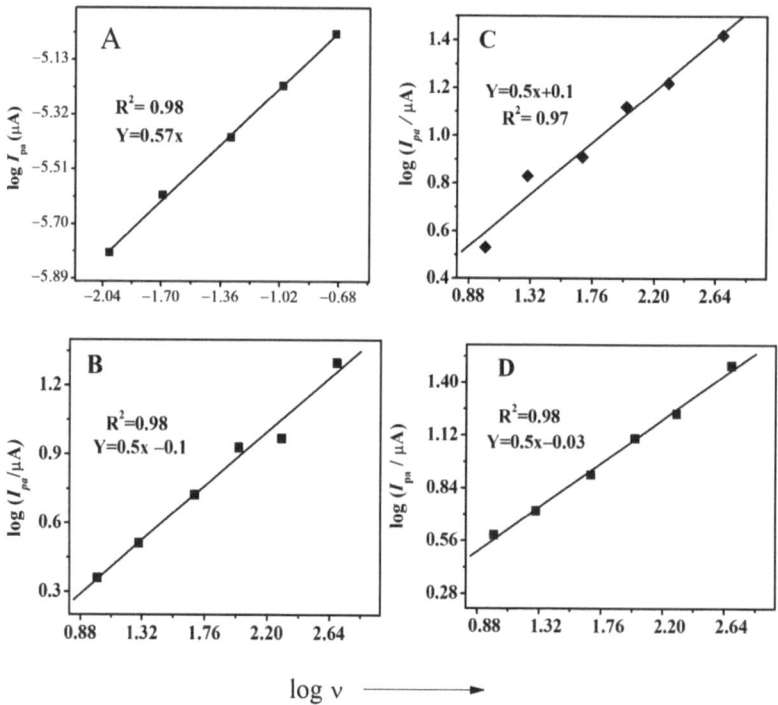

Figure 3. Plot of log I_{pa} vs. log ν for (**A**) FBA (**B**) FA (**C**) CFA (**D**) FMA.

The effect of the scan rate for FMA is displayed in Figure 4, where the much greater positive potential shift observed for the 2a anodic signal with increasing scan rate supports the irreversible nature of this oxidation step. The values of D for all derivatives (peak 1a) are displayed in Table 1 and were found to be in the range of 10^{-7} cm^2 s^{-1}.

The Nicholson method [44] has been used to calculate the rate constant k_{sh} at the electrode. Peak separation (ΔE_p) between the anodic and cathodic peak from the background subtracted voltammogram is used to evaluate ψ using the classic relation given by Nicholson between ΔE_p value of 0.07 V and the relation is presented in Figure 3 of Nicholson's paper [44]. The value of Ψ = 1.51 was then employed to determine the value of k_{sh} using Equation (1).

$$\Psi = \frac{k_{sh}}{(\alpha D)^{1/2}} \tag{1}$$

where α is for charge transfer coefficient, which has been calculated using the relation α = nFν/RT (ν being the scan rate, 1.46 V/s for this experiment) and was found to be α = 5.69 (Table 2). D is the diffusion coefficient, and its value was obtained from the study of the scan rate effect of FBA and was found to be 5.645 × 10^{-7} cm^2 s^{-1} (see Table 1). By putting the values of α, D and Ψ in Equation (1), the heterogeneous rate constant k_{sh} was calculated giving the value of 6.74 × 10^{-4} cm s^{-1} for FBA. The values of k_{sh} calculated for all other derivatives are listed in Table 2.

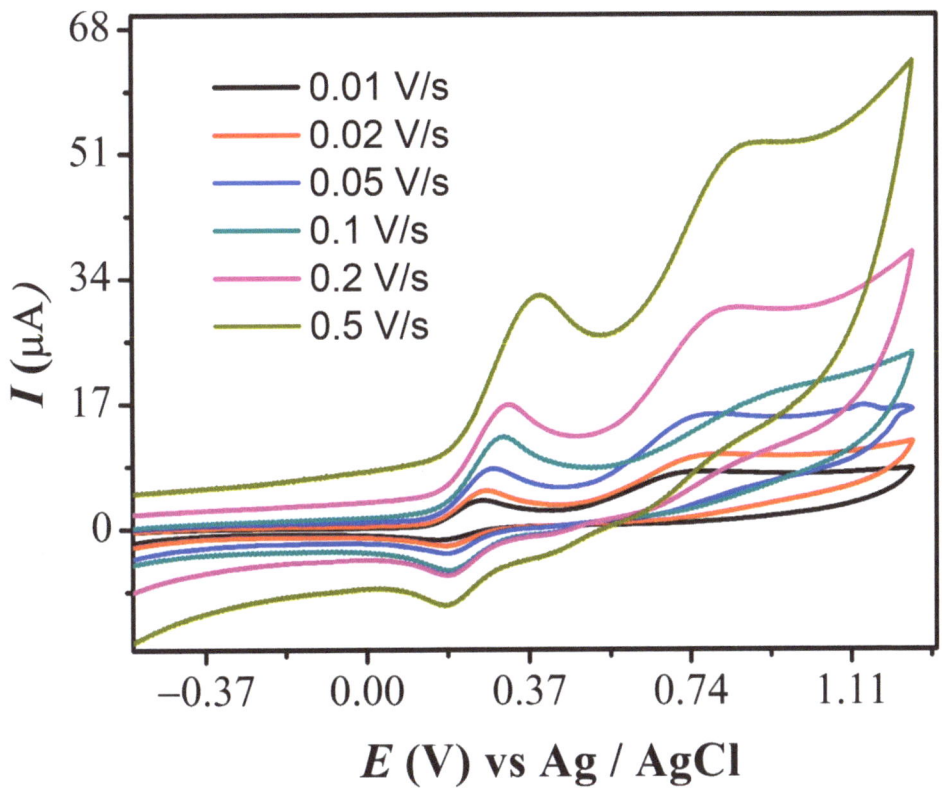

Figure 4. CVs of FMA showing scan rates effect.

Table 2. Summary of important electrochemical parameters obtained for all compounds.

Compound	ΔE_p (V)	ψ	$k_{sh} \times 10^{-4}$ (cm/sec)	$\alpha = (E_{1/2}-E_{pc})/E_{pa}-E_{pc}$
FMA	115	0.394	8.9	0.476
CFA	150	0.212	4.8	0.453
FA	105	0.496	8.7	1.1
FBA	0.073	1.51	6.7	5.69

3.2. Differential Pulse Voltammetry

Differential pulse voltammetry (DPV) was employed to study the effect of pH and determine the number of electrons and protons involved in the redox processes. DPV is a very sensitive technique to reduce background charging currents. The waveform in DPV is a succession of pulses, where a baseline potential is a particular period of time earlier than the application of a potential pulse. The medium effect was successfully studied in the pH range from 3–12 using DPV. The shifting of anodic peaks towards less positive potential with increasing pH projects the convenience of the oxidation process in basic media [43,45]. For FBA, only one anodic signal was observed in accordance with CV, which displayed slight shifting towards positive potential with increasing pH (Figure 5A). The width at half peak height, $W_{1/2}$, value close to 90 mV and literature review of structurally similar ferrocenyl derivatives confirmed one electron involvement in its oxidation [46–48]. Participation of protons in the oxidation step were calculated by plotting E_p vs. pH plot

(Figure 5B), using relation $E = E^O - m/n \times 0.059$ pH where m is the number of protons and n is number of electrons involved in the redox process [46–48]. The slope value deviating from 25.6 mV per pH unit suggested the m = 0.4, indicating almost no proton involvement in this oxidation step [46–48].

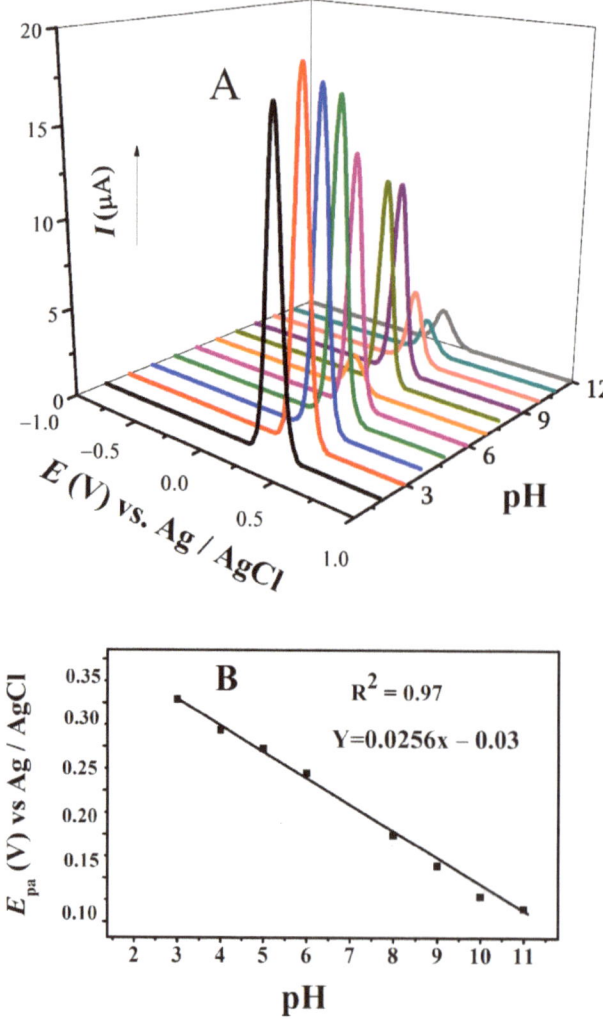

Figure 5. (**A**) DPV of FBA obtained at 0.1 V/s in different pH media and (**B**) plot of E_{pa} vs. pH of FBA.

Variation in current has also been observed with the changing pH, owing to medium effect. The variation in voltammetric response reflects that the change in pH can alter the biochemical pathways, as they depend on the redox potential of compounds.

The DPV plots of FMA vs pH are displayed in Figure 6A,B. Anodic peak 1a does not show any significant shift with pH, however, 2a has a large shift towards negative potential with increasing pH and displayed ease of oxidation of -NH$_2$ group of FMA in a basic environment. The variation of E_{pa2} with pH for FMA, with a slope of 58 mV per pH unit close to 59 mV per pH unit, suggests that the oxidation of the -NH$_2$ group for this and other amino ferrocenyl derivatives involve the same number of protons as electrons. Rising pH is also accompanied by the changing current values for 2a. This variation in redox potentials and currents confirmed the medium sensitivity, complementing CV results.

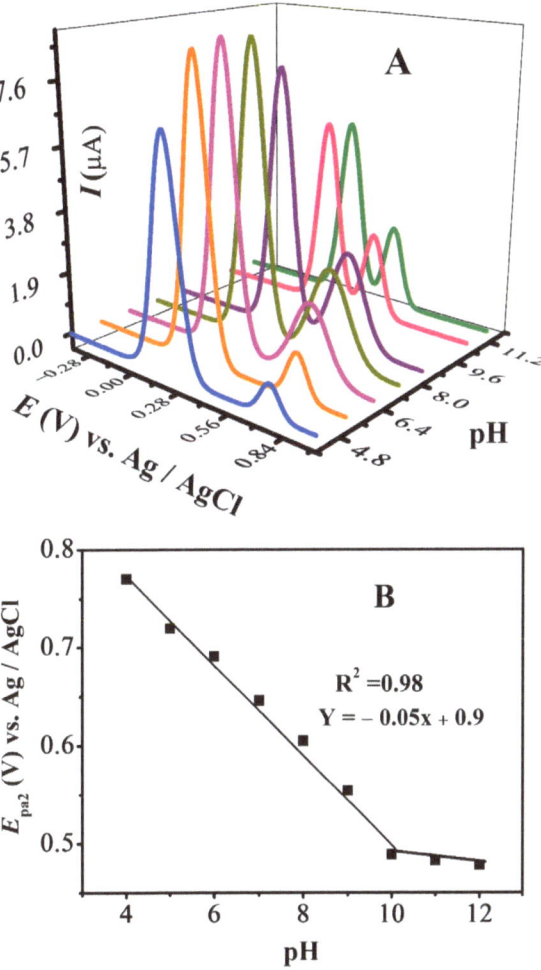

Figure 6. (**A**) DPVs obtained at 0.1 V/s in different pH media and (**B**) plot of E_{pa} vs. pH of FMA.

The $W_{1/2}$ FMA, CFA and FA are approximately equal to 90 mV (Figure S2A,B) and a literature review of structurally related derivatives suggests one electron involvement in oxidation of these ferrocenyl compounds [35]. The pk_a is determined to be at pH 10, as evidenced from the plot of E_{pa2} vs. pH. No significant peak shifting is observed for FA and CFA, with changing pH values indicating the involvement of only electrons in the oxidation process.

3.3. Square Wave Voltammetry

SWV is the most sensitive technique, as compared to other electrochemical techniques, because of greater pace of analysis, reserved consumption of electro-active species, rather than DPV and minimized problems of electrode surface poisoning [49]. The main advantage of SWV is the simultaneous recording of oxidation and reduction currents in one scan. This improvement thus helps in confirming the reversibility, irreversibility or quasi-reversibility of any electrochemical processes. Thus, SWV was employed to confirm the reversibility of redox peaks and to calculate k_{sh} using diffusion coefficient values obtained from CV.

The quasi-reversible nature of the redox peaks of FBA and FA was confirmed by the inequality of magnitudes of I_f and I_r [46–48], as observed in SWV (see Figure 7 and Figure S3A,B). The effect of consecutive scans, by recording successive SWVs in the same solution for FBA without cleaning the GCE surface, also depicts the reversibility or non-reversibility of the redox process. In case of FBA, the peak intensity of the oxidation peak almost remained constant after the first scan without cleaning the electrode surface (see Figure 8A), displaying the electrochemical stability of ferrocene/ferrocenium [37] couple, complementing the CV results. Whereas for FMA, CFA and FA, intensity of the oxidation peak decreased significantly with the number of scans, owing to the adsorption of the oxidation product, making the redox process non-reversible (Figure 8B).

The effect of different concentrations was evaluated at pH 7 for all compounds. The results for FBA and FA are displayed in Figure 9A,B (see also Figure S4A,B). Decrease in the analyte concentration was followed by the decrease in anodic current intensity at 0.1 V/s. Values of k_{sh} for all compounds listed in Table 3 were calculated using Reinmuths expression [50]. The values in the order of 10^{-4} cm/s supports the quasi-reversible nature of their oxidation process, in accordance with the reported criterion [50].

Table 3. D and k_{sh} values for quasi-reversible and irreversible systems for all studied ferrocenyl derivatives.

Compounds	Signal 1a		Signal 2a	
	D (cm^2/s)	k_{sh} (cm/s)	D (cm^2/s)	k_{sh} (cm/s)
FMA	9.03×10^{-7}	8.90×10^{-4}	8.92×10^{-6}	4.66×10^{-4}
CFA	9.18×10^{-7}	8.47×10^{-4}	6.79×10^{-6}	1.40×10^{-4}
FA	5.53×10^{-7}	5.45×10^{-4}	5.752×10^{-6}	1.30×10^{-4}
FBA	5.64×10^{-7}	6.74×10^{-4}	-	-

The standard deviation of concentration vs. peak current provided the values of the limit of detection (LOD) and limit of quantification (LOQ), using the relations LOD = 3s/m (slope = µA M^{-1}) and LOQ = 10 s/m where 's' is the standard deviation of the intercept and 'm' is the slope of the related calibration plot as listed in Table 4 [51] (Figure 9).

Table 4. The values of LOD and LOQ calculated for all compounds.

Compounds	s	R^2	m	LOD (µA) (1a)	LOD (µA) (2a)	LOQ (µA) (1a)	LOQ (µA) (2a)
FMA	0.432	0.97	7.10	0.183	0.112	0.61	0.028
CFA	0.269	0.97	5.72	0.141	0.113	0.54	0.094
FA	0.209	0.97	3.67	0.171	0.119	0.56	0.390
FBA	0.238	0.94	4.58	0.113	-	0.52	-

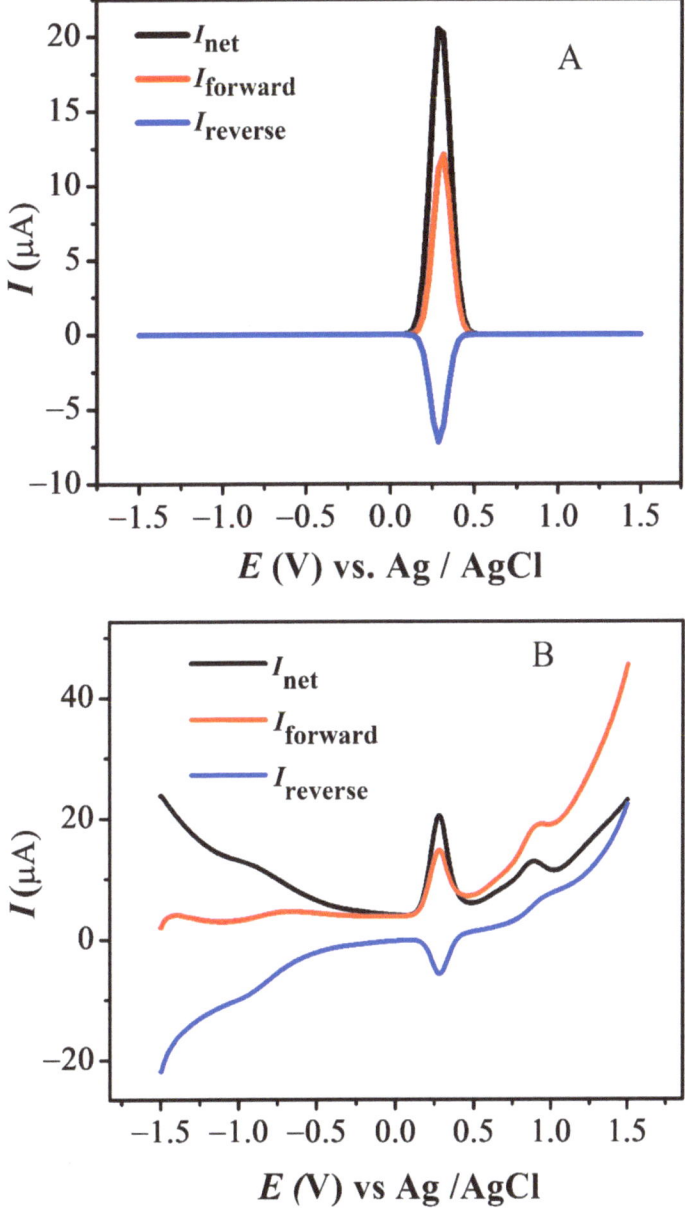

Figure 7. SWV indicating net, forward and reverse current for (**A**) FBA and (**B**) FA.

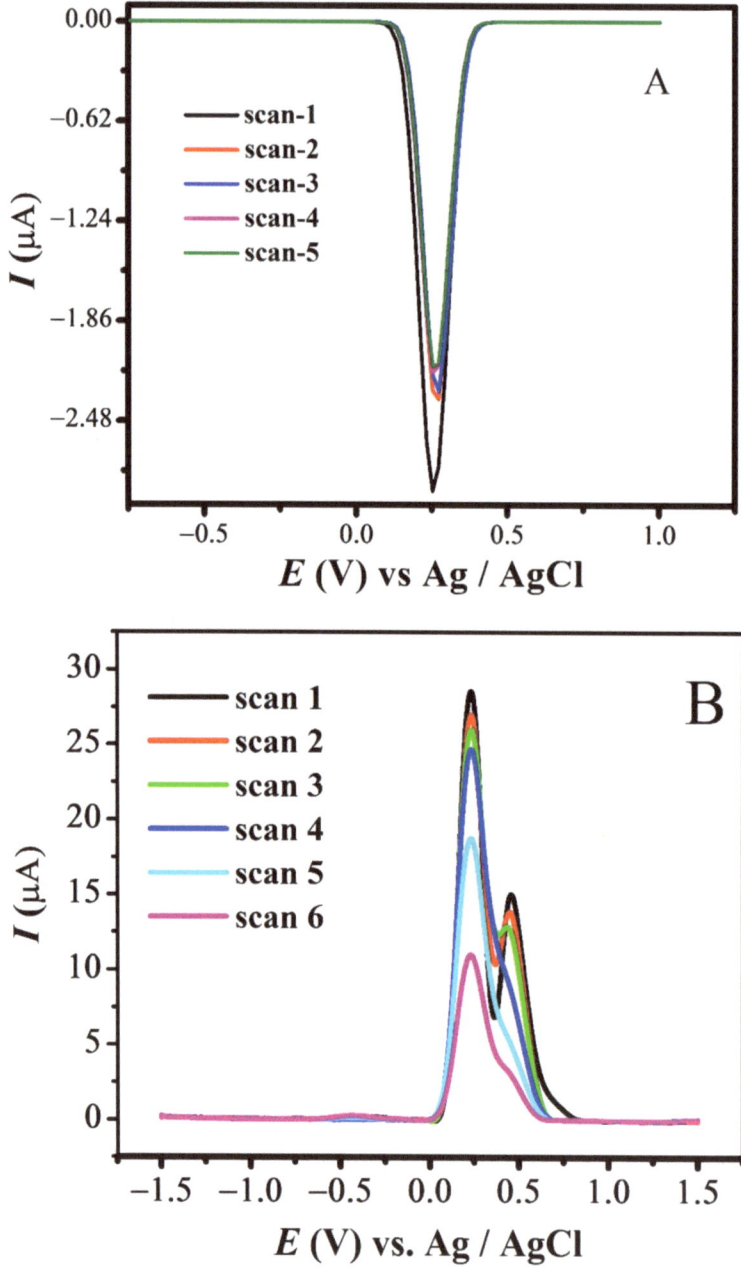

Figure 8. SWV Consecutive scans of (**A**) FBA and (**B**) FMA without cleaning electrode.

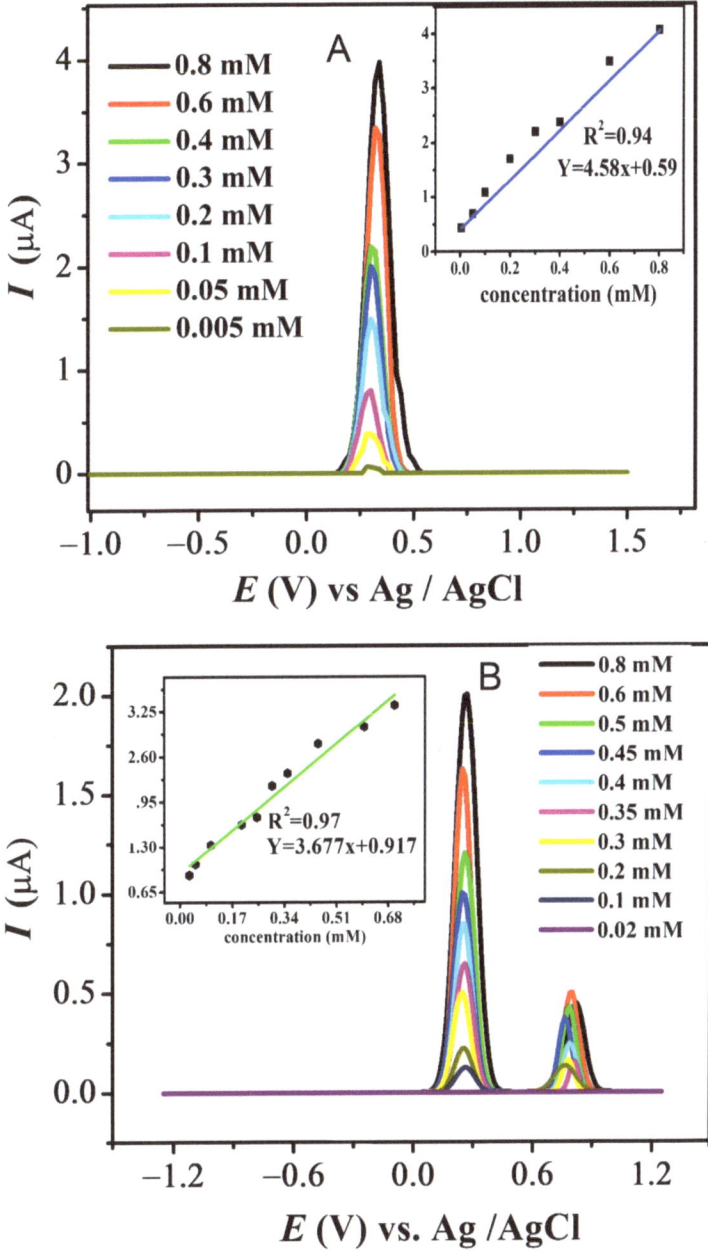

Figure 9. Concentration effect recorded in pH-7 at scan rate 0.1 V/s and inset plot of concentration (mM) vs. peak current (μA) for (**A**) FBA and (**B**) FA.

4. Proposed Redox Mechanism

Based on the information obtained from voltammetric measurements, we have proposed the redox mechanism of all the selected compounds on a carbon electrode surface. For FBA, peak 1a and 1c is attributed to redox behavior of the ferrocene nucleus, supported by various research publications [36]. E_p, $W_{1/2}$, ΔE_p and I_a/I_c current ratio, D and k_{sh} values revealed one electron, quasi-reversible, diffusion-controlled oxidation of ferrocene moiety as shown in Scheme 2.

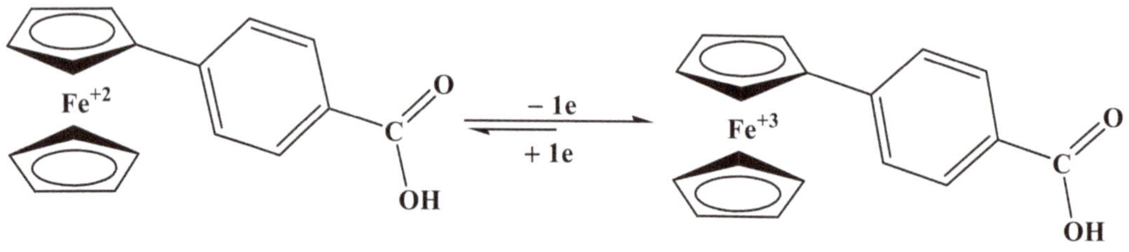

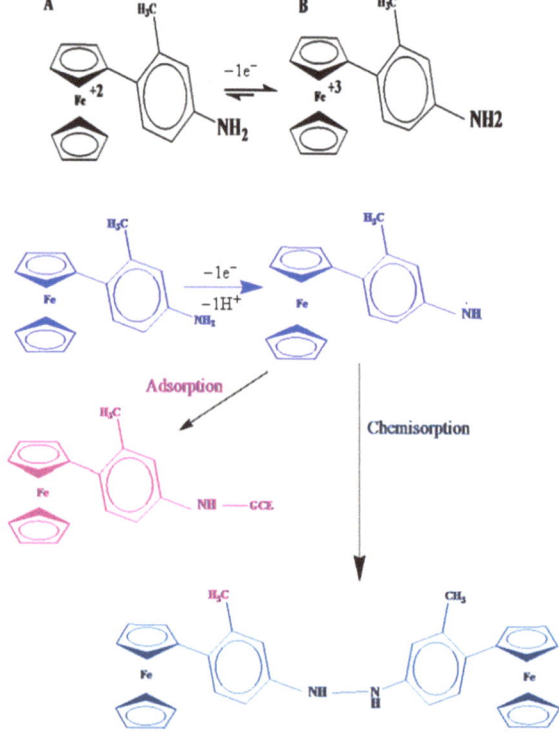

Scheme 2. Proposed redox mechanism of FBA at pH 3 to 10 for oxidation/reduction peak.

For the other three aniline containing ferrocenyl compounds, (CFA, FMA and FA), redox signals (1a and 1c) are representative of the Fe^{+2}/Fe^{+3} reversible couple. The second chemically irreversible oxidation peak (2a) is attributed to the oxidation of -NH_2 group, by the loss of 1e$^-$ and 1H$^+$ in FMA and loss of 1e$^-$ with no proton involvement in CFA and FA. The pk_a value of 11 for FMA (2a) shows protonation/deprotonation at this pH. Decrease in the original 2a signal (see Figure 5B) with scan numbers and the appearance of a new voltammetric signal corresponded to the electro polymerization of amine group, similar to what happens in the electrochemically induced electro polymerization of 4- N,N –Dimethyl amino phenyl group (DMAPP), attached to 1-ferrocenyl-prop-2-en-1-one. Scheme 3 shows the redox mechanism for one of these derivatives, FMA.

Scheme 3. Proposed redox mechanism of 4-Ferrocenyl,3-methyl aniline (FMA).

5. Conclusions

Detailed electrochemistry of four ferrocene derivatives was investigated using cyclic voltammetry, differential pulse voltammetry and square wave voltammetry. Cyclic voltammetry of ferrocenyl derivatives revealed that they displayed well-defined redox behavior of ferrocene moiety. A 2nd irreversible oxidation signal was also observed in addition to the conventional ferrocene redox couple, associated with attached -NH_2 groups. The effect of different scan rates was employed to determine the diffusion coefficient of the compounds. Differential pulse voltammetry established that redox signals were found to be sensitive to pH and was employed to determine the number of electrons and protons involved in the process. Physical parameters like k_{sh}, LOD and LOQ have been successfully determined through square wave voltammetry. Moreover, the reversibility of redox peaks was also evaluated through SWV, showing the quasi-reversible nature of redox peaks associated with ferrocene moiety in compounds. A redox mechanism was proposed based on the results of CV, DPV and SWV, demonstrating $1e^-$ oxidation of ferrocene moiety and $1e^-/1H^+$ oxidation of -NH_2 groups attached with FMA. The peak potentials of CFA and FA did not shift with pH, therefore, indicating the involvement of one electron for both oxidation peaks. Results obtained from all three voltammetry techniques complemented well with each other and proposed a mechanism provided insight into the action mechanism of these derivatives in biological systems.

Supplementary Materials: The following supporting information can be downloaded at: https://www.mdpi.com/article/10.3390/c8030045/s1, Figure S1. Effect of scan rate using cyclic voltammetry for (A) FMA (B) CFA (C) FA. Figure S2. DPV of (A) CFA and (B) FA at various pH values. Figure S3. SWV of (A) CFA and (B) FA. Figure S4. Concentration effect of (A) CFA and (B) FMA studied using SWV to determine LOQ and LOD.

Author Contributions: S.A. performed the experiments and wrote the manuscript. S.M. contributed to the interpretation of results and explanation of the redox mechanism. A.S. supervised the research project. All authors have read and agreed to the published version of the manuscript.

Funding: This research received no external funding.

Institutional Review Board Statement: Not applicable.

Informed Consent Statement: Not applicable.

Data Availability Statement: The data presented in this study are available within the article and supplementary materials.

Acknowledgments: Afzal Shah gratefully acknowledges the research facilities provided by Quaid-i-Azam University, Islamabad, Pakistan.

Conflicts of Interest: The authors declare no conflict of interest.

References

1. Koch, H.; Jørgensen, P.; Helgaker, T. The molecular structure of ferrocene. *J. Chem. Phys.* **1996**, *104*, 9528–9530. [CrossRef]
2. Fabbrizzi, L. The ferrocenium/ferrocene couple: A versatile redox switch. *ChemTexts* **2020**, *6*, 22. [CrossRef]
3. Bhatt, V. *Essentials of Coordination Chemistry: A Simplified Approach with 3D Visuals*; Academic Press: Cambridge, MA, USA, 2015.
4. Saenger, W. Cyclodextrin inclusion compounds in research and industry. *Angew. Chem. Int. Ed. Engl.* **1980**, *19*, 344–362. [CrossRef]
5. Bersier, P.M.; Bersier, J.; Klingert, B. Electrochemistry of cyclodextrins and cyclodextrin inclusion complexes. *Electroanalysis* **1991**, *3*, 443–455. [CrossRef]
6. Heinze, K.; Lang, H. *Ferrocene—Beauty and Function*; ACS Publications: Washington, WA, USA, 2013; Volume 32, pp. 5623–5625.
7. Amer, W.A.; Wang, L.; Amin, A.M.; Ma, L.; Yu, H. Recent progress in the synthesis and applications of some ferrocene derivatives and ferrocene-based polymers. *J. Inorg. Organomet. Polym. Mater.* **2010**, *20*, 605–615. [CrossRef]
8. Corra, S.; Curcio, M.; Baroncini, M.; Silvi, S.; Credi, A. Photoactivated artificial molecular machines that can perform tasks. *Adv. Mater.* **2020**, *32*, 1906064. [CrossRef]
9. Wang, C.-H.; Chen, K.-J.; Wu, T.-H.; Chang, H.-K.; Tsuchido, Y.; Sei, Y.; Chen, P.-L.; Horie, M. Ring rotation of ferrocene in interlocked molecules in single crystals. *Chem. Sci.* **2021**, *12*, 3871–3875. [CrossRef]
10. Kaifer, A.E.; Gómez-Kaifer, M. *Supramolecular Electrochemistry*; John Wiley & Sons: Hoboken, NJ, USA, 2008.

11. Shim, N.Y.; Bernards, D.A.; Macaya, D.J.; DeFranco, J.A.; Nikolou, M.; Owens, R.M.; Malliaras, G.G. All-plastic electrochemical transistor for glucose sensing using a ferrocene mediator. *Sensors* **2009**, *9*, 9896–9902. [CrossRef]
12. Cunningham, L.; Benson, A.; Guiry, P.J. Recent developments in the synthesis and applications of chiral ferrocene ligands and organocatalysts in asymmetric catalysis. *Org. Biomol. Chem.* **2020**, *18*, 9329–9370. [CrossRef]
13. Altun, A.; Apetrei, R.-M.; Camurlu, P. Reagentless amperometric glucose biosensors: Ferrocene-tethering and copolymerization. *J. Electrochem. Soc.* **2020**, *167*, 107507. [CrossRef]
14. Soon, G.H.; Deasy, M.; Dempsey, E. An Electrochemical Evaluation of Novel Ferrocene Derivatives for Glutamate and Liver Biomarker Biosensing. *Biosensors* **2021**, *11*, 254. [CrossRef]
15. Guven, N.; Apetrei, R.-M.; Camurlu, P. Next step in 2nd generation glucose biosensors: Ferrocene-loaded electrospun nanofibers. *Mater. Sci. Eng. C* **2021**, *128*, 112270. [CrossRef] [PubMed]
16. Beitollahi, H.; Khalilzadeh, M.A.; Tajik, S.; Safaei, M.; Zhang, K.; Jang, H.W.; Shokouhimehr, M. Recent advances in applications of voltammetric sensors modified with ferrocene and its derivatives. *ACS Omega* **2020**, *5*, 2049–2059. [CrossRef] [PubMed]
17. Zhang, M.; Ma, C.; Du, D.; Xiang, J.; Yao, S.; Hu, E.; Liu, S.; Tong, Y.; Wong, W.Y.; Zhao, Q. Donor–acceptor metallopolymers containing ferrocene for brain inspired memristive devices. *Adv. Electron. Mater.* **2020**, *6*, 2000841. [CrossRef]
18. Salas, M.; Gordillo, B.; González, F.J. Current measurements as a tool to characterise the H-bonding between 1-ferrocenylmethylthy-mine and 9-octyladenine: A voltammetric and chronoamperometric analysis. *J. Electroanal. Chem.* **2004**, *574*, 33–39. [CrossRef]
19. Camm, K.D.; Furtado, S.J.; Gott, A.L.; McGowan, P.C. Synthesis and structural studies of bis-amino-functionalised ferrocene salts and ferrocenium salts. *Polyhedron* **2004**, *23*, 2929–2936. [CrossRef]
20. Padeste, C.; Steiger, B.; Grubelnik, A.; Tiefenauer, L. Molecular assembly of redox-conductive ferrocene–streptavidin conjugates—towards bio-electrochemical devices. *Biosens. Bioelectron.* **2004**, *20*, 545–552. [CrossRef]
21. Zhang, F.-F.; Wan, Q.; Wang, X.-L.; Sun, Z.-D.; Zhu, Z.-Q.; Xian, Y.-Z.; Jin, L.-T.; Yamamoto, K. Amperometric sensor based on ferrocene-doped silica nanoparticles as an electron transfer mediator for the determination of glucose in rat brain coupled to in vivo microdialysis. *J. Electroanal. Chem.* **2004**, *571*, 133–138. [CrossRef]
22. Asaftei, S.; Walder, L. Covalent layer-by-layer type modification of electrodes using ferrocene derivatives and crosslinkers. *Electrochim. Acta* **2004**, *49*, 4679–4685. [CrossRef]
23. Pandey, P.; Upadhyay, S.; Upadhyay, A. Electrochemical sensors based on functionalized ormosil-modified electrodes—Role of ruthenium and palladium on the electrocatalysis of nadh and ascorbic acid. *Sens. Actuators B Chem.* **2004**, *102*, 126–131. [CrossRef]
24. Pandey, P.; Upadhyay, S.; Shukla, N.; Sharma, S. Studies on the electrochemical performance of glucose biosensor based on ferrocene encapsulated ORMOSIL and glucose oxidase modified graphite paste electrode. *Biosens. Bioelectron.* **2003**, *18*, 1257–1268. [CrossRef]
25. Pandey, P.; Upadhyay, S.; Tiwari, I.; Sharma, S. A Novel Ferrocene-Encapsulated Palladium-Linked Ormosil-Based Electrocatalytic Biosensor. The Role of the Reactive Functional Group. *Electroanal. Int. J. Devoted Fundam. Pract. Asp. Electroanal.* **2001**, *13*, 1519–1527. [CrossRef]
26. Gritzner, G.; Kuta, J. Recommendations on reporting electrode potentials in nonaqueous solvents (Recommendations 1983). *Pure Appl. Chem.* **1984**, *56*, 461–466. [CrossRef]
27. Abraham, K. Directions in secondary lithium battery research and development. *Electrochim. Acta* **1993**, *38*, 1233–1248. [CrossRef]
28. Nawaz, H.; Akhter, Z.; Yameen, S.; Siddiqi, H.M.; Mirza, B.; Rifat, A. Synthesis and biological evaluations of some Schiff-base esters of ferrocenyl aniline and simple aniline. *J. Organomet. Chem.* **2009**, *694*, 2198–2203. [CrossRef]
29. Kerman, K.; Mahmoud, K.A.; Kraatz, H.-B. An electrochemical approach for the detection of HIV-1 protease. *Chem. Commun.* **2007**, *37*, 3829–3831. [CrossRef]
30. Sarhan, A.A.; Ibrahim, M.S.; Kamal, M.M.; Mitobe, K.; Izumi, T. Synthesis, cyclic voltammetry, and UV–Vis studies of ferrocene-dithiafulvalenes as anticipated electron-donor materials. *Mon. Chem.-Chem. Mon.* **2009**, *140*, 315–323. [CrossRef]
31. Lal, B.; Kanwal, A.; Altaf, A.A.; Badshah, A.; Asghar, F.; Akhter, S.; Ullah, S.; Khan, S.I.; Tahir, M.N. Synthesis, crystal structure, spectral and electrochemical characterization, DNA binding and free radical scavenging studies of ferrocene-based thioureas. *J. Coord. Chem.* **2019**, *72*, 2376–2392. [CrossRef]
32. Asghar, F.; Munir, S.; Fatima, S.; Murtaza, B.; Patujo, J.; Badshah, A.; Butler, I.S.; Taj, M.B.; Tahir, M.N. Ferrocene-functionalized anilines as potent anticancer and antidiabetic agents: Synthesis, spectroscopic elucidation, and DFT calculations. *J. Mol. Struct.* **2022**, *1249*, 131632. [CrossRef]
33. Asghar, F.; Badshah, A.; Fatima, S.; Zubair, S.; Butler, I.S.; Tahir, M.N. Biologically active meta-substituted ferrocenyl nitro and amino complexes: Synthesis, structural elucidation, and DFT calculations. *J. Organomet. Chem.* **2017**, *843*, 48–61. [CrossRef]
34. Nawaz, S.; Asghar, F.; Patujo, J.; Fatima, S.; Murtaza, B.; Munir, S.; Naz, M.; Badshah, A.; Butler, I.S. New ferrocene-integrated multifunctional guanidine surfactants: Synthesis, spectroscopic elucidation, DNA interaction studies, and DFT calculations. *New J. Chem.* **2022**, *46*, 185–198. [CrossRef]
35. Muller, T.J.; Conradie, J.; Erasmus, E. A spectroscopic, electrochemical and DFT study of para-substituted ferrocene-containing chalcone derivatives: Structure of FcCOCHCH (p-tBuC6H4). *Polyhedron* **2012**, *33*, 257–266. [CrossRef]
36. Cardona, R.A.; Hernández, K.; Pedró, L.E.; Otaño, M.R.; Montes, I.; Guadalupe, A.R. Electrochemical and spectroscopical characterization of ferrocenyl chalcones. *J. Electrochem. Soc.* **2010**, *157*, F104. [CrossRef]
37. Terki, B.; Lanez, T. Anodic behaviour investigation of (ferrocenylmethyl) trimethylammonium cation. *Ann. Sci. Technol.* **2007**, *1*, 6.

38. Kennedy, K.G.; Miles, D.T. Electrochemistry of ferrocene-modified monolayer-protected gold nanoclusters at reduced temperatures. *J. Undergrad. Chem. Res.* **2004**, *4*, 145.
39. Neghmouche, N.; Lanez, T. Electrochemical properties of ferrocene in aqueous and organic mediums at glassy carbon electrode. *Recent Trends Phys. Chem. Int. J.* **2013**, *1*, 1–3.
40. Seiwert, B.; Karst, U. Ferrocene-based derivatization in analytical chemistry. *Anal. Bioanal. Chem.* **2008**, *390*, 181–200. [CrossRef]
41. Pournaghi-Azar, M.; Ojani, R. Catalytic oxidation of ascorbic acid by some ferrocene derivative mediators at the glassy carbon electrode. Application to the voltammetric resolution of ascorbic acid and dopamine in the same sample. *Talanta* **1995**, *42*, 1839–1848. [CrossRef]
42. Shah, A.; Nosheen, E.; Munir, S.; Badshah, A.; Qureshi, R.; Muhammad, N.; Hussain, H. Characterization and DNA binding studies of unexplored imidazolidines by electronic absorption spectroscopy and cyclic voltammetry. *J. Photochem. Photobiol. B Biol.* **2013**, *120*, 90–97. [CrossRef]
43. Nosheen, E.; Shah, A.; Badshah, A.; Hussain, H.; Qureshi, R.; Ali, S.; Siddiq, M.; Khan, A.M. Electrochemical oxidation of hydantoins at glassy carbon electrode. *Electrochim. Acta* **2012**, *80*, 108–117. [CrossRef]
44. Nicholson, R.S. Theory and application of cyclic voltammetry for measurement of electrode reaction kinetics. *Anal. Chem.* **1965**, *37*, 1351–1355. [CrossRef]
45. Shah, A.; Nosheen, E.; Qureshi, R.; Yasinzai, M.M.; Lunsford, S.K.; Dionysiou, D.D.; Siddiq, M.; Badshah, A.; Ali, S. Electrochemical characterization, detoxification and anticancer activity of didodecyldimethylammonium bromide. *Int. J. Org. Chem.* **2011**, *1*, 183. [CrossRef]
46. Munir, S.; Shah, A.; Zafar, F.; Badshah, A.; Wang, X.; Rehman, Z.-U.; Hussain, H.; Lunsford, S.K. Redox behavior of a derivative of vitamin K at a glassy carbon electrode. *J. Electrochem. Soc.* **2012**, *159*, G112. [CrossRef]
47. Munir, S.; Shah, A.; Rauf, A.; Badshah, A.; Hussain, H.; Ahmad, Z. Redox behavior of juglone in buffered aq.: Ethanol media. *C. R. Chim.* **2013**, *16*, 1140–1146. [CrossRef]
48. Munir, S.; Shah, A.; Rauf, A.; Badshah, A.; Lunsford, S.K.; Hussain, H.; Khan, G.S. Redox behavior of a novel menadiol derivative at glassy carbon electrode. *Electrochim. Acta* **2013**, *88*, 858–864. [CrossRef]
49. Golea, D.; Diculescu, V.; Enache, A.; Butu, A.; Tugulea, L.; Brett, A.O. Electrochemical evaluation of dsDNA—Liposomes interactions. *Dig. J. Nanomater. Biostruct.* **2012**, *7*, 1333–1342.
50. Shah, A.; Khan, A.M.; Qureshi, R.; Ansari, F.L.; Nazar, M.F.; Shah, S.S. Redox behavior of anticancer chalcone on a glassy carbon electrode and evaluation of its interaction parameters with DNA. *Int. J. Mol. Sci.* **2008**, *9*, 1424–1434. [CrossRef]
51. Yardim, Y.; Şentürk, Z. Voltammetric behavior of indole-3-acetic acid and kinetin at pencil-lead graphite electrode and their simultaneous determination in the presence of anionic surfactant. *Turk. J. Chem.* **2011**, *35*, 413–426. [CrossRef]

Article

Power Generation Characteristics of Polymer Electrolyte Fuel Cells Using Carbon Nanowalls as Catalyst Support Material

Takayuki Ohta [1,*], Hiroaki Iwata [1], Mineo Hiramatsu [1], Hiroki Kondo [2] and Masaru Hori [2]

[1] Department of Electrical and Electronic Engineering, Meijo University, 1-501 Shiogamaguchi, Tempaku, Nagoya 468-8502, Japan
[2] Center for Low-Temperature Plasma Sciences, Nagoya University, Furo, Chikusa, Nagoya 464-8603, Japan
* Correspondence: tohta@meijo-u.ac.jp

Abstract: We evaluated the power generation characteristics of a polymer electrolyte fuel cell (PEFC) composed of Pt-supported carbon nanowalls (CNWs) and a microporous layer (MPL) of carbon black on carbon paper (CP) as catalyst support materials. CNWs, standing vertically on highly crystallizing graphene sheets, were synthesized on an MPL/CP by plasma-enhanced chemical vapor deposition (PECVD) using inductively coupled plasma (ICP). Pt nanoparticles were supported on the CNW surface using the liquid-phase reduction method. The three types of voltage loss, namely those due to activated polarization, resistance polarization, and diffusion polarization, are discussed for the power generation characteristics of the PEFC using the Pt/CNWs/MPL/CP. The relationship between the height or gap area of the CNWs and the voltage loss of the PEFC is demonstrated, whereby the CNW height increased with the extension of growth time. The three-phase interface area increased with the increase in the CNW height, resulting in mitigation of the loss due to activated polarization. The gap area of the CNWs varied when changing the CH_4/H_2 gas ratio. The loss due to diffusion polarization was reduced by enlarging the gap area, due to the increased diffusion of fuel gas and discharge of water. The secondary growth of the CNWs caused the three-phase interface area to decrease as a result of platinum aggregation, impedance of the supply of ionomer dispersion solution to the bottom of the CNWs, and inhibition of fuel gas and water diffusion, which led to the loss of activated and diffuse polarizations. The voltage losses can be mitigated by increasing the height of CNWs while avoiding secondary growth.

Keywords: polymer electrolyte fuel cell; carbon nanowalls; plasma-enhanced chemical vapor deposition

Citation: Ohta, T.; Iwata, H.; Hiramatsu, M.; Kondo, H.; Hori, M. Power Generation Characteristics of Polymer Electrolyte Fuel Cells Using Carbon Nanowalls as Catalyst Support Material. *C* **2022**, *8*, 44. https://doi.org/10.3390/c8030044

Academic Editors: Olena Okhay and Gil Goncalves

Received: 7 July 2022
Accepted: 22 August 2022
Published: 27 August 2022

Publisher's Note: MDPI stays neutral with regard to jurisdictional claims in published maps and institutional affiliations.

Copyright: © 2022 by the authors. Licensee MDPI, Basel, Switzerland. This article is an open access article distributed under the terms and conditions of the Creative Commons Attribution (CC BY) license (https://creativecommons.org/licenses/by/4.0/).

1. Introduction

Polymer electrolyte fuel cells (PEFCs) have been applied as power sources in cogeneration systems and vehicles due to their compact size and low operating temperature. However, the improvement of their power generation efficiency and ensuring the high durability of electrode material are needed. The structure of the fuel cell is sandwich-like, where there is a catalyst layer (CL) and gas diffusion layer (GDL) on either side of the proton exchange membrane in addition to another catalyst layer. The CL is essential for accelerating the oxygen reduction reaction (ORR), which is slow. The morphological features (shape/size and porosity) of the catalyst determines its degree of dispersion and accessibility to protons and oxygen/hydrogen. The GDL is also an important component in the PEFC, and the main function of the GDL is to transfer water and gas. Accordingly, the interaction with the catalyst can affect the catalyst electronic energy, thereby affecting the kinetics. Carbon nanoparticle-based platinum catalysts have been widely used, which comprise platinum nanoparticles (2–5 nm) supported on carbon black nanoparticles (20–30 nm) covered with a nanothin film of ionomer. The modification of carbon support materials for Pt and Pt-alloy cathode catalysts has been reported to successfully improve the performance and durability of PEFCs. The criteria for use as a support material are as follows: (1) sufficient electrical conductivity; (2) large surface area; (3) high resistance to electrochemical corrosion; (4) suitable porosity and a porous structure; (5) strong stability in acidic or alkaline media; (6) adequate proton conductivity; (7) sufficient compatibility with electrodes; (8) adequate water handling to avoid flooding; and (9) strong interaction between the support and the catalyst. The following have been reported as new carbon-based materials: carbon black [1–12], mesoporous carbon [13–19], carbon nanotubes (CNTs) [10,12,13,20–28], hollow graphite spheres [29,30], carbon cloth [31], graphene [32–38], carbon nanofibers (CNFs) [9,13,21,39–43], carbon aerogels [13,44], carbon xerogel [45–49], carbon nanocoils (CNCs) [5,50–55], fullerene [56–58], carbon nano-onions [59,60], carbon nanohorn [61,62], and polymer-based nanohybrids [63]. In addition, the following noncarbon-supported materials have been developed: Pt/Ta-SnO$_2$ [64], Pt–Co/C [65], Fe–N–C catalyst [66], and silica-coated Pt [67].

In recent years, carbon nanowalls (CNWs) were discovered, and fundamental research on them has been progressing [68–71]. CNWs have been grown using various chemical vapor deposition (CVD) methods, and several applications using CNWs have been reported, such as fuel cells [72–77], biofuel cells [78], strain sensors [79], electrochemical sensors [80–82], scaffolds [83], and substrates for surface-assisted laser desorption/ionization mass spectrometry [84,85]. CNWs are composed of stacked graphene sheets and are stranded perpendicular to the substrate. CNWs are expected to be alternative catalyst support materials due to their high electric conductivity, very large specific surface area, and physically robust shape due to having a high aspect ratio. The electrochemical characterization or high durability of CNWs for a fuel cell has been reported [77–81]. A test cell unit using a Pt-supported CNW/carbon fiber paper was fabricated and the V–I curve characterized for assessment of proton exchange membrane (PEM) fuel cell application. However, the PEM fuel cell exhibits a voltage drop to 0.2 V due to activated polarization, because no ionomer binder is incorporated in the catalyst layers. The three-phase boundary region near the surface of the catalyst layers and the proton conduction between the catalyst layer and membrane are both essential for improving the power generation efficiency. Since CNWs have a unique structure, the gap area and height of CNWs can be controlled by changing the deposition conditions.

In this study, CNWs were synthesized on a microporous layer (MPL) of carbon black on carbon paper (CP) by plasma-enhanced chemical vapor deposition (PECVD) using inductively coupled plasma (ICP) to improve the power generation of the PEFC. The relationship between the CNW height or gap area and the PEFC power generation characteristics was investigated.

2. Materials and Methods
2.1. Plasma-Enhanced Chemical Vapor Deposition for CNW Growth

Figure 1 shows a schematic diagram of the ICP reactor used for CNW growth. RF (13.56 MHz) power was applied to the coil antenna. Two types of CNWs with different gap spaces between walls were synthesized on an MPL of carbon black on the CP (MPL/CP) (SGL carbon GmbH; GDL 29BC, thickness: 235 µm, void fraction: 40–41%). Condition (1) for the small gap was 550 W RF power, 20/44 sccm Ar/CH$_4$ gas flow rates, 22 mTorr total pressure, and 720 °C growth temperature. Condition (2) for the large gap was 600 W RF power, 20/9/8 sccm Ar/CH$_4$/H$_2$ gas flow rates, 16 mTorr total pressure, and 720 °C growth temperature.

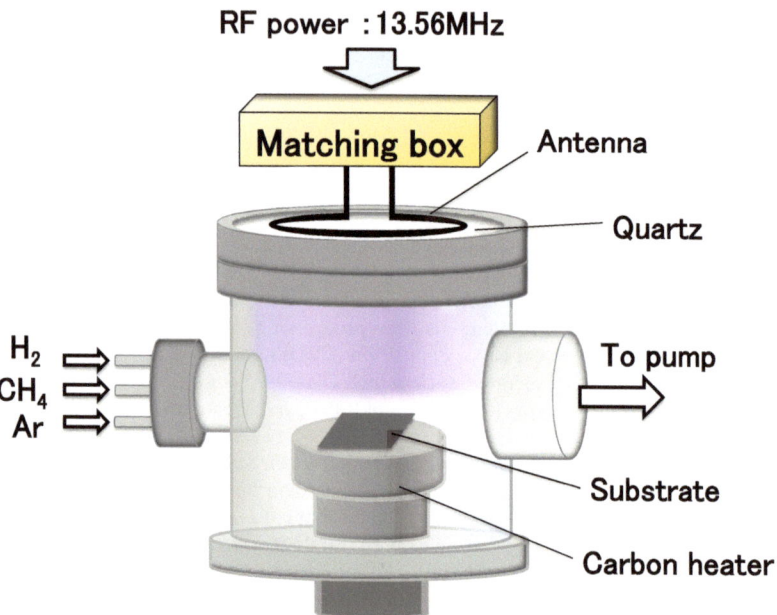

Figure 1. Schematic diagram of the ICP reactor.

2.2. Structure of a Single PEFC Using CNWs

Figure 2 shows a schematic diagram of the single PEFC, which was assembled with a membrane electrode assembly (MEA) and separators. The MEA consisted of the Pt-supported CNWs on MPL/CP, ionomer dispersion solution, and non-treated ionomer membrane (Nafion 117) for proton exchange. The surface of the CNWs on the MPL/CP was treated with atmospheric pressure plasma to endow hydrophilicity. The Ar gas flow rate was 2.0 sccm, the applied voltage to the electrodes was 6 kV, and the treatment distance was 5 mm. Pt nanoparticles were supported on the surface of the CNWs/MPL/CP by the liquid-phase reduction method using chloroplatinic acid solution (H_2PtCl_6, 8 wt.% in H_2O) and $NaBH_4$ as the reducing agents. A total volume of 0.1 mL ionomer dispersion solution (Nafion DE202) diluted with 2-propanol was applied dropwise on the Pt/CNWs/MPL/CP. The MEA was fabricated by hot-pressing at 135 °C with a pressing pressure of 5.0 MPa and pressing time of 90 s. The MEA with an electrode area of 5 cm^2 was assembled to the standard single PEFC (Electro Chem, Inc., EFC-05-02, Woburn, MA, USA). The power generation characteristics were investigated using a fuel cell test system (Scribner Associates Inc., AutoPEM, Southern Pines, NC, USA) at an anode H_2 gas flow rate of 0.1 L/m, cathode air flow rate of 0.2 L/m, and a temperature of 80 °C.

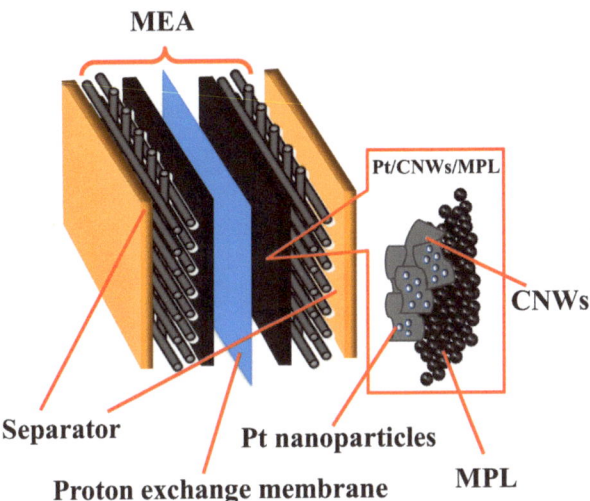

Figure 2. Schematic diagram of the single PEFC using CNWs.

3. Results and Discussion

3.1. CNWs on MPL/CP

Figure 3 shows the height of the CNWs as a function of the growth time for condition (1). The height of the CNWs increased from 1.7 µm to 7.2 µm as the growth time increased. The growth rate was estimated as 28 nm/min. Figure 4 shows the SEM images of the CNWs for various growth times. The CNWs were directly formed on the MPL of carbon black as shown in Figure 4a. The growth mechanism of CNWs on the substrates was detailed in a previous report [80]. The growth was as follows: (1) A very thin amorphous carbon layer formed on the substrate. (2) A nucleation site was formed there by ion irradiation. (3) Nanoislands were formed by aggregating carbon radicals at the nucleation site. (4) The nanoislands were irradiated with ions, and the absorption of carbon radicals was enhanced. (5) Nanographene grew preferentially in the height direction to form carbon nanowalls. In this study, the flake-like CNWs grew on the MPL, and some branches grew from the flake-like CNWs after 3 h.

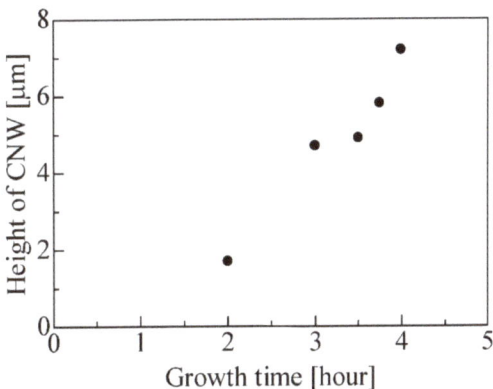

Figure 3. Height of CNW as a function of growth time.

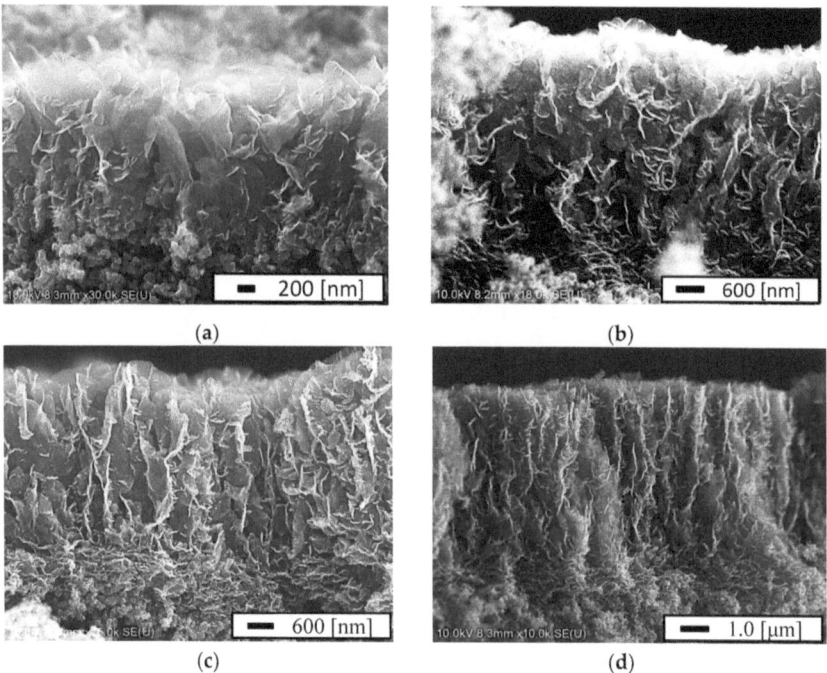

Figure 4. Cross-sectional view of the SEM images of the CNWs at a growth time of (**a**) 2 h, (**b**) 3 h, (**c**) 3.5 h, and (**d**) 4 h.

3.2. Relationship between CNW Height and Power Generation Characteristics

Figure 5 shows the voltage–current density (V–J) curve of the single PEFC for the various growth times for deposition under condition (1). Three gradients, due to activated polarization, resistance polarization, and diffusion polarization, were observed on the PEFC using Pt/CNW/MPL/CP. The maximum current density increased with the increase in the growth time up to 3.5 h, since the reaction area at the three-phase boundary region among Pt on CNW, ionomer dispersion solution, and supplied fuel gas increased with the increase in the CNW height. The maximum current density, however, decreased after more than 3.5 h, because the dispersion of the Pt and ionomer dispersion solution to the bottom of CNWs was suppressed due to the secondary growth (branches) from the sidewalls of the flake-like CNWs.

The magnitude of the three types of voltage loss, activated polarization, resistance polarization, and diffusion polarization, was evaluated with consideration to the power generation characteristics and how improvements could be achieved by reducing this loss [86]. Figure 6 shows the voltage loss due to activated polarization for the various growth times. Activated polarization is the potential difference beyond the value of equilibrium needed to generate currents and is dependent on the energy activation of a redox reaction and its reaction area, and it can be calculated using the following equations:

$$\eta_a = a + b \log i \tag{1}$$

$$a = -\frac{2.303RT}{\alpha nF} \log i_0 \tag{2}$$

$$b = \frac{2.303RT}{\alpha nF} \log i \tag{3}$$

$$i_0 = nFAck_0 \exp\left(\frac{-E_a}{RT}\right) \tag{4}$$

where i is the total current (A), a is the coefficient of migration, R is the gas constant (8.314 J/K mol), T is the temperature (K), i_0 is the exchange current (A), n is the number of electrons involved in the reaction (mol), F is the Faraday constant (96,485 C/mol), E_a is the activation energy (J/mol), k_0 is the reaction rate constant (s^{-1}), and A is the reaction area (m^2).

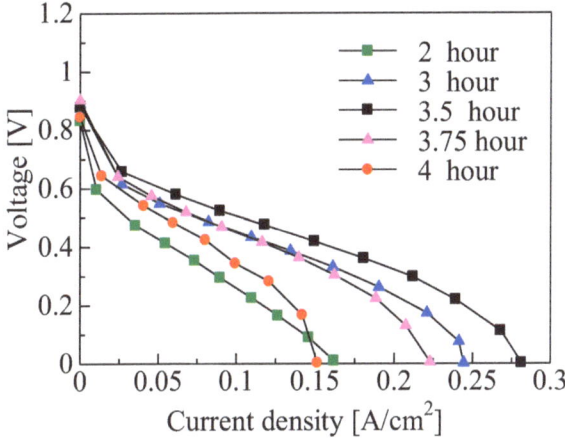

Figure 5. Voltage–current density (V–J) curve of single PEFC cells of Pt/CNW/MPL/CP for various growth times of the CNWs.

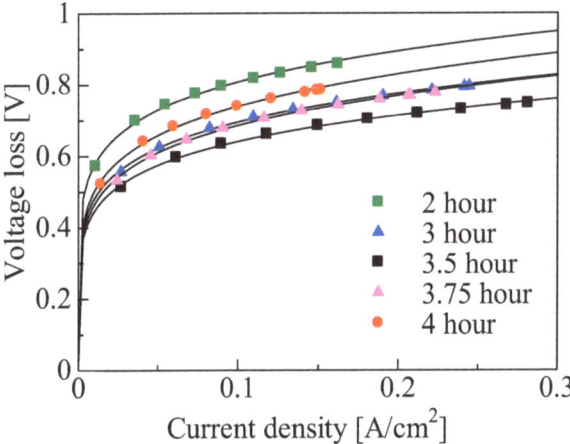

Figure 6. Activated polarization for various growth times.

The loss due to activated polarization decreased as the growth time was increased from 2 to 3.5 h, followed by an increase from 3.5 to 4 h. The activation energy remained constant in this study because the catalytic activity did not change for any of the experimental conditions. As shown in Figure 3, the three-phase interface region increased with the increase in CNW height observed from 2 to 3.5 h, resulting in an increase in reaction area A. However, the Pt particles aggregated due to CNW secondary growth from 3.5 to 4 h. Accordingly, the decrease in the three-phase interface led to the loss of activated polarization.

Figure 7 shows the voltage loss due to resistance polarization for the various growth times. Resistance polarization is a part of electrode polarization arising from an electric current through an ohmic resistance within the electrode or the electrolyte. The resistance polarization can be calculated using the following equation:

$$\eta_r = jRa \qquad (5)$$

where j is the current density (A/cm^2) and R is the internal resistance of the fuel cell (Ωcm^2). The ionic conductivity of the electrolyte, the electrical conductivity of the CNWs, and the contact resistance did not change regardless of the CNW growth time.

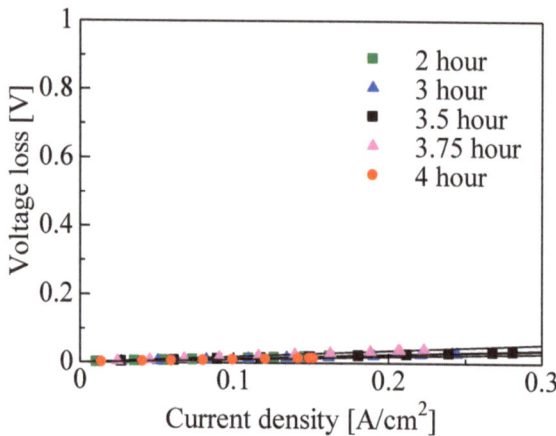

Figure 7. Resistance polarization for various growth times.

Figure 8 shows the voltage loss due to diffusion polarization for the various growth times. The diffusion polarization of an electrode is a result of the formation of a diffusion layer for fuel gas at the three-phase boundary or the water as a byproduct of an oxygen reduction reaction. Diffusion polarization can be calculated using the following equation:

$$\eta_d = -b \log \frac{C_{OX}}{C°_{OX}} \qquad (6)$$

where C_{OX} is the concentration of oxygen at the three-phase boundary and $C°_{OX}$ is the concentration of bulk oxygen. The loss due to diffusion polarization decreased with the increase in growth time from 2 to 3.5 h, since the oxygen concentration (C_{OX}) at the three-phase boundary increased along with the CNW height. However, from 3.5 to 4 h, the loss due to diffusion polarization decreased. The supply of fuel gas to the three-phase interface or the discharge of water molecules as the byproduct of the oxygen reduction reaction would be suppressed by CNW secondary growth. It was found that the best power generation characteristics were obtained at the growth time of 3.5 h, as shown in Figure 5, due to an increase in the three-phase interface area and the improvements such as the increased diffusion of fuel gas and water molecules.

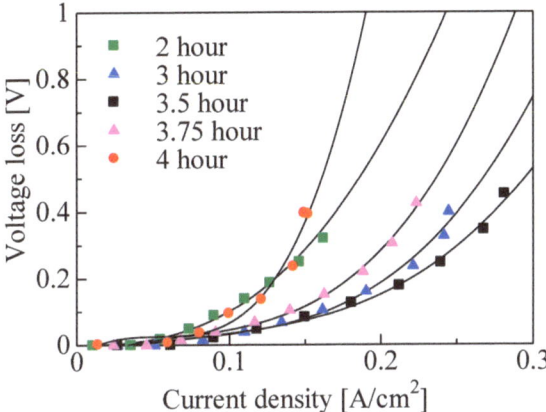

Figure 8. Diffusion polarization for various growth times.

3.3. Relationship between CNW Gap Area and Power Generation Characteristics

Figure 9 shows plane or cross-sectional SEM images of the CNWs for different gap areas. The CNWs with a small gap area (0.37 μm^2) were deposited under condition (1), and those with a large gap area (0.79 μm^2) were obtained under condition (2). The growth time to achieve a height of 1.6 μm was 2 h for condition (1) and 70 min for condition (2) since the influence of the secondary growth can be suppressed under relatively low height. Here, the gap area of CNWs was evaluated from the area enclosed by the walls as their reciprocal [87].

Figure 9. (**a**,**c**) Plane view and (**b**,**d**) cross-sectional view of the SEM images of CNWs grown on MPL with (**a**,**b**) a small gap area (0.37 μm^2) and (**c**,**d**) a large gap area (0.79 μm^2).

Figure 10 shows the V–J curve of a single PEFC for the different gap areas. Three gradients, due to activated polarization, resistance polarization, and diffusion polarization, were observed on the large gap area. The voltage of approximately 0.91 V at a current density of 0 mA/cm^2 was obtained for a large gap area under relatively low height. Moreover, the maximum current density for a large gap area was larger than that with a small gap area and was increased from approximately 0.15 to 0.3 A/cm^2 at the same height. Figure 11 shows the voltage loss due to activated polarization for the different gap areas. The voltage loss due to the activated polarization for the large gap was smaller than that for the small gap. The difference was approximately 0.2 V regardless of the current density. These results indicate that the three-phase interface region increased with the increase in gap area since the supply of platinum or ionomer dispersion solution to the bottom of the CNWs may have been enhanced. Figure 12 shows the voltage loss due to diffusion polarization for the different gap areas. The voltage loss due to diffusion polarization for the large gap was improved to be approximately 4 times smaller than that for the small gap. This result indicates that the supply of fuel gas to the three-phase interface and the discharge of water molecules were drastically improved by enlarging the gap of CNWs.

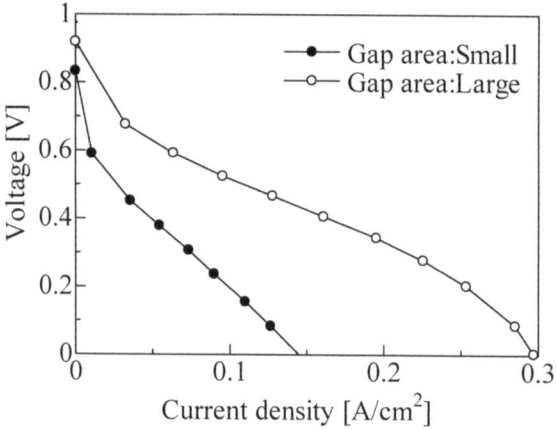

Figure 10. Voltage–current density (V–J) curve of single PEFC cells of Pt/CNWs/MPL/CP for different gap areas.

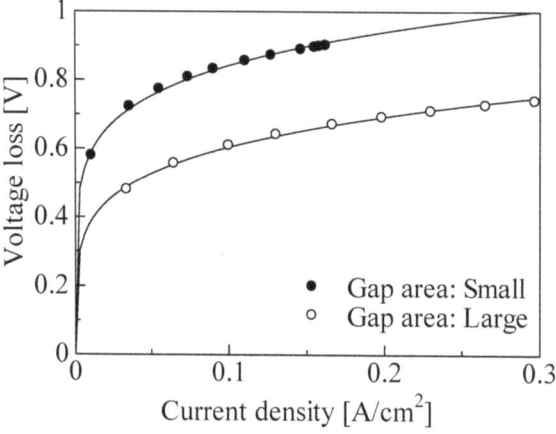

Figure 11. Activated polarization for different gap areas.

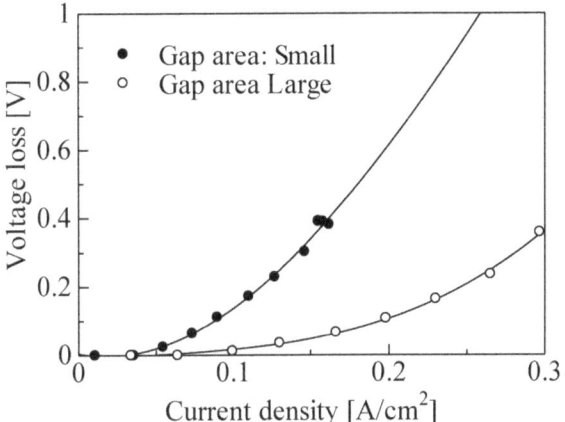

Figure 12. Diffusion polarization for different gap areas.

The voltage loss due to activated polarization and diffusion polarization was improved by an increase in gap area, although the surface area of the CNWs was decreased. The improvement in the diffusion polarization indicates that the supply of fuel gas to the three-phase interface and the discharge of water molecules were enhanced. Accordingly, the loss due to the activated polarization was improved, since the effective area of the three-phase interface was increased by the supply of fuel gas. Moreover, the Pt particle and the ionomer dispersion solution would be supplied to the bottom of CNWs in the case of the large gap. It was found that the power generation characteristics were improved by enlarging the gap of CNWs, since the loss due to activated polarization and diffusion polarization was mitigated.

4. Conclusions

CNWs as catalyst support materials were synthesized on an MPL/CP by ICP-PECVD. The power generation of the single PEFC using Pt/CNWs/MPL/CP as an MEA was demonstrated. The power generation characteristics of the PEFC were evaluated based on the three types of voltage loss due to activated polarization, resistance polarization, and diffusion polarization.

- An increase in the height of the CNWs increased the three-phase interface area with a reduction in the loss due to activated polarization.
- An increase in the gap area of the CNWs resulted in improvements due to increases in fuel gas diffusion and water discharge, and the loss due to diffusion polarization was reduced.
- The secondary growth of the CNWs caused a reduction in the three-phase interface area due to platinum aggregation, impedance of the supply of ionomer dispersion solution to the bottom of CNWs, and inhibited fuel gas and water diffusion, which led to the loss of activated and diffuse polarizations.

The voltage losses can be mitigated by increasing the height of the CNWs and expanding the gap area. The data obtained in this study are useful for the fabrication of PEFCs using CMWs.

Author Contributions: Conceptualization, investigation and writing—original draft preparation, T.O., data curation, H.I.; supervision, resources and project administration, H.K., M.H. (Masaru Hori) and M.H. (Mineo Hiramatsu). All authors have read and agreed to the published version of the manuscript.

Funding: This work was partly supported by the project for promoting research in Meijo University.

Institutional Review Board Statement: Not applicable.

Informed Consent Statement: Not applicable.

Data Availability Statement: Not applicable.

Conflicts of Interest: The authors declare no conflict of interest.

References

1. Clingerman, M.L.; Weber, E.H.; King, J.A.; Schulz, K.H. Synergistic effect of carbon fillers in electrically conductive nylon 6,6 and polycarbonate based resins. *Polym. Compos.* **2002**, *23*, 911–924. [CrossRef]
2. Rimbu, G.A.; Jackson, C.L.; Scott, K. Platinum/carbon/polyaniline based nanocomposites as catalysts for fuel cell technology. *J. Optoelectron. Adv. Mater.* **2006**, *8*, 611–616.
3. Tang., S.; Sun, G.; Qi, J.; Sun, S.; Guo, J.; Xin, Q. Review of new carbon materials as catalyst supports in direct alcohol fuel cells. *Chin. J. Catal.* **2010**, *3*, 12–17. [CrossRef]
4. Lavacchi, A.; Miller, H.; Vizza., F. *Nanostructure Science and Technology*; Springer: New York, NY, USA, 2013.
5. Celorrio, V.; Flórez-Montaño, J.; Moliner, R.; Pastor, E.; Lázaro, M.J. Fuel cell performance of Pt electrocatalysts supported on carbon nanocoils. *Int. J. Hydrogen Energy.* **2014**, *39*, 5371–5377. [CrossRef]
6. Kaluža, L.; Larsen, M.J.; Zdražil, M.; Gulková, D.; Vít, Z.; Šolcová, O.; Soukup, K.; Koštejn, M.; Bonde, J.L.; Maixnerová, L.; et al. Highly loaded carbon black supported Pt catalysts for fuel cells. *Catal. Today* **2015**, *256*, 375–383. [CrossRef]
7. Fujii, K.; Ito, M.; Sato, Y.; Takenaka, S.; Kishida, M. Performance and durability of carbon black-supported Pd catalyst covered with silica layers in membrane electrode assemblies of proton exchange membrane fuel cells. *J. Power Sources* **2015**, *279*, 100–106. [CrossRef]
8. Lee, W.H.; Seo, J.; Lee, T.; Kim, H. Preparation of a self-assembled organosilane coating on carbon black as a catalyst support in polymer electrolyte membrane fuel cells. *J. Power Sources* **2015**, *274*, 1140–1146. [CrossRef]
9. Li, M.; Wu, X.; Zeng, J.; Hou, Z.; Liao, S. Heteroatom doped carbon nanofibers synthesized by chemical vapor deposition as platinum electrocatalyst supports for polymer electrolyte membrane fuel cells. *Electrochim. Acta* **2015**, *182*, 351–360. [CrossRef]
10. Geraldes, A.N.; Furtunato da Silva, D.; Martins da Silva, J.C.; Antonio de Sá, O.; Spinacé, E.V.; Neto, A.O.; Coelho dos Santos, M. Palladium and palladium–Tin supported on multi wall carbon nanotubes or carbon for alkaline direct ethanol fuel cell. *J. Power Sources.* **2015**, *275*, 189–199. [CrossRef]
11. Zhou, K.; Li, T.; Han, Y.; Wang, J.; Chen, J.; Wang, K. Optimizing the hydrophobicity of GDL to improve the fuel cell performance. *RSC Adv.* **2021**, *11*, 2010. [CrossRef]
12. Quan, D.L.; Le, P.H. Enhanced methanol oxidation activity of PtRu/C$_{100-x}$MWCNTs$_x$ (x = 0–100 wt.%) by controlling the composition of C-MWCNTs support. *Coatings* **2021**, *11*, 571. [CrossRef]
13. Antolini, E. Carbon supports for low-temperature fuel cell catalysts. *Appl. Catal. B Environ.* **2009**, *88*, 1–24. [CrossRef]
14. Liu, H.-J.; Cui, W.-J.; Jin, L.-H.; Wang, C.-X.; Xia, Y.-Y. Preparation of three-dimensional ordered mesoporous carbon sphere arrays by a two-step templating route and their application for supercapacitors. *J. Mater. Chem.* **2009**, *19*, 3661–3667. [CrossRef]
15. Tang, Z.-H.; He, X.; Song, Y.; Liu, L.; Guo, Q.-G.; Yang, J.-H. Properties of mesoporous carbons prepared from different carbon precursors using nanosize silica as a template. *N. Carbon Mater.* **2010**, *25*, 465–469. [CrossRef]
16. Li, Y.; Fu, Z.-Y.; Su, B.-L. Hierarchically structured porous materials for energy conversion and storage. *Adv. Funct. Mater.* **2012**, *22*, 4634–4667. [CrossRef]
17. Kuppan, B.; Selvam, P. Platinum-supported mesoporous carbon (Pt/CMK-3) as anodic catalyst for direct methanol fuel cell applications: The effect of preparation and deposition methods. *Prog. Nat. Sci. Mater. Int.* **2012**, *22*, 616–623. [CrossRef]
18. Bruno, M.M.; Viva, F.A.; Petruccelli, M.A.; Corti, H.R. Platinum supported on mesoporous carbon as cathode catalyst for direct methanol fuel cells. *J. Power Sources* **2015**, *278*, 458–463. [CrossRef]
19. Hung, C.-T.; Liou, Z.-H.; Veerakumar, P.; Wu, P.-H.; Liu, T.-C.; Liu, S.-B. Ordered mesoporous carbon supported bifunctional PtM (M = Ru, Fe, Mo) electrocatalysts for a fuel cell anode. *Chin. J. Catal.* **2016**, *37*, 43–53. [CrossRef]
20. Knupp, S.L.; Li, W.; Paschos, O.; Murray, T.M.; Snyder, J.; Haldar, P. The effect of experimental parameters on the synthesis of carbon nanotube/nanofiber supported platinum by polyol processing techniques. *Carbon* **2008**, *46*, 1276–1284. [CrossRef]
21. Monthioux, M. *Carbon Meta-Nanotubes: Synthesis, Properties and Applications*; Wiley: Hoboken, NJ, USA, 2011.
22. Rahsepar, M.; Pakshir, M.; Nikolaev, P.; Piao, Y.; Kim, H. A combined physicochemical and electrocatalytic study of microwave synthesized tungsten mono-carbide nanoparticles on multiwalled carbon nanotubes as a co-catalyst for a proton-exchange membrane fuel cell. *Int. J. Hydrogen Energy* **2014**, *39*, 15706–15717. [CrossRef]
23. Chiang, Y.-C.; Hsieh, M.-K.; Hsu, H.-H. The effect of carbon supports on the performance of platinum/carbon nanotubes for proton exchange membrane fuel cells. *Thin Solid Film.* **2014**, *570*, 221–229. [CrossRef]
24. Weng, B.; Xu, F.; Wu, Z.; Li, Z. Hydrogen generation from LiBH4 solution catalyzed by multiwalled carbon nanotubes supported Co-B nanocatalysts for a portable micro proton exchange membrane fuel cell application. *Int. J. Hydrogen Energy.* **2014**, *39*, 14942–14948. [CrossRef]
25. Jin, H.; Zhu, L.; Bing, N.; Wang, L.; Wang, L. No cytotoxic nitrogen-doped carbon nanotubes as efficient metal-free electrocatalyst for oxygen reduction in fuel cells. *Solid State Sci.* **2014**, *30*, 21–25. [CrossRef]

26. Kitahara, T.; Nakajima, H.; Okamura, K. Gas diffusion layers coated with a microporous layer containing hydrophilic carbon nanotubes for performance enhancement of polymer electrolyte fuel cells under both low and high humidity conditions. *J. Power Sources* 2015, *283*, 115–124. [CrossRef]
27. Ortiz-Herrera, J.C.; Cruz-Martínez, H.; Solorza-Feria, O.; Medina, D.I. Recent progress in carbon nanotubes support materials for Pt-based cathode catalysts in PEM fuel cells. *Int. J. Hydrogen Energy* 2022. (In Press) [CrossRef]
28. Nasrabadi, M.K.; Ebrahimi-Moghadam, A.; Kumar, R.; Nabipour, N. Electrochemical performance improvement of the catalyst of the methanol microfuel cell using carbon nanotubes. *Int. J. Chem. Eng.* 2021, *2021*, 8894768. [CrossRef]
29. Kelly, B.T. *Physics of Graphite*; Applied Science Publishers: London, UK, 1981; p. 291.
30. Galeano, C.; Meier, J.C.; Peinecke, V.; Bongard, H.; Katsounaros, I.; Topalov, A.A.; Lu, A.; Mayrhofer, K.J.; Schüth, F. Toward highly stable electrocatalysts via nanoparticle pore confinement. *J. Am. Chem. Soc.* 2012, *134*, 20457–20465. [CrossRef]
31. Vivekananthan, J.; Masa, J.; Chen, P.; Xie, K.; Muhler, M.; Schuhmann, W. Nitrogen-doped carbon cloth as a stable self-supported cathode catalyst for air/H2-breathing alkaline fuel cells. *Electrochim. Acta* 2015, *182*, 312–319. [CrossRef]
32. Jha, N.; Jafri, R.I.; Rajalakshmi, N.; Ramaprabhu, S. Graphene-multi walled carbon nanotube hybrid electrocatalyst support material for direct methanol fuel cell. *Int. J. Hydrogen Energy* 2011, *36*, 7284–7290. [CrossRef]
33. Lee, J.; Kim, K.; Park, W.I.; Kim, B.-H.; Park, J.H.; Kim, T.-H. Uniform graphene quantum dots patterned from self-assembled silica nanodots. *Nano Lett.* 2012, *12*, 6078–6083. [CrossRef]
34. Zhang, X.; Yuan, W.; Duan, J.; Zhang, Y.; Liu, X. Graphene nanosheets modified by nitrogen-doped carbon layer to support Pt nanoparticles for direct methanol fuel cell. *Microelectron. Eng.* 2015, *141*, 234–237. [CrossRef]
35. Li, L.; Hu, L.; Li, J.; Wei, Z. Enhanced stability of Pt nanoparticle electrocatalysts for fuel cells. *Nano Res.* 2015, *8*, 418–440. [CrossRef]
36. Yang, H.N.; Lee, D.C.; Park, K.W.; Kim, W.J. Platinum—Boron doped grapheme intercalated by carbon black for cathode catalyst in proton exchange membrane fuel cell. *Energy* 2015, *89*, 500–510. [CrossRef]
37. Song, C.; Gui, Y.; Xing, X.; Zhang, W. Well-dispersed chromium oxide decorated reduced graphene oxide hybrids and application in energy storage. *Mater. Chem. Phys.* 2016, *173*, 460–466. [CrossRef]
38. Lazar, O.-A.; Marinoiu, A.; Raceanu, M.; Pantazi, A.; Mihai, G.; Varlam, M.; Enachescu, M. Reduced graphene oxide decorated with dispersed gold nanoparticles: Preparation, characterization and electrochemical evaluation for oxygen reduction reaction. *Energies* 2020, *13*, 4307. [CrossRef]
39. Yamada, T.; Yabutani, H.; Saito, T.; Yang, C.Y. Temperature dependence of carbon nanofiber resistance. *Nanotechnology* 2010, *21*, 265707. [CrossRef] [PubMed]
40. Oh, Y.; Kim, S.-K.; Peck, D.-H.; Jang, J.-S.; Kim, J.; Jung, D.-H. Improved performance using tungsten carbide/carbon nanofiber based anode catalysts for alkaline direct ethanol fuel cells. *Int. J. Hydrogen Energy* 2014, *39*, 15907–15912. [CrossRef]
41. Thamer, B.M.; El-Newehy, M.H.; Barakat, N.A.M.; Abdelkareem, M.A.; Al-Deyab, S.S.; Kim, H.Y. Influence of nitrogen doping on the catalytic activity of niincorporated carbon nanofibers for alkaline direct methanol fuel cells. *Electrochim. Acta* 2014, *142*, 228–239. [CrossRef]
42. Giorgi, L.; Salernitano, E.; Dikonimos Makris, T.; Gagliardi, S.; Contini, V.; De Francesco, M. Innovative electrodes for direct methanol fuel cells based on carbon nanofibers and bimetallic PtAu nanocatalysts. *Int. J. Hydrogen Energy* 2014, *39*, 21601–21612. [CrossRef]
43. Zainoodin, A.M.; Kamarudin, S.K.; Masdar, M.S.; Daud, W.R.W.; Mohamad, A.B.; Sahari, J. High power direct methanol fuel cell with a porous carbon nanofiber anode layer. *Appl. Energy* 2014, *113*, 946–954. [CrossRef]
44. Pekala, R.W.; Farmer, J.C.; Alviso, C.T.; Tran, T.D.; Mayer, S.T.; Miller, J.M.; Dunn, B. Carbon aerogels for electrochemical applications. *J. Non-Cryst. Solids* 1998, *225*, 74–80. [CrossRef]
45. Job, N.; Marie, J.; Lambert, S.; Berthon-Fabry, S.; Achard, P. Carbon xerogels as catalyst supports for PEM fuel cell cathode. *Energy Convers. Manag.* 2008, *49*, 2461–2470. [CrossRef]
46. Sharma, C.; Kulkarni, M.; Sharma, A.; Madou, M. Synthesis of carbon xerogel particles and fractal-like structures. *Chem. Eng. Sci.* 2009, *64*, 1536–1543. [CrossRef]
47. Liu, B.; Creager, S. Carbon xerogels as Pt catalyst supports for polymer electrolyte membrane fuel-cell applications. *J. Power Sources* 2010, *195*, 1812–1820. [CrossRef]
48. Calderón, J.C.; Mahata, N.; Pereira, M.F.R.; Figueiredo, J.L.; Fernandes, V.R.; Rangel, C.M.; Calvillo, L.; Lázaro, M.J.; Pastor, E. Pt–Ru catalysts supported on carbon xerogels for PEM fuel cells. *Int. J. Hydrogen Energy* 2012, *37*, 7200–7211. [CrossRef]
49. Gao, X.; Omosebi, A.; Landon, J.; Liu, K. Surface charge enhanced carbon electrodes for stable and efficient capacitive deionization using inverted adsorption—Desorption behavior. *Energy Environ. Sci.* 2015, *8*, 897–909. [CrossRef]
50. Hayashida, T.; Pan, L.; Nakayama, Y. Mechanical and electrical properties of carbon tubule nanocoils. *Physica B* 2002, *323*, 352–353. [CrossRef]
51. Sevilla, M.; Sanchís, C.; Valdés-Solís, T.; Morallón, E.; Fuertes, A.B. Highly dispersed platinum nanoparticles on carbon nanocoils and their electrocatalytic performance for fuel cell reactions. *Electrochim. Acta* 2009, *54*, 2234–2238. [CrossRef]
52. Jafri, R.I.; Rajalakshmi, N.; Ramaprabhu, S. Nitrogen-doped multi-walled carbon nanocoils as catalyst support for oxygen reduction reaction in proton exchange membrane fuel cell. *J. Power Sources* 2010, *195*, 8080–8083. [CrossRef]
53. Shaikjee, A.; Coville, N.J. The synthesis, properties and used of carbon materials with helical morphology. *J. Adv. Res.* 2012, *3*, 195–223. [CrossRef]

54. Suda, Y.; Ozaki, M.; Tanoue, H.; Takikawa, H.; Ue, H.; Shimizu, K. Supporting PtRu catalysts on various types of carbon nanomaterials for fuel cell applications. *J. Phys. Conf. Ser.* **2013**, *433*, 012008. [CrossRef]
55. Tang, S.; Huangfu, H.; Dai, Z.; Sui, L.; Zhu, Z. Preparation of Fe-N-carbon nanocoils as catalyst for oxygen reduction reaction. *Int. J. Electrochem. Sci.* **2015**, *10*, 7180–7191.
56. Mohammad, K.; Forouzan, A.; Hamid Reza Lotfi Zadeh, Z.; Omid, S.; Ezzatollah, N. Electrocatalytic performance of Pt/Ru/Sn/W fullerene electrode for methanol oxidation in direct methanol fuel cell. *J. Fuel Chem. Technol.* **2013**, *41*, 91–95. [CrossRef]
57. Zhang, Q.; Bai, Z.; Shi, M.; Yang, L.; Qiao, J.; Jiang, K. High-efficiency palladium nanoparticles supported on hydroxypropyl-b-cyclodextrin modified fullerene [60] for ethanol oxidation. *Electrochim. Acta* **2015**, *177*, 113–117. [CrossRef]
58. Rambabu, G.; Bhat, S.D. Sulfonated fullerene in SPEEK matrix and its impact on the membrane electrolyte properties in direct methanol fuel cell. *Electrochim. Acta* **2015**, *176*, 657–669.
59. Borgohain, R.; Yang, J.; Selegue, J.P.; Kim, D.Y. Controlled synthesis, efficient purification, and electrochemical characterization of arc-discharge carbon nano-onions. *Carbon* **2014**, *66*, 272–284. [CrossRef]
60. Dhand, V.; Prasad, J.S.; Rao, M.V.; Bharadwaj, S.; Anjaneyulu, Y.; Jain, P.K. Flame synthesis of carbon nano onions using liquefied petroleum gas without catalyst. *Mater. Sci. Eng. C* **2013**, *33*, 758–762. [CrossRef]
61. Zhu, S.; Xu, G. Single-walled carbon nanohorns and their applications. *Nanoscale* **2010**, *2*, 2538–2549. [CrossRef]
62. Unni, S.M.; Bhange, S.N.; Illathvalappil, R.; Mutneja, N.; Patil, K.R.; Kurungot, S. Nitrogen-induced surface area and conductivity modulation of carbon nanohorn and its function as an efficient metal-free oxygen reduction electrocatalyst for anion-exchange membrane fuel cells. *Small* **2015**, *11*, 352–360. [CrossRef]
63. Ghosh, S.; Das, S.; Mosquera, M.E.G. Conducting polymer-based nanohybrids for fuel cell application. *Polymers* **2020**, *12*, 2993. [CrossRef]
64. Takahashi, K.; Koda, R.; Kakinuma, K.; Uchida, M. Improvement of cell performance in low-pt-loading PEFC cathode catalyst layers with Pt/Ta-SnO$_2$ prepared by the electrospray method. *J. Electrochem. Soc.* **2017**, *164*, F235. [CrossRef]
65. Tan, Y.; Matsui, H.; Ishiguro, N.; Uruga, T.; Nguyen, D.-N.; Sekizawa, O.; Sakata, T.; Maejima, N.; Higashi, K.; Dam, H.C.; et al. Pt−Co/C cathode catalyst degradation in a polymer electrolyte fuel cell investigated by an infographic approach combining threedimensional spectroimaging and unsupervised learning. *J. Phys. Chem. C* **2019**, *123*, 18844–18853. [CrossRef]
66. Osmieri, L.; Cullen, D.A.; Chung, H.T.; Ahluwalia, R.K.; Neyerlin, K.C. Durability evaluation of a Fe–N–C catalyst in polymer electrolyte fuel cell environment via accelerated stress tests. *Nano Energy* **2020**, *78*, 105209. [CrossRef]
67. Inoue, G.; Takenaka, S. Design of interfaces and phase interfaces on cathode catalysts for polymer electrolyte fuel cells. *Chem. Lett.* **2021**, *50*, 136–143. [CrossRef]
68. Hiramatsu, M.; Hori, M. *Carbon Nanowalls*; Springer: Wien, Austria, 2010. [CrossRef]
69. Hiramatsu, M.; Kondo, H.; Hori, M. Graphene Nanowalls. In *New Progress on Graphene Research*; Gong, J.R., Ed.; Intech Open Ltd.: London, UK, 2013; Chapter 9; pp. 235–260. [CrossRef]
70. Ichikawa, T.; Shimizu, N.; Ishikawa, K.; Hiramatsu, M.; Hori, M. Synthesis of isolated carbon nanowalls via high-voltage nanosecond pulses in conjunction with CH$_4$/H$_2$ plasma enhanced chemical vapor deposition. *Carbon* **2020**, *161*, 403. [CrossRef]
71. Yerlanuly, Y.; Christy, D.; Nong, N.-V.; Kondo, H.; Alpysbayeva, B.; Nemkayeva, R.; Kadyr, M.; Ramazanov, T.; Gabdullin, M.; Batryshev, D.; et al. Synthesis of carbon nanowalls on the surface of nanoporous alumina membranes by RI-PECVD method. *Appl. Surf. Sci.* **2020**, *523*, 146533. [CrossRef]
72. Shin, S.-C.; Yoshimura, A.; Matsuo, T.; Mori, M.; Tanimura, M.; Ishihara, A.; Ota, K.; Tachibana, M. Carbon nanowalls as platinum support for fuel cells. *J. Appl. Phys.* **2011**, *110*, 104308. [CrossRef]
73. Ashikawa, A.; Yoshie, R.; Kato, K.; Miyazawa, K.; Murata, H.; Hotozuka, K.; Tachibana, M. Pt nanoparticles supported on carbon nanowalls with different domain sizes, for oxygen reduction reaction. *J. Appl. Phys.* **2015**, *118*, 214303. [CrossRef]
74. Hiramatsu, M.; Mitsuguchi, S.; Takeyoshi, H.; Kondo, H.; Hori, M.; Kano, H. Fabrication of carbon nanowalls on carbon fiber paper for fuel cell application. *Jpn. J. Appl. Phys.* **2013**, *52*, 01AK03. [CrossRef]
75. Imai, S.; Kondo, H.; Hyungjun, C.; Kano, H.; Ishikawa, K.; Sekine, M.; Hiramatsu, M.; Ito, M.; Hori, M. High-durability catalytic electrode composed of Pt nanoparticles-supported carbon nanowalls synthesized by radical-injection plasma-enhanced chemical vapor deposition. *J. Phys. D* **2017**, *50*, 40LT01. [CrossRef]
76. Imai, S.; Kondo, H.; Hyungjun, C.; Ishikawa, K.; Tsutsumi, T.; Sekine, M.; Hiramatsu, M.; Hori, M. Pt nanoparticle-supported carbon nanowalls electrode with improved durability for fuel cell applications using C$_2$F$_6$/H$_2$ plasma-enhanced chemical vapor deposition. *Appl. Phys. Express* **2018**, *12*, 015001. [CrossRef]
77. Imai, S.; Kondo, H.; Hyungjun, C.; Ishikawa, K.; Tsutsumi, T.; Sekine, M.; Hiramatsu, M.; Hori, M. Effects of three-dimensional structure on electrochemical oxygen reduction characteristics of Pt-nanoparticle-supported carbon nanowalls. *J. Phys. D* **2019**, *52*, 105503. [CrossRef]
78. Lehmann, K.; Yurchenko, O.; Urban, G. Carbon nanowalls for oxygen reduction reaction in bio fuel cells. *J. Phys. Conf. Ser.* **2014**, *557*, 012008. [CrossRef]
79. Slobodian, P.; Riha, P.; Kondo, H.; Cvelbar, U.; Olejnik, R.; Matyas, J.; Sekine, M.; Hori, M. Transparent elongation and compressive strain sensors based on aligned carbon nanowalls embedded in polyurethane. *Sens. Actuator A Phys.* **2020**, *306*, 111946. [CrossRef]
80. Tomatsu, M.; Hiramatsu, M.; Foord, J.S.; Kondo, H.; Ishikawa, K.; Sekine, M.; Takeda, K.; Hori, M. Hydrogen peroxide sensor based on carbon nanowalls grown by plasma-enhanced chemical vapor deposition. *Jpn. J. Appl. Phys.* **2017**, *56*, 06HF03. [CrossRef]

81. Tomatsu, M.; Hiramatsu, M.; Kondo, H.; Ishikawa, K.; Tsutsumi, T.; Sekine, M.; Hori, M. Electrochemical Reaction in Hydrogen Peroxide and Structural Change of Platinum Nanoparticle-Supported Carbon Nanowalls Grown Using Plasma-Enhanced Chemical Vapor Deposition. *C* **2019**, *5*, 7. [CrossRef]
82. Slobodian, P.; Cvelbar, U.; Riha, P.; Olejnik, R.; Matyas, J.; Filipič, G.; Watanabe, H.; Tajima, S.; Kondo, H.; Sekine, M.; et al. High sensitivity of a carbon nanowall-based sensor for detection of organic vapours. *RSC Adv.* **2015**, *5*, 90515. [CrossRef]
83. Ichikawa, T.; Tanaka, S.; Kondo, H.; Ishikawa, K.; Tsutsumi, T.; Sekine, M.; Hori, M. Effect of electrical stimulation on proliferation and bone-formation by osteoblast-like cells cultured on carbon nanowalls scaffolds. *Appl. Phys. Express* **2019**, *12*, 025006. [CrossRef]
84. Ohta, T.; Ito, H.; Ishikawa, K.; Kondo, H.; Hiramatsu, M.; Hori, M. Atmospheric pressure plasma-treated carbon nanowalls surface-assisted laser desorption/ionization time-of-flight mass spectrometry (CNW-SALDI-MS). *C* **2019**, *5*, 40. [CrossRef]
85. Sakai, R.; Ichikawa, T.; Kondo, H.; Ishikawa, K.; Shimizu, N.; Ohta, T.; Hiramatsu, M.; Hori, M. Effects of carbon nanowalls (CNWs) substrates on soft ionization of low-molecular-weight organic compounds in surface-assisted laser desorption/ionization mass spectrometry (SALDI-MS). *Nanomaterials* **2021**, *11*, 262. [CrossRef]
86. Vielstich, W.; Gasteiger, H.A.; Lamm, A. *Handbook of Fuel Cells—Fundamentals, Technology and Applications*; John Wiley & Sons, Ltd.: Hoboken, NJ, USA, 2003.
87. Schneider, C.A.; Rasband, W.S.; Eliceiri, K.W. NIH image to imageJ: 25 years of image analysis. *Nat. Methods* **2012**, *9*, 671–675. [CrossRef] [PubMed]

Article

Development of Disposable and Flexible Supercapacitor Based on Carbonaceous and Ecofriendly Materials

Giovanni G. Daniele [1], Daniel C. de Souza [1], Paulo Roberto de Oliveira [1], Luiz O. Orzari [1,2], Rodrigo V. Blasques [1,2], Rafael L. Germscheidt [3], Emilly C. da Silva [4], Leandro A. Pocrifka [4], Juliano A. Bonacin [3] and Bruno C. Janegitz [1,*]

1. Laboratory of Sensors, Nanomedicine and Nanostructured Materials, Federal University of São Carlos, Araras 13600-970, São Paulo, Brazil; giovannigomesdaniele@gmail.com (G.G.D.); cardoso.ds@outlook.com (D.C.d.S.); paulo.oliveira@ufscar.br (P.R.d.O.); l.o.orzari@gmail.com (L.O.O.); blasques@live.com (R.V.B.)
2. Department of Physics, Chemistry and Mathematics, Federal University of São Carlos, Sorocaba 18052-780, São Paulo, Brazil
3. Institute of Chemistry, University of Campinas, Campinas 13083-859, São Paulo, Brazil; r226597@dac.unicamp.br (R.L.G.); jbonacin@unicamp.br (J.A.B.)
4. Laboratory of Electrochemistry and Energy, Chemistry Graduate Program of Federal University of Amazonas, Manaus 69067-005, Amazonas, Brazil; e.millycruz@hotmail.com (E.C.d.S.); pocrifka@gmail.com (L.A.P.)
* Correspondence: brunocj@ufscar.br

Abstract: A novel flexible supercapacitor device was developed from a polyethylene terephthalate substrate, reused from beverage bottles, and a conductive ink based on carbon black (CB) and cellulose acetate (CA). The weight composition of the conductive ink was evaluated to determine the best mass percentage ratio between CB and CA in terms of capacitive behavior. The evaluation was performed by using different electrochemical techniques: cyclic voltammetry, obtaining the highest capacitance value for the device with the 66.7/33.3 wt% CB/CA in a basic H_2SO_4 solution, reaching 135.64 F g^{-1}. The device was applied in potentiostatic charge/discharge measurements, achieving values of 2.45 Wh kg^{-1} for specific energy and around 1000 W kg^{-1} for specific power. Therefore, corroborated with electrochemical impedance spectroscopy assays, the relatively low-price proposed device presented a suitable performance for application as supercapacitors, being manufactured from reused materials, contributing to the energy storage field enhancement.

Keywords: disposable device; supercapacitor; carbon material; carbon black; cellulose acetate

1. Introduction

The energy issue has been one of the great discussions around the world due to the depletion of fossil fuels, which has attracted the search for new sources of energy conversion and storage. This issue arises from the constant demand for fast, economically efficient, and ecologically correct technological devices, to meet the world's growing population's desire for frequent information [1–3]. In such industry, electrochemical energy conversion systems function as key technologies, allowing the management of renewable energy while diminishing the pollution caused by other sources, such as the greenhouse effect [2]. Common batteries are widely researched, discussed, and employed, especially in electric or hybrid vehicles [2,4,5]. On the other hand, they have some lifetime and power issues, such as low power density, requiring a longer charge and/or discharge time, and capacitors can operate at short charge-discharge times (<0.1 s) resulting in low production and maintenance costs. As demonstrated by Libich et al. [1], electrochemical capacitors, also known as supercapacitors, present high energy densities, close to that of batteries, while also bringing the higher power densities of pure metal capacitors. Moreover, they have been gaining prominence due to relatively low-cost, renewable materials usage, and high-efficiency energy storage capability [6–8]. It is also known that even at high discharge rates, their efficiency and reversibility exceed 90% [9,10], further suggesting that supercapacitors could be ideal for renewable energy storage.

Hence, technological development has opened new applications and, in this scenario, flexible supercapacitors have attracted great interest in the industry. This class of materials meets the technological requirements for advances in the field of wearable, compact, and portable electronics, such as flexible displays on phones, health tracking devices, computers, televisions, electronic textiles, artificial electronic skin, and distributed sensors [11–15]. Furthermore, the key challenge for the conception of the commercial supercapacitors is the development of devices with excellent mechanical flexibility and adjustable dimensions along with decreasing production costs [16,17].

In this context, various carbon materials have been applied as electrode materials for electric double-layer capacitors, and among them, carbon black (CB) stands out due to the structure of the graphitic domains in the concentric particles, ensuring adequate electrical conduction, combined with the porosity of the carbonaceous material, having the capacity to store energy in the form of electrical charges [18–20].

In addition, CB can form stable dispersions of polymeric matrices [21], which is applied for the fabrication of 2D-printed electrodes, using a technology known as screen-printing [22,23]. This technology consists of the application of a conductive ink over a screen, delimiting the form of the desired electrodes on the surface of an inert substrate. This conductive ink is conventionally made by the dispersion of electrically conductive particles on a polymeric matrix, and, among renewable options, cellulose acetate (CA) could be an interesting choice as it is a biodegradable polymer of biological sources, synthesized from cellulose with relatively low production cost [24–26]. The literature also presents screen-printed systems as possible disposable materials [27–29], due to their reduced production cost, when compared to conventional electrodes. This, however, also demands an ecofriendly inert substrate, such as reusable beverage bottles (polyethylene terephthalate, PET) [27] and sandpaper [30]; or recyclable materials, such as paper [31], waterproof paper [32], fiberboard [33], and polyvinyl chloride [34]. The selection of electrode materials directly affects the electrochemical performance, physical properties, and further applications of the device. Therefore, the development of flexible supercapacitors is very important for different technologies, making essential the search for cheaper and renewable materials, with optimal operational parameters. In this work, we propose the development of a disposable and flexible supercapacitor, using the screen-printing technique, with a conductive and lab-made ink of CB and CA, deposited over a PET substrate, presenting and discussing its behaviors, characteristics, and efficiencies in different pH media.

2. Materials and Methods

2.1. Reagents and Solutions

All chemicals used in this work were purchased from Sigma-Aldrich (Taufkirchen city, Bavaria state, Germany) and/or Fluka (Muskegon, MI, USA), with analytical grade. In addition, all aqueous solutions used for electrochemical characterizations, including the supporting electrolytes, sulfuric acid (H_2SO_4), potassium hydroxide (KOH), and sodium sulfate (Na_2SO_4), were prepared by an ultrapure water system Synergy® (type 1) from Merck Millipore, Merck Group (Burlington, MA, USA), with resistivity > 18 MΩ cm. For ink preparation, VULCAN® XC-72 powder was provided by CABOT (Boston city, Massachusetts state, USA), and substrates were obtained by reusing different beverage bottles.

2.2. Apparatus

The CB-based electrodes were prepared and assembled with the aid of Silhouette Studio 4.4® (Silhouette, Brazil) software, used to elaborate the capacitor mask design, and command the Silhouette Cameo 3® cutting machine, Silhouette (São Paulo city, São Paulo state, Brazil). For the ink preparation, a magnetic stirrer and heater (Biomixer AM-10), and an ultrasonic bath purchased from Cristófoli (Campo Mourão, Paraná state, Brazil) were used.

The electrochemical characterizations were performed in a Potentiostat/Galvanostat (Metrohm, Autolab/PGSTAT204) with an expansion module for Electrochemical Impedance Spectroscopy (EIS) FRA32M performance. Managed by NOVA software 2.1.4 version (Utrecht, The Netherlands) at the open circuit potential (OCP) with sine wave type, 10 mV amplitude, in the frequency range of 0.1–10 MHz. Morphological characterizations were conducted using SEM (Scanning Electronic Microscope) images obtained in a Vega-3 LMU Tescan electronic microscope, in 5.0 kV voltage acceleration and 50 P low vacuum mode. The electrochemical impedance spectroscopy (EIS) data were obtained in three different 1.0 mol L^{-1} electrolytes: H_2SO_4, KOH, and Na_2SO_4. The data was obtained with the following parameters: applied potential corresponding to the OCP of each system, obtained after 300 s stabilization, being −26 mV for H_2SO_4 and −56 mV for KOH; potential amplitude of 10 mV, sine waves type throughout all analysis, and 10 frequency increments per decade, in the range of 1.0×10^5 to 1.0×10^{-2} Hz.

2.3. Production of the Ink and Manufacture of the Disposable Electrodes

To guarantee optimal ink formulation and material dispersion, two solvent solutions were prepared, with a mixture of N, N-Dimethylformamide (DMF), and acetone (ACE) of distinct volume %. The solvent to best disperse CA had a 52.5/47.5% ratio, while the solvent that better dispersed CB presented a ratio of 42.4/57.6%. The CA dispersion was prepared by adding 250 mg of the acetate in the mentioned solvent, and subsequently stirring it for 10 min at room temperature. Meanwhile, the CB dispersion was produced by adding a desired amount of CB to the corresponding DMF/ACE mixture and placing it in an ultrasonic bath, for 30 min, to enhance homogenization. Finally, these two dispersions were mixed in different proportions, following another ultrasonic bath for 30 min. After, the final dispersion was magnetically stirred for 10 min, finishing the conductive ink. The final CB:CA proportions are presented in Table 1.

To assemble the electrode systems, the polyethylene terephthalate (PET) substrates, obtained by reusing beverage bottles, were cut, properly washed, and polished to enhance ink adhesion. After, adhesive paper sheets were cut with the cutting machine with a proper design, based on the three-electrodes system. These adhesive masks were attached to the PET surface, and the ink was applied over them. After 10 min of drying, a second ink layer was produced, and it was left to dry at 19 °C. After the paper masks were removed and the system was ready to use. Figure 1 illustrates the whole process.

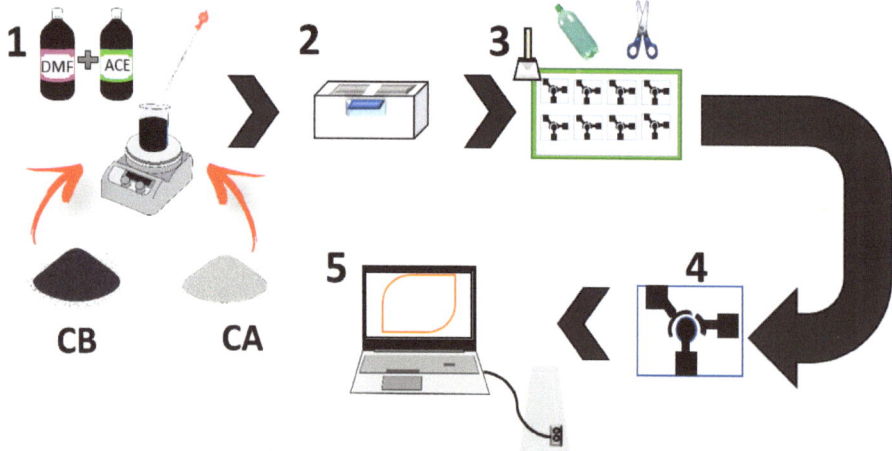

Figure 1. Experimental procedure representation: (1) elaboration of the CB and CA base dispersions; (2) ink preparation; (3) electrode assembly with screen printing technique over PET substrates; (4) finalized capacitor electrode ready for (5) electrochemical analyses.

Table 1. CB/CA proportions at the different prepared inks.

Formulation N°	CB (mg)	CA (mg)	CB/CA (wt%/wt%)
1	125	250	33.3/66.7
2	250	250	50.0/50.0
3	375	250	60.0/40.0
4	500	250	66.7/33.3

3. Results

3.1. Electrochemical and Morphological Characterizations

To understand the CB/CA ink capacitive behavior, different characterization procedures should be investigated. Therefore, the first electrochemical study performed was a profile determining the cyclic voltammetry (CV) technique, aiming to analyze and compare the four different ink formulations, based on each specific capacitance (SC) response at different electrolytes: an acid (H_2SO_4), a saline (Na_2SO_4), and a basic (KOH) media, with respective 1.0 mol L^{-1} concentration. Properties such as active surface area, porosity, and conductivity are crucial points in the development of electrochemical capacitors, and this is directly influenced by the amount of carbonaceous material used [35]. Thus, the voltammetric profiles in Figure 2A–C, obtained with different CB/CA compositions, demonstrate that, majorly, the capacitive current is predominant, even though a few electrodes presented redox interactions, limiting their use as capacitive devices. This highly capacitive behavior indicates that the materials perform energy storage through an expansion in the electrical double layer, storing charges directly on the double-layer [36]. This class of devices is commonly called an expanded double-layer capacitor (EDLC) [37,38].

This characteristic profile observed is possibly caused by the presence of discrete pores, typical of carbonaceous materials, where the carbon atoms of the CB are compacted at the edges of the channel [39]. In Figure 2, the voltammograms demonstrate an approximately rectangular voltammetric profile, a characteristic near-ideal capacitor behavior, which is justified by diffusion processes. Additionally, this study shows that the variation in current flowing through the circuit increases linearly with the CB concentration, indicating that the larger amounts of CB enhance the capacitive behavior. Probably, this functions to increase the number of active sites, due to the higher atomic density, contributing to the double layer formation process [38]. It is also interesting to highlight the presence of anodic and cathodic peaks in Figure 2C, mainly for the 60.0/40.0 and 50.0/50.0 % (w/w). These phenomena have been attributed to the result of electroactive surface functional groups redox reactions, such as phenols, ketones, and carboxylic structures [40,41]. This process could generate pseudocapacitance, but with uncertain benefits, as the self-discharge rate of such groups is higher than the overall surface [40,42]. Herewith, the ink containing the highest quantity of CB (66.7/33.3%, w/w) was chosen to continue this work.

Thus, with the previous charge data acquisition used for the CV technique, the resultant charge data influence, at each electrolyte, with different scan rates (ν), guided by SC values would be estimated by the Equation (1):

$$SC = \frac{\Delta Q}{m \times \Delta V} \quad (1)$$

where SC is expressed in (F g^{-1}); ΔQ is the charge variation in terms of C s^{-1}; m is the electrode mass, which was estimated at around 0.134 mg for the chosen formulation; and ΔV is the working potential used, represented in V. Therefore, the resultant SC values vs. different ν (mV s^{-1}) are shown in Figure 2D, for each electrolyte. The resultant dispersion, in general, presented a capacitance decrease with the increase in ν. This was expected behavior, especially because of the velocity of surface formation processes, including surface resistance, double-layer thickness, and charge diffusion as well, which are all processes in progress during data collection. It is important to highlight, however, that the CB/CA electrode demonstrated a more stable *SC* vs. ν curve in the acidic electrolyte, with a decrease of only 75%, when compared to the other media (93% for KOH and 87% for Na_2SO_4).

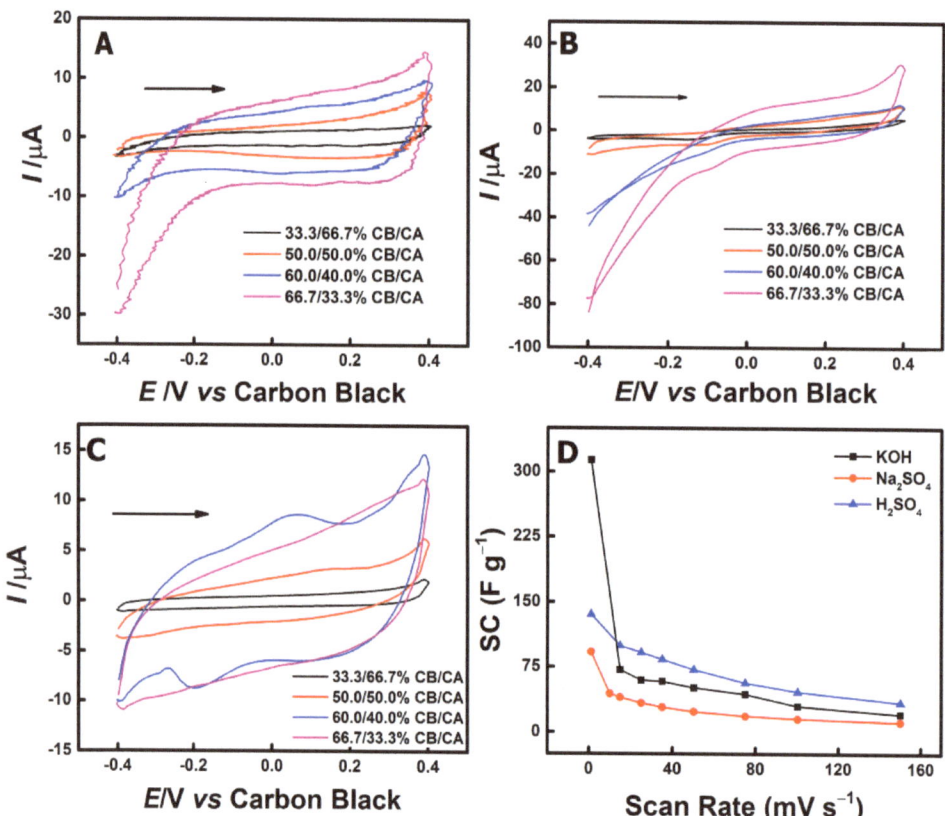

Figure 2. Cyclic voltammograms of different CB/CA formulations: 33.3/66.7 (**black**), 50.0/50.0 (**red**), 60.0/40.0 (**blue**), and 66.7/33.3% (**pink**); in (**A**) 1.0 mol L^{-1} H_2SO_4, (**B**) 1.0 mol L^{-1} KOH, and (**C**) 1.0 mol L^{-1} Na_2SO_4; ν = 10 mV s^{-1} (**D**) SC vs. scan rate plots for 66.7/33.3% CB/CA at equimolar 1.0 mol L^{-1} H_2SO_4 (▲); Na_2SO_4 (●) and KOH (■) solutions.

The EIS technique configures a very important tool for mapping processes in the electrode interface, which may include redox reactions, adsorption, and forced mass transfer processes. It responds with the function of the relation between frequency and a given potential, in contrast with the most used electroanalytical methods. This technique is useful for electrochemical devices, especially energy storage systems, such as batteries, capacitors, and supercapacitors, especially as they need more comprehensive information about resistivity and charge transfer of electrode materials.

In this study (Figure 3A), it was observed that the acidic profile presents a low-frequency region almost in parallel to the imaginary axis, which characterizes an ideal capacitive behavior due to minimal material resistance effects being observed in lower frequencies. This could imply that the double-layer ion transport has ceased, achieving equilibrium, and only dielectric behavior is being observed [43,44]. This behavior is described by a [R(RQ)C] equivalent circuit. For the KOH medium, the electrode shows a profile containing deviations from ideal capacitor behavior, especially by the formation of two semicircles, probably caused by proper material hybrid capacitive/resistive behavior, better described with the [R(RQ)([RW]C)] equivalent circuit. The first semi-circle is, potentially, due to the device geometry [43], while the second could originate from the surface redox reactions, as presented in Figure 2B. The literature proposes a few explanations for such reactions: the device could be functionalized by the basic electrolyte, adsorbing hydroxyl ions [45], or promoting rise to pseudocapacitance by oxygen-involved redox processes [46].

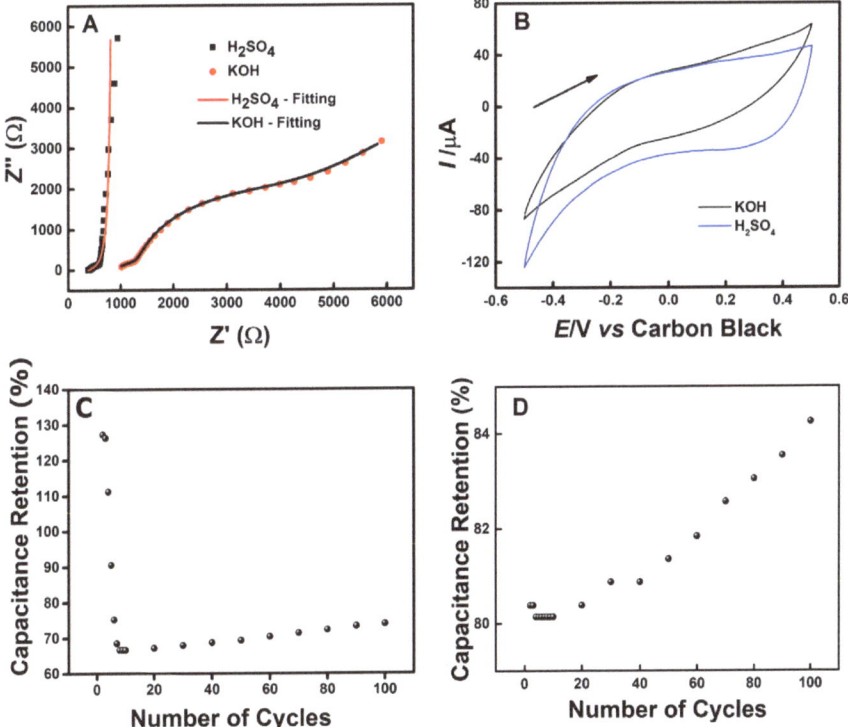

Figure 3. (**A**) Nyquist diagrams for H_2SO_4 (■) and KOH (•). (**B**) Cyclic voltammograms obtained in OCP by 66.7/33.3% CB/CA electrode, in equimolar 1.0 mol L^{-1} H_2SO_4 (—) and KOH (—); ν = 10 mV s^{-1}. Capacitance retention dispersion plots of 100 cycles in (**C**) 1.0 mol L^{-1} KOH and (**D**) 1.0 mol L^{-1} H_2SO_4.

Each system was then evaluated for its behavior against the OCP values used in EIS, and the profiles are presented in Figure 3B. As can be noted, both presented characteristic voltammograms with the only capacitive current generation. It could also be highlighted that the acidic system is considered a more rectangular shape than the basic one, implying a higher energy storage capability. By increasing the number of cycles in CV to 100, the capacitance retention of each system was evaluated against previously obtained data (100%). In alkaline solution (Figure 3C), the device rapidly presented an increase in capacitive current, equivalent to 130%, but quickly dropped to around 65%. The device then linearly increased the retention to 74% after all scans. This behavior implies a non-stable surface. In acidic solutions (Figure 3D), the retentions started at around 80%, increasing to 84% after all cycles, suggesting higher stability. This data also corroborates the use of acidic electrolytes to properly employ the proposed device as a supercapacitor.

Morphological characterizations were required to better understand the proposed material and confirm its electrochemical behavior. Therefore, SEM was carried out for the optimal 66.7/33.3% (w/w) CB/CA ink without CB (Figure 4A–C), without CA (Figure 4D–F), and in its complete form (Figure 4G–I). In the images without CB (only CA), it is possible to observe a roughish polymeric structure, with small, intertwined wing-like structures. For the ink without CA (only CB), purely carbonaceous structures can be observed, with high roughness and flake-like formation. Additionally, this sample presented a high cracking rate, as can be observed in Figure 4D right side, probably due to the electrostatic interaction between CB agglomerates being destroyed as the solvents are eliminated. In the last sample, it is possible to observe the roughness of the whole surface, with pores distributed throughout the film structure, corroborating with the data discussed previously. Additionally, the carbonaceous structure of CB is still observed, and without cracks, proving a homogeneous distribution of CB in CA. This active distribution of amorphous carbon can be associated with the overall material conductivity.

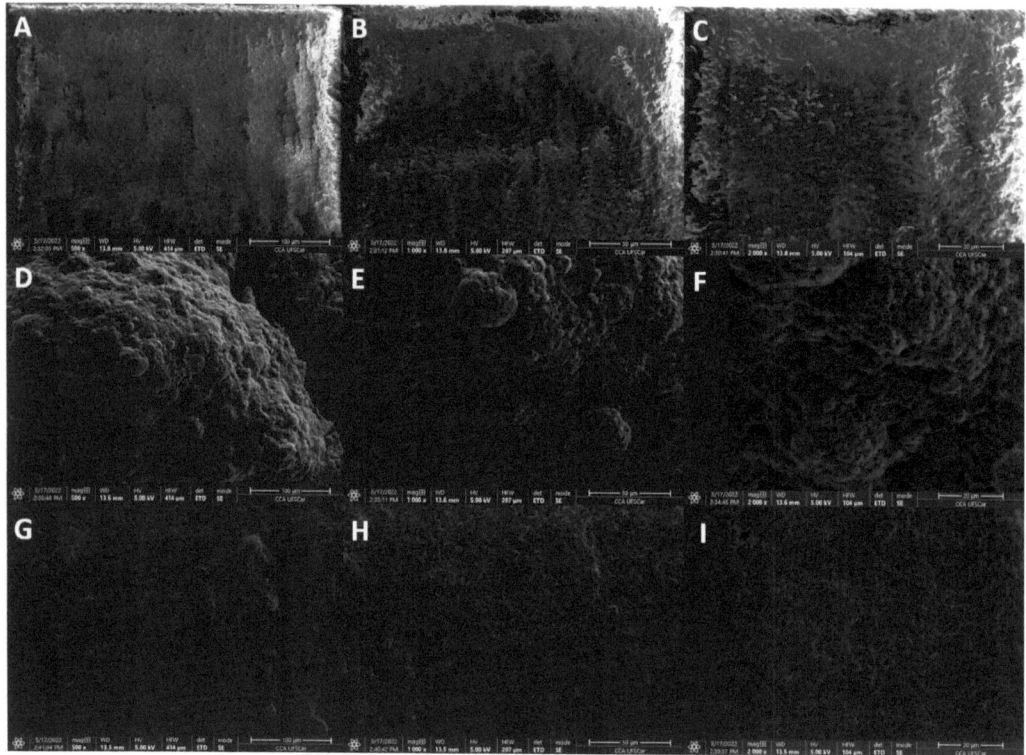

Figure 4. SEM images of the 66.67% CB/CA ink without CB (**A–C**), without CA (**D–F**) and the complete ink (**G–I**) at (**A,D,G**) 500×, (**B,E,H**) 1000×, and (**C,F,I**) 2000× magnification.

3.2. Potentiostatic Charge/Discharge Evaluation

Charge and discharge tests were carried out by performing a potentiostatic charge/discharge technique (PCD). These tests are extremely important for a more complex analysis of the behavior of the device in an applied situation, by inserting the material under charging and discharging phenomena, at distinct current density (CD) values, through time. Thus, it is expected to observe profiles with a triangular shape, that results from the device's double-layer charge/discharge response. This process depends on the charge movement rate through each material gram. As such, this behavior is principally affected by resistance and diffusion layer processes in the electrolyte/electrode interface.

Therefore, the CD values were selected randomly for the PCD procedure and the chosen values were 0.5, 1.0, 2.0, 3.0, and 4.0 A g^{-1}, and their resulting profiles for the CB ink are presented in Figure 5. In Figure 5A, the charge and discharge relationship at H_2SO_4 presented the discussed triangular profile, as expected for a capacitor. Nonetheless, plotline inclination suggested a smooth charge transition. This type of profile indicates a closer relationship with supercapacitor behavior, which conventionally shows relatively high energy and power densities with smooth rate transitions.

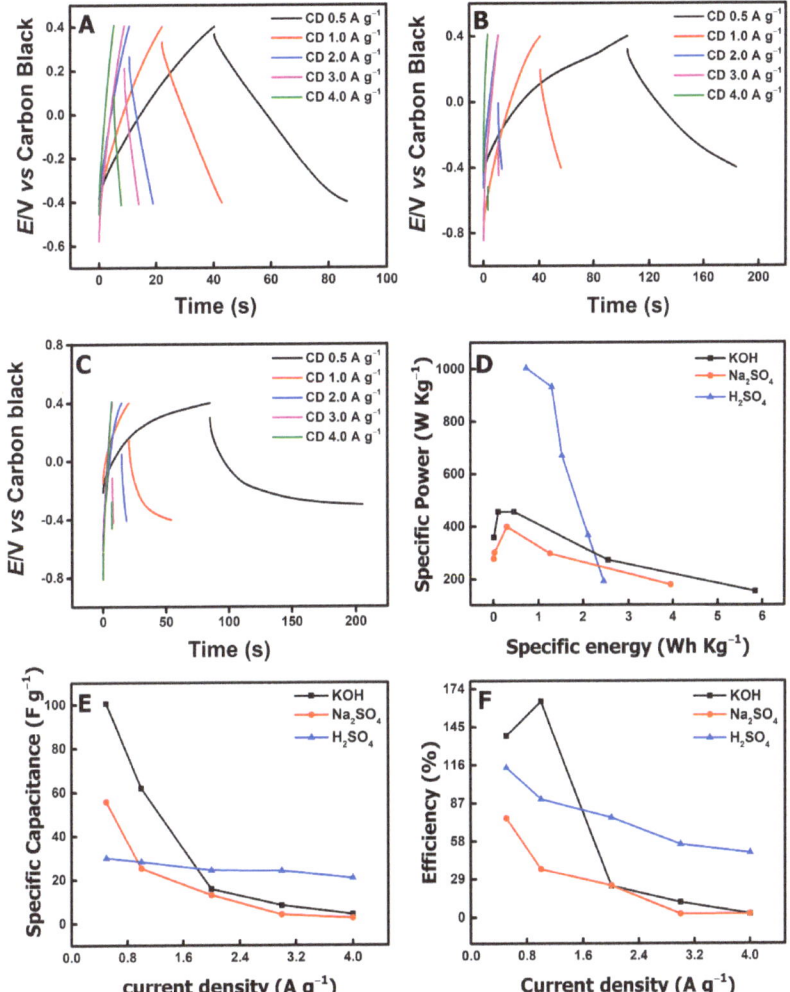

Figure 5. E vs. t plots of CB/CA PCD profiles in 1.0 mol L^{-1} (**A**) H_2SO_4, (**B**) Na_2SO_4, (**C**) KOH, at different CD values: 0.5, 1.0, 2.0, 3.0, and 4.0 A g^{-1}. (**D**) SP vs. SE (Ragone plots) for CB/CA in different and equimolar 1.0 mol L^{-1} electrolytes: H_2SO_4 (▲), Na_2SO_4 (•), and KOH (■). (**E**) Dispersion plot of PCDs SC vs. different current densities (0.5, 1.0, 2.0, 3.0, and 4.0 A g^{-1}) for CB/CA in different and equimolar 1.0 mol L^{-1} electrolytes: H_2SO_4 (▲), Na_2SO_4 (•), and KOH (■). (**F**) Capacitor efficiency, in equimolar 1.0 mol L^{-1} H_2SO_4 (▲), Na_2SO_4 (•), and KOH (■), for CB/CA devices.

Hereupon, profiles for Na_2SO_4 presented in Figure 5B also demonstrated the triangular tendency through CD rising, but started to show some alterations in shape resolution, probably due to process alterations at the double layer, like a pseudocapacitance rising phenomenon [47]. This tendency is accentuated at KOH profiles (Figure 5C), especially at discharge lines, where its definition intensifies the pseudocapacitance growth hypothesis, added to a possible decrease in activity, due to the higher pH, since water-based electrolytes were used [48,49]. Table 2 shows the calculated rate capacitance for each of the aforementioned systems, in which the alkaline performance starts as the highest, but is diminished to ~4% of its initial value. In saline solution the observed behavior is similar, while in H_2SO_4 only a third of the initial rate capacitance is observed to be lost with the increase in current density, further demonstrating higher stability for the device in such media. With these data, it is possible to conclude that the charge and discharge time increases with saline and alkaline solutions, while the acidic electrolyte seems to have more attractive features for this supercapacitor, including a flatter profile and lower PCD time.

Table 2. Rate capacitance obtained by CB/CA in different 1.0 mol L^{-1} electrolytes and current densities.

Electrolyte	Rate Capacitance (F g^{-1})				
	0.5 A g^{-1}	1.0 A g^{-1}	2.0 A g^{-1}	3.0 A g^{-1}	4.0 A g^{-1}
H_2SO_4	30.01	28.23	24.38	24.13	20.7
KOH	100.6	64.85	15.70	8.280	4.242
Na_2SO_4	55.79	25.29	13.10	4.172	2.590

3.3. Energy and Power Density Calculations

Energy and power density values present a relevant form to cross and compare data between different energy storage devices, especially through the Ragone plot (Figure 5D), which are composed of specific power (*SP*) and specific energy (*SE*) magnitudes. Therefore, the Ragone graphic was plotted by utilizing two equations, one for the device-specific energy (*SE*, Equation (2))

$$SE = \frac{I \times \Delta V \times \Delta t}{2 \times m} \quad (2)$$

where *SPe* is expressed in Wh Kg^{-1}; *I* the PCDs discharge current in A; Δt the discharge time in seconds (s); *m* the working electrode mass g; ΔV is voltage variation through the discharge process in *V*; and the other is for its specific power (*SP*, Equation (3)):

$$SP = \frac{I \times \Delta V}{2 \times m} \quad (3)$$

where *SPw* is in terms of W Kg^{-1}. The graphic obtained by plotting *SP* vs. *SE* presents a comparison of the electrode performance in different media, and shows a very interesting behavior between the two axes, especially for the acidic medium, which get closer to a storage device behavior, with high *SP* per *SE*.

To better understand the *SC* behavior of the device, obtained by PCD, the *SC* values were converted with a different *SC* formula, which considers the proper time data obtained by PCD assays, expressed by Equation (4):

$$SC = \frac{I \times \Delta t}{m \times \Delta V} \quad (4)$$

where *SC* is the PCD *SC* in terms of F g^{-1}; *I* is the applied current in terms of A; Δt *is* the discharge time (s); *m* is the work electrode mass (g); and ΔV is the voltage variation through the discharge process in *V*. Herewith, the *SC* calculated values were plotted against the previously chosen CD values, for each electrolyte studied, as can be seen in Figure 5E.

At low CD values, the saline and alkaline electrolytes presented higher *SC* values than the acidic ones. Although, with the CD increase, this tendency decays and, in general, the acidic medium presented more stability throughout the *SC* variation. This behavior shows an interesting feature that justifies a supercapacitor-like performance, in sulfuric acid, especially because of its flat and stable PCD profile.

Figure 5F shows the device efficiency plot in the function of the selected different current densities, calculated by Equation (5):

$$n = \frac{Dt}{Ct} \quad (5)$$

where n is the estimated efficiency in percentage (%); Dt is the PCD discharge time in seconds (s); and Ct, the PCDs charge time, also in seconds. This plot shows the device efficiency at the selected electrolyte. This is an important indicator to corroborate device performance over a range of current densities. Unlike before, the device presents more stable efficiencies in acidic and alkaline media, but with a considerably higher value for the first one. This further justifies the previous hypothesis about this device super capacitive behavior. Whereas, saline medium showed a more linear behavior with the previous tests, presenting certain instabilities under the studied current densities, compromising the device in optimized supercapacitor applications.

3.4. Flexibility Study

Flexibility is, as discussed previously, an important attribute for capacitors and supercapacitors, enabling the production of different and adaptable technologies. To evaluate it, tests were carried out by twisting the electrode and subsequently collecting its response with CV at 50 mV s^{-1}, with a torsion interval of 20 torsions, for a total of 200. This process was repeated twice, with different configurations for data obtention, one with the device still twisted and the other with the device laid flat, to better understand the effect of such strain in the ink. The voltammograms were recorded in 1.0 mol L^{-1} H$_2$SO$_4$ solution, from -0.5 to 0.5 V, and the data was collected at 0.0 V for all torsions. As can be seen in Figure 6A,B, both obtained profiles presented stable profiles across all cycles, with little difference between the curved and the flat electrode system, as can be noted in Figure 6C. As the PET substrate has a geometrically curved shape, due to the circumference of beverage bottles, the forced flattering of electrodes puts the ink under increased stress, diminishing the current obtained. Besides that, the electrode shows considerable stability under mechanical stress, effectively working as a flexible device.

Similar supercapacitors were chosen for comparison with the values of specific capacitance, power, and energy, which are comparable to the literature (Table 3). The data collected suggests the interesting performance of the developed device in terms of specific energy and capacitance, implying its use as an energy storage device. The majority of the selected works describe different capacitor, micro-capacitor, and supercapacitor systems, usually utilizing relatively higher-cost materials with complex processes such as vapor-liquid-solid chemical vapor deposition [50], and hydrogel synthesis [51]. On the other hand, the proposed work used a simple synthesis method, while still employing relatively low-cost and ecologically friendly materials.

Table 3. Comparison between this work and others present in literature for supercapacitor characteristics.

Types of Cell	Composition	Specific Capacitance	Specific Power	Specific Energy	Works
Three electrodes	Highly doped silicon nanowires	46 µF cm^{-2}	1.6 mF cm^{-2}	Not specified	[50]
Two electrodes	N-doped Graphene/ PANI hydrogel	584.7 mF cm^{-2}	$-$ 0.5 mW cm^{-2}	~0.1 mWh cm^{-2}	[51]
Two electrodes	PANI-N/CNT	341.0 F g^{-1}	40.0 W kg^{-1}	11.5 Wh kg^{-1}	[52]
Two electrodes	ACTIVATED CARBON-PTFE	2.1 mF cm^{-2}	Non-specified	Non-specified	[23]
Two electrodes	PANI-MLG	110 mF cm^{-2}	0.15 mW cm^{-2}	~0.01 mWh cm^{-2}	[6]
Two electrodes	PANI/cellulose/Ag	310.9 F g^{-1}	238.2 W kg^{-1}	0.7 Wh kg^{-1}	[13]

Table 3. Cont.

Types of Cell	Composition	Specific Capacitance	Specific Power	Specific Energy	Works
Two electrodes	PANI-MLG	451.0 F g^{-1}	3610.0 W kg^{-1}	17.0 Wh kg^{-1}	[6]
Three electrodes	Vanadium nitride and nickel oxide thin film	1.85 mF cm^{-2}	28 mW cm^{-2}	0.7 µWh cm^{-2}	[53]
Two electrodes	IPCN-alkaline carbon metal	200 F g^{-1}	Non-specified	Non-specified	[54]
Three electrodes	Carbon black-cellulose acetate	313.53 F g^{-1}	455.691 W kg^{-1}	5.84 Wh kg^{-1}	This work
		168.05 mF cm^{-2}	0.24437 mW cm^{-2}	0.00313 mWh cm^{-2}	This work

Highly doped silicon nanowires: silicon nanowire-based electrodes for micro-capacitor applications, elaborated with highly n and p type doping in silicon nanowires grown with the gold catalyst in chemical vapor deposition (CVD) reactor. *N-doped Graphene/PANI hydrogel*: N-type doped PANI hydrogel/graphene oxide (GO) composite electrode synthesized by hydrothermal process, for flexible supercapacitor assemble. *PANI-N/CNT*: electrode based on activated carbon nanoparticles doped with nitrogen, obtained from PANI nanoparticles pyrolysis for capacitor application. *ACTIVATED CARBON-PTFE*: Ink-jet printed electrode with an ink formulation obtained by mixing activated carbon powder, polytetrafluoroethylene (PTFE) polymer, ethylene glycol, and surfactant. Subsequently printing over a silicon substrate for micro-capacitor applications. *PANI-MLG*: PANI modified multilayer graphene electrodes. *IPCN-alkaline carbon metal*: interconnected porous carbon nanosheets with controllable pore size.

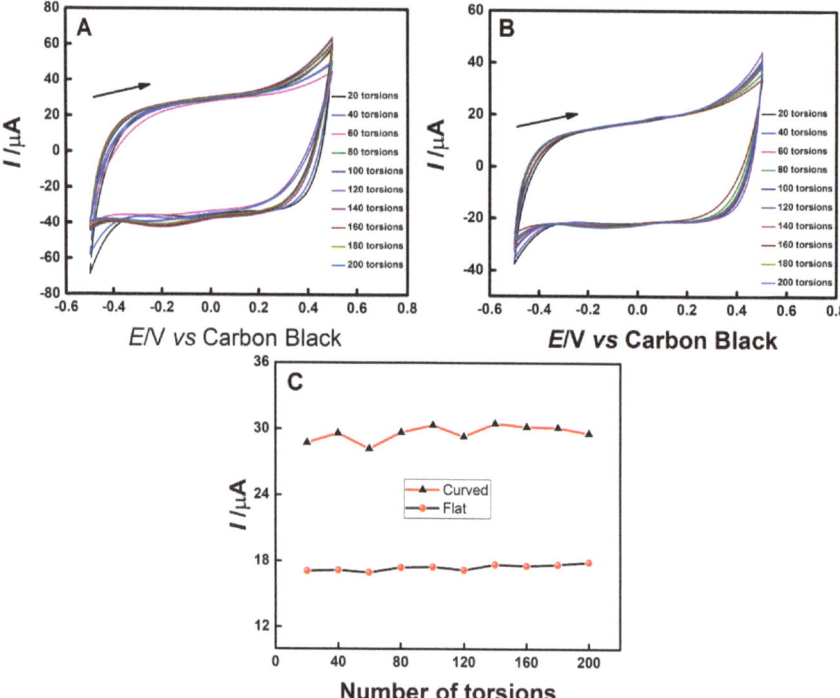

Figure 6. Cyclic voltammograms of the flexibility study on the (**A**) curved device and (**B**) flat device, in 1.0 mol L^{-1} H$_2$SO$_4$; ν = 50 mV s^{-1}. (**C**) I vs. the number of torsions dispersion plots for (▲) curved and (●) flat electrodes. Data was collected at 0.0 V.

4. Conclusions

The production of a flexible supercapacitor electrode was demonstrated in this paper, using a polyethylene terephthalate (PET) substrate. The mass quantity of CA and carbon black was tested in different aqueous media to optimize the capacitance response. The device is potentially disposable as it is obtained from eco-friendly and relatively low-cost materials, compared to others employed in electrochemical capacitors development. The optimal composition was achieved with 66.7/33.3 wt% CB/CA, and its storage mechanism is predominantly by the electrical double layer, as evidenced by cyclic voltammetry, presenting an outstanding capacitive behavior on acidic media, demonstrating the absence of redox reactions in this condition. In the PCD technique, the device presented more defined charge/discharge graphics, leading to more capacitive behaviors in both acidic and basic solutions. In these conditions, specific power of 1000 W Kg^{-1} in the acidic medium was obtained. Through the EIS technique, it was possible to observe a highly capacitive behavior in acidic media and a hybrid capacitive/resistive profile in the basic electrolyte. The device also presented 84% capacitance retention after 100 cycles and is quite stable after 200 torsions. All these data suggest that the device is indeed optimal in lower pH systems, but some energy storage can still be obtained in higher pH electrolytes. The proposed device also demonstrated interesting charge-storage capacity and cyclic stability, especially in acidic solutions, affirming its path as an alternative environmentally friendly supercapacitor with simple fabrication.

Author Contributions: Conceptualization, G.G.D. and B.C.J.; methodology, G.G.D., P.R.d.O. and B.C.J.; software, G.G.D., D.C.d.S. and L.O.O.; validation, D.C.d.S., L.O.O., P.R.d.O., R.V.B., R.L.G., E.C.d.S., L.A.P., J.A.B. and B.C.J.; formal analysis, G.G.D. and D.C.d.S.; investigation, G.G.D., P.R.d.O. and D.C.d.S.; resources, B.C.J.; data curation, G.G.D., D.C.d.S., L.O.O., R.V.B., R.L.G. and E.C.d.S.; writing—original draft preparation, G.G.D. and D.C.d.S.; writing—review and editing, L.O.O., R.V.B., R.L.G., E.C.d.S., L.A.P., J.A.B. and B.C.J.; visualization, L.O.O. and B.C.J.; supervision, B.C.J.; project administration, G.G.D. and B.C.J.; funding acquisition, B.C.J. All authors have read and agreed to the published version of the manuscript.

Funding: The authors are grateful for the financial support of the Brazilian Funding Agencies. This work was supported in part by Coordenação de Aperfeiçoamento de Pessoal de Nível Superior (CAPES) (financial code 001 and CAPES 09/2020 Epidemias 88887.504861/2020-00) and the Conselho Nacional de Desenvolvimento Científico e Tecnológico (CNPq, 303338/2019-9) and Fundação de Amparo à Pesquisa do Estado de São Paulo, FAPESP (2017/23960-0, 2017/11986-5, 2019/23342-0).

Conflicts of Interest: The authors declare no conflict of interest.

References

1. Libich, J.; Máca, J.; Vondrák, J.; Čech, O.; Sedlaříková, M. Supercapacitors: Properties and applications. *J. Energy Storage* **2018**, *17*, 224–227. [CrossRef]
2. Badwal, S.P.S.; Giddey, S.S.; Munnings, C.; Bhatt, A.I.; Hollenkamp, A.F. Emerging electrochemical energy conversion and storage technologies. *Front. Chem.* **2014**, *2*, 79. [CrossRef] [PubMed]
3. Rapisarda, M.; Marken, F.; Meo, M. Graphene oxide and starch gel as a hybrid binder for environmentally friendly high-performance supercapacitors. *Commun. Chem.* **2021**, *4*, 169. [CrossRef]
4. Faggioli, E.; Rena, P.; Danel, V.; Andrieu, X.; Mallant, R.; Kahlen, H. Supercapacitors for the energy management of electric vehicles. *J. Power Sources* **1999**, *84*, 261–269. [CrossRef]
5. Kouchachvili, L.; Yaici, W.; Entchev, E. Hybrid battery/supercapacitor energy storage system for the electric vehicles. *J. Power Sources* **2018**, *374*, 237–248. [CrossRef]
6. Augusto, G.D.S.; Scarminio, J.; da Silva, P.R.C.; de Siervo, A.; Rout, C.S.; Rouxinol, F.; Gelamo, R.V. Flexible metal-free supercapacitors based on multilayer graphene electrodes. *Electrochim. Acta* **2018**, *285*, 241–253. [CrossRef]
7. Lekakou, C.; Moudam, O.; Markoulidis, F.; Andrews, T.; Watts, J.F.; Reed, G.T. Carbon-based fibrous EDLC capacitors and supercapacitors. *J. Nanotechnol.* **2011**, *2011*, 409382. [CrossRef]
8. Kandalkar, S.; Dhawale, D.; Kim, C.-K.; Lokhande, C. Chemical synthesis of cobalt oxide thin film electrode for supercapacitor application. *Synth. Met.* **2010**, *160*, 1299–1302. [CrossRef]
9. Chmiola, J.; Largeot, C.; Taberna, P.-L.; Simon, P.; Gogotsi, Y. Monolithic carbide-derived carbon films for micro-supercapacitors. *Science* **2010**, *328*, 480–483. [CrossRef]
10. Simon, P.; Gogotsi, Y. Materials for electrochemical capacitors. In *Nanoscience Technology: A Collection of Reviews from Nature Journals*; World Scientific Publishing: Singapore, 2010; pp. 320–329.
11. Palchoudhury, S.; Ramasamy, K.; Gupta, R.K.; Gupta, A. Flexible supercapacitors: A materials perspective. *Front. Mater.* **2019**, *5*, 83. [CrossRef]

12. Xie, P.; Yuan, W.; Liu, X.; Peng, Y.; Yin, Y.; Li, Y.; Wu, Z. Advanced carbon nanomaterials for state-of-the-art flexible supercapacitors. *Energy Storage Mater.* **2021**, *36*, 56–76. [CrossRef]
13. Khosrozadeh, A.; Darabi, M.A.; Xing, M.; Wang, Q. Flexible electrode design: Fabrication of freestanding polyaniline-based composite films for high-performance supercapacitors. *ACS Appl. Mater. Interfaces* **2016**, *8*, 11379–11389. [CrossRef] [PubMed]
14. Cheng, J.; Zhao, B.; Zhang, W.; Shi, F.; Zheng, G.P.; Zhang, D.; Yang, J. High-Performance Supercapacitor Applications of NiO-Nanoparticle-Decorated Millimeter-Long Vertically Aligned Carbon Nanotube Arrays via an Effective Supercritical CO_2-Assisted Method. *Adv. Funct. Mater.* **2015**, *25*, 7381–7391. [CrossRef]
15. Cheng, J.; Chen, S.; Chen, D.; Dong, L.; Wang, J.; Zhang, T.; Jiao, T.; Liu, B.; Wang, H.; Kai, J.J.; et al. Editable asymmetric all-solid-state supercapacitors based on high-strength, flexible, and programmable 2D-metal–organic framework/reduced graphene oxide self-assembled papers. *J. Mater. Chem. A* **2018**, *6*, 20254–20266. [CrossRef]
16. Yu, L.; Hu, L.; Anasori, B.; Liu, Y.-T.; Zhu, Q.; Zhang, P.; Gogotsi, Y.; Xu, B. MXene-bonded activated carbon as a flexible electrode for high-performance supercapacitors. *ACS Energy Lett.* **2018**, *3*, 1597–1603. [CrossRef]
17. Wang, S.; Liu, N.; Su, J.; Li, L.; Long, F.; Zou, Z.; Jiang, X.; Gao, Y. Highly stretchable and self-healable supercapacitor with reduced graphene oxide based fiber springs. *Acs Nano* **2017**, *11*, 2066–2074. [CrossRef]
18. Ban, S.; Malek, K.; Huang, C.; Liu, Z. A molecular model for carbon black primary particles with internal nanoporosity. *Carbon* **2011**, *49*, 3362–3370. [CrossRef]
19. Silva, T.; Moraes, F.C.; Janegitz, B.C.; Fatibello-Filho, O. Electrochemical biosensors based on nanostructured carbon black: A review. *J. Nanomater.* **2017**, *2017*, 4571614. [CrossRef]
20. Delgado, K.P.; Raymundo-Pereira, P.A.; Campos, A.M.; Oliveira, O.N., Jr.; Janegitz, B.C. Ultralow Cost Electrochemical Sensor Made of Potato Starch and Carbon Black Nanoballs to Detect Tetracycline in Waters and Milk. *Electroanalysis* **2018**, *30*, 2153–2159. [CrossRef]
21. Phillips, C.; Al-Ahmadi, A.; Potts, S.-J.; Claypole, T.; Deganello, D. The effect of graphite and carbon black ratios on conductive ink performance. *J. Mater. Sci.* **2017**, *52*, 9520–9530. [CrossRef]
22. Garcia, D.M.E.; Pereira, A.S.T.M.; Almeida, A.C.; Roma, U.S.; Soler, A.B.A.; Lacharmoise, P.D.; Ferreira, I.M.D.M.; Simão, C.C.D. Large-Area Paper Batteries with Ag and Zn/Ag Screen-Printed Electrodes. *ACS Omega* **2019**, *4*, 16781–16788. [CrossRef] [PubMed]
23. Pech, D.; Brunet, M.; Taberna, P.-L.; Simon, P.; Fabre, N.; Mesnilgrente, F.; Conédéra, V.; Durou, H. Elaboration of a microstructured inkjet-printed carbon electrochemical capacitor. *J. Power Sources* **2010**, *195*, 1266–1269. [CrossRef]
24. Rodríguez, F.J.; Galotto, M.J.; Guarda, A.; Bruna, J.E. Modification of cellulose acetate films using nanofillers based on organoclays. *J. Food Eng.* **2012**, *110*, 262–268. [CrossRef]
25. Wsoo, M.A.; Shahir, S.; Bohari, S.P.M.; Nayan, N.H.M.; Razak, S.I.A. A review on the properties of electrospun cellulose acetate and its application in drug delivery systems: A new perspective. *Carbohydr. Res.* **2020**, *491*, 107978. [CrossRef] [PubMed]
26. Xu, Y.; Hao, Q.; Mandler, D. Electrochemical detection of dopamine by a calixarene-cellulose acetate mixed Langmuir-Blodgett monolayer. *Anal. Chim. Acta* **2018**, *1042*, 29–36. [CrossRef] [PubMed]
27. Andreotti, I.A.D.A.; Orzari, L.O.; Camargo, J.R.; Faria, R.C.; Marcolino-Junior, L.H.; Bergamini, M.F.; Gatti, A.; Janegitz, B.C. Disposable and flexible electrochemical sensor made by recyclable material and low cost conductive ink. *J. Electroanal. Chem.* **2019**, *840*, 109–116. [CrossRef]
28. Kondo, T.; Sakamoto, H.; Kato, T.; Horitani, M.; Shitanda, I.; Itagaki, M.; Yuasa, M. Screen-printed diamond electrode: A disposable sensitive electrochemical electrode. *Electrochem. Commun.* **2011**, *13*, 1546–1549. [CrossRef]
29. Kadara, R.O.; Jenkinson, N.; Banks, C.E. Characterization and fabrication of disposable screen printed microelectrodes. *Electrochem. Commun.* **2009**, *11*, 1377–1380. [CrossRef]
30. Rocha, D.S.; Duarte, L.C.; Silva-Neto, H.A.; Chagas, C.L.; Santana, M.H.; Filho, N.R.A.; Coltro, W.K. Sandpaper-based electrochemical devices assembled on a reusable 3D-printed holder to detect date rape drug in beverages. *Talanta* **2021**, *232*, 122408. [CrossRef]
31. Li, H.; Wang, W.; Lv, Q.; Xi, G.; Bai, H.; Zhang, Q. Disposable paper-based electrochemical sensor based on stacked gold nanoparticles supported carbon nanotubes for the determination of bisphenol A. *Electrochem. Commun.* **2016**, *68*, 104–107. [CrossRef]
32. Camargo, J.R.; Andreotti, I.A.; Kalinke, C.; Henrique, J.M.; Bonacin, J.A.; Janegitz, B.C. Waterproof paper as a new substrate to construct a disposable sensor for the electrochemical determination of paracetamol and melatonin. *Talanta* **2020**, *208*, 120458. [CrossRef] [PubMed]
33. Orzari, L.; Andreotti, I.A.D.A.; Bergamini, M.F.; Marcolino-Junior, L.H.; Janegitz, B.C. Disposable electrode obtained by pencil drawing on corrugated fiberboard substrate. *Sens. Actuators B Chem.* **2018**, *264*, 20–26. [CrossRef]
34. Carvalho, J.H.; Gogola, J.L.; Bergamini, M.F.; Marcolino-Junior, L.H.; Janegitz, B.C. Disposable and low-cost lab-made screen-printed electrodes for voltammetric determination of L-dopa. *Sens. Actuators Rep.* **2021**, *3*, 100056. [CrossRef]
35. Liu, B.; Shioyama, H.; Jiang, H.-L.; Zhang, X.; Xu, Q. Metal-organic framework (MOF) as a template for syntheses of nanoporous carbons as electrode materials for supercapacitor. *Carbon* **2010**, *48*, 456–463. [CrossRef]
36. Danaee, I.; Jafarian, M.; Forouzandeh, F.; Gobal, F.; Mahjani, M. Mahjani, Electrochemical impedance studies of methanol oxidation on GC/Ni and GC/NiCu electrode. *Int. J. Hydrogen Energy* **2009**, *34*, 859–869. [CrossRef]
37. Krause, A.; Kossyrev, P.; Oljaca, M.; Passerini, S.; Winter, M.; Balducci, A. Electrochemical double layer capacitor and lithium-ion capacitor based on carbon black. *J. Power Sources* **2011**, *196*, 8836–8842. [CrossRef]

38. Ma, X.; Song, X.; Yu, Z.; Li, S.; Wang, X.; Zhao, L.; Zhao, L.; Xiao, Z.; Qi, C.; Ning, G.; et al. S-doping coupled with pore-structure modulation to conducting carbon black: Toward high mass loading electrical double-layer capacitor. *Carbon* **2019**, *149*, 646–654. [CrossRef]
39. Khamkeaw, A.; Asavamongkolkul, T.; Perngyai, T.; Jongsomjit, B.; Phisalaphong, M. Interconnected Micro, Meso, and Macro Porous Activated Carbon from Bacterial Nanocellulose for Superior Adsorption Properties and Effective Catalytic Performance. *Molecules* **2020**, *25*, 4063. [CrossRef]
40. Toupin, M.; Bélanger, D.; Hill, I.R.; Quinn, D. Performance of experimental carbon blacks in aqueous supercapacitors. *J. Power Sources* **2005**, *140*, 203–210. [CrossRef]
41. Frackowiak, E.; Beguin, F. Carbon materials for the electrochemical storage of energy in capacitors. *Carbon* **2001**, *39*, 937–950. [CrossRef]
42. Conway, B.E. *Electrochemical Supercapacitors: Scientific Fundamentals and Technological Applications*; Springer Science & Business Media: Berlin/Heidelberg, Germany, 2013.
43. Orazem, M.E.; Tribollet, B. *Electrochemical Impedance Spectroscopy*; Whily: Hoboken, NJ, USA, 2008; pp. 383–389.
44. Lasia, A. *Electrochemical Impedance Spectroscopy and Its Applications, Modern Aspects of Electrochemistry*; Springer: Berlin/Heidelberg, Germany, 2002; pp. 143–248.
45. Nasibi, M.; Golozar, M.A.; Rashed, G. Nano zirconium oxide/carbon black as a new electrode material for electrochemical double layer capacitors. *J. Power Sources* **2012**, *206*, 108–110. [CrossRef]
46. Bobacka, J.; Lewenstam, A.; Ivaska, A. Electrochemical impedance spectroscopy of oxidized poly (3, 4-ethylenedioxythiophene) film electrodes in aqueous solutions. *J. Electroanal. Chem.* **2000**, *489*, 17–27. [CrossRef]
47. Zhang, C.; Xie, Y.; Wang, J.; Pentecost, A.; Long, D.; Ling, L.; Qiao, W. Effect of graphitic structure on electrochemical ion intercalation into positive and negative electrodes. *J. Solid State Electrochem.* **2014**, *18*, 2673–2682. [CrossRef]
48. Mitchell, J.B. *Understanding the Role of Structural Water for High Power Electrochemical Energy Storage in Tungsten Oxide Hydrates*; North Carolina State University: Raleigh, NC, USA, 2021.
49. Orellana, K.P.D. Low Cost, Carbon-Based Micro-and Nano-Structured Electrodes for High Performance Supercapacitors. Ph.D. Thesis, Clemson University, Clemson, SC, USA, 2016.
50. Thissandier, F.; Le Comte, A.; Crosnier, O.; Gentile, P.; Bidan, G.; Hadji, E.; Brousse, T.; Sadki, S. Highly doped silicon nanowires based electrodes for micro-electrochemical capacitor applications. *Electrochem. Commun.* **2012**, *25*, 109–111. [CrossRef]
51. Zou, Y.; Zhang, Z.; Zhong, W.; Yang, W. Hydrothermal direct synthesis of polyaniline, graphene/polyaniline and N-doped graphene/polyaniline hydrogels for high performance flexible supercapacitors. *J. Mater. Chem. A* **2018**, *6*, 9245–9256. [CrossRef]
52. Zhou, J.; Zhu, T.; Xing, W.; Li, Z.; Shen, H.; Zhuo, S. Activated polyaniline-based carbon nanoparticles for high performance supercapacitors. *Electrochim. Acta* **2015**, *160*, 152–159. [CrossRef]
53. Eustache, E.; Frappier, R.; Porto, R.L.; Bouhtiyya, S.; Pierson, J.-F.; Brousse, T. Asymmetric electrochemical capacitor microdevice designed with vanadium nitride and nickel oxide thin film electrodes. *Electrochem. Commun.* **2013**, *28*, 104–106. [CrossRef]
54. Lee, J.; Lee, Y.A.; Yoo, C.-Y.; Yoo, J.J.; Gwak, R.; Cho, W.K.; Kim, B.; Yoon, H. Self-templated synthesis of interconnected porous carbon nanosheets with controllable pore size: Mechanism and electrochemical capacitor application. *Microporous Mesoporous Mater.* **2018**, *261*, 119–125. [CrossRef]

Review

Recent Advances on Capacitive Proximity Sensors: From Design and Materials to Creative Applications

Reza Moheimani [1], Paniz Hosseini [2], Saeed Mohammadi [1,3] and Hamid Dalir [2,*]

1. Birck Nanotechnology Center, Purdue University, West Lafayette, IN 47907, USA; rezam@purdue.edu (R.M.); saeedm@purdue.edu (S.M.)
2. Department of Mechanical and Energy Engineering, Indiana University—Purdue University, Indianapolis, IN 46202, USA; phossei@iupui.edu
3. School of Electrical and Computer Engineering, Purdue University, West Lafayette, IN 47907, USA
* Correspondence: hdalir@purdue.edu

Abstract: Capacitive proximity sensors (CPSs) have recently been a focus of increased attention because of their widespread applications, simplicity of design, low cost, and low power consumption. This mini review article provides a comprehensive overview of various applications of CPSs, as well as current advancements in CPS construction approaches. We begin by outlining the major technologies utilized in proximity sensing, highlighting their characteristics and applications, and discussing their advantages and disadvantages, with a heavy emphasis on capacitive sensors. Evaluating various nanocomposites for proximity sensing and corresponding detecting approaches ranging from physical to chemical detection are emphasized. The matrix and active ingredients used in such sensors, as well as the measured ranges, will also be discussed. A good understanding of CPSs is not only essential for resolving issues, but is also one of the primary forces propelling CPS technology ahead. We aim to examine the impediments and possible solutions to the development of CPSs. Furthermore, we illustrate how nanocomposite fusion may be used to improve the detection range and accuracy of a CPS while also broadening the application scenarios. Finally, the impact of conductance on sensor performance and other variables that impact the sensitivity distribution of CPSs are presented.

Keywords: active materials; capacitive-based sensor; nanocomposites; proximity sensors

Citation: Moheimani, R.; Hosseini, P.; Mohammadi, S.; Dalir, H. Recent Advances on Capacitive Proximity Sensors: From Design and Materials to Creative Applications. *C* **2022**, *8*, 26. https://doi.org/10.3390/c8020026

Academic Editors: Olena Okhay and Gil Goncalves

Received: 31 March 2022
Accepted: 21 April 2022
Published: 5 May 2022

Publisher's Note: MDPI stays neutral with regard to jurisdictional claims in published maps and institutional affiliations.

Copyright: © 2022 by the authors. Licensee MDPI, Basel, Switzerland. This article is an open access article distributed under the terms and conditions of the Creative Commons Attribution (CC BY) license (https://creativecommons.org/licenses/by/4.0/).

1. Introduction

Flexible electronics have recently experienced extensive applications in the Internet of Things (IoT), human–machine interfaces, robotics, safety protection, human motion tracking, and healthcare systems, among other applications. With the IoT increasingly making its way into thousands of homes, the area of flexible electronic wearable gadgets has also entered a new phase [1]. On the one hand, traditional ceramic/metallic sensors suffer from the limitations of fragile materials, limited size, high cost, and a limited detecting range. On the other hand, the fast advancement of visualization technology, as well as the enormous market for smart devices, has raised the bar for flexible and wearable sensors, requiring excellent performance, mass manufacturing, low weight, and flexibility, among other characteristics [2]. Additionally, flexible wearable technologies hold tremendous promise for health monitoring and nursing applications. Clearly, flexible sensors with great performance will be vital for the implementation of flexible wearable devices [3]. As such, flexible nanocomposites may open the gates for implantable and stretchy devices.

Numerous flexible and stretchy sensors have been created for a variety of applications using micro-electro-mechanical system (MEMS) technology. Strain sensors, for example, detect body motion [4], tactile sensors monitor three-axis item handling/manipulation [5], and proximity sensors assist in detecting objects within a certain range [6]. Among all these applications, proximity detection is a key task that aims to identify information about objects that are physically near another object without requiring physical contact. Numerous electronic platforms and industrial equipment need the use of proximity sensors. These sensors are critical components of systems ranging from human–machine interfaces to health care, smart homes, shipping industry, and soft robotics [7–10]. In addition to detecting external stimuli, proximity sensors are anticipated to detect instantaneous and continuous activities such as vibration, inertia, shear force, and normal force [11,12] to meet the requirements of smooth multifunctional interactions. These applications are highly demanded to be expanded in the near future.

To sense objects, a proximity sensor often measures change in either an electrostatic field or some form of electromagnetic field. Hence, they can be categorized into capacitive [13], electrostatic [14], magnetic [15], electromagnetic radiation (infrared) [16], and light (visual) sensors [17]. Another way to classify proximity sensors is to categorize them according to their operating principles as capacitive [18], piezoresistive [19], triboelectric [20], piezoelectric [21], and photo-detecting devices. In comparison to triboelectric and piezoelectric sensors, capacitive and piezoresistive sensors have received substantial research and many similarities in their mechanics are apparent [22]. Piezoelectric and triboelectric sensors, on the other hand, have the apparent benefits of not requiring an external power source and being more responsive to dynamic stimuli. The main characteristics of each proximity detection technique are summarized in Table 1.

Table 1. A comparison of the most frequently used proximity detection techniques.

Sensing Technique	Detected Objects	Sensing Element	Operational Range	Standard Detective Circuit	Tangible Limitation
Optical	Non-conductive and conductive	Lighting resource	Frequency and condition dependent	Converter (V–I)	Lenses and object preparation needed
Ultrasonic	Non-conductive and conductive	Sound producer	Frequency and condition dependent	Digital to analog converter or Sensor modules	Object dependent
Inductive	Only Conductive	Metal coil	Coil size dependent	Impedance analyzer, LCR oscillator,	-
Capacitive	Non-conductive and conductive	Conductive electrode	Conductive electrode fabrication/size dependent	Charge amplifier, RC low pass filter, capacitance meter	-

Inductive sensing can detect conductive and/or ferromagnetic objects, and the maximum scanning range is typically about equal to the sensing coil's diameter. While both optical and ultrasonic techniques are capable of detecting non-conductive and conductive items, implementing a complicated light source or a sound wave actuator on fabric is challenging. Additionally, the detection range depends on the object's surface polish and material quality. Because the capacitive proximity sensors detect both conductive and non-conductive objects, they are well-suited to identify humans and passive objects. Additionally, the capacitive technique is easy to set up and usually requires fewer components than other sensing systems. It is the most suited mechanism for printed implementation since it allows for the use of a simple conductive electrode of any shape as the sensing element. A basic sensing element of any shape is useful for creative applications since the sensor may be customized to any artistic shape required by the designer. Capacitive proximity sensors (CPSs) have been used in a variety of applications to date, including determining the aging of composite insulators [23], estimating the permittivity and thickness of dielectric plates and shells [24], measuring tire strain in automobiles [25], force sensing in biomedical applications [26], liquid level detection [27], and harvest yield monitoring [28].

CPSs, in comparison to other proximity sensors, have a number of advantageous qualities, including affordable construction cost, low energy consumption, broad monitoring range, excellent dynamic response, and a flexible and changeable structural design [29].

Additionally, CPSs may respond more strongly to static stimuli, and their mechanisms and production methods are relatively straightforward [30]. CPSs have gained considerable attention among researchers in the past few years, as Figure 1 suggests.

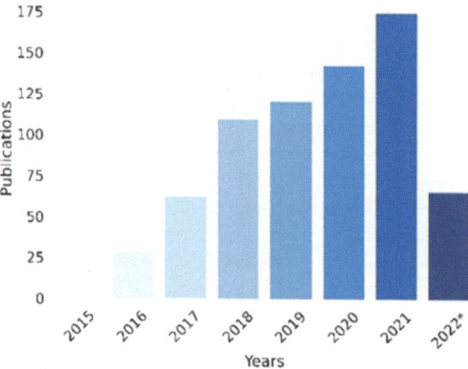

Figure 1. Number of published proximity-sensor-related papers during recent years.

The numbers in Figure 1 were obtained after a thorough exploration of search engines such as the Elsevier Online Library, Google Scholar, Web of Science, IEEE digital, and a few publications such as Wiley, Taylor & Francis, etc. We searched through peer-reviewed journals, technical bulletins, textbooks, and dissertations. The following keywords were used to compile the databases through an iterative process of research: "proximity sensors", "piezoelectric and triboelectric sensors", "displacement sensing", "proximity detection", and "occupancy-based control in HVAC". This research aims to provide standard terminology and taxonomy that will serve as a unifying framework for all engineering disciplines involved in the design and construction of capacitive proximity sensors.

Given the importance and numerous applications of CPSs, there is a lack of a thorough review study to highlight the recent progress and publications. Ref. [31], which is the closest match to the present study, was published in 2019 and many investigations have been performed since then, necessitating the need for a fresh review article. Ever since, a substantial body of research has been conducted on this topic, necessitating a need for an up-to-date review paper examining the most current state-of-the-art procedures. Nonetheless, extensive research on the variations of materials and technologies used for CPSs and their applications under various criteria has not been adequately addressed. The majority of previous literature is focused on closed-form mathematical modeling of CPSs, ignoring the materials employed in the production process. We want to bridge this gap by addressing current advancements in the manufacturing of CPSs. An overview of modeling approaches with their strengths and weaknesses, categorization of construction methods, design parameters, and constraints imposed by different applications are among some of the highlights of this review article. This study is rather different from previous publications as it addresses these issues and incorporates more recent publications and it provides a detailed picture of many factors related to proximity sensors. The evaluated articles are examined from various angles to illuminate their limitations across several aspects. This article will also assist in highlighting the present state of research on CPSs and any research gaps that have yet to be discovered.

2. Capacitive Sensing: Principals and Applications

A capacitive sensor functions similarly to a standard capacitor. A metal plate on the detecting face of the sensor is electrically coupled to an oscillator circuit, while the target to be detected serves as the second plate of the capacitor. In contrast to inductive sensors, which generate a magnetic field, capacitive sensors generate an electrostatic field. External capacitance between the internal sensor plate and target plate contributes to the oscillator tank impedance. As the target gets closer to the sensors, the increase of the oscillator tank capacitance decreases the oscillation frequency until it hits a threshold value and trigger the output. Figure 2 shows the schematic of a capacitive proximity sensor.

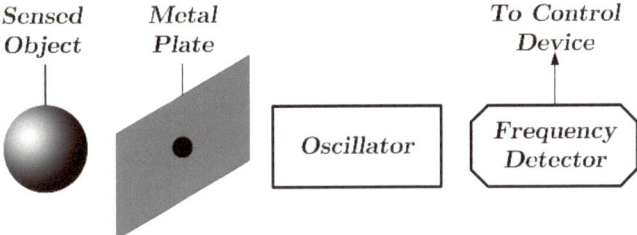

Figure 2. Schematic of a capacitive proximity sensor.

Capacitive sensing follows the principle of an interaction between a material under test and the probing electric field. A sensor-electrode-generated electric field penetrates a sensed object, causing electric displacement inside the tested material to counteract the applied field. The displacement pitch modifies the stored charge between those electrodes, thereby altering the inter-electrode capacitive field, which is used to infer the material properties, such as permittivity, conductivity, and their distributions, and ultimately to derive system variables such as temperature and humidity. As such, a CPS is usually made up of two electrically conducting electrodes that are connected through a potential difference to produce an alternating electrostatic field. When a targeting object moves near to the sensor, this field is disrupted, and the change in capacitance indicates the item's vicinity (self-capacitance if one electrode is utilized, as seen in Figure 3 on the right, or mutual capacitance if two electrodes are used, as shown in the left side of Figure 2). Thus, CPTs are capable of detecting the existence of any solid, metallic, or nonmetallic object by altering the capacitance of the sensor [32]. It is worth noting that different sensor combinations, geometries, and designs have been customized to meet various application requirements.

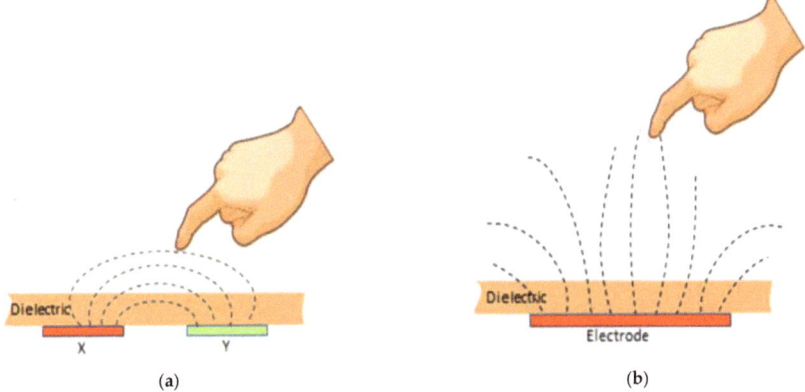

Figure 3. Schematic of two basic CPT sensors [31]: (**a**) mutual capacitance sensor, and (**b**) self-capacitance sensor.

Generally, a driving electrode is used to provide an electrical stimulation, whereas a sensing electrode is used to collect data. Typically, the frequency of the electrical stimulus, and hence the produced alternating electric field, are restricted. For example, the Agilent 4294A precision impedance analyzer operates between 35 and 105 MHz. Therefore, the relative permittivity and conductivity are only characterized in this frequency range [33].

Table 2 shows a few applications for the capacitive sensors implemented in previous literature. The capacitance of the circuit changes as the sensible object moves away from the electrodes. While capacitive displacement sensing may be used to measure distance, displacement, and position, the detection range is limited by the size and dielectric constant of the detected object. Additionally, by putting multiple electrodes in a regular pattern, one can discern the object's position, movement direction, and certain interaction intents expressed by the movement trajectories of a human body. Numerous applications for this sensing mode have been suggested, including electrical capacitance tomography [34], capacitive voltage sensors [35], capacitive humidity sensors [36], and capacitive gas sensors [37]. For example, electrical capacitance tomography is largely utilized for non-invasive imaging, with capacitance measurements used to determine the dielectric permittivity distribution inside inaccessible domains [34]. Additionally, based on the underlying features of the electrodes and certain transformation associations between the electrode distortion and the observed physical variables, it is easy to infer the force, acceleration, or actions of the subject human as the muscle moves from the displacements. The capacitance changes generated by electrode deformation are used to determine acceleration [35], angles [36], force [37], displacement [38], and muscle action for interaction [39].

Table 2. Some of the practical applications associated with capacitive sensors.

Reference	Application
[34]	Electrical capacitance tomography
[35]	Capacitive voltage sensors
[36]	Capacitive humidity sensors
[37]	Capacitive gas sensors
[38]	Displacement detection
[39]	Muscle action for interaction

In Refs. [40–43], CPSs are offered as a viable cost-effective and nondestructive alternative to optical sensors for a broad variety of applications involving the examination of geometrical and physical properties. Their sensitivity, on the other hand, is affected by moisture and temperature. Additionally, parasitic capacitance and noise from external disturbances may affect the response of capacitive sensors, necessitating proper shielding of the device and design of the readout circuits. Capacitive proximity-sensing methods have been extensively used for the nondestructive evaluation of materials with poor conductivity [44].

Capacitive sensing methods have also been utilized to monitor the structure healthy state of a concrete slab retrofitted with composites by detecting the local fluctuation in the dielectric characteristics of the materials under test [45]. The proximity capacitive approach was used by El-Dakhakhn et al. [46] to identify empty cells in grouted masonry buildings. A concentric coplanar capacitive sensor was designed for quantitative material property assessment of multilayered dielectrics [47], and the suggested approach for detecting water incursion in random constructions was experimentally confirmed. Additionally, the outer insulating layer of electric wires is often made of a low-conductivity substance. Capacitive sensing methods have been used to characterize the insulation qualities of cables. Chen et al. [48] developed a capacitive probe for determining the permittivity of wire insulation, and tests established the possibility of determining the state of wiring insulation deterioration using quantitative capacitive approaches. The introduction of proximity-coupled interdigital sensors to detect insulation degradation in power system cables confirms that the proximity capacitive approach is sensitive to the existence of holes and water trees in a power line cable [49]. Sheldon et al. [50] designed an interdigital capacitive sensor to detect aircraft wire aging damage and experimentally confirmed the capacitance fluctuation caused by aviation fluid immersion. Figure 4 shows a few CPSs used in traditional applications, as well as creative industries.

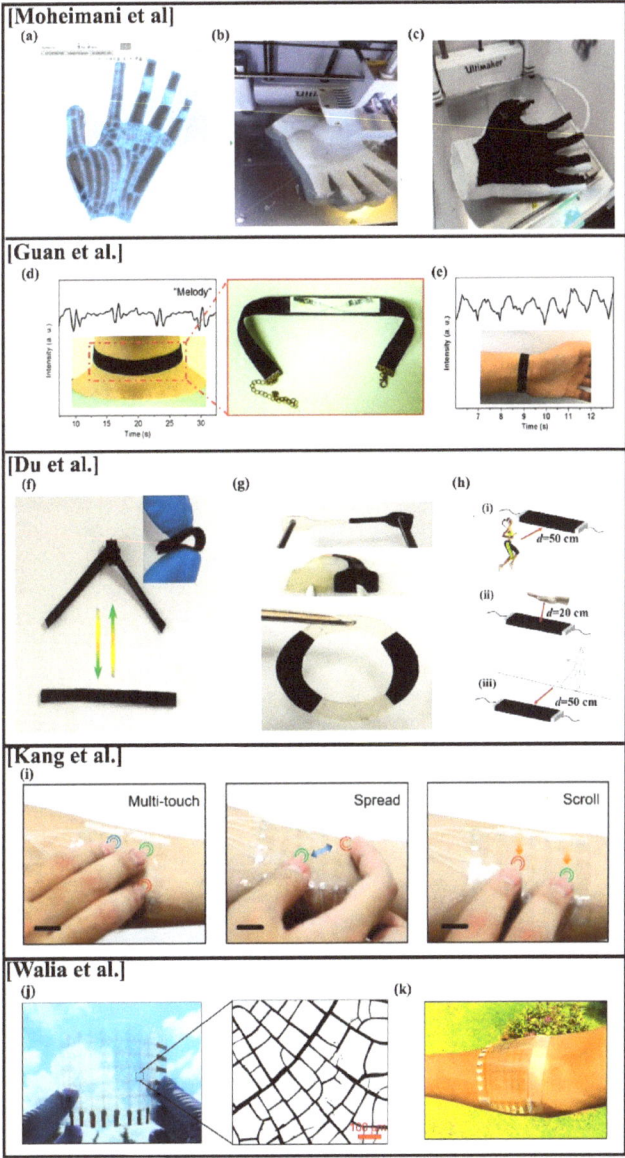

Figure 4. Application of proximity sensors proposed by Refs. [12,15,19,24,47]: (**a**) solid-shell curvy model, (**b**) 3D-printed thermoplastic polyurethane (TPU)/PVA model, (**c**) 3D printing a shell model before wiring, (**d**) signal of relative capacitance as measured by a choker sensor while the phrase "melody" is repeated four times, (**e**) recording pulse signal when a fiber-sensor is attached to the wrist, (**f**) optical images demonstrating the resulting flexible conductive films, (**g**) PAM (white) and PAM-FGO (black) fragments in optical photos of the healed specimens, (**h**) the PAM-FGO film-based proximity sensors allow for remote monitoring of human movements, (**i**) image of a graphene electrode-based wearable capacitive touch sensor, (**j**) optoelectronic characteristics of capacitive sensor, and (**k**) bendability and wearability of the proposed sensor.

3. Design, Materials, and Fabrication of CPSs

The most often used electrode configurations for capacitive sensing are planar parallel plate electrodes and co-planar electrodes (also called capacitive proximity sensors). Capacitive proximity sensors operate on the basis of the fringing electric field effect. In comparison to the traditional parallel plate capacitor, capacitive proximity sensors have several advantages, including one-sided access (the other side can be open to the environment), easy control of signal strength via dimension changes, multiple physical implications in the same structure (magnetic, acoustic, and electric), and a broad frequency range of operation. As a result, they are extensively employed in a variety of sectors, including humidity sensing, monitoring material qualities, chemical sensing, biosensing, and sensing of electrical insulating properties [51,52].

According to research, the electrode designs and parameters have a significant effect on capacitive sensor performance metrics such as signal intensity, diameter, sensitivity, and signal-to-noise ratio, which all impact the detecting capabilities of capacitive proximity sensors. Numerous improvements have been made to the performance of capacitive proximity sensors [53,54]. Various sensor designs were explored, including square-shaped, maze, spiral, and comb patterns. It was found that complex sensor patterns may increase the effective electrode area, hence increasing sensor signal and sensitivity [55]. Rivadeneyra et al. [56] designed a serpentine structure that combines meandering and interdigitated electrodes to increase signal strength and sensitivity. For humidity measurement, a capacitive sensor with an interdigital electrode arrangement and an enhanced height was built [57]. In comparison to the conventional interdigital electrode sensor, the suggested sensor demonstrated increased sensitivity because of the confinement of horizontal electric field lines in the polyimide sensing layer. Syaifudin et al. [58,59] investigated the effect of electrode configuration on capacitive proximity-sensor performance and discovered that the optimal number of negative electrodes between two adjacent positive electrodes can increase chemical detection sensitivity. A few petal-like electrode devices were constructed for water detection in an automated windshield system [60].

To enhance the flexibility and wearability of CPSs, flexible electrodes and dielectric layers made of polymer elastomers such as polyethylene terephthalate (PET) [61], polydimethylsiloxane (PDMS) [62], polyvinyl alcohol (PVA) [63], polyimide (PI) [64], polyvinylidene fluoride (PVDF), and Eco-flex [65] are frequently used [66]. Pressure sensitivity of these flexible substrates still needs improvement. As a result, a straightforward approach for introducing conductive fillers into the dielectric layer of polymer elastomers has been widely researched [67]. Due to the percolation threshold hypothesis [68], the inclusion of a conductive filler may raise the dielectric constant under applied pressure, resulting in a change in capacitance [69]. Additionally, the sensor's noncontact detection mode enables it to recognize and monitor the form and location of an item without a physical touch and interaction with the surrounding environment, highlighting the sensor's particular capabilities. Zhang et al. [70] produced a stretchy dual-mode sensor array capable of detecting a 4% relative capacitance fluctuation at a noncontact distance of 10 cm. Sarwar et al. [71] described a transparent touch sensor developed on a hydrogel electrode with a maximum capacitance corresponding difference of 15% in absolute value. Due to the structure and material properties of the classic film- or resin-based capacitive sensors, their low air permeability is incompatible with sweat evaporation for wearable electronic application, thus impeding long-term deployment of such devices for wearable electronics. As a result, flexible capacitive sensors with a high degree of breathability are still required to increase comfort and durability.

Textile-based capacitive sensors have been claimed to increase the permeability of flexible capacitive sensors owing to their low weight, flexibility, deformability, comfort, and softness [72,73]. By depositing PDMS on the surface of the conductive fiber as a dielectric layer and vertically stacking the two PDMS-coated fibers, Lee et al. [74] created a highly sensitive CPS. Chen et al. [75] electrospun the nylon dielectric constant and electrode to create a CPS that can detect human joint motion with high accuracy. As a result, textile-based CPSs are favored for achieving durability, flexibility, multifunctional sensing capabilities, and comfort all at the same time, making them an important research topic in the field of flexible and wearable CPSs.

Nonetheless, the construction process of some CPSs is barely documented, particularly for fabric-based CPSs with multifunctional sensing. Table 3 summarizes some of the studies that report the construction processes of their proposed sensors in a detailed manner.

Table 3. Summary description of few CPS studies with manufacturing details, CDC (capacitance-to-digital converter), PCB (printed circuit board), and LCR (inductance, capacitance, resistance).

Reference	Application	Shape/Size	Measuring Range	Shielding	Error or Resolution	Method of Measurement
[44]	Inductive and capacitive sensors integration	20 mm × 5 mm	10 mm	×	-	Resonance
[45]	Ultrasonic and capacitive integration	60 mm × 30 mm × 0.1 mm	200 mm	✓	30 mm	CDC:AD7143
[63]	Two arrays of 16 × 16 electrodes	5 mm × 100 mm Rectangle	170 mm	✓	-	200 kHz charging circuit
[64]	Woven-polyester fabric, printed on a standard	180 mm × 180 mm × 3 mm Spiral	80 mm	×	0.5 mm	CDC:MTCH112
[67]	Temperature and capacitive sensor combined	7 mm × 3 mm × 0.1 mm Rectangle	17 mm	✓	-	CDC:AD7746
[73]	Moving target detection CPS	310 mm × 190 mm	60 mm	✓	5 mm	LPF, C/V circuit
[76]	Symmetrical distribution of electrodes—circular shape	45 mm × 45 mm Circular	55 mm	✓	0.3–4.6%	LCR meter
[47]	Equally distributed sensors (120°)	85 mm × 40 mm Single	336 mm	✓	3.3 mm	Neural Network

In this context, a 3D honeycomb, which consists of a supporting yarn layer and two independent mesh-knitted textiles, has been suggested because of its great wearing comfort and compression, making it a viable material for flexible CPSs [76]. In this context, Ref. [77] described the construction of a bimodal fabric-based CPS using a practical and affordable manufacturing process. The suggested sensor was made out of a 3D honeycomb fabric dielectric surface (weight: 220 g/m^2, thickness: 3 mm) and bottom and top conductive Ni-plated woven electrodes, giving it a high sensitivity for noncontact detection with a detection distance of 10 cm and a maximum relative capacitance change of 15%. Furthermore, at the short distance range ($\leq$3 cm), a maximum sensitivity of 0.022 cm^{-1} is obtained. When a hand is hovered at various distances, the capacitance fluctuations are consistently maintained, proving the proposed sensor's stable noncontact detecting response. Furthermore, when the finger is hung above the sensor array unit, the capacitance change rate of the corresponding sensor unit is 9%. This demonstrates that the proposed fabric-based sensor array can precisely detect the finger and provides outstanding noncontact spatial response.

In another study, the authors described a hierarchically porous silver nanowire-bacterial cellulose fiber that can be used to sense both the pressure and closeness of human fingers [15]. The conductive fiber was made by continuous wet spinning at a speed of 20 m/min, and it had a diameter of 52 cm, an electrical conductivity of 1.3×10^4 s/cm, a tensile strength of 198 MPa, and an elongation strain of 3% at break. To create the fiber sensor element, which is thinner than a human hair, the fibers were coaxially coated with a 10 cm thick poly(dimethyl siloxane) dielectric elastomer. The capacitance variations between the conductive cores in response to closeness were monitored using two fiber-based sensors that were arranged diagonally. The suggested sensor was found to be very sensitive to objects up to 29 cm away. In addition, the fiber may be simply sewed into clothes as pleasant and stylish heartbeat and voice-pulse sensors. A fiber sensor array may be used to play music and correctly identify the closeness of an item without the need for a touchpad. A two-by-two array was also shown for detecting faraway objects in two- and three-dimensional space.

A recent study looked at the usage of a noninvasive omnidirectional CPS developed in-house as a possible option for human–machine interaction applications [78]. The performance of the proposed sensor was compared to that of infrared, time-of-flight, and ultrasonic sensors, all of which are routinely employed in comparable applications. The suggested sensor is based on a heterodyning approach employed in theremin, which consists of two digital oscillators, one of which is coupled to a sensing (conductive) plate (primary/sensing). If a grounded item enters the detecting range, the oscillator's total capacitance rises, affecting the frequency of the produced square wave. In our tests, we employed a 10×10 cm^2 thin copper PCB square as a sensor plate. The detecting plate is linked to the main oscillator, which has a fixed 10 kΩ resistor in the feedback line. Any grounded item put within the sensing range will add capacitance to this oscillator, causing it to shift frequency by some amount. A measuring system is built to verify the performance, as shown in Figure 5. The camera is used to compute the displacement of the user's hand from the sensor module in this arrangement. They concluded that the sensing mechanism may be extended to all directions surrounding the detecting element by using an omnidirectional CPSs, such as the one described in their study. In addition, the suggested capacitive sensor's range (4–11 cm) is greater than that of comparable capacitive sensing technologies. The power usage was further decreased to 5 mW by duty-cycling the power supply while still allowing 50 readings per second to be acquired.

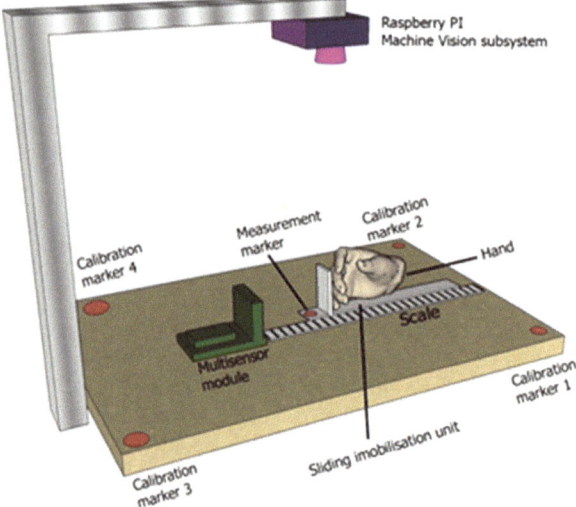

Figure 5. Measurement setup proposed by Ref. [78] for CPSs.

To advance contactless measurement, some research has been conducted on incorporating the proximity-sensing function of nanofillers into applications for other kinds of sensors, such as touch, pressure, and strain sensors, which are gaining growing popularity in wearable electronics [79–88]. Table 4 summarizes the key characteristics of modern CPSs coated on a variety of flexible polymeric substrates using nanostructured particles/fillers. As previously stated, only a few flexible nanocomposite polymeric CPSs with a broad detection range exist.

Table 4. Summary description of a few CPS studies with manufacturing details. A summary of the fundamental properties of flexible capacitance-type proximity sensors enhanced with nanomaterials. PET stands for polyethylene terephthalate that is ultrathin FPCB is for flexible printed circuit board. CNC stands for cellulose nanocrystals. GO is for graphene oxide, PDMS stands for polydimethylsiloxane, AgNWs stands for silver nanowires, CMC stands for carbon microcoils, MWCNT stands for multiwall carbon nanotube, and ms stands for millisecond.

Reference	Active Materials and Substrate	Response Time (1 pF)	Shape/Size	Sensitivity $[\frac{\Delta C}{c_0} mm^{-1}]$	Operational Range	Error or Resolution	Other Features
[15]	AgNW–BC/PDMS fibers	<75 ms	53 µm Diameter Thickness 10 µm	0.19 (Skin)	30 cm	1 mm	Proximity and Pressure, bacterial cellulose coated with PDMS, Wet spinning for compressibility
[16]	CNC-m-rGO-epoxy GO (conducting particles)	-	1 × 2 cm² Thickness 0.16 mm	7.8 (Skin) 0.0 (Copper and plastic rod)	0.6 cm	0.6 mm	Durability of the touch sensor (100 cycles at the distance of 0.02 cm), Excellent stability and repeatability, Average recovery time (3 s)
[10,82]	PET-PDMS-AgNWs PDMS (dielectric layer) AgNWS (electrodes)	<40 ms	7.5 × 2.5 cm² 1 mm Thickness	0.06–0.12 (Skin)	9 cm [10] 14 cm [82]	4.8 mm	All pressure sensing Reversibility (up to 100 kPa) [82] and (50% strain) [10], Stability (2 h), Bending stability (310 cycles and r_b = 3 cm) [10]
[24]	Graphene, acrylic PET, PET (mesh-structured), Graphene (electrodes), Acrylic polymer (dielectric layer)	<60 ms	6 × 4 cm² 8 × 8 Channels 0.03 mm	0.66 (iron) 0.10 (skin)	1 cm (Iron) 7 cm (Skin)	5 mm	Touch sensing Searchability~9–16% (r_b = 0.15 cm)
[33]	AgNW-PDMS AgNWS (electrodes) PDMS (dielectric layer)	<40 ms	800 × 2500 µm² 3 µm Thickness	0.16 (Skin)	15 cm	5 mm	Durability (200 cycles in 100 kPa), All pressure sensing Reversibility
[63]	PCB	-	0.7 × 0.3 cm² 16 × 16 sensor array	0.18 (Steel)	8 cm	1% output frequency	switched-charge amplifier and sampling/filtering for noise rejection,
[67]	CMC-MWCNT-silicone CMC (elastomer composite sheet)	-	3.3 × 3.3 cm² FPCB electrode layer Thickness 0.6 mm	0.10 (Copper)	6 cm	2 mm	Inductive and capacitive sensing modes, Repeatability and reversibility (5 cycles), Durability (3000 cycles for 150 kPa), Maximum detection 1.5%
[74]	PDMS Copper Electrodes	-	600 × 600 µm² 16 × 16 capacitor array	0.5 (Plastics, PVC, Acrylic, HDPE)	17 cm	0.5 mm	Dual-mode functioning custom circuit board, many possible variations of the electrode configuration
[78]	-	<16 ms	-	-	5–10 cm	-	Energy efficient using duty-cycling power supply, boot-up self-adjustment mechanism via digital potentiometer, Wide variety of applications
[81]	Fabinks TC-C4001, Polyester woven	<30 ms	0.4 × 0.4 cm² Thickness 30 µm	0.79 (Skin)	0.1–40 cm	0.5 mm	76% less conductive ink via loop design, Microchip MTCH112 to simplify the circuit
[75]	Micro-Electro-Mechanical	-	sensors size 500 × 50 µm² electrode width of 10 µm	0.8 (Skin)	10–10,000 µm	0.48 fF/µm	Conductor or nonconductor measuring, batch fabricated via Micro-Electro-Mechanical, Micro sensor size, Capable of measuring permittivity
[87]	PET-PDMS CP coating	<100 ms	10 × 10 cm² 15 grid lines, Effective line width 3.3 mm	0.5 (Skin)	13 cm	-	High flexibility (2 cm), Transparent (90%), Can detect several stimuli, Pressure touch, Minimal noise
[88]	TPU-CNT	<30 ms	6 × 2 cm² Thickness 0.5 mm	0.3 (Brass, Skin)	2–22 cm	0.3 mm	Excellent detection range, Reasonable flexibility and durability

In Refs. [19,79], a new form of conductive flexible film is developed using self-healing polyazomethine (PAM) and functionalized graphene oxide (FGO) as conductive fillers. PAM-FGO conductive films were made by using imine bonds to crosslink the PAM polymer chains with the FGO. With a skin-like elongation of 200 percent and elastic moduli of 0.75 MPa, the conductive films produced demonstrated excellent flexibility. Furthermore, the papered conductive films showed an excellent intrinsic self-healing capability that benefited from dynamic covalent interactions, with an important role in enhancing mechanical properties and electrical conductivity as high as 95 percent, respectively, after healing the fractured sample for 24 h at 25 °C. Importantly, strain sensors based on the PAM-FGO conductive film showed ultrahigh sensing sensitivity, with GF up to 641, and could detect large-scale human movements and delicate physical signals with accuracy. Because of its ultrasensitive capabilities, proximity sensors based on the organic film field-effect transistor architecture could record human movements from a distance of up to 1 m. This research can pave the way for the creation of multifunctional soft materials. Sensors based on the FGO-PAM film have a wide range of applications in wearable electronics, health diagnostics, human–machine interface, and security protection.

Ref. [80] describes the dispenser printing of a CPS on a woven cloth made entirely of polyester. Three electrode designs, spiral, filled, and loop, are reviewed and contrasted to determine a trade-off between ink consumption and maximum detection distance. To facilitate operation, a simple detecting circuit based on a proximity sensor chip is constructed. Three patterns with identical exterior dimensions were used to investigate the effect of sensing electrode design on performance. A commercially available CPS integrated circuit (IC), the Microchip MTCH112, is chosen to obtain the proximity sensor functionality. The MTCH112 supports a maximum input capacitance of 41 pF and may be set to offer two independent CPSs channels or one sensor channel plus an active guard electrode to mitigate the effects of external electrical noise. The linearity (maximum detection distance correlation coefficient with the proximity sensor width) of the sensing circuit is determined to be 0.8 after testing printed CPS with varied diameters ranging from 1 cm to 40 cm. A broad detection range indicates that the proximity sensor is capable of interacting with humans in big-scale creative applications.

Using an extremely transparent material, Ref. [81] investigated the continuous response of pressure and proximity sensing. The sensor functions by detecting the capacitance change between two transparent silver nanowire electrodes. A sandwich construction was used to construct the capacitive sensor. The substrate was a polyethylene terephthalate (PET) sheet with parallel electrode stripes printed on it. Two of these PET substrates were connected together using an elastic polydimethylsiloxane dielectric layer with orthogonal top and bottom electrode stripes (row-and-column electrodes). Throughout their tests, the authors encountered a constant issue: the development of ghost points for multi-point object recognition, which occur from virtual intersections in places other than the actual locations of the items. To examine the sensor's reaction performance, the capacitance fluctuations as a finger approached and moved away for four cycles was depicted; the device demonstrated excellent reversibility and stability with a rapid response, allowing for accurate proximity detection.

Refs. [74,82] describe a modular dual-mode capacitive sensor for robot applications that combines touch and proximity-sensing capabilities into a single platform. The sensor is composed of a PDMS (polydimethylsiloxane) mechanical framework and a mesh of numerous copper electrode strips. The mesh is constructed from 16 top and 16 bottom copper strips that are crossed to produce a 16-capacitor array. The suggested sensor is capable of switching between tactile and proximity sensing modes or vice versa by simply rearranging the electrode connections. Through simulation of numerous two-dimensional models, the capacitance shift produced by an approaching item was calculated. For proximity measurement, we evaluated a variety of materials ranging from conducting metals to a human hand. The built sensor was capable of detecting a human hand up to 170 mm distant. Additionally, the authors successfully showed the viability of the suggested sensor operating in dual modes in real time by using a custom-designed PCB, and a data-collecting pad.

Ref. [83] demonstrated an electronic skin equipped with CPSs. They incorporated carbon nanotube forests, which serve as the sensing element, into a transparent and flexible artificial skin through a simple and inexpensive manufacturing technique. The electronic skin exhibited a high capacitance and sensitivity, allowing for easy detection of proximity using specialized laboratory equipment and commercially accessible on-chip circuits. At room temperature, the capacitance between the electronic skin and two CNT forest sensing devices was continually monitored. They plotted capacitance against the vertical distance between the grounded electrode and the electronic skin, as seen in Figure 6. The maximum vertical gap considered in this study is 10 mm.

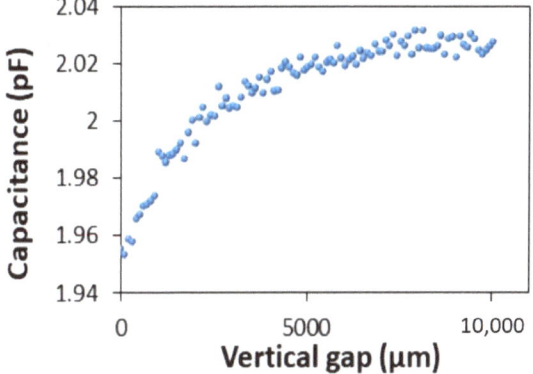

Figure 6. Measured capacitance versus the vertical gap of the proposed CPS in Ref. [83].

In reference [24,84], a CPS with excellent sensing capabilities in both contact and non-contact modes is described. This is made possible by the use of graphene and a thin device architecture. The resultant graphene-based three-dimensional sensor is directly attached to deformable human body parts such as the palms and forearms and demonstrated high stretchability (15%) and reasonable sensing performance in noncontact modes (22 dB SNR at a 7 cm distance). They demonstrated a three-dimensional mapping graph of relative capacitance changes across three distinct distances between the item and sensor in this setting (0.5, 1, and 1.5 cm). The depicted surfaces represent the results of the mapping of relative capacitance changes from the 8 × 8 capacitive sensor array. Their findings are shown in Figure 7.

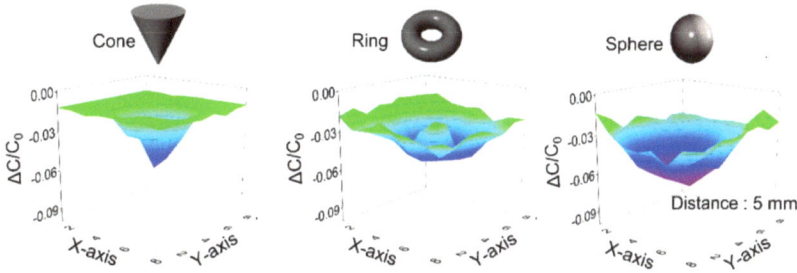

Figure 7. 3D representation of capacitance changes for three shapes: cone, ring, and sphere [24].

Ref. [85] describes a flexible CPS with great transmittance and sensitivity. An inter-digital capacitance (IDC) structure is developed using a polydimethylsiloxane (PDMS) sensing layer to cover indium tin oxide (ITO) electrodes interdigitated on a polyethylene-terephthalate (PET) substrate. They studied the pressure and proximity sensing capabilities of the constructed IDC-based CPS sensor, resulting in a maximum sensing distance of 200 mm. Figure 8 illustrates the manufacturing process of the proposed sensor. To construct the sensor, a PET sheet with a thickness of 0.125 mm was employed as the flexible substrate, which was then coated with ITO film. Second, laser etching was used to form the coplanar ITO interdigital electrodes. Third, the Dow Corning 184 prepolymer was mixed with the curing agent at a 10:1 weight ratio for 15 min and then spin-coated on the ITO electrodes for 31 s at a speed of 1500 rpm. Next, the sensor was degassed for 31 min in a vacuum desiccator to eliminate any air bubbles embedded in the sensitive film. It was then cured at 70 °C for over an hour to generate the sensor's final shape.

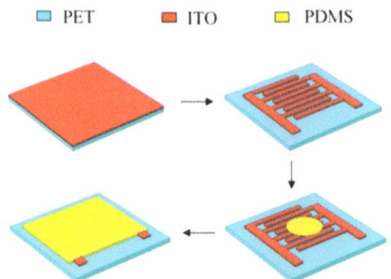

Figure 8. Construction process of IDC CPS sensor proposed by Ref. [85].

A dual-mode array sensor based on carbon micro-coils (CMC) on a soft dielectric elastomer surface layer is presented in [86]. It uses variations in inductance to determine the distance to an item. Numerous tests on dielectric substrates, electrode structures, and target objects are carried out by varying the electrical impedance created by CMC when excited with an alternating current at a dominating excitation frequency of 90+ kHz. The authors proved the performance of their 10×10 proximity tactile sensor, which is able to detect a 30 mg droplet at a maximum distance of 15 cm from the metal object.

In Refs. [47,87], the crackling templating approach was used to fabricate a flexible CPS panel. The metal-mesh electrode was pixelated into interlocking diamond patterns on flexible PET using a mask created by laser-printing toner. The dielectric material was a thin (30 m) PDMS layer. Although their suggested sensor is versatile and innovative, they do not discuss the specific properties of proximity sensing.

A polymer-based sensor with a nanostructure composite sensing element has been constructed for use in healthcare and automotive applications [88,89]. A probing station was used to apply distances ranging from 2 to 20 cm. To determine the maximum change in capacitance, the samples were saturated with 6 V direct current to eliminate the tunneling effect. Additionally, a 25 mV alternating current swiping signal was used to determine the film's capacitance at various frequencies. The CPS has a detection range of 12 cm and a precision of 0.3 percent/mm. Tunneling and fringing effects are studied to explain substantial capacitance shifts as sensing mechanisms. Percolation threshold investigation of various TPU/CNT concentrations revealed that nanocomposites containing 2% carbon nanotubes had outstanding sensing capabilities, achieving maximum detection accuracy with the least amount of noise. In this context, Figure 9 summarizes the platform setup, the sensor construction, and the results associated with some of the references studied so far. A modeling of capacitive proximity sensors using the Laplace's equation has been analytically solved and proven to be a fast and reliable technique to obtain the capacitance of a CPS [90].

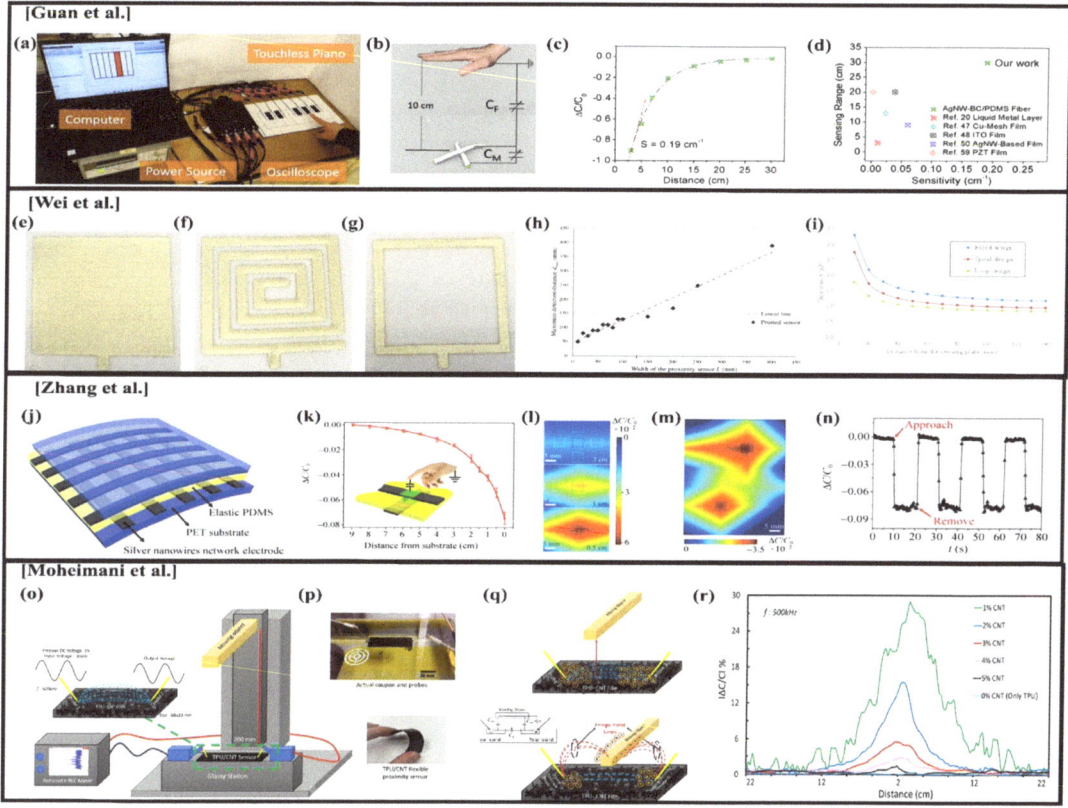

Figure 9. Measurement setup and the results proposed by Refs. [15,79,81,88]: (**a**) touchless piano being played optically with a human finger hovering 40 mm above the keyboard, (**b**) schematic representation of sensor sensing, (**c**) increase in relative capacitance with response to an approaching object, (**d**) performance comparison of the proposed setup with previous works, (**e**) printed proximity sensors on fabric with filled design, (**f**) printed proximity sensors on fabric with spiral design, (**g**) printed proximity sensors on fabric with loop design, (**h**) maximum detection distance as a function of the proximity sensor width, (**i**) capacitance change simulation via COMSOL, (**j**) sandwich structure of CPS with the AgNWs stripes serving as the row and column electrodes, placed orthogonally on the top and bottom PET substrates, (**k**) capacitance variation as a function of vertical distance from the intersection, (**l**) contour graphics depicting the estimated capacitance change profile of a center pixel and its four closest neighbors with varying degrees of sensitivity, (**m**) proximity sensing depiction of two metal bars, the intersections of which are denoted by dashed black boxes, (**n**) relative capacitance change vs. response time during four cycles, (**o**) a schematic representation of a TPU/carbon nanotube proximity sensor configuration, (**p**) noise minimization using semi-planner 45° probes, and (**q**) mutual capacitance becomes apparent when an item is moved near to the sensor. Shunting the initial electric lines results in a highly strong and distributed fring field between the object, film, and probes, resulting in a dramatic decrease in capacitance. (**r**) Comparing the maximum sensitivity of various weight percentages of carbon nanotubes (CNTs).

4. Opportunities and Challenges

Although CPSs have grown in popularity over the past decade because of their attractive features, several issues have remained persistent. The following issues can be further explored to forward the progress of CPSs:

As the literature survey suggests, the integration of CNTs into TPU have the potential to outperform other flexible CPS designs. However, fabricating a flexible nanocomposite tailored to desired performance and functionality is not addressed primarily because of the lack of basic understanding of microstructure formation from the molecular level to higher scales. Integrating CNTs into polymer composites often involves time/cost-inefficient processes that lead to inhomogeneous dispersion of CNTs, weak interfacial bonding between polymer and CNT, and damage to the sidewalls of CNTs that alter their intrinsic properties. Therefore, there is a need for a transformative strategy in the processing and manufacturing of CNT-based polymer nanocomposite flexible capacitive sensors to overcome these limitations. However, there are few reported applications of incorporation of CNTs in TPU-based nanocomposites for proximity sensing.

Meanwhile, because the detection range of CPSs is highly dependent on the size of the electrodes, they must have a particular size to allow for a good detection at a significant distance. In the absence of sufficient contact distance, the sensors' use may be restricted. Typical methods for boosting the detection range include expanding the electrode size and optimizing the interface circuit's performance, although these measures have only a marginal impact. Additionally, a sensor array system requires a compromise between resolution and detection range. Recent research provides intriguing examples, such as combining it with another detection approach, such as ultrasonic and inductance sensors, or fusing several detection techniques (e.g., CMUT). This method may provide high-quality sensing and a wide detection range. In the meantime, a quicker reaction time of the capacitance measuring circuit is required to fulfill the demand for real-time performance, since the working electrodes must flip numerous times throughout the measurement period.

Last but not least, CPSs are sensitive to their surroundings (changes in ambient variables, such as temperature, humidity, or illumination; or changes in the presence or location of interior items), which may impair sensor-data accuracy. Post-processing raw sensor data, rather than the converted discrete distance, may effectively decrease sensor-data variability and noise caused by deployment-specific environmental variables. As a result, significant effort will be required to address these critical concerns and challenges. Regarding future study, the authors believe that the combination of object distance characteristics with machine/deep learning methodologies is one of the most promising and leading research topics.

5. Conclusions

In this study, the current level of knowledge about the role of CPSs operating in the displacement sensing mode is summarized. CPSs are simple, inexpensive, and very energy efficient, and may be installed under any nonconductive shielding. It is necessary to describe and categorize these applications as has been done in this review article by addressing the most pertinent literature on the approaches to CPS applications. Another objective of this article is to further stimulate academics and developers to explore broader application possibilities. It is possible to expand the investigation of CPSs in the areas of object detection and human interaction in the future. Particularly since they do not compromise user privacy and require touch, the application of capacitive displacement sensing in intelligent devices cannot be overlooked and has considerable future potential. Meanwhile, capacitive technology continues to confront a number of obstacles, including production constraints, a limited detection distance, imprecise measurement precision, and interference from the environment, necessitating more discoveries and developments.

Author Contributions: Conceptualization, methodology, investigation, resources, and writing—original draft preparation, R.M., P.H.; supervision and project administration, S.M., H.D.; funding acquisition, H.D. All authors have read and agreed to the published version of the manuscript.

Funding: The authors would like to express their gratitude towards the National Science Foundation Small Business Technology Transfer (NSF STTR) (#2036490) for sponsoring this research; the National Science Foundation Major Research Instrumentation Program (NSF-MRI) (#1229514) for supporting this research for FESEM. This work is also sponsored in part by the Indiana 21st Century Fund and the Indiana Economic Development Corporation as well as Indiana University's Research Support Funds Grant (RSFG). Any opinions, findings, and conclusions or recommendations expressed in this material are those of the author(s) and do not necessarily reflect the views of the National Science Foundation.

Institutional Review Board Statement: Not applicable.

Informed Consent Statement: Not applicable.

Data Availability Statement: The data that support the findings of this study are available from the corresponding authors upon reasonable request.

Conflicts of Interest: The authors declare no conflict of interest.

References

1. Chen, Y.L.; Huang, Y.; Shih, F.; Chou, T.; Chien, T.L.; Chen, R.; Fang, W. A dual sensing modes capacitive tactile sensor for proximity and tri-axial forces detection. In Proceedings of the IEEE 35th International Conference on Micro Electro Mechanical Systems Conference (MEMS), Tokyo, Japan, 9–13 January 2022; pp. 710–713.
2. Mondal, I.; Ganesha, M.K.; Singh, A.K.; Kulkarni, G.U. Inkjet printing aided patterning of transparent metal mesh for wearable tactile and proximity sensors. *Mater. Lett.* **2022**, *15*, 131724. [CrossRef]
3. Lee, H.; Mandivarapu, J.K.; Ogbazghi, N.; Li, Y. Real-time interface control with motion gesture recognition based on non-contact capacitive sensing. *arXiv* **2022**, arXiv:2201.01755.
4. Ge, C.; Yang, B.; Wu, L.; Duan, Z.; Li, Y.; Ren, X.; Jiang, L.; Zhang, J. Capacitive sensor combining proximity and pressure sensing for accurate grasping of a prosthetic hand. *ACS Appl. Electron. Mater.* **2022**, *4*, 869–877. [CrossRef]
5. Zhang, Z.; Li, J.; Yu, B.; Huang, D.; Wang, Q.; Li, Z. Low-cost, flexible annular interdigital capacitive sensor (Faics) with carbon black-pdms sensitive layer for proximity and pressure sensing. In Proceedings of the IEEE 35th International Conference on Micro Electro Mechanical Systems Conference (MEMS), Tokyo, Japan, 9–13 January 2022; pp. 35–38.
6. Fleming, A.J. A review of nanometer resolution position sensors: Operation and performance. *Sens. Actuators A Phys.* **2013**, *190*, 106–126. [CrossRef]
7. George, B.; Tan, Z.; Nihtianov, S. Advances in capacitive, eddy current, and magnetic displacement sensors and corresponding interfaces. *IEEE Trans. Ind. Electron.* **2017**, *64*, 9595–9607. [CrossRef]
8. Grosse-Puppendahl, T.; Holz, C.; Cohn, G.; Wimmer, R.; Bechtold, O.; Hodges, S.; Smith, J.R. Finding common ground: A survey of capacitive sensing in human-computer interaction. In Proceedings of the CHI Conference on Human Factors in Computing Systems (ACM CHI), Gaithersburg, ML, USA, 6–11 May 2017; pp. 3293–3315.
9. Sivayogan, T. Design and Development of a Contactless Planar Capacitive Sensor. Master's Thesis, University of Toronto, Toronto, ON, Canada, 2013.
10. Kulkarni, M.R.; John, R.A.; Rajput, M.; Tiwari, N.; Yantara, N.; Nguyen, A.C.; Mathews, N. Transparent flexible multifunctional nanostructured architectures for non-optical readout, proximity, and pressure sensing. *ACS Appl. Mater. Interfaces* **2017**, *9*, 15015–15021. [CrossRef]
11. Tholin-Chittenden, C.; Abascal, J.F.P.-J.; Soleimani, M. Automatic parameter selection of image reconstruction algorithms for planar array capacitive imaging. *IEEE Sens. J.* **2018**, *18*, 6263–6272. [CrossRef]
12. Moheimani, R.; Agarwal, M.; Dalir, H. 3D-printed flexible structures with embedded deformation/displacement sensing for the creative industries. In *AIAA Scitech 2021 Forum*; Aerospace Research Central: New York, NY, USA, 2021; p. 0534.
13. Zeinali, S.; Homayoonnia, S.; Homayoonnia, G. Comparative investigation of interdigitated and parallel-plate capacitive gas sensors based on Cu-BTC nanoparticles for selective detection of polar and apolar VOCs indoors. *Sens. Actuators B Chem.* **2019**, *278*, 153–164. [CrossRef]
14. Chen, W.P.; Zhao, Z.G.; Liu, X.W.; Zhang, Z.X.; Suo, C.G. A capacitive humidity sensor based on multi-wall carbon nanotubes (MWCNTs). *Sensors* **2009**, *9*, 7431–7444. [CrossRef]
15. Guan, F.; Xie, Y.; Wu, H.; Meng, Y.; Shi, Y.; Gao, M.; Zhang, Z.; Chen, S.; Chen, Y.; Wang, H.; et al. Silver nanowire–bacterial cellulose composite fiber-based sensor for highly sensitive detection of pressure and proximity. *ACS Nano* **2020**, *14*, 15428–15439. [CrossRef]
16. Sadasivuni, K.K.; Kafy, A.; Zhai, L.; Ko, H.U.; Mun, S.; Kim, J. Transparent and flexible cellulose nanocrystal/reduced graphene oxide film for proximity sensing. *Small* **2015**, *11*, 994–1002. [CrossRef] [PubMed]
17. Watzenig, D.; Fox, C. A review of statistical modelling and inference for electrical capacitance tomography. *Meas. Sci. Technol.* **2009**, *20*, 052002. [CrossRef]
18. Kim, H.; Kim, G.; Kim, T.; Lee, S.; Kang, D.; Hwang, M.S.; Chae, Y.; Kang, S.; Lee, H.; Park, H.G.; et al. Transparent, flexible, conformal capacitive pressure sensors with nanoparticles. *Small* **2018**, *14*, 1703432. [CrossRef] [PubMed]

19. Du, Y.; Yu, G.; Dai, X.; Wang, X.; Yao, B.; Kong, J. Highly stretchable, self-healable, ultrasensitive strain and proximity sensors based on skin-inspired conductive film for human motion monitoring. *ACS Appl. Mater. Interfaces* **2020**, *12*, 51987–51998. [CrossRef]
20. Zhu, M.; Shi, Q.; He, T.; Yi, Z.; Ma, Y.; Yang, B.; Chen, T.; Lee, C. Self-powered and self-functional cotton sock using piezoelectric and triboelectric hybrid mechanism for healthcare and sports monitoring. *ACS Nano* **2019**, *13*, 1940–1952. [CrossRef]
21. Kumar, V.; Kumar, A.; Han, S.S.; Park, S.-S. RTV silicone rubber composites reinforced with carbon nanotubes, titanium-di-oxide and their hybrid: Mechanical and piezoelectric actuation performance. *Nano Mater. Sci.* **2020**, *3*, 233–240. [CrossRef]
22. Min, S.D.; Wang, C.; Park, D.S.; Park, J.H. Development of a textile capacitive proximity sensor and gait monitoring system for smart healthcare. *J. Med. Syst.* **2018**, *42*, 76. [CrossRef]
23. Sarwar, M.S.; Dobashi, Y.; Preston, C.; Wyss, J.K.M.; Mirabbasi, S.; Madden, J.D.W. Bend, stretch, and touch: Locating a finger on an actively deformed transparent sensor array. *Sci. Adv.* **2017**, *3*, e1602200. [CrossRef]
24. Kang, M.; Kim, J.; Jang, B.; Chae, Y.; Kim, J.H.; Ahn, J.H. Graphene-based three-dimensional capacitive touch sensor for wearable electronics. *ACS Nano* **2017**, *11*, 7950–7957. [CrossRef]
25. Arshad, A.; Khan, S.; Alam, A.H.M.Z.; Kadir, K.A.; Tasnim, R.; Ismail, A.F. A capacitive proximity sensing scheme for human motion detection. In Proceedings of the IEEE International Instrumentation and Measurement Technology Conference (I2MTC), Turin, Italy, 22–25 May 2017; pp. 1–5.
26. Hosseini, M.; Zhu, G.; Peter, Y.-A. A new formulation of fringing capacitance and its application to the control of parallel-plate electrostatic micro actuators. *Analog Integr. Circuits Signal Process.* **2007**, *53*, 119–128. [CrossRef]
27. Wang, B.; Long, J.; Teo, K. Multi-channel capacitive sensor arrays. *Sensors* **2016**, *16*, 150. [CrossRef] [PubMed]
28. Muhlbacher-Karrer, S.; Mosa, A.H.; Faller, L.-M.; Ali, M.; Hamid, R.; Zangl, H.; Kyamakya, K. A driver state detection system—Combining a capacitive hand detection sensor with physiological sensors. *IEEE Trans. Instrum. Meas.* **2017**, *66*, 624–636. [CrossRef]
29. Durukan, M.B.; Cicek, M.O.; Doganay, D.; Gorur, M.C.; Çınar, S.; Unalan, H.E. Multifunctional and Physically Transient Supercapacitors, Triboelectric Nanogenerators, and Capacitive Sensors. *Adv. Funct. Mater.* **2022**, *32*, 2106066. [CrossRef]
30. Maxwell, J.C. *A Treatise on Electricity and Magnetism*; Oxford University Press: Oxford, UK, 1881.
31. Ye, Y.; Zhang, C.; He, C.; Wang, X.; Huang, J.; Deng, J. A review on applications of capacitive displacement sensing for capacitive proximity sensor. *IEEE Access* **2019**, *8*, 45325–45342. [CrossRef]
32. Baxter, L.K. *Capacitive Sensors: Design and Applications*; Institute of Electrical and Electronics Engineers Inc.: New York, NY, USA, 1997.
33. Yao, S.; Zhu, Y. Wearable multifunctional sensors using printed stretchable conductors made of silver nanowires. *Nanoscale* **2014**, *6*, 2345–2352. [CrossRef]
34. Abdelhamid, H.; Morsy, O.E.; El-Shibiny, A.; Abdelbaset, R. Detection of foodborne pathogens using novel vertical capacitive sensors. *Alex. Eng. J.* **2022**, *61*, 3873–3882. [CrossRef]
35. Ahmadi, A.; Nabipour, M.; Taheri, S.; Mohammadi-Ivatloo, B.; Vahidinasab, V. A New False Data Injection Attack Detection Model for Cyberattack Resilient Energy Forecasting. *IEEE Trans. Ind. Inform.* **2022**. [CrossRef]
36. Taheri, S.; Akbari, A.; Ghahremani, B.; Razban, A. Reliability-based energy scheduling of active buildings subject to renewable energy and demand uncertainty. *Therm. Sci. Eng. Prog.* **2022**, *28*, 101149. [CrossRef]
37. Pourteimoor, S.; Haratizadeh, H. Performance of a fabricated nanocomposite-based capacitive gas sensor at room temperature. *J. Mater. Sci. Mater. Electron.* **2017**, *28*, 18529–18534. [CrossRef]
38. Merdassi, A.; Yang, P.; Chodavarapu, V. A wafer level vacuum encapsulated capacitive accelerometer fabricated in an unmodified commercial MEMS process. *Sensors* **2015**, *15*, 7349–7359. [CrossRef]
39. Fulmek, P.L.; Wandling, F.; Zdiarsky, W.; Brasseur, G.; Cermak, S.P. Capacitive sensor for relative angle measurement. *IEEE Trans. Instrum. Meas.* **2002**, *51*, 1145–1149. [CrossRef]
40. Jindal, S.K.; Agarwal, Y.K.; Priya, S.; Kumar, A.; Raghuwanshi, S.K. Design and analysis of MEMS pressure transmitter using Mach–Zehnder interferometer and artificial neural networks. *IEEE Sens. J.* **2018**, *18*, 7150–7157. [CrossRef]
41. Zeng, T.; Lu, Y.; Liu, Y.; Yang, H.; Bai, Y.; Hu, P.; Li, Z.; Zhang, Z.; Tan, J. A capacitive sensor for the measurement of departure from the vertical movement. *IEEE Trans. Instrum. Meas.* **2016**, *65*, 458–466. [CrossRef]
42. Arjomandi-Nezhad, A.; Ahmadi, A.; Taheri, S.; Fotuhi-Firuzabad, M.; Moeini-Aghtaie, M.; Lehtonen, M. Pandemic-Aware Day-Ahead Demand Forecasting using Ensemble Learning. *IEEE Access* **2022**. [CrossRef]
43. Kwon, O.-K.; An, J.-S.; Hong, S.-K. Capacitive touch systems with styli for touch sensors: A review. *IEEE Sens. J.* **2018**, *18*, 4832–4846. [CrossRef]
44. George, B.; Zangl, H.; Bretterklieber, T.; Brasseur, G. A combined inductive–capacitive proximity sensor for seat occupancy detection. *IEEE Trans. Instrum. Meas.* **2010**, *59*, 1463–1470. [CrossRef]
45. Xia, F.; Campi, F.; Bahreyni, B. Tri-mode capacitive proximity detection towards improved safety in industrial robotics. *IEEE Sens. J.* **2018**, *18*, 5058–5066. [CrossRef]
46. Derakhshan, R.; Ramiar, A.; Ghasemi, A. Numerical investigation into continuous separation of particles and cells in a two-component fluid flow using dielectrophoresis. *J. Mol. Liq.* **2020**, *310*, 113211. [CrossRef]
47. Walia, S.; Mondal, I.; Kulkarni, G.U. Patterned Cu-Mesh-based transparent and wearable touch panel for tactile, proximity, pressure, and temperature sensing. *ACS Appl. Electron. Mater.* **2019**, *1*, 1597–1604. [CrossRef]

48. Chen, T.; Bowler, N. Analysis of a capacitive sensor for the evaluation of circular cylinders with a conductive core. *Meas. Sci. Technol.* **2012**, *23*, 045102. [CrossRef]
49. Yang, M. Development of Electrical Tomography Systems and Application for Milk Flow Metering. Ph.D. Thesis, The University of Manchester, Manchester, UK, 2007.
50. Sheldon, R.T.; Bowler, N. An interdigital capacitive sensor for nondestructive evaluation of wire insulation. *IEEE Sens. J.* **2014**, *14*, 961–970. [CrossRef]
51. Yang, W.Q. Hardware design of electrical capacitance tomography systems. *Meas. Sci. Technol.* **1996**, *7*, 225–232. [CrossRef]
52. Yang, W.Q.; Peng, L.H. Image reconstruction algorithms for electrical capacitance tomography. *Meas. Sci. Technol.* **2003**, *14*, R1–R13. [CrossRef]
53. Zeothout, J.; Boletis, A.; Bleuler, H. High performance capacitive position sensing device for compact active magnetic bearing spindles. *JSME Int. J. Ser. B Fluids Therm. Eng.* **2003**, *46*, 900–907. [CrossRef]
54. Igreja, R.; Dias, C.J. Analytical evaluation of the interdigital electrodes capacitance for a multi-layered structure. *Sens. Actuators A Phys.* **2004**, *112*, 291–301. [CrossRef]
55. Huang, S.M.; Plaskowski, A.B.; Xie, C.G.; Beck, M.S. Tomographic imaging of two-component flow using capacitance sensors. *J. Phys. E Sci. Instrum.* **1989**, *22*, 173–177. [CrossRef]
56. Rivadeneyra, A.; Fernandez-Salmeron, J.; Agudo-Acemel, M.; Lopez-Villanueva, J.A.; Capitan-Vallvey, L.F.; Palma, A.J. Printed electrodes structures as capacitive humidity sensors: A comparison. *Sens. Actuators A Phys.* **2016**, *244*, 56–65. [CrossRef]
57. Lee, J.W.; Min, D.J.; Kim, J.; Kim, W. A 600-dpi capacitive fingerprint sensor chip and image-synthesis technique. *IEEE J. Solid-State Circuits* **1999**, *34*, 469–475.
58. Mohd Syaifudin, A.R.; Mukhopadhyay, S.C.; Yu, P.L. Modelling and fabrication of optimum structure of novel interdigital sensors for food inspection. *Int. J. Numer. Model. Electron. Netw. Devices Fields* **2012**, *25*, 64–81. [CrossRef]
59. Zia, A.I.; Syaifudin, A.M.; Mukhopadhyay, S.C.; Yu, P.L.; Al-Bahadly, I.H.; Gooneratne, C.P.; Kosel, J.; Liao, T.S. Electrochemical impedance spectroscopy based MEMS sensors for phthalates detection in water and juices. *J. Phys. Conf. Ser.* **2013**, *439*, 012026. [CrossRef]
60. Khawaja, K.; Seneviratne, L.; Althoefer, K. Wheel-tooling gap measurement system for conform extrusion machinery based on a capacitive sensor. *J. Manuf. Sci. Eng.* **2010**, *127*, 394–401. [CrossRef]
61. Addabbo, T.; Bertocci, F.; Fort, A.; Mugnaini, M.; Panzardi, E.; Vignoli, V.; Cinelli, C. A clearance measurement system based on on component multilayer tri-axial capacitive probe. *Measurement* **2018**, *124*, 575–581. [CrossRef]
62. Petchmaneelumka, W.; Phankamnerd, P.; Rerkratn, A.; Riewruja, V. Capacitive sensor readout circuit based on sample and hold method. *Energy Rep.* **2022**, *8*, 1012–1018. [CrossRef]
63. Kim, Y.S.; Cho, S.I.; Shin, D.H.; Lee, J.; Baek, K.H. Single chip dual plate capacitive proximity sensor with high noise immunity. *IEEE Sens. J.* **2013**, *14*, 309–310. [CrossRef]
64. Lee, H.-K.; Chang, S.-I.; Yoon, E. A capacitive proximity sensor in dual implementation with tactile imaging capability on a single flexible platform for robot assistant applications. In Proceedings of the 19th IEEE International Conference on Micro Electro Mechanical Systems, Istanbul, Turkey, 22–26 January 2006; pp. 606–609.
65. Kohama, T.; Tsuji, S. Tactile and proximity sensor by 3D tactile sensor using self-capacitance measurement. In Proceedings of the IEEE SENSORS, Busan, Korea, 1–4 November 2015; pp. 1–4.
66. Qiu, S.; Huang, Y.; He, X.; Sun, Z.; Liu, P.; Liu, C. A dual-mode proximity sensor with integrated capacitive and temperature sensing units. *Meas. Sci. Technol.* **2015**, *26*, 105101. [CrossRef]
67. Nguyen, T.D.; Kim, T.; Han, H.; Shin, H.Y.; Nguyen, C.T.; Phung, H.; Choi, H.R. Characterization and optimization of flexible dual mode sensor based on Carbon Micro Coils. *Mater. Res. Express* **2018**, *5*, 015604. [CrossRef]
68. Lu, Y.; Bai, Y.; Zeng, T.; Li, Z.; Zhang, Z.; Tan, J. Coplanar capacitive sensor for measuring horizontal displacement in joule balance. In Proceedings of the Conference on Precision Electromagnetic Measurements (CPEM 2016), Ottawa, ON, Canada, 10–15 July 2016; pp. 1–2.
69. Lu, X.; Li, X.; Zhang, F.; Wang, S.; Xue, D.; Qi, L.; Wang, H.; Li, X.; Bao, W.; Chen, R. A novel proximity sensor based on parallel plate capacitance. *IEEE Sens. J.* **2018**, *18*, 7015–7022. [CrossRef]
70. Taheri, S.; Talebjedi, B.; Laukkanen, T. Electricity demand time series forecasting based on empirical mode decomposition and long short-term memory. *Energy Eng. J. Assoc. Energy Eng.* **2021**, *118*, 1577–1594. [CrossRef]
71. Nguyen, N.T.; Sarwar, M.S.; Preston, C.; Le Goff, A.; Plesse, C.; Vidal, F.; Cattan, E.; Madden, J.D. Transparent stretchable capacitive touch sensor grid using ionic liquid electrodes. *Extrem. Mech. Lett.* **2019**, *33*, 100574. [CrossRef]
72. Long, J.; Wang, B. A metamaterial-inspired sensor for combined inductive-capacitive detection. *Appl. Phys. Lett.* **2015**, *106*, 074104. [CrossRef]
73. Böhmländer, D.; Doric, I.; Appel, E.; Brandmeier, T. Video camera and capacitive sensor data fusion for pedestrian protection systems. In Proceedings of the 11th Workshop on Intelligent Solutions in Embedded Systems (WISES), Pilsen, Czech Republic, 10–11 September 2013; pp. 1–7.
74. Lee, H.K.; Chang, S.I.; Yoon, E. Dual-mode capacitive proximity sensor for robot application: Implementation of tactile and proximity sensing capability on a single polymer platform using shared electrodes. *IEEE Sens. J.* **2009**, *9*, 1748–1755. [CrossRef]
75. Chen, Z.; Luo, R.C. Design and implementation of capacitive proximity sensor using microelectromechanical systems technology. *IEEE Trans. Ind. Electron.* **1998**, *45*, 886–894. [CrossRef]

76. Ye, Y.; Deng, J.; Shen, S.; Hou, Z.; Liu, Y. A novel method for proximity detection of moving targets using a large-scale planar capacitive sensor system. *Sensors* **2016**, *16*, 699. [CrossRef] [PubMed]
77. Ye, X.; Tian, M.; Li, M.; Wang, H.; Shi, Y. All-fabric-based flexible capacitive sensors with pressure detection and non-contact instruction capability. *Coatings* **2022**, *12*, 302. [CrossRef]
78. Čoko, D.; Stančić, I.; Dujić Rodić, L.; Čošić, D. TheraProx: Capacitive proximity sensing. *Electronics* **2022**, *11*, 393. [CrossRef]
79. Wei, Y.; Torah, R.; Li, Y.; Tudor, J. Dispenser printed capacitive proximity sensor on fabric for applications in the creative industries. *Sens. Actuators A Phys.* **2016**, *247*, 239–246. [CrossRef]
80. Luo, N.; Dai, W.; Li, C.; Zhou, Z.; Lu, L.; Poon, C.C.Y.; Chen, S.-C.; Zhang, Y.; Zhao, N. Flexible piezoresistive sensor patch enabling ultralow power cuffless blood pressure measurement. *Adv. Funct. Mater.* **2016**, *26*, 1178–1187. [CrossRef]
81. Zhang, B.; Xiang, Z.; Zhu, S.; Hu, Q.; Cao, Y.; Zhong, J.; Zhong, Q.; Wang, B.; Fang, Y.; Hu, B.; et al. Dual functional transparent film for proximity and pressure sensing. *Nano Res.* **2014**, *7*, 1488–1496. [CrossRef]
82. Alagi, H.; Navarro, S.E.; Mende, M.; Hein, B. A versatile and modular capacitive tactile proximity sensor. In Proceedings of the 2016 IEEE Haptics Symposium (HAPTICS), Philadelphia, PA, USA, 8–11 April 2016; pp. 290–296.
83. Ben-Yasharand, G.; Ya'akobovitz, A. Electronic skin with embedded carbon nanotubes proximity sensors. *IEEE Trans. Electron Devices* **2021**, *68*, 4098–4103. [CrossRef]
84. Addabbo, T.; Fort, A.; Mugnaini, M.; Rocchi, S.; Vignoli, V. A heuristic reliable model for guarded capacitive sensors to measure displacements. In Proceedings of the IEEE International Instrumentation and Measurement Technology Conference (I2MTC) Proceedings, Pisa, Italy, 11–14 May 2015; pp. 1488–1491.
85. Bu, D.; Li, S.Q.; Sang, Y.M.; Qiu, C.J. High transparency flexible sensor for pressure and proximity sensing. *Mod. Phys. Lett. B* **2018**, *32*, 1850394. [CrossRef]
86. Nguyen, T.D.; Han, H.S.; Shin, H.Y.; Nguyen, C.T.; Phung, H.; Van Hoang, H.; Choi, H.R. Highly sensitive flexible proximity tactile array sensor by using carbon micro coils. *Sens. Actuators A Phys.* **2017**, *266*, 166–177. [CrossRef]
87. Hein, B.; Li, X.; Alagi, H. Two Examples of Using Machine Learning for Processing Sensor Data of Capacitive Proximity Sensors. Available online: https://arxiv.org/ftp/arxiv/papers/1903/1903.08495.pdf (accessed on 1 December 2019).
88. Moheimani, R.; Aliahmad, N.; Aliheidari, N.; Agarwal, M.; Dalir, H. Thermoplastic polyurethane flexible capacitive proximity sensor reinforced by CNTs for applications in the creative industries. *Sci. Rep.* **2021**, *11*, 1104. [CrossRef] [PubMed]
89. Moheimani, R.; Gonzalez, M.; Dalir, H. An integrated nanocomposite proximity sensor: Machine learning-based optimization, simulation, and experiment. *Nanomaterials* **2022**, *12*, 1269. [CrossRef] [PubMed]
90. Moheimani, R.; Pasharavesh, A.; Agarwal, M.; Dalir, H. Mathematical model and experimental design of nanocomposite proximity sensors. *IEEE Access* **2020**, *8*, 153087–153097. [CrossRef]

MDPI AG
Grosspeteranlage 5
4052 Basel
Switzerland
Tel.: +41 61 683 77 34

C Editorial Office
E-mail: carbon@mdpi.com
www.mdpi.com/journal/carbon

Disclaimer/Publisher's Note: The title and front matter of this reprint are at the discretion of the Guest Editors. The publisher is not responsible for their content or any associated concerns. The statements, opinions and data contained in all individual articles are solely those of the individual Editors and contributors and not of MDPI. MDPI disclaims responsibility for any injury to people or property resulting from any ideas, methods, instructions or products referred to in the content.